Essentials of Stem Cell Biology

Essentials of Stem Cell Biology

Editor: Jack Collins

FOSTER
ACADEMICS

www.fosteracademics.com

www.fosteracademics.com

FA FOSTER
ACADEMICS

Cataloging-in-Publication Data

Essentials of stem cell biology / edited by Jack Collins.
    p. cm.
Includes bibliographical references and index.
ISBN 978-1-63242-506-5
1. Stem cells. 2. Cells. 3. Cytology. I. Collins, Jack.
QH588.S83 E87 2017
616.027 74--dc23

Foster Academics,
118-35 Queens Blvd., Suite 400,
Forest Hills, NY 11375, USA

ISBN 978-1-63242-506-5 (Hardback)

Printed and bound in the United States of America.

Contents

Preface

Stem cells are undifferentiated cells that can divide to reproduce themselves. This book on stem cell biology discusses the application of the stem cells for various purposes such as tissue regeneration, limb/organ transplantation, blood generation, etc. Stem cell therapy and bone marrow transplantation are two effective treatments that harvest stem cells from particular areas. Diabetes, heart disease and spinal cord injury are major diseases that are treated through stem cell therapy. This book is a complete source of knowledge on the present status of this important field. For all readers who are interested in stem cell biology, the case studies included in this book will serve as excellent guide to develop a comprehensive understanding.

All of the data presented henceforth, was collaborated in the wake of recent advancements in the field. The aim of this book is to present the diversified developments from across the globe in a comprehensible manner. The opinions expressed in each chapter belong solely to the contributing authors. Their interpretations of the topics are the integral part of this book, which I have carefully compiled for a better understanding of the readers.

At the end, I would like to thank all those who dedicated their time and efforts for the successful completion of this book. I also wish to convey my gratitude towards my friends and family who supported me at every step.

Editor

Intraspinal bone-marrow cell therapy at pre- and symptomatic phases in a mouse model of amyotrophic lateral sclerosis

Fernanda Gubert[1*], Ana B. Decotelli[1], Igor Bonacossa-Pereira[1], Fernanda R. Figueiredo[1], Camila Zaverucha-do-Valle[2], Fernanda Tovar-Moll[3,4], Luísa Hoffmann[1], Turan P. Urmenyi[1], Marcelo F. Santiago[1] and Rosalia Mendez-Otero[1]

Abstract

Background: Amyotrophic lateral sclerosis (ALS) is a progressive neurological disease that selectively affects the motor neurons. The details of the mechanisms of selective motor-neuron death remain unknown and no effective therapy has been developed. We investigated the therapy with bone-marrow mononuclear cells (BMMC) in a mouse model of ALS (SOD1^{G93A} mice).

Methods: We injected 10^6 BMMC into the lumbar portion of the spinal cord of SOD1^{G93A} mice in presymptomatic (9 weeks old) and symptomatic (14 weeks old) phases. In each condition, we analyzed the progression of disease and the lifespan of the animals.

Results: We observed a mild transitory delay in the disease progression in the animals injected with BMMC in the presymptomatic phase. However, we observed no increase in the lifespan. When we injected BMMC in the symptomatic phase, we observed no difference in the animals' lifespan or in the disease progression. Immunohistochemistry for NeuN showed a decrease in the number of motor neurons during the course of the disease, and this decrease was not affected by either treatment. Using different strategies to track the BMMC, we noted that few cells remained in the spinal cord after transplantation. This observation could explain why the BMMC therapy had only a transitory effect.

Conclusion: This is the first report of intraspinal BMMC therapy in a mouse model of ALS. We conclude this cellular therapy has only a mild transitory effect when performed in the presymptomatic phase of the disease.

Keywords: Amyotrophic lateral sclerosis, Bone marrow cell therapy, Neuronal degeneration

Background

Amyotrophic lateral sclerosis (ALS) is a fatal neurodegenerative disease involving the death of motor neurons in the spinal cord, brain stem, and motor cortex. Approximately 90 % of ALS cases are sporadic, and the remaining 10 % are inherited (familial). A mutation in the Cu/Zn superoxide dismutase 1 (SOD1) gene is responsible for 20 % of the familial forms [1]. In rodents, overexpression of the human SOD1 mutant protein leads to progressive motor-neuron degeneration, which mimics the pathological progression observed in humans [2, 3].

Several lines of evidence suggest that nonneuronal cells play an important role in ALS pathogenesis, since both the astrocytes and the microglia expressing the mutated SOD1 are toxic to motor neurons and contribute significantly to the disease progression [4–8]. In addition, immune cells such as T lymphocytes have also been implicated in ALS, and an increase in the number of these cells in the affected areas of the central nervous system (CNS) has been described both in humans and in mouse models [9–11]. The CNS-infiltrating CD4 T lymphocytes seem to communicate with the microglia, and this dialog directs the microglia to a neuroprotective profile [12, 13].

* Correspondence: fegubert@biof.ufrj.br
[1]Instituto de Biofísica Carlos Chagas Filho, Centro de Ciências da Saúde, Sala G2-028, Universidade Federal do Rio de Janeiro, Cidade Universitária, RJ 21941-902 Rio de Janeiro, Brazil
Full list of author information is available at the end of the article

Currently, the only treatment for ALS is Riluzole®, Sanofi S.A., (Gentilly, Paris, France), which increases the patient's lifespan by only 3 months [14]. Alternatively, cell therapies with different types of stem cells have been suggested as a possible therapeutic approach. Bone marrow-derived stem cells have been used in several models of neurological disease, including ALS [15–19]. In several of these studies, beneficial effects were observed; the suggested mechanism of action includes release of trophic factors and local or systemic modulation of the immune response.

In addition to different cell types, different routes of injection have been tested in animal ALS models; for instance, intravenous, intrathecal, intramuscular, and direct delivery in the spinal cord [16, 20–22]. In the majority of studies using animal ALS models, the cells were injected in the presymptomatic phase of the disease. For clinical translation, however, cell therapy during the symptomatic phase could be more meaningful. In this respect, a few clinical trials with bone marrow mononuclear cells or mesenchymal stem cells in ALS patients showed that these therapies are safe, although the efficacy of these therapies still needs to be investigated [23, 24].

In this study, we investigated the potential therapeutic benefit of bone marrow-derived mononuclear cells (BMMC) injected directly into the lumbar spinal cord of ALS mice in the presymptomatic or symptomatic phases. BMMC have been broadly tested in preclinical models of traumas or neurodegenerative diseases in the nervous system, showing positive results. These cells are easy to isolate and require minimum manipulation prior to transplant, becoming a good candidate for cell therapies. Using different models of neurological diseases, our group showed in previous publications that BMMC transplantation results in neuronal protection, axonal regeneration, decreases microglia activation, and improves functional recovery [17, 25, 26]. More importantly, this is the first study using this approach, and we report here that the BMMC therapy delayed the animals' functional outcome only when administered in the presymptomatic phase.

Methods
Animals
All of the experiments were conducted in accordance with the recommendations of the National Institutes of Health Guide for the Care and Use of Laboratory Animals, and were approved by the Committee for the Use of Experimental Animals of our institution (permit number: IBCCF213-09/16). The strain of mice used was B6SJL-Tg(SOD1-G93A)1Gur (SOD1^{G93A}), developed by Gurney in 1994 [2], which carries the mutant allele human SOD1 containing the Gly 93 → Ala substitution. The colony was maintained by crossing transgenic male founders with wild-type female mice. The number of human SOD1 transgenic copies was assessed as described in the Jackson Laboratory manual (https://www.jax.org/strain/002726#jump-nav-5). The breeding pairs were donated by the ALS Foundation through Dr R Brown (University of Massachusetts, Worcester, USA).

Isolation and labeling of mononuclear bone marrow cells
Bone marrow was obtained from the femur and tibia of adult B6SJL mice of both genders. BMMC were isolated by density gradient (Histopaque 1083; Sigma, St. Louis, MO, USA) and washed three times with phosphate-buffered saline (PBS). Cells were suspended in saline (250,000 cells/µl) and injected into the lumbar portion of the spinal cord as described in the next section.

In some animals the BMMC were labeled, before transplantation, with a fluorescent marker (CellTrace™ Far Red DDAO-SE; Life Technologies, Carlsbad, CA, USA) or with iron nanoparticles (superparamagnetic iron nanoparticles (SPION), FeraTrack™; Miltenyi Biotec, Bergisch Gladbach, Germany). For the fluorescent labeling, BMMC were incubated with Cell Trace™ (1:1000) diluted in Dulbecco's modified Eagle's medium (DMEM)–F12 (Life Technologies) for 40 minutes at 37 ° C, washed five times with PBS, and then suspended in saline.

For the SPION labeling, BMMC were incubated with FeraTrack™ contrast particles and FeraTrack™ loading reagent for 3 hours at 37 °C/5 % CO_2, as described previously by Zaverucha-do-Valle et al. [27]. After incubation, the cells were washed three times with PBS and suspended in saline. The labeled cells were injected in week 9 or week 14. Cells were tracked in vivo by magnetic resonance imaging (MRI) measurements at 1 and 7 days after the injection and in the end stage of the disease.

Images were acquired with a 7 T magnetic resonance scanner (7 T/400 horizontal Varian scanner; Agilent Technologies, Santa Clara, CA, USA). Four proton-density fast spin-echo imaging sequences (repetition time = 2000 milliseconds; echo spacing = 10.5–12 milliseconds: matrix = 192 × 192; slice thickness = 0.5 mm; 7–10 continuous slices, no gap; 16 averages, axial field of view (FOV) = 25 mm × 25 mm, coronal FOV = 30 mm × 30 mm, sagittal FOV = 40 mm × 30 mm, total sagittal FOV = 50 mm × 30 mm) were used to investigate the spinal cord region.

Prior to image analysis, datasets were anonymized. For each dataset, all images were visually inspected for important artifacts. Data were processed using MRIcron software (Columbia, SC, USA).

Injection
SOD1^{G93A} animals were injected with BMMC or vehicle (saline) at 9 or 14 weeks old. Mice were anesthetized with xylazine (15 mg/kg; Vetbrands and ketamine (150 mg/kg; Vetbrands) (Goiânia, GO, Brazil) intraperitoneally. The animals were immobilized and the spine was exposed. The

vertebrae were carefully separated using two fine tweezers in order to reveal the lumbar spinal cord (L4–L5). The BMMC (10^6 cells) or saline were injected intraparenchymally with a glass micropipette connected to the nanoinjector (Nanoinject II; Drummond Scientific Company, Broomall, PA, USA) at the rate of 1 μl/minute for a total volume of 4 μl. After recovery from anesthesia, the animals from both groups were returned to the animal facility and kept in cages with food and water ad libitum. To assess the survival and functional outcome, the animals were divided into five groups: BMMC treated in week 9 ($n = 22$), saline injected in week 9 ($n = 24$), BMMC treated in week 14 ($n = 22$), saline injected in week 14 ($n = 22$), and wild-type ($n = 16$). Equal numbers of male and female were used in each group.

Functional evaluation

Functional tests were performed weekly in all experimental groups by blinded investigators. The tests performed were the rotarod, hanging wire, and motor score. In the rotarod test, the animals were placed in a rolling rod (maximum of 180 seconds) with an accelerating speed from 20 to 35 rpm. In the hanging-wire test, the animals were placed on the wire lid from their housing cage, where they remained upside down until they fell (maximum time 90 seconds). The longest period that the animal remained in the rotarod test and hanging-wire test was recorded after three trials.

In the motor score, the animals were graded as follows: 4 = no sign of motor dysfunction, 3 = hind-limb tremors when suspended by the tail and collapse or partial collapse of leg extension, 2 = evident motor dysfunction, 1 = paralysis in at least one limb, and 0 = not able to right itself within 30 seconds when placed on its side.

The statistical analyses were performed with the two-way analysis of variance (ANOVA) and Bonferroni post-test in GraphPad Prism 4.02 (GraphPad Software, Inc., La Jolla, CA, USA). For survival comparison, the log-rank test was used. Statistical significance was considered when $p \leq 0.05$. All data are presented as mean ± standard error of the mean (SEM).

Tissue preparation and immunohistochemistry

Animals ($n = 5/6$ animals per group) were perfused through the heart with ice-cold saline, followed by a solution of 4 % paraformaldehyde in 0.1 M phosphate buffer, pH 7.4, via the ascending aorta. The spinal cords were then removed, and cryoprotected by immersion in 20 % sucrose followed by 30 % sucrose in phosphate buffer. The spinal cords were sectioned transversely at 20 μm on a cryostat (CM 1850; Leica, Wetzlar, Germany) and collected on gelatin-coated slides.

For immunohistochemistry, sections were washed three times with PBS with 0.3 % Triton X-100, incubated with 5 % normal goat serum (NGS; Sigma) in PBS for 30 minutes, and then incubated with the primary antibodies monoclonal mouse anti-NeuN (1:200, #MAB 377; Chemicon, Temecula, CA, USA) and polyclonal rabbit anti-Iba1 (1:400, #019-19741; Wako, Richmond, VA, USA) overnight at 4 °C. After the primary incubations, the appropriate secondary Cy3®-conjugated or Alexa® 488-conjugated antibodies (Jackson ImmunoResearch, West Grove, PA, USA and Invitrogen Inc., Carlsbad, CA, respectively) were used. In some cases, the sections were incubated with TO-PRO®-3 (1:1000; Invitrogen) for nuclei staining. Sections were mounted with VectaShield® (Vector, Burlingame, CA, USA) and analyzed using confocal microscopes (LSM 510 META or LSM 510 META NLO; Zeiss GmbH, Oberkochen, Germany).

Motor-neuron and microglia quantification

For motor-neuron quantification we stained the lumbar spinal cord with NeuN antibody. We inspected every fifth transverse section of a total of 10 sections from the anterior horn of the lumbar spinal cord (20 μm). Counts were made using a 20× objective on a Zeiss Axiovert 200 M microscope equipped with fluorescence optics. The size of NeuN-positive cells in the anterior horn was quantified; cells featuring cell bodies with cross-sectional area ≥ 250 μm^2 were considered motor neurons. Smaller NeuN-positive cells were considered interneurons.

For microglia quantification, we stained the spinal cord with Iba1 antibody. We quantified every fifth section of a total of 20 transverse sections from the anterior horn of the lumbar spinal cord (20 μm). Counts were made using a 40× objective on a Zeiss Axiovert 200 M microscope with fluorescence optics.

Statistical analysis was performed by one-way ANOVA and Tukey's multiple comparison test in GraphPad Prism 4.02 (GraphPad Software, Inc., La Jolla, CA, USA). Statistical significance was considered when $p \leq 0.05$. All data are presented as mean ± SEM.

Results

Effect of BMMC treatment on motor functional outcome

To assess the effect of BMMC transplanted directly into the lumbar spinal cord, we analyzed the disease functional outcome using the rotarod, hanging-wire, and motor-score tests. We divided the animals into two groups: presymptomatic animals, which were injected with saline ($n = 24$) or BMMC ($n = 22$) in week 9; and symptomatic animals, which received the injection only in week 14 (saline, $n = 22$; BMMC, $n = 22$). These groups were also compared with wild-type animals of equivalent ages ($n = 16$).

In the rotarod test, compared with the wild-type mice, the animals treated with BMMC in the presymptomatic phase started to show motor deficits in week 13, while in the untreated group this decline was already apparent in week 11 (Fig. 1a). In the hanging-wire test, we also observed that the saline group started to show loss of function 2 weeks before the treated group (Fig. 1c). In the motor score analyses, we observed a delay in the functional loss in the BMMC-treated animals in week 9

(Fig. 1e). This group showed deficits in week 15 compared with wild-type mice, while saline injected animals showed a deficit already in week 14. However, we did not observe differences between the BMMC-injected and saline-injected animals.

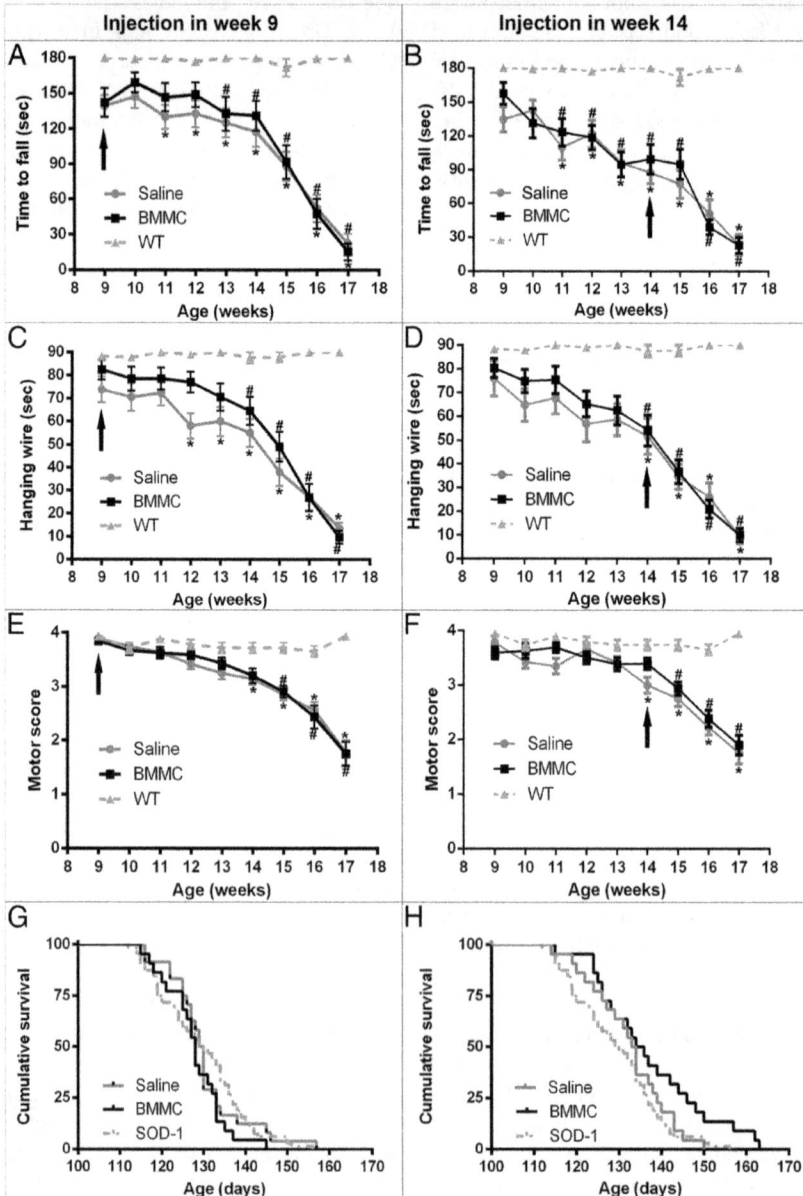

Fig. 1 Functional outcome and survival of SOD-1^{G93A} mice after BMMC therapy. **a, b** Rotarod test, **c, d** hanging-wire test, **e, f** motor-score test, and **g, h** cumulative survival. SOD-1^{G93A} mice were injected with BMMCs or saline at two different time points: presymptomatic phase (9 weeks old **a, c, e, g**) and symptomatic phase (14 weeks old **b, d, f, h**). The animals that received the treatment in week 9 showed a 2-week delay in the functional loss in the rotarod and hanging-wire tests. In the motor-score analyses, the saline-injected animals in week 9 showed a decline in week 14, while the BMMC-treated animals showed a decline in week 15. No differences were observed between the saline-injected and BMMC-injected animals in week 9 in the cumulative survival. The therapy in week 14 did not show an effect in any analysis performed. *Arrows* indicate week of treatment. *WT vs. saline (p <0.05), #WT vs. BMMC (p <0.05). *BMMC* bone marrow-derived mononuclear cells, *SOD1* Cu/Zn superoxide dismutase 1, *WT* wild-type

The animals treated with BMMCs in the symptomatic phase showed motor deficit as early as at the injection day when measured by all functional tests; and these deficits were never recovered or stabilized (Fig. 1b, d, f).

To determine whether the treatment could affect differently males and females, we performed a new analysis by gender. In the presymptomatic animals, we could observe

that female mice treated with BMMC showed a better performance in all tests described than the saline-injected female mice, when we compared both of them with the wild-type group (Fig. 2b, d, f). Male animals treated in week 9 showed a difference only in the hanging-wire test, where we could observe a delay in the functional loss (Fig. 2c). In the symptomatic animals, we observed a

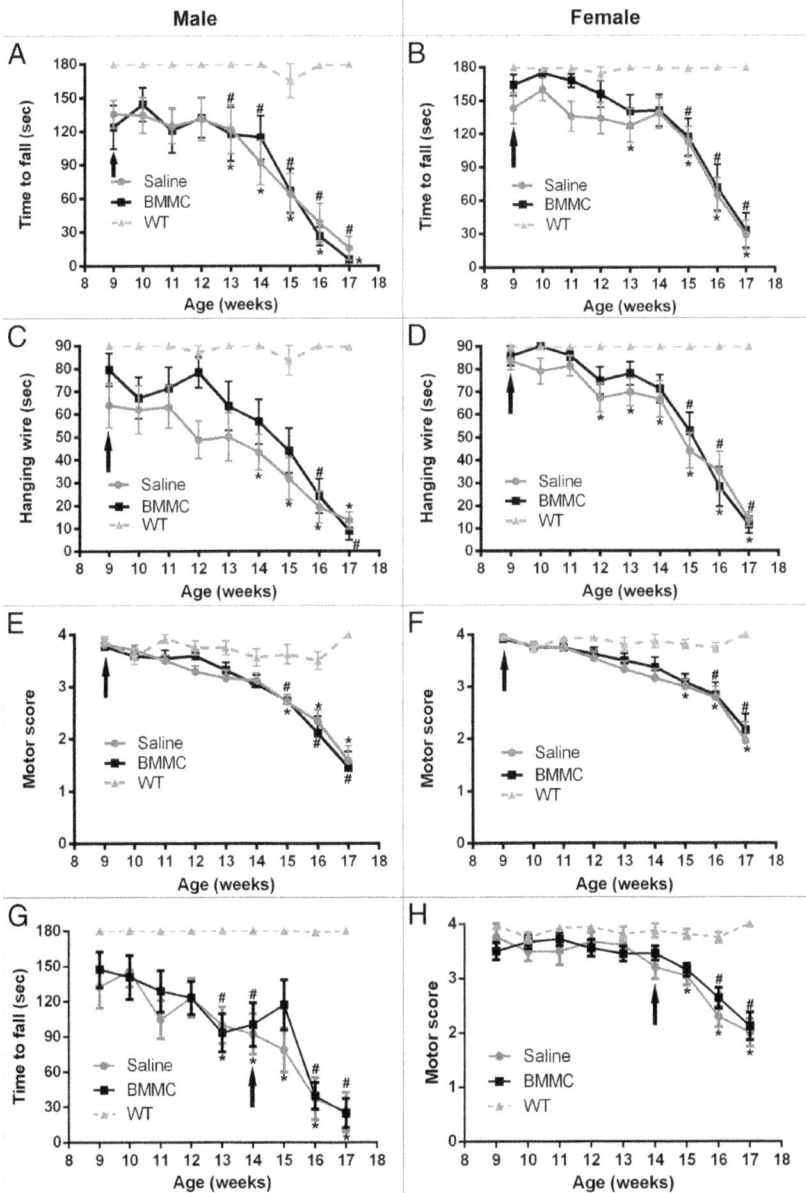

Fig. 2 Gender effect in the functional outcome of SOD-1^G93A mice after BMMC therapy. **a, b, g** Rotarod test, **c, d** hanging-wire test, and **e, f, h** motor-score test. Female SOD-1^G93A animals that received the treatment in week 9 showed a delay in the functional loss in all tests performed **b, d, f**. Male animals treated in week 9 showed a delay only in the hanging-wire test **c**. The therapy in week 14 improved the rotarod performance in males **g** and delayed the deficit analyzed by motor score grade in BMMC-injected females **h**. *Arrows* indicate week of treatment. *WT vs. saline (p <0.05), #WT vs. BMMC (p <0.05). BMMC bone marrow-derived mononuclear cells, WT wild-type

difference in the treated male animals only for the rotarod test (Fig. 2g). In this test, we observed that 1 week after treatment the animals improved their performance (week 15; Fig. 2g). However, this effect was not sustained in the following weeks. Moreover, when we compared both female groups with wild-type females we observed a 1-week delay in the manifestation of motor-score deficits in the BMMC animals when compared with the saline group (Fig. 2h).

There were no differences in survival between the BMMC-treated animals in the presymptomatic phase or the BMMC-treated animals in the symptomatic phase compared with the untreated controls (BMMC presymptomatic: 127.86 ± 7.05 days, saline presymptomatic: 130.33 ± 9.12 days, $p = 0.303$; BMMC postsymptomatic: 137.27 ± 13 days, saline postsymptomatic: 132.45 ± 9.323 days, $p = 0.11$) (Fig. 1g, h). However, we observed an increase in survival of BMMC-treated animals, in week 14, compared with noninjected mice (SOD1 non-injected: 129.54 ± 10.4; $n = 64$; male:female = 1:1; $p < 0.05$) (Fig. 1h). Also, it is important to notice that only BMMC-treated animals in the symptomatic phase (two of a total of 22; male:female = 1:1) survived longer than 160 days. We did not observe any gender effect in the

survival in animals treated in week 9 (male BMMC presymptomatic: 125.90 ± 7.84 days, male saline presymptomatic: 130.46 ± 7.81 days, $p = 0.06$; female BMMC presymptomatic: 130.07 ± 6.43 days, female saline presymptomatic: 130 ± 10.38 days, $p = 0.92$). We also did not observe a difference by gender in the animals treated in week 14 (male BMMC postsymptomatic: 134.45 ± 14.55 days, male saline postsymptomatic: 129 ± 9.90 days, $p = 0.35$; female BMMC postsymptomatic: 138.53 ± 10.98 days, female saline postsymptomatic: 134.58 ± 8.4 days, $p = 0.33$).

Neuronal survival and microglia activation

To determine whether the BMMC injection could have any local effect on the spinal cord, we analyzed the neuronal survival and microglia activation in the anterior horn of the lumbar spinal cord. Neuronal survival was analyzed using NeuN staining, and motor neurons were distinguished from interneurons by size (see Materials and methods; Fig. 3). We analyzed the spinal cord of the animals injected in the presymptomatic phase in week 12, 3 weeks after the injection. At this age we could already observe a significant decrease in the number of motor neurons in the lumbar spinal cord compared with

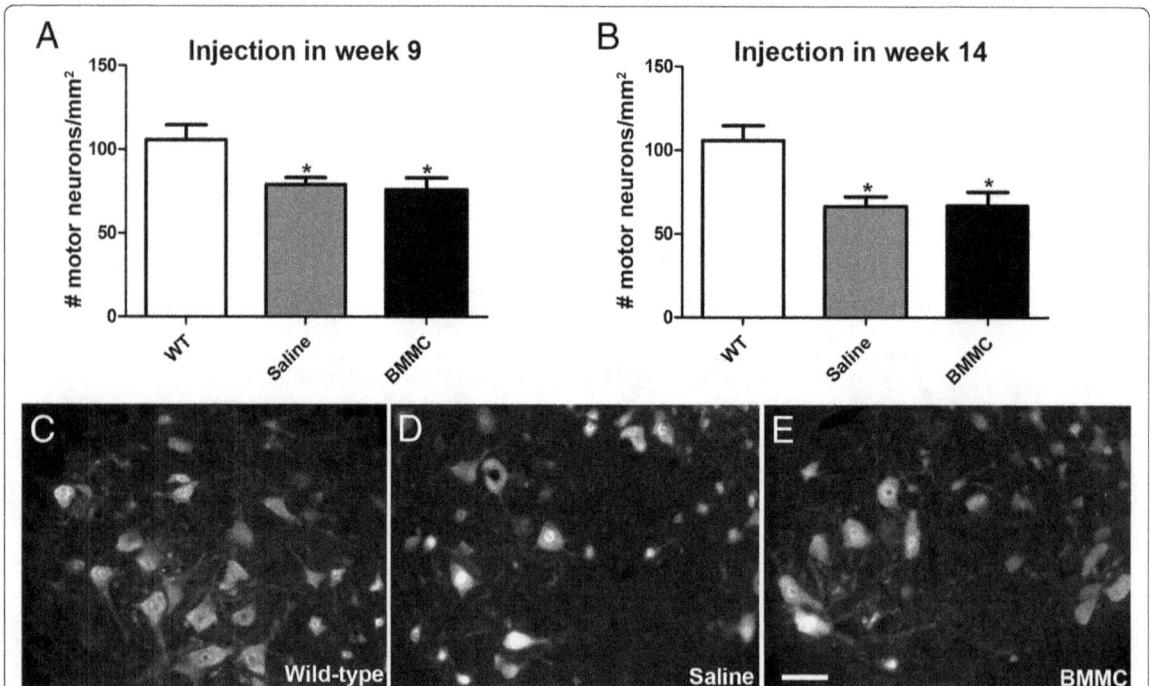

Fig. 3 Quantification of motor neurons in the anterior horn of the spinal cord. The number of NeuN-positive cells was quantified in week 12 in the presymptomatic injected animals **a** and in week 15 in the symptomatic injected animals **b**. Both saline-injected and BMMC-injected mice showed a decrease in the number of motor neurons compared with the wild-type mice at the time point analyzed. There was no difference in the number of motor neurons after the treatment with BMMC injected in week 9 or in week 14 **c–e**. Representative images of NeuN expression in wild-type mice **c**, saline-injected animals in week 9 **d**, and BMMC-injected animals in week 9 **e**. *$p < 0.01$. Scale bar: 50 μm. *BMMC* bone marrow-derived mononuclear cells, *WT* wild-type

the wild-type animals. These data are consistent with the motor-function analyses, which showed a decline in the untreated animals in week 12. The BMMC treatment in the presymptomatic phase did not affect the motor-neuron survival (saline injected in week 9: 79.02 ± 11.3 cells/mm^2, $n = 6$; BMMC injected in week 9: 76.05 ± 17.54 cells/mm^2, $n = 5$; wild-type: 105.8 ± 19.75 cells/mm^2, $n = 4$) (Fig. 3a).

In the symptomatic treatment protocol, we analyzed the spinal cord in week 15, 1 week after the BMMC or saline injection. As expected, we observed a significant reduction in the number of motor neurons in the untreated animals compared with the wild-type; however, the treatment did not prevent this loss (saline injected in week 14: 66.3 ± 14.69 cells/mm^2, $n = 7$; BMMC injected in week 14: 66.96 ± 23.04 cells/mm^2, $n = 8$) (Fig. 3b). Importantly, the number of interneurons observed in the anterior horn did not change in the transgenic animals compared with the wild-type animals at the time points analyzed (saline injected in week 9: 132.8 ± 74.64 cells/mm^2, $n = 6$; BMMC injected in week 9: 167.3 ± 87.53 cells/mm^2, $n = 6$; saline injected in week 14: 159.3 ± 73.61 cells/mm^2, $n = 8$; BMMC injected in week 14: 142.2 ± 41.99 cells/mm^2, $n = 8$; wild-type: 168.6 ± 102.7 cells/mm^2, $n = 7$) (Additional file 1: Figure S1). At the end stage of the disease, we observed a more extensive loss of motor neurons in the anterior horn. In this case we also did not observe any preservation of motor neurons after the BMMC injection, in either the presymptomatic or symptomatic phase (saline injected in week 9: 24.03 ± 6.33

cells/mm^2, $n = 4$; BMMC injected in week 9: 25.63 ± 14.9 cells/mm^2, $n = 4$; saline injected in week 14: 26.88 ± 10.92 cells/mm^2, $n = 5$; BMMC injected in week 14: 14.68 ± 11.66 cells/mm^2, $n = 5$) (Additional file 1: Figure S1).

Microglia cells in the anterior horn were stained with Iba1. In the presymptomatic treated animals, we observed an increase in the number of microglia in week 12 in the untreated compared with the wild-type animals (Fig. 4). The BMMC animals treated in the presymptomatic phase showed a similar increase (saline injected in week 9: 539.6 ± 149.1 cells/mm^2, $n = 7$; BMMC injected in week 9: 495.5 ± 126.5 cells/mm^2, $n = 6$; wild-type: 172.2 ± 116.7 cells/mm^2, $n = 6$) (Fig. 4d).

The animals treated in the symptomatic phase showed a similar pattern as the presymptomatic groups: both saline and BMMC groups, at 15 weeks of age, had an increased number of microglia in the lumbar anterior horn compared with the wild-type animals (saline injected in week 14: 705.0 ± 190.0 cells/mm^2, $n = 8$; BMMC injected in week 14: 861.3 ± 166.1 cells/mm^2, $n = 8$) (Fig. 4g). However, we did not observe a difference between the treated and untreated groups.

At the end stage of the disease, the number of microglia increased greatly compared with the wild-type animals; however, there was no difference between the untreated animals and the BMMC-injected animals in the presymptomatic or symptomatic phase (saline injected in week 9: 877.56 ± 160.8 cells/mm^2, $n = 4$; BMMC injected in week 9: 1016.36 ± 147.1 cells/mm^2, $n = 5$; saline injected in week

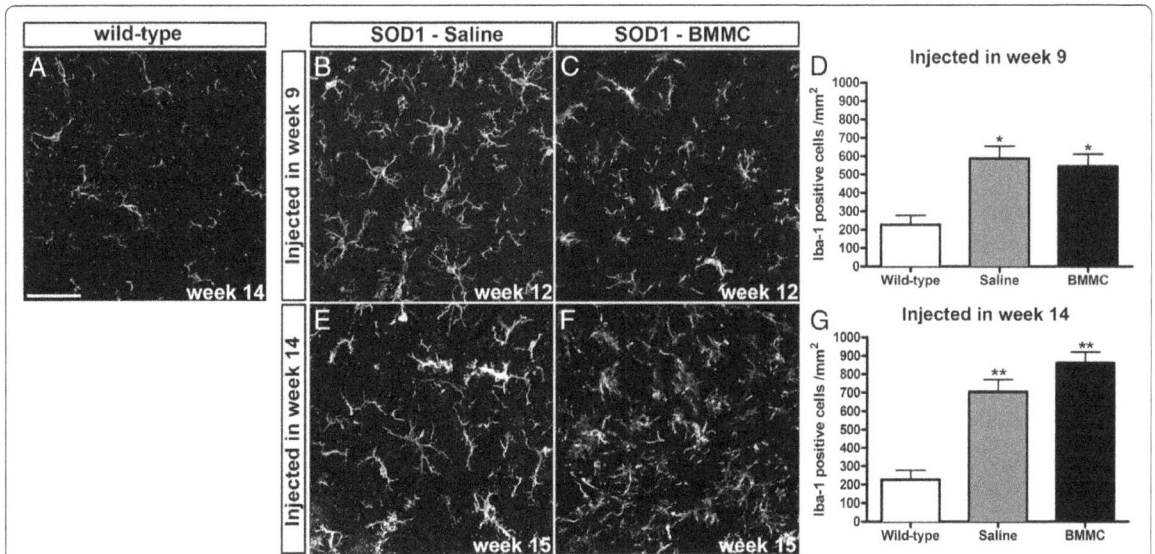

Fig. 4 Quantification of microglial cells in the anterior horn of the spinal cord. The number of Iba1-positive cells was quantified in week 12 in the presymptomatic injected animals **b–d** and in week 15 in the symptomatic injected animals **e–g**. Iba1-positive cells in the wild-type animals are shown in **a**. The saline-injected and BMMC-injected mice showed an increase in the number of Iba1-positive cells compared with the wild-type mice. There was no difference in the number of microglia cells after the treatment with BMMC injected in week 9 or week 14. $*p < 0.01$, $**p < 0.001$. Scale bar: 50 μm. *BMMC* bone marrow-derived mononuclear cells, *SOD1* Cu/Zn superoxide dismutase 1

14: 744.2 ± 323.8 cells/mm^2, $n = 5$; BMMC injected in week 14: 824.5 ± 333.7 cells/mm^2, $n = 6$).

Fate of BMMC after spinal cord injection

In this study, we used different techniques in order to investigate the fate of the injected cells. In the first approach, we used cells from transgenic mice carrying the enhanced green fluorescent protein (eGFP) under the control of the actin promoter. The fate of BMMC-eGFP transplanted in presymptomatic animals was analyzed at 4, 7, and 65 (end stage) days after the injection. In sections of spinal cord tissue we observed many eGFP cells 4 days after the injection, distributed on the dorsal–ventral axis. However, the number of these cells decreased drastically by 7 days, and at the end stage we could not find any transplanted cells (Fig. 5). Analyzing the distribution of microglia (Iba1-positive cells) in the same animals, we observed that 4 days after injection many BMMC were found in association with microglia (Fig. 5a, arrows). Seven days after injection, the few BMMC observed were not clearly associated with microglia (Fig. 5b′).

Alternatively, we also labeled the cells with the fluorescent marker CellTrace™ before injection. Seven days after injection of BMMC in a symptomatic animal, it was possible to observe CellTrace™-labeled cells in the spinal cord sections. However, we did not observe the usual cytoplasmatic pattern of labeling, and only a few cells showed the staining associated with cell nuclei (Fig. 6). Analyzing microglia in these animals, we observed a high concentration of microglia in the injection site (square in Fig. 6a and at higher magnification in Fig. 6b–d), where many cells exhibited a round phagocyte morphology similar to the transplanted cells (Fig. 6b–d, arrows).

To further investigate the fate of the transplanted cells, we labeled them with SPION prior to the injection and we imaged the spinal cord region in vivo with MRI for several weeks in the same animal. We were able to detect the injected cells as a hypointense signal in the spinal cord, and this signal was present at all stages examined (Fig. 7, arrows). Importantly, even 72 days after the injection (end stage of the animal injected in week 9) the signal was still present at the injection site. Histological analyses of these animals were performed in order to identify whether the signal could be correlated with cell bodies or was due to free iron. We used differential and interferential contrast to observe the SPION and Iba1 to identify microglia/macrophages. The majority of the SPION staining was located in the ventral horn, and most of the iron seemed be associated with Iba1-positive cells (Fig. 7, arrowheads). However, it was possible to find some SPION that was not associated with the Iba1 staining (Fig. 7, arrows).

Discussion

ALS is a fatal neurodegenerative disease for which there is no available treatment, and patients survive only 3–5 years on average after the symptoms start. ALS is a multifactorial disease, and different cell types seem to play important roles in the physiopathology. The need for an effective treatment has stimulated many researchers to test cell therapy. In our study, we injected BMMC into a mouse ALS model during the presymptomatic and symptomatic phases of the disease, to test its therapeutic potential.

Bone marrow cell therapy seems to be a promising treatment for ALS. This approach has shown good results in preclinical studies of other neurological diseases or traumas in the CNS, such as brain ischemia or optic-nerve crush [17, 25, 28]. The BMMC fraction constitutes a heterogeneous population formed mainly by hematopoietic cells, such as monocytes, lymphocytes, and granulocytes, and a small percentage of mesenchymal and hematopoietic stem cells, as described previously by our group [19]. BMMC treatment has proven to increase nerve regeneration and neural stem cell proliferation, and to reduce neuronal death and microglia activation [17, 19, 26, 29]. The main mechanism attributed to these beneficial effects is the capacity to release growth factors and cytokines [30, 31]. Besides this, the ALS model demonstrated that both wild-type monocytes and T lymphocytes (both cell types present in BMMC) delay disease progression in the mouse model, and could contribute to the beneficial effect of this therapy [4, 12].

BMMC therapy has already been tested in clinical trials for different neurological diseases [32, 33]. In ALS patients there is a clinical trial injecting these cells into the spinal cord, demonstrating that this approach is feasible and safe [23]. However, it is also important to analyze the efficacy of this therapy. Preclinical studies, as the present one, could contribute to the design of phase 2/3 clinical trials to test efficacy, such as through analysis of the best time points, dose, and routes of administration.

In this work, we performed different functional tests and histopathological analyses of the SOD1^{G93A} animals treated with BMMC in two distinctive time points. In order to analyze the functional deficit, we use three distinctive tests—rotarod, hanging wire, and motor score. Comparing the gender-matched SOD1^{G93A} animals' performance in these tests with the wild-type animals' performance, we could observe when the deficits start in our animals and we were able to observe a delay in the beginning of the motor loss in the BMMC-treated animals in the presymptomatic phase of the disease—although we were not able to see significant differences when we compared untreated and treated SOD1^{G93A} groups. Treatment in the symptomatic phase

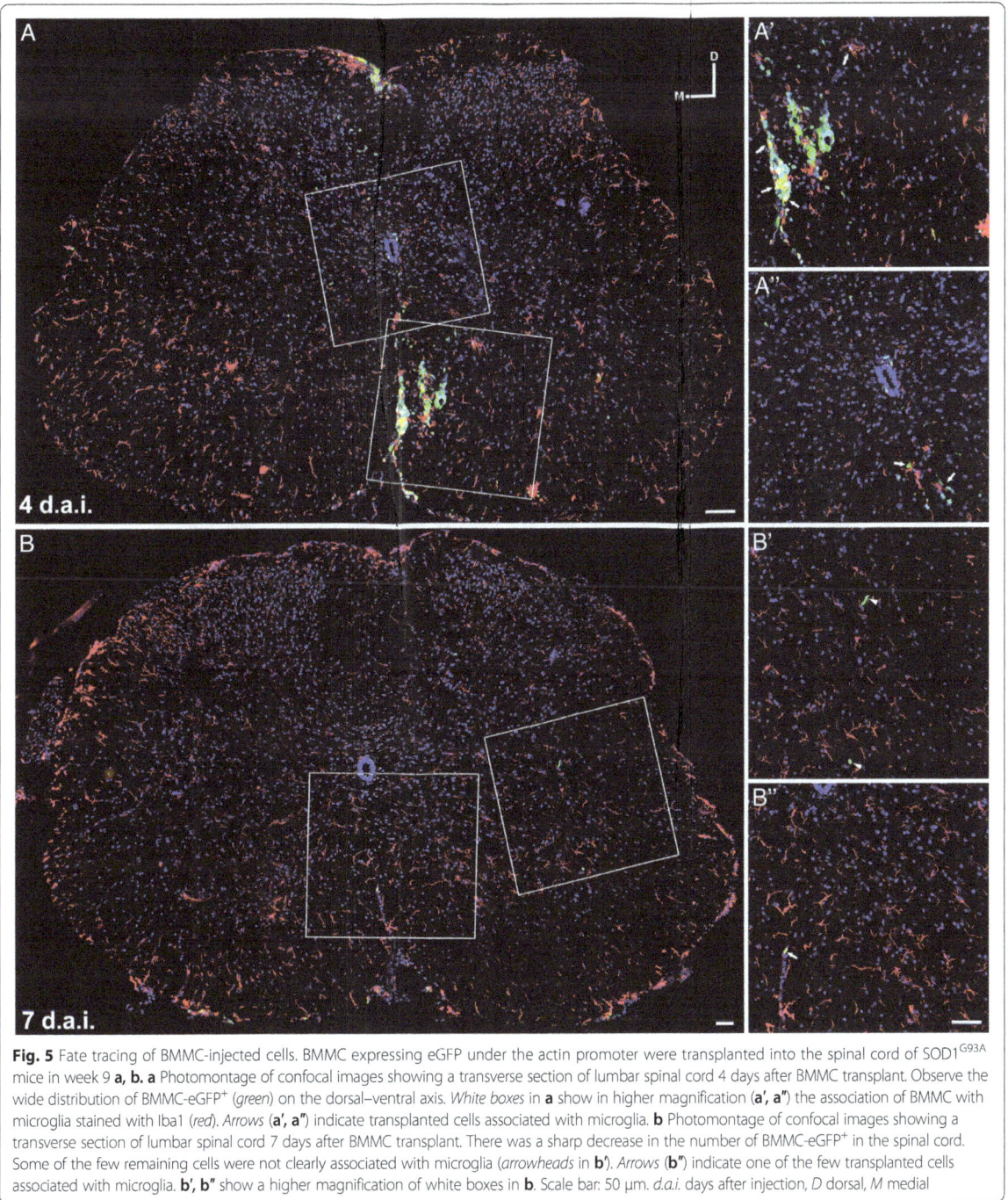

Fig. 5 Fate tracing of BMMC-injected cells. BMMC expressing eGFP under the actin promoter were transplanted into the spinal cord of SOD1^{G93A} mice in week 9 **a, b. a** Photomontage of confocal images showing a transverse section of lumbar spinal cord 4 days after BMMC transplant. Observe the wide distribution of BMMC-eGFP$^+$ (*green*) on the dorsal–ventral axis. *White boxes* in **a** show in higher magnification (**a', a''**) the association of BMMC with microglia stained with Iba1 (*red*). *Arrows* (**a', a''**) indicate transplanted cells associated with microglia. **b** Photomontage of confocal images showing a transverse section of lumbar spinal cord 7 days after BMMC transplant. There was a sharp decrease in the number of BMMC-eGFP$^+$ in the spinal cord. Some of the few remaining cells were not clearly associated with microglia (*arrowheads* in **b'**). *Arrows* (**b''**) indicate one of the few transplanted cells associated with microglia. **b', b''** show a higher magnification of white boxes in **b**. Scale bar: 50 μm. *d.a.i.* days after injection, *D* dorsal, *M* medial

had no effect. Neither therapy had an effect on survival. These results are different from most of the studies that have used bone marrow cells in rodent ALS models. Nevertheless, there are important differences between our study and others that have been reported.

In mouse ALS models, wild-type bone marrow cells were used to replace the bone marrow from irradiated SOD1^{G93A} animals. The cells were transplanted in the pre-symptomatic phase, migrated, and invaded the spinal cord during the course of the disease. Under this approach, the results were contradictory: one group observed an increase

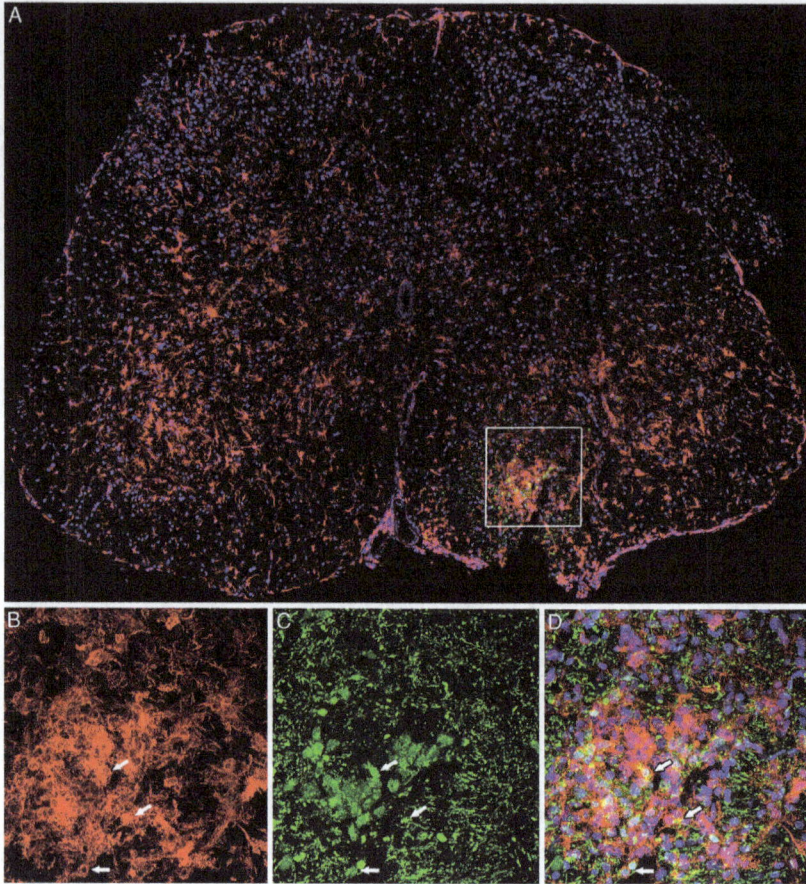

Fig. 6 Fate tracing of BMMC-injected cells stained with CellTrace™. BMMC stained with the fluorescent cell marker CellTrace™ (*green*) were transplanted into the spinal cord of SOD1[G93A] mice in week 14. Photomontage of confocal images showing a transverse section of lumbar spinal cord 7 days after BMMC transplant. We observed a dispersed pattern of CellTrace™ staining in the ventral horn. Microglia were stained with Iba1 (*red*). Observe the high concentration of microglia at the injection site (*white box* in **a**). **b–d** Higher magnification of the area delimited by the white box in **a**, showing the association of BMMC with microglia stained with Iba1. *Arrows* in **b–d** indicate microglia phagocyte profiles interacting with the transplanted cells. Scale bar: 50 µm

in lifespan, while another found no positive effect [34, 35]. These conflicting results could be due to the number of cells that reached the damaged area. In our study, we injected BMMC directly into the lumbar spinal cord of presymptomatic SOD1[G93A] mice, and we observed a positive result only in the functional outcome but not in the lifespan. However, since we performed a single local injection and we did not observe migration of these cells outside the injection site, it is possible that multiple injections are required to observe a more substantial positive effect in the functional outcome and survival. To our knowledge, this is the first report that injects BMMC into the spinal cord of SOD1[G93A] mice.

The number of injected cells could also be crucial to obtain a positive outcome. Previous works using 10^5 mesenchymal stem cells or 2×10^4 neural stem cells

observed an increase in the lifespan of ALS mice [20, 36]. Pastor et al. [37] injected 10^6 bone marrow cells into the spinal cord parenchyma of mdf/ocd mice, and observed a beneficial functional effect. However, in the present work, using the same amount of BMMC in the SOD1[G93A] mice model, we only saw a slight positive effect in the presymptomatic injected animals. One possible explanation is that although we injected a significant amount of cells, BMMC do not seem to persist in the spinal cord for a long period.

It is possible that the injection into the spinal cord could stimulate more inflammation, worsening the situation and leading to a decrease in the survival time. However, we also did not see a decrease in survival comparing the saline-injected animals with SOD1[G93A] mice that did not undergo any procedure (Additional file 2:

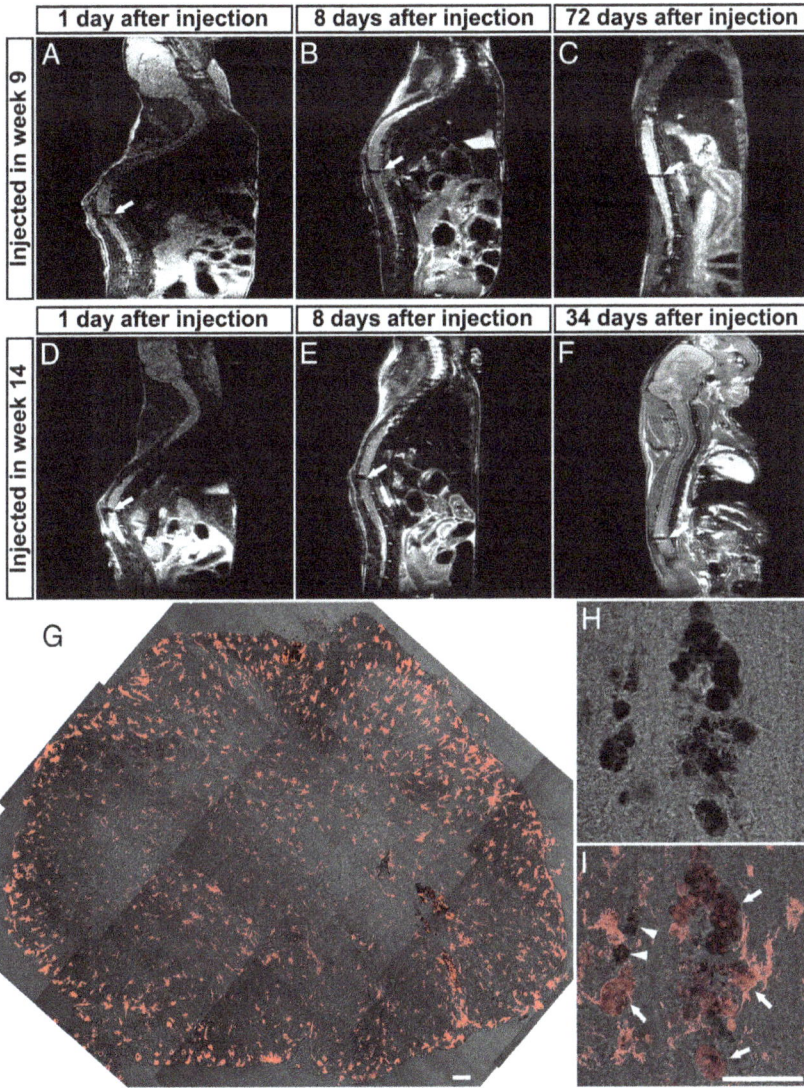

Fig. 7 In vivo fate tracing of BMMC-injected cells. BMMC were labeled with SPION and injected into the lumbar spinal cord in week 9 **a–c** or in week 14 **d–f**. Representative images of in vivo MRI in longitudinal planes of the spinal cord at different time points. *Arrows* indicate hypointense (*black*) areas corresponding to SPION-labeled cells in the spinal cord. Labeled cells were found at day 1 **a, d**, day 8 **b, e**, and at the end stage of the disease, which corresponds to 72 days after presymptomatic injection **c** and 34 days after symptomatic injection **f. g** Transverse section of the spinal cord from the animal analyzed by MRI in **f**. SPION staining is revealed by differential and interference contrast (dark signal in **g–i**). Iba1 was used to identify microglia/macrophages (*red*). We observed only faint staining throughout the spinal cord, with most of the SPION staining located in the ventral horn. Some SPION staining was associated with Iba1-positive cells (*arrowheads* in **i**) and some with Iba1-negative cells (*arrows* in **i**). Scale bar: 50 μm

Figure S2). Moreover, considering that other groups have previously tested this approach, as already described, the strategy of injecting cells directly into the spinal cord seems to be feasible.

We observed an increase in survival in the animals treated in week 14 when compared with noninjected mice. However, we did not observe this difference between saline and treated mice in week 14. It is important to highlight that most of saline-injected and BMMC-injected mice in week 14 were littermates, which decreases possible variability. We also observed that only in the BMMC-treated mice in the symptomatic phase did a couple of animals survive more than 160 days. Thus, it is possible that by increasing the number of cells or the number of injection sites we would observe a positive effect of the therapy.

In this work, we analyzed whether BMMC therapy could affect differently male and females. It is well known that in this animal model females show a longer lifespan than males [38]. This difference could be the result of a protective role of estrogen [39]. In our study, we observe a discrete influence of gender in the functional tests. In the presymptomatic injection, the therapy seems to be more effective in females than males, since BMMC-treated female mice showed a delay in functional loss in all tests performed, while treated males showed this only in the hanging-wire test. In the symptomatic injection, when we separate the groups by gender, we were able to see an improvement or stabilization in the functional outcome, especially in males for the rotarod test. Others studies had demonstrated a distinguished response by gender in many therapies. For instance, Kondo et al. [40] demonstrated that lumbar-injected glial-rich neural progenitors improved survival only in males. Morita et al. [41], however, showed that intrathecal transplanted mesenchymal stem cells decrease disease duration only in females. So far, there is still no solid explanation for how distinguished therapies could act differently in males and females.

In this study, we tried different techniques in order to trace the fate of the injected cells. Using MRI, we could observe the BMMC injected in the same animal during the course of the disease. Until the end stage, these cells seemed to remain close to the site of injection. This low migration rate could be the reason why we observed an effect only in the functional test, but not in the lifespan, since the neurodegeneration occurs in the entire spinal cord, not only in the lumbar region. In order to overcome this possible problem, it could be interesting to test multiple injections into different regions of the spinal cord.

Importantly, we could only observe BMMC until the end stage of the disease by using SPION labeling. When we injected eGFP-BMMC or BMMC labeled with Cell-Trace™, we could not track the cells until the end stage. A possible explanation is that the fluorescent stain is less stable, and during the procedures to analyze the spinal cord this stain was lost. However, when we used a specific antibody against eGFP, we still could not find the cells. When we analyzed the BMMC labeled with SPION in cryosections of the spinal cord, we could see that many labeled cells were in close apposition to Iba1-positive cells. It could be argued that most of the BMMC were phagocytized by resident microglia, and these cells retained the SPION label. However, the BMMC population includes monocytes, which could survive in the spinal cord and differentiate into macrophages that also express Iba1. Nevertheless, we also found SPION-labeled cells that did not stain with Iba1, suggesting that at least a part of the BMMC population was not phagocytized.

Many groups have tried different drugs to reduce the disease progression and extend the lifespan in animal ALS models. However, although some of them showed positive effects in the animal test, the results did not translate to clinical practice [42]. One of the main concerns at present is that most of the preclinical studies treat the animals in the presymptomatic phase. In this period, the animals do not show functional deficits or significant motor neuronal death. This contrasts with the clinical trials, where the patients already show significant loss of function. In this study we injected BMMC at two different time points, in the presymptomatic phase (9 weeks old) and in the symptomatic phase (14 weeks old). We observed a positive effect in the functional tests when we injected the BMMC in the presymptomatic phase, but not in the symptomatic phase. It is possible that the environment in the symptomatic phase is already too damaged to recover with this therapy, and it is important to take this into consideration before attempting to translate this treatment to human patients.

Conclusion

The results presented here demonstrate that BMMC therapy, used in the presymptomatic phase, could retard the loss of function in SOD1^{G93A} mice. An increase in the number of injection sites could improve this outcome. However, when BMMC were injected in the symptomatic phase we could not see any effect; this indicates that we should look for another type of cell or even combine pharmacological therapy with cell therapy to potentiate the results in the symptomatic phase.

Additional files

Additional file 1: Figure S1. Quantification of neurons in the anterior horn of the spinal cord. The number of interneurons (NeuN-positive cells with cross-sectional area ≤250 μm^2) was analyzed in week 12 in the presymptomatic injected animals **a** and in week 15 in the symptomatic injected animals **b**. There was no difference in the number of interneurons in the SOD1^{G93A} mice compared with the wild-type animals. The number of motor neurons was analyzed in the end stage of the disease in the presymptomatic injected animals **c** and in the symptomatic injected animals **d**. Both saline-injected and BMMC-injected mice showed a decrease in the number of motor neurons compared with the wild-type mice at the time point analyzed. ***p <0.001. (TIF 745 kb)

Additional file 2: Figure S2. Survival do SOD-1^{G93A} mice. **a, b** Comparison of survival of SOD-1^{G93A} mice that did not suffer any surgical procedure with SOD-1^{G93A} mice injected with saline at two different time points: presymptomatic phase (9 weeks old) and symptomatic phase (14 weeks old). No differences were observed between the groups. (TIF 288 kb)

Abbreviations
ALS: Amyotrophic lateral sclerosis; ANOVA: Analysis of variance; BMMC: Bone marrow-derived mononuclear cells; CNS: Central nervous system; DMEM: Dulbecco's modified Eagle's medium; eGFP: Enhanced green fluorescent protein; FOV: Field of view; MRI: Magnetic resonance imaging; NGS: Normal goat serum; PBS: Phosphate-buffered saline; SEM: Standard

error of the mean; SOD1: Cu/Zn superoxide dismutase 1;
SPION: Superparamagnetic iron nanoparticles.

Competing interests
The authors declare that they have no competing interests.

Authors' contributions
FG was responsible for conception and design, acquisition of data, analysis and interpretation of the data, and development and writing of the manuscript. ABD performed functional tests, BMMC extraction, BMMC staining, histology procedures, motor-neuron quantification, and interpretation of experimental results and revised the manuscript. IB-P performed functional tests, BMMC extraction, and interpretation of experimental results and revised the manuscript. FRF performed functional tests, BMMC extraction, histology procedures, and motor-neuron quantification and revised the manuscript. CZ-d-V performed BMMC staining, MRI analysis, interpretation of experimental results, and manuscript writing. FT-M performed MRI acquisition, interpretation of experimental results, and drafting and revision of the manuscript. LH performed animal copy number analysis and interpretation of experimental results and revised the manuscript. TPU performed animal copy number design analysis and interpretation of experimental results and revised the manuscript. MFS performed fluorescent analysis and contributed to the interpretation of experimental results, and development and writing of the manuscript.RM-O directed the project and contributed to the general administration, interpretation of experimental results, and development and writing of the manuscript. All authors read and approved the manuscript.

Acknowledgements
The authors thank Felipe Marins, Camila Teixeira, and Fernanda Meireles Ferreira for technical assistance, and Robert Brown Jr and the ALS Foundation for providing the mutant animals.
This study was supported by grants from the Conselho Nacional de Desenvolvimento Científico e Tecnológico (www.cnpq.br), Coordenação de Aperfeiçoamento de Pessoal de Nível Superior (www.capes.gov.br), Fundação Carlos Chagas Filho de Amparo à Pesquisa do Estado do Rio de Janeiro (www.faperj.br), and the Brazilian Ministry of Health (www.saude.gov.br). The funders had no role in study design, data collection and analysis, decision to publish, or preparation of the manuscript.

Author details
[1]Instituto de Biofísica Carlos Chagas Filho, Centro de Ciências da Saúde, Sala G2-028, Universidade Federal do Rio de Janeiro, Cidade Universitária, RJ 21941-902 Rio de Janeiro, Brazil. [2]Evandro Chagas National Institute of Infectious Diseases (INI), Oswaldo Cruz Foundation, Avenida Brasil 4365Maguinhos RJ 21040-900 Rio de Janeiro, Brazil. [3]Institute of Biomedical Sciences and National Center of Structural Biology and Bioimaging, CENABIO, Centro de Ciências da Saúde, Universidade Federal do Rio de Janeiro, Cidade Universitária, RJ 21941-902 Rio de Janeiro, Brazil. [4]Instituto D'Or de Pesquisa e Educação (IDOR), Rua Diniz Cordeiro 30Botafogo RJ 22281-100 Rio de Janeiro, Brazil.

References
1. Rosen DR, Siddique T, Patterson D, Figlewicz DA, Sapp P, Hentati A, et al. Mutations in Cu/Zn superoxide dismutase gene are associated with familial amyotrophic lateral sclerosis. Nature. 1993;362(6415):59–62.
2. Gurney ME. Transgenic-mouse model of amyotrophic lateral sclerosis. N Engl J Med. 1994;331(25):1721–2.
3. Nagai M, Aoki M, Miyoshi I, Kato M, Pasinelli P, Kasai N, et al. Rats expressing human cytosolic copper-zinc superoxide dismutase transgenes with amyotrophic lateral sclerosis: associated mutations develop motor neuron disease. J Neurosci. 2001;21(23):9246–54.
4. Beers DR, Henkel JS, Xiao Q, Zhao W, Wang J, Yen AA, et al. Wild-type microglia extend survival in PU.1 knockout mice with familial amyotrophic lateral sclerosis. Proc Natl Acad Sci U S A. 2006;103(43):16021–6.
5. Boillee S, Yamanaka K, Lobsiger CS, Copeland NG, Jenkins NA, Kassiotis G, et al. Onset and progression in inherited ALS determined by motor neurons and microglia. Science. 2006;312(5778):1389–92.
6. Di Giorgio FP, Carrasco MA, Siao MC, Maniatis T, Eggan K. Non-cell autonomous effect of glia on motor neurons in an embryonic stem cell-based ALS model. Nat Neurosci. 2007;10(5):608–14.
7. Nagai M, Re DB, Nagata T, Chalazonitis A, Jessell TM, Wichterle H, et al. Astrocytes expressing ALS-linked mutated SOD1 release factors selectively toxic to motor neurons. Nat Neurosci. 2007;10(5):615–22.
8. Haidet-Phillips AM, Hester ME, Miranda CJ, Meyer K, Braun L, Frakes A, et al. Astrocytes from familial and sporadic ALS patients are toxic to motor neurons. Nat Biotechnol. 2011;29(9):824–8.
9. Beers DR, Zhao W, Liao B, Kano O, Wang J, Huang A, et al. Neuroinflammation modulates distinct regional and temporal clinical responses in ALS mice. Brain Behav Immun. 2011;25(5):1025–35.
10. Engelhardt JI, Tajti J, Appel SH. Lymphocytic infiltrates in the spinal cord in amyotrophic lateral sclerosis. Arch Neurol. 1993;50(1):30–6.
11. Alexianu ME, Kozovska M, Appel SH. Immune reactivity in a mouse model of familial ALS correlates with disease progression. Neurology. 2001;57(7):1282–9.
12. Beers DR, Henkel JS, Zhao W, Wang J, Appel SH. CD4+ T cells support glial neuroprotection, slow disease progression, and modify glial morphology in an animal model of inherited ALS. Proc Natl Acad Sci U S A. 2008;105(40):15558–63.
13. Henkel JS, Beers DR, Zhao W, Appel SH. Microglia in ALS: the good, the bad, and the resting. J Neuroimmune Pharmacol. 2009;4(4):389–98.
14. Morrison KE. Therapies in amyotrophic lateral sclerosis-beyond riluzole. Curr Opin Pharmacol. 2002;2(3):302–9.
15. Ohnishi S, Ito H, Suzuki Y, Adachi Y, Wate R, Zhang J, et al. Intra-bone marrow-bone marrow transplantation slows disease progression and prolongs survival in G93A mutant SOD1 transgenic mice, an animal model mouse for amyotrophic lateral sclerosis. Brain Res. 2009;1296:216–24.
16. Corti S, Nizzardo M, Nardini M, Donadoni C, Salani S, Simone C, et al. Systemic transplantation of c-kit + cells exerts a therapeutic effect in a model of amyotrophic lateral sclerosis. Hum Mol Genet. 2010;19(19):3782–96.
17. Zaverucha-do-Valle C, Gubert F, Bargas-Rega M, Coronel JL, Mesentier-Louro LA, Mencalha A, et al. Bone marrow mononuclear cells increase retinal ganglion cell survival and axon regeneration in the adult rat. Cell Transplant. 2011;20(3):391–406.
18. Uccelli A, Milanese M, Principato MC, Morando S, Bonifacino T, Vergani L, et al. Intravenous mesenchymal stem cells improve survival and motor function in experimental amyotrophic lateral sclerosis. Mol Med. 2012;18:794–804.
19. Gubert F, Zaverucha-do-Valle C, Figueiredo FR, Bargas-Rega M, Paredes BD, Mencalha AL, et al. Bone-marrow cell therapy induces differentiation of radial glia-like cells and rescues the number of oligodendrocyte progenitors in the subventricular zone after global cerebral ischemia. Stem Cell Res. 2013;10(2):241–56.
20. Vercelli A, Mereuta OM, Garbossa D, Muraca G, Mareschi K, Rustichelli D, et al. Human mesenchymal stem cell transplantation extends survival, improves motor performance and decreases neuroinflammation in mouse model of amyotrophic lateral sclerosis. Neurobiol Dis. 2008;31(3):395–405.
21. Zhou C, Zhang C, Zhao R, Chi S, Ge P, Zhang C. Human marrow stromal cells reduce microglial activation to protect motor neurons in a transgenic mouse model of amyotrophic lateral sclerosis. J Neuroinflammation. 2013;10:52.
22. Suzuki M, McHugh J, Tork C, Shelley B, Hayes A, Bellantuono I, et al. Direct muscle delivery of GDNF with human mesenchymal stem cells improves motor neuron survival and function in a rat model of familial ALS. Mol Ther. 2008;16(12):2002–10.
23. Blanquer M, Moraleda JM, Iniesta F, Gomez-Espuch J, Meca-Lallana J, Villaverde R, et al. Neurotrophic bone marrow cellular nests prevent spinal motoneuron degeneration in amyotrophic lateral sclerosis patients: a pilot safety study. Stem Cells. 2012;30(6):1277–85.
24. Karussis D, Karageorgiou C, Vaknin-Dembinsky A, Gowda-Kurkalli B, Gomori JM, Kassis I, et al. Safety and immunological effects of mesenchymal stem cell transplantation in patients with multiple sclerosis and amyotrophic lateral sclerosis. Arch Neurol. 2010;67(10):1187–94.
25. de Vasconcelos Dos Santos A, da Costa RJ, Diaz Paredes B, Moraes L, Jasmin, Giraldi-Guimaraes A, et al. Therapeutic window for treatment of cortical ischemia with bone marrow-derived cells in rats. Brain Res. 2010;1306:149–58.
26. Ramos AB, Vasconcelos-Dos-Santos A, de Souza SAL, Rosado-de-Castro PH, da Fonseca LMB, Gutfilen B, et al. Bone-marrow mononuclear cells reduce neurodegeneration in hippocampal CA1 layer after transient global ischemia in rats. Brain Res. 2013;1522:1–11.
27. Zaverucha-do-Valle C, Mesentier-Louro L, Gubert F, Mortari N, Padilha AB, Paredes BD, et al. Sustained effect of bone marrow mononuclear cell therapy in axonal regeneration in a model of optic nerve crush. Brain Res. 2014;1587:54–68.

28. Rosado-de-Castro PH, Pimentel-Coelho PM, da Fonseca LM, de Freitas GR, Mendez-Otero R. The rise of cell therapy trials for stroke: review of published and registered studies. Stem Cells Dev. 2013;22(15):2095–111.

29. Kocsis JD, Honmou O. Bone marrow stem cells in experimental stroke. Prog Brain Res. 2012;201:79–98.

30. Borlongan CV, Glover LE, Tajiri N, Kaneko Y, Freeman TB. The great migration of bone marrow-derived stem cells toward the ischemic brain: therapeutic implications for stroke and other neurological disorders. Prog Neurobiol. 2011;95(2):213–28.

31. Leal MM, Costa-Ferro ZS, Souza BS, Azevedo CM, Carvalho TM, Kaneto CM, et al. Early transplantation of bone marrow mononuclear cells promotes neuroprotection and modulation of inflammation after status epilepticus in mice by paracrine mechanisms. Neurochem Res. 2014;39(2):259–68.

32. Rosado-de-Castro PH, Schmidt Fda R, Battistella V, de Lopes de Souza SA, Gutfilen B, Goldenberg RC, et al. Biodistribution of bone marrow mononuclear cells after intra-arterial or intravenous transplantation in subacute stroke patients. Regen Med. 2013;8(2):145–55.

33. Sharma AK, Sane HM, Paranjape AA, Gokulchandran N, Nagrajan A, D'sa M, et al. The effect of autologous bone marrow mononuclear cell transplantation on the survival duration in amyotrophic lateral sclerosis—a retrospective controlled study. Am J Stem Cells. 2015;4(1):50–65.

34. Corti S, Locatelli F, Donadoni C, Guglieri M, Papadimitriou D, Strazzer S, et al. Wild-type bone marrow cells ameliorate the phenotype of SOD1-G93A ALS mice and contribute to CNS, heart and skeletal muscle tissues. Brain. 2004; 127(Pt 11):2518–32.

35. Solomon JN, Lewis CA, Ajami B, Corbel SY, Rossi FM, Krieger C. Origin and distribution of bone marrow-derived cells in the central nervous system in a mouse model of amyotrophic lateral sclerosis. Glia. 2006;53(7):744–53.

36. Corti S, Locatelli F, Papadimitriou D, Del Bo R, Nizzardo M, Nardini M, et al. Neural stem cells LewisX+ CXCR4+ modify disease progression in an amyotrophic lateral sclerosis model. Brain. 2007;130(Pt 5):1289–305.

37. Pastor D, Viso-Leon MC, Jones J, Jaramillo-Merchan J, Toledo-Aral JJ, Moraleda JM, et al. Comparative effects between bone marrow and mesenchymal stem cell transplantation in GDNF expression and motor function recovery in a motorneuron degenerative mouse model. Stem Cell Rev. 2012;8(2):445–58.

38. Scott S, Kranz JE, Cole J, Lincecum JM, Thompson K, Kelly N, et al. Design, power, and interpretation of studies in the standard murine model of ALS. Amyotroph Lateral Scler. 2008;9(1):4–15.

39. Choi CI, Lee YD, Gwag BJ, Cho SI, Kim SS, Suh-Kim H. Effects of estrogen on lifespan and motor functions in female hSOD1 G93A transgenic mice. J Neurol Sci. 2008;268(1-2):40–7.

40. Kondo T, Funayama M, Tsukita K, Hotta A, Yasuda A, Nori S, et al. Focal transplantation of human iPSC-derived glial-rich neural progenitors improves lifespan of ALS mice. Stem Cell Reports. 2014;3(2):242–9.

41. Morita E, Watanabe Y, Ishimoto M, Nakano T, Kitayama M, Yasui K, et al. A novel cell transplantation protocol and its application to an ALS mouse model. Exp Neurol. 2008;213(2):431–8.

42. Benatar M. Lost in translation: treatment trials in the SOD1 mouse and in human ALS. Neurobiol Dis. 2007;26(1):1–13.

2

Relevance of HCN2-expressing human mesenchymal stem cells for the generation of biological pacemakers

Ieva Bruzauskaite[1], Daiva Bironaite[1,2*], Edvardas Bagdonas[1], Vytenis Arvydas Skeberdis[3], Jaroslav Denkovskij[1], Tomas Tamulevicius[4], Valentinas Uvarovas[5] and Eiva Bernotiene[1]

Abstract

Background: The transfection of human mesenchymal stem cells (hMSCs) with the hyperpolarization-activated cyclic nucleotide-gated ion channel 2 (HCN2) gene has been demonstrated to provide biological pacing in dogs with complete heart block. The mechanism appears to be the generation of the ion current (I_f) by the HCN2-expressing hMSCs. However, it is not clear how the transfection process and/or the HCN2 gene affect the growth functions of the hMSCs. Therefore, we investigated survival, proliferation, cell cycle, and growth on a Kapton® scaffold of HCN2-expressing hMSCs.

Methods: hMSCs were isolated from the bone marrow of healthy volunteers applying a selective cell adhesion procedure and were identified by their expression of specific surface markers. Cells from passages 2–3 were transfected by electroporation using commercial transfection kits and a pIRES2-EGFP vector carrying the pacemaker gene, mouse HCN2 (mHCN2). Transfection efficiency was confirmed by enhanced green fluorescent protein (EGFP) fluorescence, quantitative real-time polymerase chain reaction (RT-qPCR) and enzyme-linked immunosorbent assay (ELISA). After hMSCs were transfected, their viability, proliferation, I_f generation, apoptosis, cell cycle, and expression of transcription factors were measured and compared with non-transfected cells and cells transfected with pIRES2-EGFP vector alone.

Results: Intracellular mHCN2 expression after transfection increased from 22.14 to 62.66 ng/mg protein ($p < 0.05$). Transfection efficiency was 45 ± 5 %. The viability of mHCN2-transfected cells was 82 ± 5 %; they grew stably for more than 3 weeks and induced I_f current. mHCN2-transfected cells had low mitotic activity (10.4 ± 1.24 % in G2/M and 83.6 ± 2.5 % in G1 phases) as compared with non-transfected cells (52–53 % in G2/M and 31–35 % in G1 phases). Transfected cells showed increased activation of nine cell cycle-regulating transcription factors: the most prominent upregulation was of AMP-dependent transcription factor ATF3 (7.11-fold, $p = 0.00056$) which regulates the G1 phase. mHCN2-expressing hMSCs were attached and made anchorage-dependent connection with other cells without transmigration through a 12.7-μm thick Kapton® HN film with micromachined 1–3 μm diameter pores.

Conclusions: mHCN2-expressing hMSCs preserved the major cell functions required for the generation of biological pacemakers: high viability, functional activity, but low proliferation rate through the arrest of cell cycle in the G1 phase. mHCN2-expressing hMSCs attached and grew on a Kapton® scaffold without transmigration, confirming the relevance of these cells for the generation of biological pacemakers.

Keywords: HCN2, Stem cells, Scaffold, Tissue regeneration, Biological pacemaker

---

* Correspondence: daibironai@gmail.com
[1]Department of Regenerative Medicine, State Research Institute Centre for Innovative Medicine, Vilnius, Lithuania
[2]Department of Pathology, Forensic Medicine and Pharmacology, Vilnius University, Faculty of Medicine, Vilnius, Lithuania
Full list of author information is available at the end of the article

Background

Electronic pacing is the state-of-the art treatment for complete heart block and a variety of other cardiac arrhythmias. While they successfully regulate heart rhythm, electronic pacemakers are not without risk, including infection, lead fracture and displacement, and potential interference from other devices. These concerns plus the need for frequent monitoring and for battery replacement have led to a search for other means—such as biological pacemakers—to provide cardiac pacing.

Some of these other means reported to provide biological pacing are virally delivered gene therapies aimed at increasing inward and or decreasing outward ionic currents during diastole, the use of transcription factors to convert mature myocytes into sinus node-like cells, and the use of embryonic stem cells (ESCs) or induced pluripotent stem cells (iPSCs) converted into a pacemaker lineage [1–5]. Another approach used mesenchymal stem cells (MSCs) and a self-inactivating HIV-based lentiviral vector for delivery of human potassium/sodium hyperpolarization-activated cyclic nucleotide-gated ion channel 2 (HCN2) into rabbit MSCs [6]. Bone marrow (BM)-derived human mesenchymal stem cells (hMSCs) transfected with the mouse HCN2 (mHCN2) gene were grafted to initiate a biological rhythm in the canine heart [7]. However, the MSCs have been shown to migrate away from the administration site within weeks of engraftment, making this a short-term solution. This problem might solved by trapping HCN2-expressing and heart rhythm-stimulating cells in cages/scaffolds thus preventing transmigration, preserving their long-term growth, sufficient nutrition and functional properties. Additionally, cell functions such as viability, stability, and renewal rate of HCN2-transfected cells are important factors.

Therefore, in the present study we explored the relevance of mHCN2-expressing hMSCs for the generation of biological pacemakers, paying particular attention to cell proliferation and cell cycle regulation and to the ability of mHCN2-transfected cells to grow and function on the porous Kapton® scaffold.

Methods

Isolation and expansion of BM-derived hMSCs

hMSCs were isolated, identified, and expanded from BM aspirates following previously described protocols [8, 9]. Briefly, BM aspirates (~6 ml) from the iliac crest of humans (up to 40 years old) were collected after obtaining written informed consent according to the Declaration of Helsinki and approval by the Lithuanian Bioethics Committee (No. 158200-14-741-257). BM donors were tested according to guidelines for the preparation of blood and blood products. Unprocessed bone marrow was seeded at a density of 12,000 mononuclear cells (MNC)/cm^2 in T-75 flasks (NUNC, Wiesbaden,

Germany) and cultured in Dulbecco's modified Eagle's medium (DMEM) containing 10 % fetal bovine serum (FBS) at 37 °C in a humidified atmosphere of 5 % CO_2. After 24 h, supernatant containing non-adherent cells was removed, cells were rinsed with phosphate-buffered saline (PBS) without Ca^{2+}/Mg^{2+} (Biochrom, UK) and growth medium was added. Additional changes of growth medium were performed every 3 days. For further experiments, cells were used from passages 2–3. Only the MSCs that corresponded to all main MSC identification criteria were included in the study.

Transfection of hMSC with the mHCN2 gene

A full-length of mHCN2 cDNA, subcloned in a pIRES2-EGFP vector (BD Biosciences Clontech, San Jose, USA), was transfected into the hMSCs by electroporation using the Lonza 4D-Nucleofector™ system (Lonza, USA). hMSCs were transfected with 3 μg pIRES2-EGFP and pIRES2-mHCN2-EGFP vectors, and the efficiency of transfection was estimated by the number of green fluorescence-emitting cells (enhanced green fluorescence protein (EGFP) excitation 488 nm, emission 509 nm). The total number of green fluorescent cells 24–48 h after the incorporation of vectors revealed 40–50 % transfection efficiency. Transfected cells were cultured in hMSC growing medium (Poietics MSCGM, BioWhittaker, Lonza) at 37 °C in a 5 % CO_2 humidified atmosphere. To select a pure population of mHCN2-expressing hMSCs, cells were exposed to 50 μM geneticin for an additional 5 days. Transfection efficiency was also confirmed by both quantitative real-time polymerase chain reaction (RT-qPCR) and enzyme-linked immunosorbent assay (ELISA).

Investigation of mHCN2 expression by RT-qPCR

pIRES-EGFP- and pIRES-mHCN2-EGFP-transfected hMSCs were lysed, and their RNA was extracted and purified using RNeasy Mini Spin columns (Qiagen, USA) according to the manufacturer's instructions. RNA concentration and purity were evaluated with a SpectraMAX i3 spectrophotometer (Molecular Devices, USA). RNA samples were treated with DNase I (Thermo Scientific™) and reversely transcribed with the Maxima®First Strand cDNA Synthesis Kit (Thermo Scientific™) according to the manufacturer's protocols. PCRs were performed using Maxima® Probe qPCR Master Mix (2×) (Thermo Scientific™) and Stratagene MX-3005P detection instrument (Agilent Technologies, USA). The TaqMan® Gene Expression Assays (Applied Biosystems, USA) for *HCN2* (human, Hs00606903_m1), *mHCN2* (mouse, Mm00468538_m1), and *ACTB* (housekeeping gene, Hs01060665_g1) were used for gene expression analysis and separation of endogenous human HCN2 from transfected mHCN2. Expression of the mHCN2 gene after transfection was compared with the level of the endogenous human HCN2 gene in hMSCs (Fig. 1a). All reactions were run in triplicate starting with a denaturation step for 10 min at 95 °C followed by 40 cycles of 15 s at 95 °C for denaturation and 60 s for annealing and extension. The gene expression ratio (pIRES-mHCN2 vs. non-transfected cells) was calculated using the $2^{-\Delta\Delta Ct}$ equation. The efficiency of mHCN2 transfection was measured 5 days after the cell growth with 50 μM geneticin.

Evaluation of mHCN2 protein expression by ELISA

Cells were lysed using three cycles of freezing-thawing. Before making measurements, all samples were kept on ice. mHCN2 expression after hMSC transfection was measured using an ELISA kit for the estimation of mouse potassium/sodium hyperpolarization-activated cyclic nucleotide-gated channel 2 (EIAab, cat. No.E15069m) following the manufacturer's instructions. Absorbance was measured at 450 nm using a Spectramax plate reader. Three groups of cell lysates were investigated: pIRES-mHCN2-EGFP-expressing (positive control), pIRES-EGFP-transfected (control with transfection reagent), and non-transfected (negative control) hMSCs. The concentration of total protein in all tested groups was measured using the Bio-Rad DC Protein Kit according to the manufacturers' instructions. Absorbance was read at 750 nm using a Spectramax plate reader. The final concentration of intracellular mHCN2 after transfection was expressed as ng/mg protein. The efficiency of mHCN2 protein expression in hMSCs was measured 5 days after cell growth with 50 μM geneticin.

Patch-clamp and dye transfer measurements

For electrophysiological recordings, glass coverslips with hMSCs were transferred to the experimental chamber with constant flow-through perfusion, and mounted on the stage of an inverted microscope (Olympus IX81). Junctional conductance between hMSCs (abutted or connected through tunneling tubes (TT)) was measured

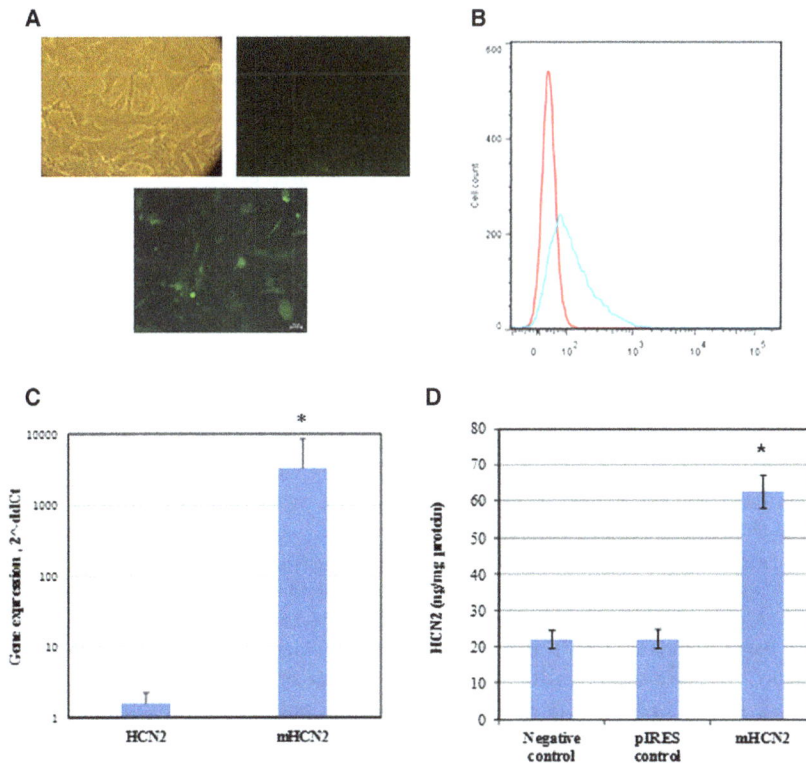

Fig. 1 Efficiency of mouse HCN2 (*mHCN2*) transfection in hMSCs cells. **a** Primary evaluation of transfection efficiency by microscopy: *upper left panel*, light microscopy picture of hMSCs; *upper right panel*, fluorescent picture of not transfected cells; *lower panel*, fluorescent picture of mHCN2-transfected cells. Cells were investigated 48 h after the seeding. Magnification 20x. **b** Evaluation of transfection efficiency by flow cytometry. *Red*, fluorescence of cells before transfection; *blue*, fluorescence of cells after transfection. One representative graph is shown. **c** Expression of the mHCN2 gene in hMSCs was estimated by RT-qPCR. HCN2 column shows the level of endogenous human HCN2 in hMSCs cells estimated with cDNA against human HCN2. **d** Level of intracellular mHCN2 protein after transfection was estimated by ELISA and expressed as ng/mg protein. Protein was measured as described in the Methods section. Data are presented as means ± SD. The increased mHCN2 expression after transfection was significant at *$p < 0.05$

using the dual whole-cell patch-clamp technique. Cells 1 and 2 of a cell pair were voltage clamped independently with the patch-clamp amplifier (MultiClamp 700B; Molecular Devices, Inc., USA) at the same holding potential ($V_1 = V_2$). Voltages and currents were digitized using the Digidata 1440A data acquisition system (Molecular Devices, Inc.) and acquired and analyzed using pClamp 10 software (Molecular Devices, Inc.). By stepping the voltage in cell 1 (ΔV_1) and keeping the other constant, junctional current was measured as the change in current in the unstepped cell 2, $I_j = \Delta I_2$. Thus, g_j was obtained from the ratio $-I_j/\Delta V_1$, where ΔV_1 is equal to transjunctional voltage (V_j), and a negative sign indicates that the junctional current measured in cell 2 is oppositely oriented to that measured in cell 1.

To examine whether cells residing on the opposite sides of Kapton® scaffold can couple through 3 μm diameter pores, non-transfected hMSCs were seeded on one side of the scaffold and 24 h later the mHCN2-transfected cells were seeded on the other side of the scaffold. After the attachment of transfected cells, DAPI dye (20 μM) was injected through the patch pipette into the mHCN2-transfected hMSC, and its transfer to the non-transfected cells residing on the other side of the scaffold was monitored by time lapse imaging at 37 °C in a humidified atmosphere of 5 % CO_2 using an incubation system (INUBG2E-ONICS; Tokai Hit, Shizuoka-ken, Japan) with an incubator mounted on the stage of the microscope equipped with an Orca-R^2 cooled digital camera (Hamamatsu Photonics K.K., Japan), fluorescence excitation system MT10 (Olympus Life Science Europa Gmbh, Hamburg, Germany), and XCELLENCE software (Olympus Soft Imaging Solutions Gmbh, München, Germany).

Patch pipettes were pulled from borosilicate glass capillary tubes with filaments. To minimize the effect of series resistance on the measurements of g_j [10], we maintained pipette resistances below 3 milli-ohms. Patch pipettes were pulled from borosilicate glass capillary tubes with filaments. Experiments were performed at room temperature in Krebs-Ringer solution (mM): NaCl, 140; KCl, 4; $CaCl_2$, 2; $MgCl_2$, 1; glucose, 5; pyruvate, 2; HEPES, 5 (pH = 7.4). Patch pipettes were filled with internal solution (mM): KCl, 130; Na aspartate, 10; MgATP, 3; $MgCl_2$, 1; $CaCl_2$, 0.2; EGTA, 2; HEPES, 5 (pH = 7.3). Patch-clamp measurements were performed 48–72 h after transfection.

Cell viability-apoptosis assay
The number of viable cells, type of cell, and cell death, and stage of apoptosis of mHCN2-expressing hMSCs after selection with geneticin were analyzed using the Muse® Annexin V & dead cell assay (Merck Millipore) following the manufacturer's instructions. This assay allows

quantitative identification of live, early and late apoptotic, and dead cells by measuring the intensity of cell fluorescence. Briefly, cells transfected with the mHCN2 gene, only, with the pIRES vector, and non-transfected hMSCs were harvested with trypsin-EDTA and suspended in DMEM containing 10 % fetal calf serum. Cell suspension (100 μl) and Muse™ annexin V & dead cell reagent (100 μl; Annexin V and 7-AAD) were thoroughly mixed. Samples were incubated for 20 min in the dark. The Muse™ Cell Analyzer (Merc Millipore) was used to measure cell fluorescence. Cells were separated into groups according to the intensity of green (Annexin; early and late apoptotic cells) and red (7-AA; dead cells) fluorescence. Cell viability was tested 5 days after the transfected cell growth with 50 μM geneticin. Non-transfected cells were grown in parallel for the same time period.

Measurements of cell proliferation and cell cycle
Proliferation measurements were performed using the CCK-8 kit (Dojindo Molecular Technologies, USA) according to the manufacturer's instructions. Briefly, after selection with geneticin, mHCN2- and pIRES-transfected hMSCs were grown for 5 days in 12-well plates. Similar numbers (25×10^3) of non-transfected hMSCs were seeded in 12-well plates and allowed to attach for 24 h. Then, the number of proliferating cells in each cell group was measured every 24 h by adding a required volume of CCK-8 reagent with subsequent incubation for 3 h and measurement of absorption at 450 nm. Proliferation rates were measured for 72 h. Non-transfected hMSCs were grown and analyzed in parallel to the transfected ones.

Cell cycle was also measured 72 h after cell growth following the manufacturer's instructions. Briefly, mHCN2- and pIRES-transfected, and non-transfected hMSCs were harvested with trypsin-EDTA. Cells were suspended in growth media containing 10 % fetal calf serum; 200 μl cells were added to each tube, centrifuged at $300 \times g$ for 5 min and washed once with PBS. Then cells were suspended in 200 μl ice cold 70 % ethanol and incubated for 3 h at −20 °C. Cells were centrifuged at $300 \times g$ for 5 min and washed once with PBS; 200 μl Muse™ Cell Cycle Reagent was added to each tube and incubated for 30 min at room temperature in the dark. The cell cycle reagent was a mixture of propidium iodide (PI) and RNAse A in respective proportions subsequently intercalating nuclear DNA. The assay allows identification and measurement of the percentage of cells in each cell cycle phase (G0/G1, S, and G2/M) according to the intensity of PI-based red fluorescence. The DNA Muse™ Cell Analyzer program was used to evaluate the results.

Investigation of transcription factors by RT-qPCR
After selection with geneticin, pIRES-EGFP, pIRES-mHCN2-EGFP-transfected, and non-transfected hMSCs

were lysed and RNA was extracted as described above. RNA was reversely transcribed with the RT^2 First Strand Kit (Qiagen, USA) according to the manufacturer's protocols. PCRs were performed using RT^2 SYBR Green ROX qPCR Mastermix (Qiagen, USA) and the Stratagene MX-3005P detection instrument (Agilent Technologies). The Human Transcription Factors RT^2 Profiler™ PCR Array (Qiagen, PAHS-075Z) was used to screen gene expression changes. Samples of three independent transfection experiments were investigated. Raw data were analyzed by the GeneGlobe Data Analysis Center platform (available online: http://www.qiagen.com/fo/shop/genes-and-pathways/data-analysis-center-overview-page/; Qiagen, USA). For the normalization of gene expression, the geometric mean of four threshold cycles (Ct) of reference genes was used. Gene expression ratio was calculated using the $2^{-\Delta\Delta Ct}$ equation. Data were statistically significant at $p < 0.05$. The expression of transcription factors in transfected cells was measured after cell growth with 50 µM geneticin for 5 days and compared with non-transfected hMSCs.

Manufacture of Kapton® scaffold

Pores were micromachined in a commercially available 12.7-µm thickness polyimide Kapton® HN (DuPont, USA) film using FemtoLab workstation (Workshop of Photonics) and second harmonic (515 nm) of Yb:KGW femtosecond laser Pharos (Light Conversion). The diameters of the resulting pores were 1–3 µm, as determined by scanning electron microscope (Quanta 200 FEG) [11] under the low-vacuum mode.

Immunocytochemistry

To detect the growth of mHCN2-transfected hMSCs and their possible transmigration through the Kapton® scaffold, 4×10^4 non-transfected hMSC were labeled with PKH26 (red fluorescent cell linker kit; Sigma-Aldrich) following the manufacturer's instructions and seeded on one side of a scaffold. The next day, 4×10^4 mHCN2-transfected hMSCs (green fluorescence of EGFP) were seeded on the other side of the scaffold, which was then fixed on a specially designed cell crown holder. After 24 h of cell growth in media at 37 °C in a humidified atmosphere with 5 % CO_2, cell were washed and fixed with 4 % paraformaldehyde and mounted in the Vectashield (Vector Labs, USA) containing DAPI for visualization of the nuclei. All samples were analyzed using a Leica TCS SP8 confocal microscope.

Statistical analyses

All statistical analyses were performed using the SPSS package (version 19.0 for Windows; SPSS Inc., Chicago, IL, USA) and considered to be significant at the 5 % level. Differences between transfected and non-transfected cells were tested by analysis of variance (ANOVA) and Student's t test. Data are presented as means ± SD.

Ethical approval

All hMSC isolation procedures were approved by the Ethics Committee of Vilnius Regional Biomedical Research, Lithuania (No. 158200-14-741-257). All volunteers gave written consent and agreed with the investigational procedure of BM.

Results

Efficiency of hMSC transfection

The efficiency of transfection with pIRES2-EGFP and pIRES2-HCN2-EGFP vectors was evaluated by fluorescence microscopy, RT-PCR, and ELISA. Data in Fig. 1a show that non-transfected cells did not have green fluorescence, whereas transfected cells had green fluorescence. The efficiency of plasmid incorporation has also been confirmed by flow cytometry (Fig. 1b). mHCN2 gene expression was also confirmed by RT-PCR (Fig. 1c). mHCN2 protein expression was investigated by ELISA and expressed as ng/mg protein (Fig. 1d). Representative profiles of cell population and apoptosis of mHCN2-transfected cells analyzed by Muse™ Annexin V and Dead cell reagent is demonstrated in Fig. 2a. The percentage of early, late, and total apoptosis in cells from three independent transfection experiments is shown in Fig. 2b. Induction of necrosis after transfection was negligible. Cells transfected with the pIRES2-EGFP and with the pIRES2-mHCN2-EGFP vectors showed similar levels of live, early, and late apoptosis, which suggests that cell apoptosis was induced by the electroporation procedure, which could damage cell membranes, rather than by the mHCN2 gene itself.

I_f and cell to cell coupling

The expression of functional mHCN2 channels and electrical coupling between abutted hMSCs (Fig. 3A) or hMSCs connected through tunneling nanotubes (TNTs) (Figs. 3B and C) were examined by dual whole-cell patch-clamp measurements. The co-culture of mHCN2-transfected (green) and control hMSCs (not fluorescent) was investigated. TNTs formed by the lamellipodium outgrowth mechanism formed the gap junction (GJ)-based electrical coupling between the hMSCs that also could be achieved by intersection of lamellipodia extensions (inset c). TNTs between hMSCs contained F-actin and Cx43 GJs at the interface of hMSC-1 and the lamellipodium ending of hMSC-2 (Fig. 3D and inset d).

Non-transfected hMSCs did not exhibit time-dependent hyperpolarization (Fig. 3E, upper panel) activated current I_f (Fig. 3E, middle panel), whereas mHCN2 expressing hMSCs showed I_f current (Fig. 3E, lower panel). In the mHCN2-transfected hMSCs, I_f was nearly

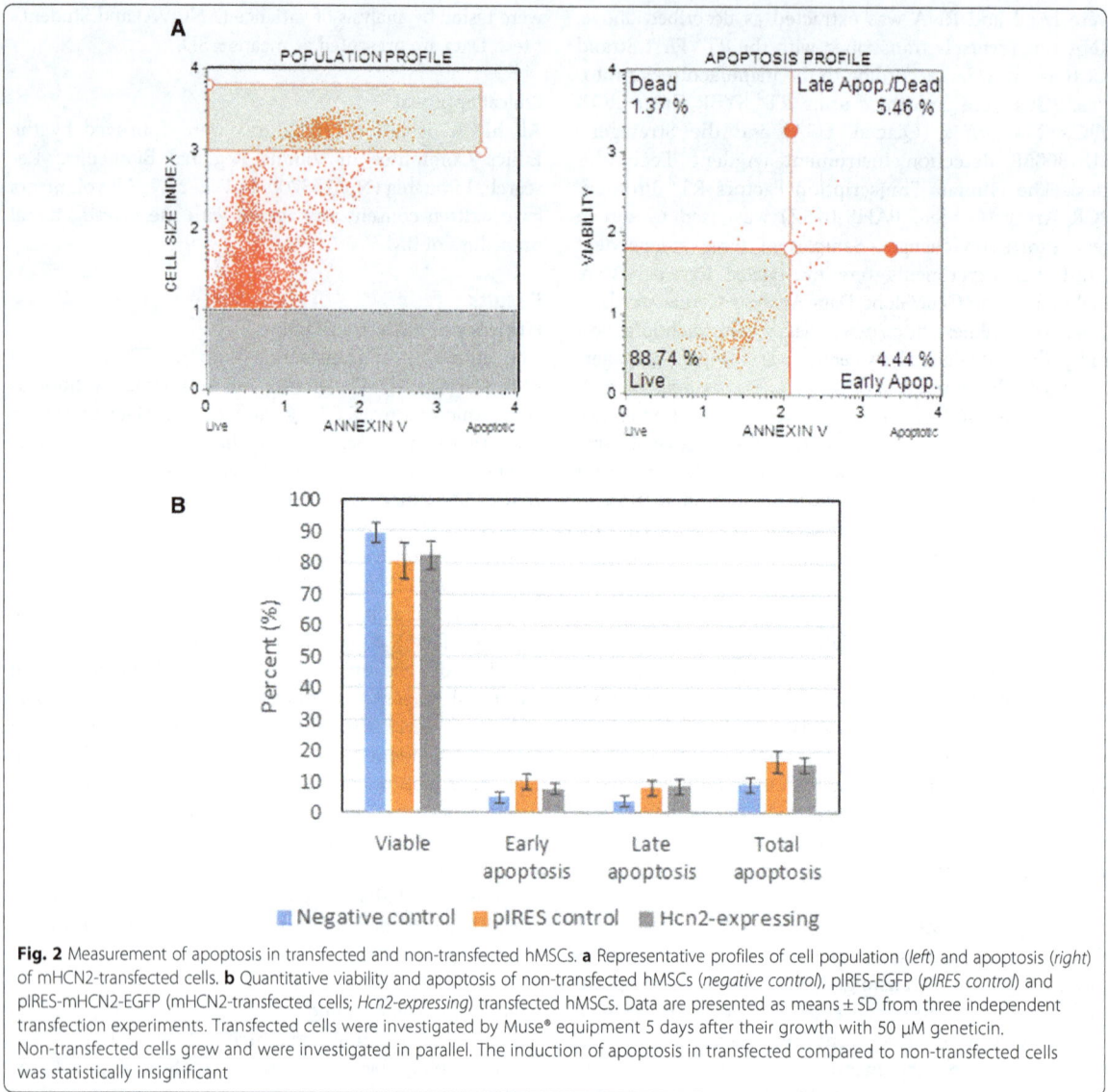

Fig. 2 Measurement of apoptosis in transfected and non-transfected hMSCs. **a** Representative profiles of cell population (*left*) and apoptosis (*right*) of mHCN2-transfected cells. **b** Quantitative viability and apoptosis of non-transfected hMSCs (*negative control*), pIRES-EGFP (*pIRES control*) and pIRES-mHCN2-EGFP (mHCN2-transfected cells; *Hcn2-expressing*) transfected hMSCs. Data are presented as means ± SD from three independent transfection experiments. Transfected cells were investigated by Muse® equipment 5 days after their growth with 50 μM geneticin. Non-transfected cells grew and were investigated in parallel. The induction of apoptosis in transfected compared to non-transfected cells was statistically insignificant

fully activated at −140 mV (504 ± 130 pA; $n = 4$) with an activation threshold of −90 mV. Fig. 3F (lower panel) displays the g_j/V_j plot obtained by measuring the I_j response (Fig. 3F, middle panel) in the hMSC-2 to the voltage ramp (Fig. 3F, upper panel) applied to the hMSC-1 with its symmetric counterpart at positive V_js. The presence of voltage gating indicates that the cells established electrical coupling by forming functional GJs. The measured g_j between six abutted cell pairs was 37.14 ± 3.61 nS and g_T between five cell pairs connected through TNTs was 1.15 ± 0.25 nS.

Cell cycle of mHCN2-expressing hMSCs

Both types of transfected cells (with empty pIRES vector and with mHCN2 gene) showed significantly

downregulated proliferation rates compared to the non-transfected hMSCs (Fig. 4a). Moreover, the mHCN2-transfected cells showed significantly lower proliferation rates compared to the pIRES-transfected cells (Fig. 4a). This finding led us to analyze the cell cycle and possible mechanisms involved in its regulation. Data in Fig. 4b show representative population and DNA content of one mHCN-2 transfection experiment, whereas data in Fig. 4c demonstrate summarized data from three independent transfection experiments compared to the non-transfected cells. mHCN2- and pIRES-transfected hMSCs were significantly concentrated in the G1 phase with low numbers of cells in G2/M phase, whereas non-transfected hMSCs had an

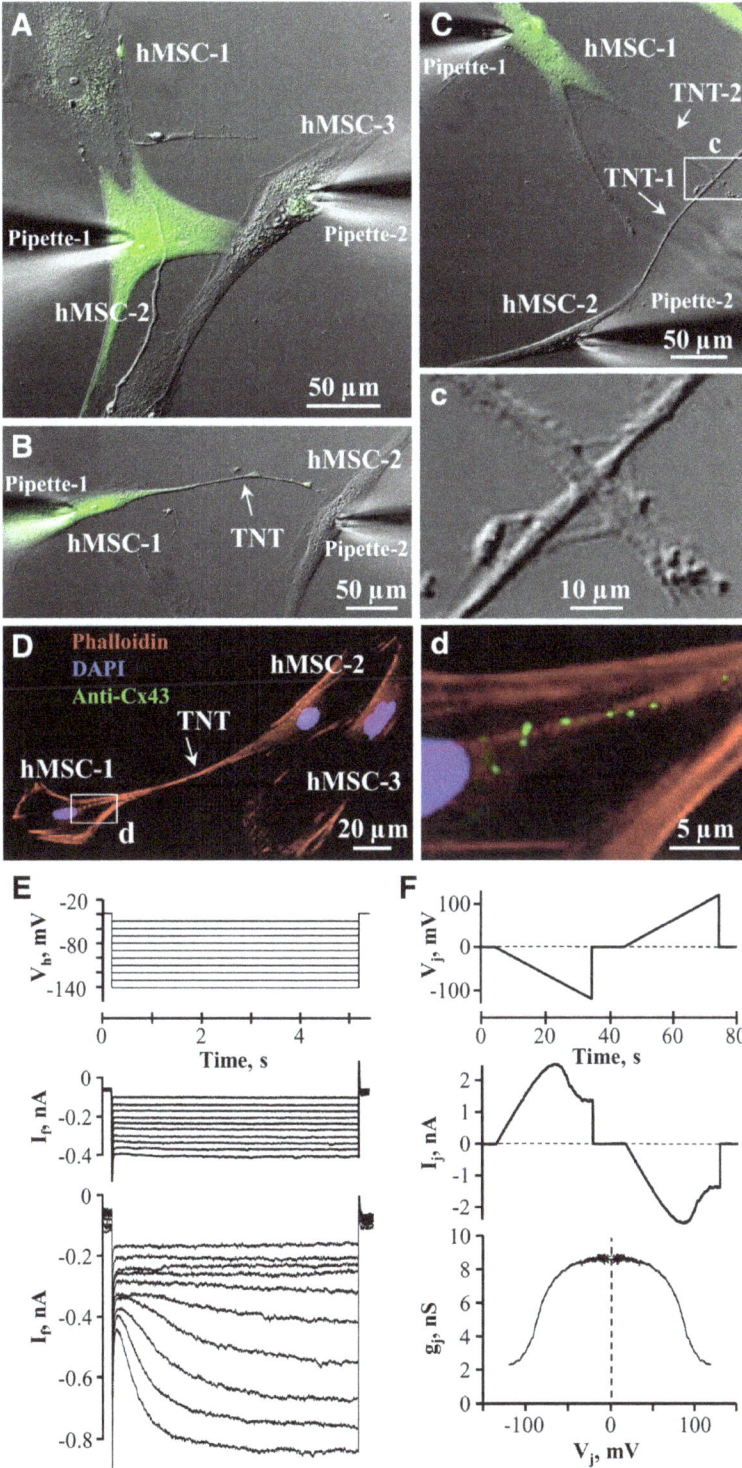

Fig. 3 (See legend on next page.)

(See figure on previous page.)
Fig. 3 Current generation by mHCN2-expressing cells and their electrical coupling. Human mesenchymal stem cells (*hMSCs*) expressing mHCN2 (*green*) abutted (**a**) and connected through tunneling nanotubes (*TNTs*) (**b, c**). (inset c) GJ-based electrical coupling between cells also could be achieved by intersection of lamellipodia extensions. **d** and inset **d** TNTs between hMSCs containing F-actin, and Cx43-based GJs between hMSC-1 and the lamellipodium of hMSC-2. I_f current was measured in mHCN2-expressing hMSCs (**e**, *lower panel*) by hyperpolarizing the cells from $V_h = -40$ mV for 5 s to voltages between −50 and −140 mV in 10 mV increments (**e**, *upper panel*). I_f current was absent in non-transfected or transfected with empty vector hMSCs (**e**, *middle panel*). **f** Typical experiment showing the measurement of electrical coupling between two hMSCs connected through TNT (shown in **b**). g_T/V_T plot (**f**, *lower panel*) was obtained by measuring the I_j response (**f**, *middle panel*) in the hMSC-2 to the voltage ramp of negative polarity from 0 to −120 mV (**f**, *upper panel*) applied to the hMSC-1 with its symmetric counterpart at positive V_js. The co-culture of mHCN2-transfected (green fluorescence) and non-transfected hMSCs (not fluorescent) was investigated

almost equal distribution between G1 and G2/M phases (Fig. 4c). The mHCN2-expressing cells showed slightly stronger downregulation of the cell cycle compared to the pIRES-transfected cells. Cell cycle of transfected cells was measured after selection by 50 μM geneticin.

Transcription factors regulating cell cycle of mHCN2-expressing hMSCs

Data from the cell cycle experiments inspired us to investigate further mechanisms regulating the cell cycle after transfection. For this purpose, we investigated an

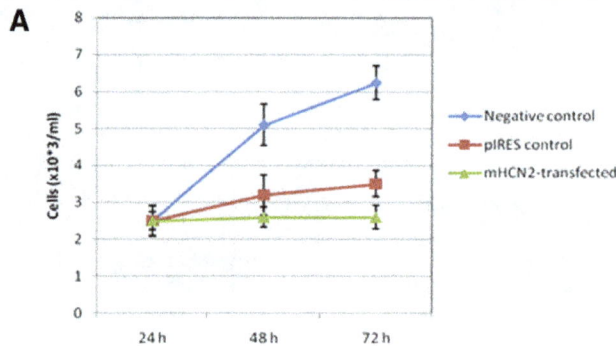

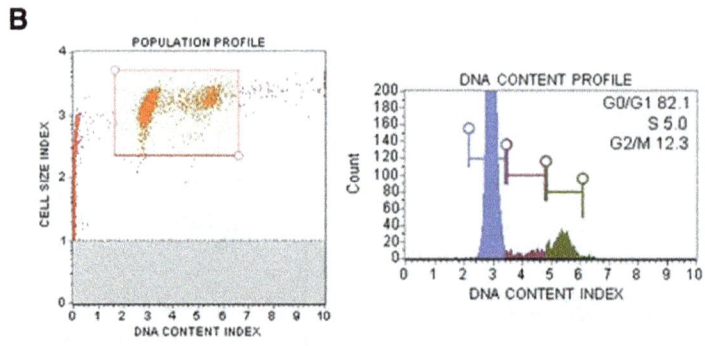

C   Percent of cells concentrated in each phase of cell cycle.

Cells	G_1	S	G2/M
mHCN2-transfected	84.6 ± 2.5*	4.0 ± 0.54	10.4 ± 1.24*
pIRES control	80.1± 1.75*	4.3 ± 1.47	13.5 ± 2.68*
Negative control	45.6 ± 2.35	5.5 ± 1.23	46.8 ± 3.31

Fig. 4 Proliferation and cell cycle of transfected and non-transfected hMSCs. **a** Proliferation rate of non-transfected (*negative control*), pIRES-EGFP-transfected (*pIRES control*) and pIRES-mHCN2-EGFP-transfected (*mHCN2-transfected*) hMSCs over 72 h. Control, pIRES-, and mHCN2-transfected cells were investigated after their growth with geneticin for 5 days. Changes in proliferation rate were significant ($p < 0.05$) compared to the non-transfected cells. **b** Representative profiles of cell population (*left*) and DNA content (*right*) of mHCN2-transfected hMSCs. **c** Summarized quantitative percentage of each cell cycle phase. Data are presented as means ± SD. *Data were significant compared to the negative control at $p < 0.05$. Proliferation of transfected cells was measured 5 days after their growth with geneticin (50 μM). Non-transfected cells grew for 72 h and were analyzed in parallel. Data are presented as means ± SD

array of transcription factors and their changes after both types of transfection (pIRES and mHCN2) compared with non-transfected hMSCs. Data presented in Fig. 5 show changes in expression of transcription factors measured in pIRES- and mHCN2-transfected cells compared to the negative control (non-transfected hMSCs). Further statistical analysis of PCR data revealed that mHCN2-transfected cells significantly upregulated nine transcription factors involved in cell cycle regulation compared to the pIRES-transfected cells (Table 1). It also demonstrates that the activation of the nine transcription factors is a result of the mHCN2 gene and not the transfection procedure. The most significantly upregulated factor in mHCN2-transfected cells was

activating transcription factor ATF3, a member of the cAMP response element-binding (CREB) protein family of transcription factors (7.11-fold, $p = 0.00056$). Other cell cycle regulating transcription factors were upregulated from 1.6- to 2.97-fold ($p < 0.05$; Table 1).

Growth of mHCN2-expressing cells on porous scaffold

Stem cells directly grafted into heart tissue or injected through the vessels may be released due to the strong contraction of heart muscle or as a response to signaling molecules. Therefore, we investigated the possibility of enclosing cells in a cage/scaffold with particular pore sizes that could be impermeable for cells but allow passing of nutrients and/or anchorage-dependent cell to cell

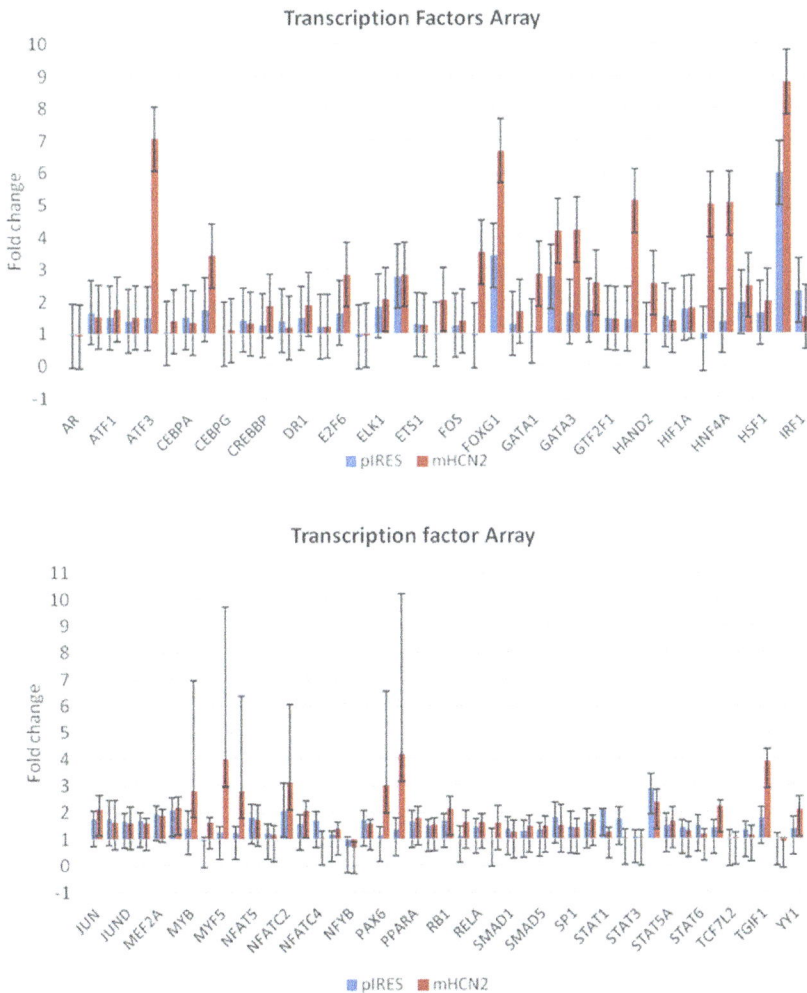

Fig. 5 Panels of changed activities of transcription factor arrays after transfection. Folds of change in activities of transcription factors after pIRES- and mHCN2-transfections were compared to non-transfected cells. Data are presented as mean values of fold-changes obtained during three independent experiments. The statistically significant activation of nine transcription factors after mHCN2 transfection compared to pIRES transfection is presented in Table 1

Table 1 Statistically evaluated impact of mouse HCN2 to the expression of transcription factors

No	Gene	Main impact to cell cycle regulation	Fold-change	P value
1.	ATF3	Delays G1 to S transition [29]	7.11	0.00056
2.	ETS2	Promotes G2/M phase [46]	2.79	0.02
3.	GTF2B	Regulates G1 phase [33]	1.9	0.01
4.	ID1	Promotes G1/S phase [37]	1.86	0.044
5.	MYC	Promotes G1/S phase [36]	2.3	0.0028
6.	NFATC3	Promotes G1/S phase [41]	1.6	0.02
7.	REL	Promotes G0 toG1 transition [45]	1.94	0.014
8.	TBP	Delays G2/M phase [24]	2.04	0.012
9.	TGIF1	Promotes G0 toG1 transition [22]	2.97	0.025

communication. Data from other authors showed that the best pore size for cell growth without transmigration through the scaffold could be up to 3 μm in diameter [12]. With this in mind, we investigated 8.9–16.5 μm thick Kapton® scaffold with 1–3 μm pores of conical shape. Data in Fig. 6 show that mHCN2-expressing cells were able to attach and grow on the Kapton® scaffold without transmigration. Confocal micrographs demonstrate that the same mHCN2-transfected cells were able to grow on one side (Fig. 6a-d) and non-transfected hMSCs stained with PKH26 on the other side of scaffold (Fig. 6e). Both types of the above-mentioned cells could grow on different sides of the scaffold without transmigration (Fig. 6f). Moreover, the mHCN2-expressing hMSCs growing on the one side of the Kapton® scaffold (cell 1 in Fig. 7a; nucleus encircled in Fig. 7c) can establish the intercellular communication with the non-transfected hMSCs growing on the other side of the scaffold (cells 2 and 3 in Fig. 7b; nuclei encircled in Fig. 7c). DAPI dye injected through the patch pipette into cell 1 was transferred to cells 2 and 3, confirming that cells residing on the opposite sides of the scaffold can couple to each other through 3-μm pores. Note that nucleus staining in the DAPI-injected cell 1 is not visible because the Kapton® scaffold is impermeable to the UV light used for excitation of DAPI fluorescence, while nuclei staining in the recipient cell 2 and cell 3 are obvious, and DAPI accumulation kinetics in these cells is shown in Fig. 7d.

Discussion

The major requirements for generating functional biological pacemakers are construction of viable, properly functioning and proliferating cells capable of generating the pacemaker current and growing on the porous scaffolds. In this study, we have demonstrated that mHCN2-transfected hMSCs expressed the mHCN2 channel protein and exhibited I_f current necessary for cardiac stimulation. However, the capacitances of transfected hMSCs varied from 100 to 200 pF, and I_f densities

in our experiments were limited to several pA/pF. This was presumably due to insufficient translation and/or not completed insertion of mHCN2 channels into the cell membrane after transfection. The mHCN2-transfected MSCs preserved their viability, generated pacemaker current, grew on the porous Kapton® scaffolds with 1–3 μm diameter pore arrays, and established intercellular communication between opposite sides of the scaffold without transmigration through the pores. The polyimide films demonstrated suitable mechanical and thermal properties, and good biocompatibility, and have already been successfully applied to a vast range of biomedical investigations [13–17]. In parallel to the formation of functionally active HCN channels, the mHCN2-transfected MSCs expressed Cx43 necessary for communication through Cx43-based GJ channels and F-actin containing TNTs necessary for the biological pacemaker functioning. The collaboration of functional GJs and HCN2 channels as a pacemaker unit in heterologous cell pairs has been shown elsewhere [18].

Importantly, data from this study show, for the first time, that the proliferation rate of mHCN2-transfected cells was significantly downregulated, mainly by the nine transcription factors that controlled the cell cycle. Our findings show mHCN2-expressing hMSCs have low proliferative activity that could be vitally important for the proper and long-term function of heart-stimulating cells. Other authors also revealed that quiescence and self renewal are critical for stem cell pool preservation and long-term engraftment potential [19]. Highly proliferating cells do not have enough time for cell renewal, and eventually their supply might become exhausted. On the other hand, given the extracellular stimuli and fulfillment of various regenerating functions, stem cells cannot be quiescent all the time [20]. Data from this study show that the balance between cell cycle and functional activity of mHCN2-transfected hMSCs is important for the biological pacemaker. The ability to modify the balance between the activation of cell cycle-regulating transcription factors and HCN channel

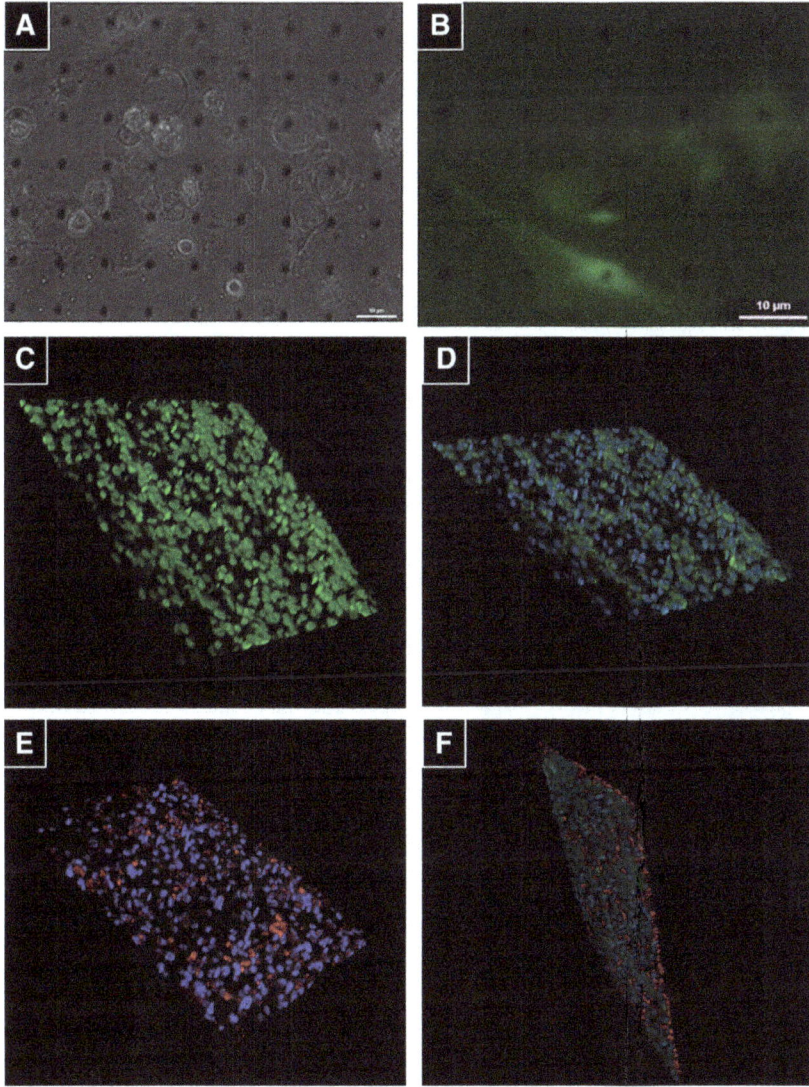

Fig. 6 Growth of mHCN2-expressing cells on Kapton® scaffold. **a** Light microscopy image of mHCN2 cells grown on the Kapton® scaffold. Magnification 20×. **b** Fluorescent image of mHCN2-transfected hMSC grown on the Kapton® scaffold. Magnification 20×. **c** Confocal three-dimensional image of mHCN2-transfected cells (*green*). Magnification 20×. *Scale bar* 10 μm. **d** Confocal three-dimensional image of mHCN2-transfected cells (*green*) and nucleus (*blue*). Magnification 20×. *Scale bar* 10 μm. **e** Confocal three-dimensional image of non-transfected hMSCs stained with PKH26 (*red*) and nucleus (*blue*). Magnification 20×. **f** Confocal three-dimensional image of mHCN2-transfected cells (*green*) on the one side of Kapton® scaffold and hMSC stained with PKH26 (*red*) on the opposite side. Nuclei were stained *blue*. Magnification 20×

proteins could be a useful tool in the field of biological pacing.

TG-interacting factor 1 (TGIF1) is a transcriptional repressor and necessary factor modulating the balance between cell quiescence, self renewal, and differentiation. TGIF1 knockdown in myeloid progenitors affected cell proliferation and induced cell cycle blocking at the G0 stage [21]. Another study also showed that TGIF1 knockout resulted in increased quiescence of hematopoietic stem cells, their self-renewal, and a tendency to reside in the G0 state [22]. On the other hand, the transcription factor c-Rel protein, a member of the NFkB transcription factor family, stimulates the cell cycle as well as the various other intracellular functions. Binding of NFkB to the cell DNA is rapidly induced by serum growth factors and stimulates the G0 to G1 transition in mouse fibroblasts [23]. Additionally, it was shown that *c-rel* is also induced by serum in quiescent fibroblasts and the level of *c-rel* transcripts decreases to nearly the basal level 3 h after stimulation [24]. This finding suggests that *c-rel* is

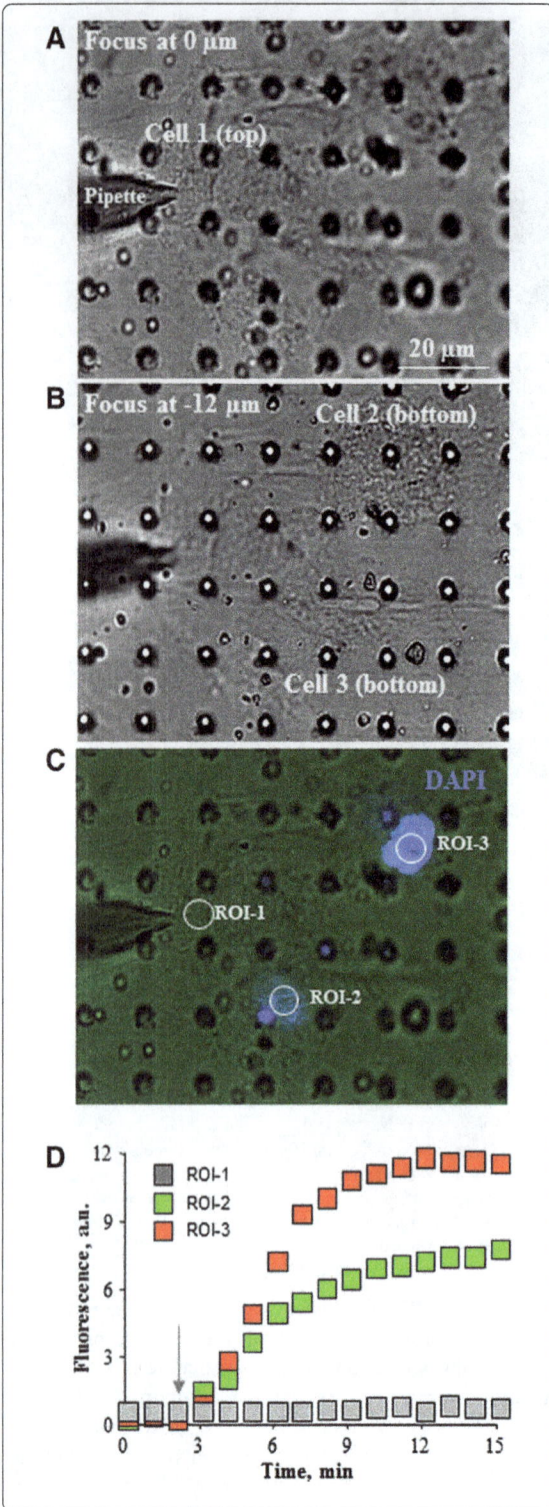

Fig. 7 Intercellular coupling of cells growing on the opposite sides of Kapton® scaffold. **a** DAPI (20 µM) was injected through the patch pipette into the mHCN2-transfected cell 1 (nucleus encircled in **c**) residing on the top side of Kapton® scaffold. **b** Non-transfected hMSCs (cell 2 and cell 3) residing on the bottom side of the scaffold (nucleus encircled in **c**). **c** Transfer of DAPI dye from the donor cell 1 to the recipient cell 2 and cell 3. Note that the fluorescence of DAPI-stained nucleus in cell 1 could not be demonstrated because of the impermeability of Kapton® scaffold to UV light. **d** DAPI accumulation kinetics measured in the regions of interest (*ROI*) situated on the nuclei of respective cells. *Arrow* indicates the onset of dye application to cell 1

involved in the transition from the G0 to G1 phase. C-Rel was also found in the S phase, whereas its level decreased when cells entered the G2 phase [25]. Our data show that increased expression of TGIF1 and c-Rel by 2.97- and 1.94-fold, respectively, can be responsible for the advancement of mHCN2-expressing cells from the G0 stage to further phases of the cell cycle.

The next step in cell cycle regulation is controlled transition of mHCN2-transfected cells into the G1 phase. This step is strongly regulated by ATF3, which belongs to the family of CREB transcription factors and is a stress-inducible gene [26]. ATF3 has an anti-apoptotic effect and inhibits adriamycin-induced apoptosis in primary cardiomyocytes [27]. Furthermore, chick embryo fibroblasts stably expressing ATF3 grew better under low serum conditions [28]. It was also shown that ATF3+/+ fibroblasts more slowly transitioned from the G1 to S phase, suggesting a growth cycle control in the G1 phase [29, 30]. Data from this study show that significant activation of ATF3 (7.11-fold, $p < 0.001$) after mHCN2 transfection in hMSCs can control the cell cycle in the G1 phase and support better cell viability. The strong activation of ATF3 might also show the formation of functionally active mHCN2 channels which are cAMP sensitive. Another factor important for the control of the G1 phase is a general transcription factor IIB (GTF2B), which stimulates transcription through the stabilization of RNA polymerase II, initiating the DNA-TBA (TATA-binding) complexes [31]. It was shown that TF2B can also act as an autoacetyltransferase, which is important for TFIIB acetylation, stabilization, and activation of cell transcription [32]. GTF2B activation is mainly observed in the G1 phase, not in the M phase, suggesting suppression of cell proliferation [33]. Our data show that when GTF2B in mHCN2-transfected hMSCs was activated 1.9-fold, G1 to S transition was delayed similarly to ATF3.

The helix-loop-helix protein ID1 is important for cell "stemness" and is scarcely expressed in normal adult differentiated tissues, whereas it is abundant in proliferating tissues [34]. Under low serum conditions, ectopic expression of ID1, but not ID2, supported proliferation of mammary epithelial cells [35]. Downregulation of ID1, similar to c-Myc, decreased expression of cyclins D1 and E. The same authors suggested that ID1 is downstream of c-Myc in regulation of cell proliferation. Consistent with this idea, both c-Myc and ID1 are necessary for sufficient G1 to S progression [36, 37]. Another transcription factor, the nuclear factor of activated T cells (NFAT), belongs to a family of transcription factors that has been foremost identified in immune cells and later on in a wide range of cell types and tissues [38, 39]. NFAT is constitutively expressed in resting cells, whereas the activation of its receptor is related to the mobilization of calcium and subsequent cell activation [40]. It was also shown that calcium signals stimulate progression of the cell cycle and promote transition of the G1/S phase [41]. Our data show that ID1, c-Myc, and NFATC3 were activated 1.86-, 2.3- and 1.6-fold ($p < 0.05$), respectively, and could be involved in transition of the G1/S phase.

The last group of transcription factors controls the G2/M phases. The TATA-binding protein (TBP) is a universal transcription factor required for the eukaryotic RNA polymerase. TBP like-null chicken cells exhibited 20 % elevated cell cycle progression due to shortening of the G2 phase [42], and TBP induces a delay in the G2/M transition which subsequently delays cell proliferation [43]. Moreover, TBP in stressed cells instead of transcription regulation preferentially binds and repairs injured DNA [44]. This study shows that 2.04-fold activation of TBP after mHCN2 transfection may delay cell entrance to the G2-M phase. On the other hand, the member of the E26 transformation-specific (ETS) family member ETS2, the winged helix-turn-helix DNA-binding domain, has been found to be a regulator of cdc2 expression necessary for the G2/M phase and better cell growth under stress conditions [45]. Moreover, ETS2 activation is necessary for trophoblast stem cell self-renewal and is a vitally important factor for the survival of mouse embryos [46]. Data from this study demonstrate that 2.79-fold ETS2 activation will slightly stimulate mHCN2-expressing cells to proliferate, and might be important for better cell survival.

Conclusions

The results of this study show that mHCN2-transfected hMSCs preserved high cell viability and functional activity necessary for cardiac stimulation: mHCN2-expressing cells had low proliferative activity due to the downregulation of the cell cycle and cell concentration in the G1 phase (~85 %); and generated I_f current and made anchorage-dependent connection with other cells without transmigration through a 12.7-μm thick Kapton® HN film with micromachined 1–3 μm diameter pores. Insertion of mHCN2 gene into hMSCs activates nine transcription factors that control each phase of the cell cycle,

subsequently downregulating cell proliferation. The strongest activation of cAMP-responsive transcription factor ATF3 suggests its particular role in the G1 phase arrest. Additionally, a strong activation of ATF3 in mHCN2-transfected cells could be a marker of formation of functionally active HCN channels which are also cAMP-dependent. This study shows that mHCN2-transfected BM-derived hMSCs are appropriate for the further generation of functional biopacemakers.

Abbreviations

AMP: adenosine monophosphate; ATF3: activating transcription factor 3; BM: bone marrow; cdc2: cell division cycle protein 2 homolog; c-Myc: v-myc myelocytomatosis viral oncogene homolog (avian); CREB: cAMP response element binding protein; c-Rel: NF-kappa-B heterodimer RELA/p65.; Cx43: connexin 43; DAPI: 4′,6-diamidino-2-phenylindole; DMEM: Dulbecco's modified Eagle's medium; EGFP: enhanced green fluorescent protein; ELISA: enzyme-linked immunosorbent assay; ESC: embryonic stem cell; ETS: E26 transformation-specific; GJ: gap junction; g$_j$: gap junction conductance; GTF2B: general transcription factor IIB; HCN2: potassium/sodium hyperpolarization-activated cyclic nucleotide-gated ion channel 2; hMSC: human mesenchymal stem cell; ID1 and ID2: DNA-binding protein inhibitors; I$_f$: funny current; mHCN2: mouse HCN2; MSC: mesenchymal stem cell; NFAT: nuclear factor of activated T cells; NFkB: nuclear factor-κB; PBS: phosphate-buffered saline; PI: propidium iodide; ROI: region of interest; RT-qPCR: quantitative real-time polymerase chain reaction; TBP: TATA-binding protein; TGIF1: TG-interacting factor 1; TNT: tunneling nanotube; TT: tunneling tubes.

Competing interests

The authors declare that they have no competing interests.

Authors' contributions

IB carried out the cell transfections and cell-based studies, participated in the writing of the manuscript, and made substantial contributions to conception and design, acquisition of data, and analysis and interpretation of data. DB carried out ELISA and protein measurements, and was involved in drafting the manuscript or revising it critically for important intellectual content. EB carried out the PCR measurements, and made substantial contributions to the acquisition of data, and analysis and interpretation of data. EB was also involved in drafting the manuscript or revising it critically for important intellectual content. VAS participated in the patch-clamp experiments, and was been involved in drafting the manuscript and revising it critically for important intellectual content. JD carried out the multiplication of plasmid, participated in the ELISA experiments, and made substantial contributions to the acquisition of data, and analysis and interpretation of data; JD was also involved in drafting the manuscript or revising it critically for important intellectual content. TT participated in the manufacturing of scaffold, and agreed to be accountable for all aspects of the work in ensuring that questions related to the accuracy or integrity of any part of the work are appropriately investigated and resolved; TT was also involved in drafting the manuscript or revising it critically for important intellectual content. VU collected BM samples, and agreed to be accountable for all aspects of the work in ensuring that questions related to the accuracy or integrity of any part of the work are appropriately investigated and resolved; VU was also involved in drafting the manuscript or revising it critically for important intellectual content. EV participated in the discussion of results, and has given final approval of the version to be published; EV was also involved in drafting the manuscript or revising it critically for important intellectual content. All authors have read and approved the manuscript.

Acknowledgement

This work was supported by ESFA project "Biocardiostim" (No. VP1-3.1-ŠMM-10-V-02-029). The authors are thankful to Prof. Ira Cohen, Department of Physiology and Biophysics, Stony Brook University, Stony Brook, New York, USA, and to Prof. Michael R. Rosen, Department of Pharmacology, Columbia University, New York, USA, for the valuable suggestions and discussions on the results presented in this paper.

Author details

[1]Department of Regenerative Medicine, State Research Institute Centre for Innovative Medicine, Vilnius, Lithuania. [2]Department of Pathology, Forensic Medicine and Pharmacology, Vilnius University, Faculty of Medicine, Vilnius, Lithuania. [3]Institute of Cardiology, Lithuanian University of Health Sciences, Kaunas, Lithuania. [4]Institute of Materials Science, Kaunas University of Technology, Kaunas, Lithuania. [5]Clinic of Rheumatology, Orthopedic and Traumatology and Reconstructive Surgery, Faculty of Medicine, Vilnius University, Vilnius, Lithuania.

References

1. Edelberg JM, Huang DT, Josephson ME, Rosenberg RD. Molecular enhancement of porcine cardiac chronotropy. Heart. 2001;86(5):559–62.

2. Miake J, Marban E, Nuss HB. Biological pacemaker created by gene transfer. Nature. 2002;419(6903):132–3. doi:10.1038/419132b.

3. Kapoor N, Liang W, Marban E, Cho HC. Direct conversion of quiescent cardiomyocytes to pacemaker cells by expression of Tbx18. Nat Biotechnol. 2013;31(1):54–62. doi:10.1038/nbt.2465.

4. Saito Y, Nakamura K, Yoshida M, Sugiyama H, Ohe T, Kurokawa J, et al. Enhancement of spontaneous activity by HCN4 overexpression in mouse embryonic stem cell-derived cardiomyocytes—a possible biological pacemaker. PLoS One. 2015;10(9):e0138193. doi:10.1371/journal.pone.0138193.

5. Semmler J, Lehmann M, Pfannkuche K, Reppel M, Hescheler J, Nguemo F. Functional expression and regulation of hyperpolarization-activated cyclic nucleotide-gated channels (HCN) in mouse iPS cell-derived cardiomyocytes after UTF1-neo selection. Cell Physiol Biochem. 2014;34(4):1199–215. doi:10.1159/000366332.

6. Zhou YF, Yang XJ, Li HX, Han LH, Jiang WP. Mesenchymal stem cells transfected with HCN2 genes by LentiV can be modified to be cardiac pacemaker cells. Med Hypotheses. 2007;69(5):1093–7. doi:10.1016/j.mehy.2007.02.032.

7. Plotnikov AN, Shlapakova I, Szabolcs MJ, Danilo Jr P, Lorell BH, Potapova IA, et al. Xenografted adult human mesenchymal stem cells provide a platform for sustained biological pacemaker function in canine heart. Circulation. 2007;116(7):706–13. doi:10.1161/CIRCULATIONAHA.107.703231.

8. Penfornis P, Pochampally R. Isolation and expansion of mesenchymal stem cells/multipotential stromal cells from human bone marrow. Methods Mol Biol. 2011;698:11–21. doi:10.1007/978-1-60761-999-4_2.

9. Sekiya I, Larson BL, Smith JR, Pochampally R, Cui JG, Prockop DJ. Expansion of human adult stem cells from bone marrow stroma: conditions that maximize the yields of early progenitors and evaluate their quality. Stem Cells. 2002;20(6):530–41. doi:10.1634/stemcells.20-6-530.

10. Wilders R, Jongsma HJ. Limitations of the dual voltage clamp method in assaying conductance and kinetics of gap junction channels. Biophys J. 1992;63:942–53.

11. Hou Q, Grijpma DW, Feijen J. Porous polymeric structures for tissue engineering prepared by a coagulation, compression moulding and salt leaching technique. Biomaterials. 2003;24(11):1937–47.

12. Bruzauskaite I, Bironaite D, Bagdonas E, Bernotiene E. Scaffolds and cells for tissue regeneration: different scaffold pore sizes-different cell effects. Cytotechnology. 2015. doi:10.1007/s10616-015-9895-4.

13. Rubehn B, Stieglitz T. In vitro evaluation of the long-term stability of polyimide as a material for neural implants. Biomaterials. 2010;31(13):3449–58. doi:10.1016/j.biomaterials.2010.01.053.

14. Rousche PJ, Pellinen DS, Pivin Jr DP, Williams JC, Vetter RJ, Kipke DR. Flexible polyimide-based intracortical electrode arrays with bioactive capability. IEEE Trans Biomed Eng. 2001;48(3):361–71. doi:10.1109/10.914800.

15. Qi Y, McAlpine MC. Nanotechnology-enabled flexible and biocompatible energy harvesting. Energy Environ Sci. 2010;3:1275–85. doi:10.1039/C0EE00137F.

16. Myllymaa S, Myllymaa K, Korhonen H, Lammi MJ, Tiitu V, Lappalainen R. Surface characterization and in vitro biocompatibility assessment of photosensitive polyimide films. Colloids Surf B Biointerfaces. 2010;76(2):505–11. doi:10.1016/j.colsurfb.2009.12.011.

17. Maenosono H, Saito H, Nishioka Y. A transparent polyimide film as a biological cell culture sheet with microstructures. J Biomater Nanobiotechnol. 2014;5(1):17–23. doi:10.4236/jbnb.2014.51003.

18. Valiunas V, Kanaporis G, Valiuniene L, Gordon C, Wang HZ, Li L, et al. Coupling an HCN2-expressing cell to a myocyte creates a two-cell pacing unit. J Physiol. 2009;587(Pt 21):5211–26. doi:10.1113/jphysiol.2009.180505.

19. Zon LI. Intrinsic and extrinsic control of haematopoietic stem-cell self-renewal. Nature. 2008;453(7193):306–13. doi:10.1038/nature07038.

20. Pittenger MF, Mackay AM, Beck SC, Jaiswal RK, Douglas R, Mosca JD, et al. Multilineage potential of adult human mesenchymal stem cells. Science. 1999;284(5411):143–7.

21. Hamid R, Brandt SJ. Transforming growth-interacting factor (TGIF) regulates proliferation and differentiation of human myeloid leukemia cells. Mol Oncol. 2009;3(5–6):451–63. doi:10.1016/j.molonc.2009.07.004.

22. Yan L, Womack B, Wotton D, Guo Y, Shyr Y, Dave U, et al. Tgif1 regulates quiescence and self-renewal of hematopoietic stem cells. Mol Cell Biol. 2013;33(24):4824–33. doi:10.1128/MCB.01076-13.

23. Baldwin Jr AS, Azizkhan JC, Jensen DE, Beg AA, Coodly LR. Induction of NF-kappa B DNA-binding activity during the G0-to-G1 transition in mouse fibroblasts. Mol Cell Biol. 1991;11(10):4943–51.

24. Bull P, Hunter T, Verma IM. Transcriptional induction of the murine c-rel gene with serum and phorbol-12-myristate-13-acetate in fibroblasts. Mol Cell Biol. 1989;9(11):5239–43.

25. Evans RB, Gottlieb PD, Bose Jr HR. Identification of a rel-related protein in the nucleus during the S phase of the cell cycle. Mol Cell Biol. 1993;13(10):6147–56.

26. Hai T, Hartman MG. The molecular biology and nomenclature of the activating transcription factor/cAMP responsive element binding family of transcription factors: activating transcription factor proteins and homeostasis. Gene. 2001;273(1):1–11.

27. Nobori K, Ito H, Tamamori-Adachi M, Adachi S, Ono Y, Kawauchi J, et al. ATF3 inhibits doxorubicin-induced apoptosis in cardiac myocytes: a novel cardioprotective role of ATF3. J Mol Cell Cardiol. 2002;34(10):1387–97.

28. Perez S, Vial E, van Dam H, Castellazzi M. Transcription factor ATF3 partially transforms chick embryo fibroblasts by promoting growth factor-independent proliferation. Oncogene. 2001;20(9):1135–41. doi:10.1038/sj.onc.1204200.

29. Lu D, Wolfgang CD, Hai T. Activating transcription factor 3, a stress-inducible gene, suppresses Ras-stimulated tumorigenesis. J Biol Chem. 2006;281(15):10473–81. doi:10.1074/jbc.M509278200.

30. Fan F, Jin S, Amundson SA, Tong T, Fan W, Zhao H, et al. ATF3 induction following DNA damage is regulated by distinct signaling pathways and over-expression of ATF3 protein suppresses cells growth. Oncogene. 2002;21(49):7488–96. doi:10.1038/sj.onc.1205896.

31. Heng HH, Xiao H, Shi XM, Greenblatt J, Tsui LC. Genes encoding general initiation factors for RNA polymerase II transcription are dispersed in the human genome. Hum Mol Genet. 1994;3(1):61–4.

32. Choi CH, Hiromura M, Usheva A. Transcription factor IIB acetylates itself to regulate transcription. Nature. 2003;424(6951):965–9. doi:10.1038/nature01899.

33. James Faresse N, Canella D, Praz V, Michaud J, Romascano D, Hernandez N. Genomic study of RNA polymerase II and III SNAPc-bound promoters reveals a gene transcribed by both enzymes and a broad use of common activators. PLoS Genet. 2012;8(11):e1003028. doi:10.1371/journal.pgen.1003028.

34. Yokota Y, Mori S. Role of Id family proteins in growth control. J Cell Physiol. 2002;190(1):21–8. doi:10.1002/jcp.10042.

35. Swarbrick A, Akerfeldt MC, Lee CS, Sergio CM, Caldon CE, Hunter LJ, et al. Regulation of cyclin expression and cell cycle progression in breast epithelial cells by the helix-loop-helix protein Id1. Oncogene. 2005;24(3):381–9. doi:10.1038/sj.onc.1208188.

36. Amati B, Alevizopoulos K, Vlach J. Myc and the cell cycle. Front Biosci. 1998;3:d250–68.

37. Norton JD. ID helix-loop-helix proteins in cell growth, differentiation and tumorigenesis. J Cell Sci. 2000;113(Pt 22):3897–905.

38. Neal JW, Clipstone NA. A constitutively active NFATc1 mutant induces a transformed phenotype in 3T3-L1 fibroblasts. J Biol Chem. 2003;278(19):17246–54. doi:10.1074/jbc.M300528200.

39. Rao A, Luo C, Hogan PG. Transcription factors of the NFAT family: regulation and function. Annu Rev Immunol. 1997;15:707–47. doi:10.1146/annurev.immunol.15.1.707.

40. Macian F, Lopez-Rodriguez C, Rao A. Partners in transcription: NFAT and AP-1. Oncogene. 2001;20(19):2476–89. doi:10.1038/sj.onc.1204386.

41. Lipskaia L, Lompre AM. Alteration in temporal kinetics of Ca2+ signaling and control of growth and proliferation. Biol Cell. 2004;96(1):55–68. doi:10.1016/j.biolcel.2003.11.001.

42. Shimada M, Nakadai T, Tamura TA. TATA-binding protein-like protein (TLP/TRF2/TLF) negatively regulates cell cycle progression and is required for the stress-mediated G(2) checkpoint. Mol Cell Biol. 2003;23(12):4107–20.

43. Um M, Yamauchi J, Kato S, Manley JL. Heterozygous disruption of the TATA-binding protein gene in DT40 cells causes reduced cdc25B phosphatase expression and delayed mitosis. Mol Cell Biol. 2001;21(7):2435–48. doi:10.1128/MCB.21.7.2435-2448.2001.

44. Vichi P, Coin F, Renaud JP, Vermeulen W, Hoeijmakers JH, Moras D, et al. Cisplatin- and UV-damaged DNA lure the basal transcription factor TFIID/TBP. EMBO J. 1997;16(24):7444–56. doi:10.1093/emboj/16.24.7444.

45. Wen SC, Ku DH, De Luca A, Claudio PP, Giordano A, Calabretta B. ets-2 regulates cdc2 kinase activity in mammalian cells: coordinated expression of cdc2 and cyclin A. Exp Cell Res. 1995;217(1):8–14. doi:10.1006/excr.1995.1057.

46. Wen F, Tynan JA, Cecena G, Williams R, Munera J, Mavrothalassitis G, et al. Ets2 is required for trophoblast stem cell self-renewal. Dev Biol. 2007;312(1):284–99. doi:10.1016/j.ydbio.2007.09.024.

3

Intravitreal administration of multipotent mesenchymal stromal cells triggers a cytoprotective microenvironment in the retina of diabetic mice

Marcelo Ezquer[1], Cristhian A. Urzua[2], Scarleth Montecino[1], Karla Leal[1], Paulette Conget[1] and Fernando Ezquer[1*]

Abstract

Background: Diabetic retinopathy is a common complication of diabetes and the leading cause of irreversible vision loss in the Western world. The reduction in color/contrast sensitivity due to the loss of neural cells in the ganglion cell layer of the retina is an early event in the onset of diabetic retinopathy. Multipotent mesenchymal stromal cells (MSCs) are an attractive tool for the treatment of neurodegenerative diseases, since they could differentiate into neuronal cells, produce high levels of neurotrophic factors and reduce oxidative stress. Our aim was to determine whether the intravitreal administration of adipose-derived MSCs was able to prevent the loss of retinal ganglion cells in diabetic mice.

Methods: Diabetes was induced in C57BL6 mice by the administration of streptozotocin. When retinal pro-damage mechanisms were present, animals received a single intravitreal dose of 2×10^5 adipose-derived MSCs or the vehicle. Four and 12 weeks later we evaluated: (a) retinal ganglion cell number (immunofluorescence); (b) neurotrophic factor levels (real-time quantitative polymerase chain reaction (RT-qPCR) and enzyme-linked immunosorbent assay (ELISA)); (c) retinal apoptotic rate (TUNEL); (d) retinal levels of reactive oxygen species and oxidative damage (ELISA); (e) electrical response of the retina (electroretinography); (f) pro-angiogenic and anti-angiogenic factor levels (RT-qPCR and ELISA); and (g) retinal blood vessels (angiography). Furthermore, 1, 4, 8 and 12 weeks post-MSC administration, the presence of donor cells in the retina and their differentiation into neural and perivascular-like cells were assessed (immunofluorescence and flow cytometry).

Results: MSC administration completely prevented retinal ganglion cell loss. Donor cells remained in the vitreous cavity and did not differentiate into neural or perivascular-like cells. Nevertheless, they increased the intraocular levels of several potent neurotrophic factors (nerve growth factor, basic fibroblast growth factor and glial cell line-derived neurotrophic factor) and reduced the oxidative damage in the retina. Additionally, MSC administration has a neutral effect on the electrical response of the retina and did not result in a pathological neovascularization.

Conclusions: Intravitreal administration of adipose-derived MSCs triggers an effective cytoprotective microenvironment in the retina of diabetic mice. Thus, MSCs represent an interesting tool in order to prevent diabetic retinopathy.

Keywords: Diabetes, Diabetic retinopathy, Multipotent mesenchymal stromal cells, Mesenchymal stem cells, Microenvironment, Cytoprotection, Retinal ganglion cells, Prevention

* Correspondence: eezquer@udd.cl
[1]Centro de Medicina Regenerativa, Facultad de Medicina Clínica
Alemana-Universidad del Desarrollo, Av. Las Condes 12438, Lo Barnechea,
Santiago 7710162, Chile
Full list of author information is available at the end of the article

Background

Diabetic retinopathy (DR) is one of the most common and frightening complications in patients with diabetes mellitus (DM) [1] and the leading cause of irreversible vision loss in developed countries [2]. In the Western world, the prevalence of DR in patients with DM is around 28 %, and strongly increases with the progression of DM [3]. Hence, almost all patients with Type 1 DM and 60 % of those with Type 2 DM will have some degree of DR after 20 years of DM evolution [4, 5].

Through the years, DR has been recognized primarily as a vascular disorder that involves pericyte loss, basement membrane thickening and endothelial dysfunction involving loss of retinal barrier integrity which leads to hemorrhage, vascular obliteration and the resulting neovascularization [6]. These events subsequently cause fibrovascular proliferation and blindness [6]. Nevertheless, it has been well documented that the hyperglycemic state adversely affects the entire neurosensory retina, and accelerates neuronal apoptosis [7, 8]. Thus, today DR is also considered a sensory neuropathy [9].

Although the molecular mechanisms by which DM induces neuronal degeneration and dysfunction are still not well understood, the role of hyperglycemia and the generation of reactive oxygen species (ROS) due to its exacerbating metabolism seem to be the most influential factor [10, 11]. ROS are mainly produced in the mitochondria through the electron transport chain, and it is well known that elevated ROS levels affect the survival and function of retinal neurons [12]. Excessive production of ROS results in the oxidative damage of several biomolecules including lipids, proteins and DNA. Retinal ganglion cells (RGCs) and glial cells are particularly sensitive to ROS-induced damage, which leads to the early death of these cells by apoptosis [13]. Additionally, it has been reported that increased ROS levels reduce the retinal levels of neurotrophic factors, including nerve growth factor (NGF), brain-derived neurotrophic factor (BDNF) and glial cell line-derived neurotrophic factor (GDNF), accelerating neuronal death [10, 12].

Nowadays, available treatments for DR are applicable in advanced stages of the disease and are primarily intended to regulate vascular changes principally mediated by the action of vascular endothelial growth factor (VEGF). These include laser treatment to destroy the hypoxic retinal cells that produce VEGF or the intravitreal administration of anti-VEGF drugs [14]. However, these therapies have only targeted the vascular pathology and have achieved limited success [15]. Therefore, the future generation of therapies that, applied at early stages of DR, can target the neural tissue eliciting a better visual prognosis are highly desirable.

Recently, the use of stem cells for the management of eye diseases has generated considerable interest [16]. Among the different type of stem cells, multipotent mesenchymal stromal cells (MSCs), also referred to as mesenchymal stem cells, appear as an ideal candidate for a cell therapy for DR because: (a) they can be obtained from different sources without major complications [17]; (b) they can be easily ex-vivo expanded [17]; (c) they can differentiate into neural cells or perivascular cells [18, 19], replacing the cells that are damaged during the course of DR; (d) they are able to secrete high levels of potent neuroprotective factors, including NGF, BDNF and GDNF [20, 21] that can reduce the apoptosis of neuronal cells in the retina; (e) they efficiently scavenge ROS, reducing the oxidative damage of the target tissues [22, 23]; and (f) MSCs have been used in cell therapy strategies to treat patients with different diseases, providing favorable outcomes without significant side effects [24, 25].

Previous work has demonstrated that the intravitreal injection of MSCs can help to ameliorate and repair different retinopathic injuries, mainly by the stabilization of retinal microvasculature [26, 27].

In this study, we evaluated whether an intravitreal administration of adipose-derived MSCs was able to prevent the loss of RGCs in diabetic mice. For this, severe diabetes was induced in C57/BL6 mice by the administration of a single high dose (200 mg/kg) of streptozotocin (STZ) [28, 29]. Diabetes progression was evaluated according to the levels of glucose, insulin and glycated hemoglobin in blood samples. Twelve weeks after diabetes induction, animals were randomly assigned into two groups: one group received a single intravitreal administration of 2×10^5 adipose-derived MSCs (DM + MSC mice) and the other group received the vehicle (DM mice). In addition, a third group of nondiabetic animals (Normal mice) was included as control. Four and 12 weeks post-MSC administration, we evaluated: (a) RGC number by immunofluorescence; (b) the levels of neuroprotective factors by real-time quantitative polymerase chain reaction (RT-qPCR) and enzyme-linked immunosorbent assay (ELISA); (c) retinal apoptotic rate by TUNEL; (d) retinal ROS levels and oxidative damage by ELISA; (e) electrical response of the retina by electroretinography; (f) pro-angiogenic and anti-angiogenic factors levels by RT-qPCR and ELISA; and (g) retinal blood vessels by angiography. Furthermore, 1, 4, 8 and 12 weeks post-MSC administration, the presence of donor cells in the retina and their differentiation into neural and perivascular-like cells were assessed by immunofluorescence and flow cytometry.

We found that the intravitreal administration of MSCs increases neurotrophic factor levels, reduces oxidative damage of the retina and prevents RGC loss in diabetic mice.

Methods

Animals

C57BL/6 and C57BL76-Tg(ACTB-EGFP)1Obs mice (Jackson Laboratory, Bar Harbor, ME, USA) were housed at constant temperature and humidity, with a 12:12 hour light–dark cycle and unrestricted access to standard diet and water. When required, animals were lightly anesthetized with sevofluorane (Abbot, Tokyo, Japan) or deeply anesthetized with ketamine (Drag Pharma, Santiago, Chile) plus xylazine (Centrovet, Santiago, Chile). All animal protocols used were approved by the Ethic Committee of Facultad de Medicina Clínica Alemana-Universidad del Desarrollo.

Diabetes induction

Ten-week-old male C57BL/6 mice were lightly anesthetized and received an intraperitoneal injection of 200 mg/kg STZ (Calbiochem, La Jolla, CA, USA) immediately after dissolving it in 0.1 M citrate buffer pH 4.5 (DM mice), or citrate buffer only (Normal mice). It has been reported that this protocol of STZ administration causes a massive cytotoxic destruction of beta-pancreatic cells generating a condition of severe hyperglycemia, which accelerates the appearance of the secondary complications associated with diabetes [28, 30].

Blood glucose quantification

Blood samples were collected from the tail vein of non-fasted alert animals, and glucose levels were determined with the glucometer system Accu-Chek Performa from Roche Diagnostic (Mannheim, Germany).

Plasma insulin quantification

Blood samples were collected from the tail vein of fasted alert animals. Plasma was recovered by centrifugation, and insulin concentrations were measured using a mouse insulin ultrasensitive ELISA kit (Mercodia, Uppsala, Sweden).

Glycated hemoglobin quantification

Blood samples were collected from the tail vein of fasted alert animals, and HbA_{1c} percentages were assessed using the DCA2000 analyzer (Bayer Corporation, Pittsburgh, PA, USA) as previously described [31].

Isolation, ex vivo expansion and characterization of MSCs

Eight- to 10-week-old female C57BL/6 or C57BL/6-Tg(ACTB-EGFP)1Obs mice were sacrificed by cervical dislocation. Epididymal fat was dissected, washed with phosphate-buffered saline (PBS) and cut into small pieces. Tissues were then digested with 1 mg/mL collagenase type II (Gibco, Grand Island, NY, USA) in PBS and incubated with agitation at 37 °C for 2 hours. At the end of digestion, 10 % fetal bovine serum (FBS; Gibco, Auckland, New Zealand) was added to neutralize collagenase. The mixture was then centrifuged at 400 g for 10 minutes to remove floating adipocytes. Pellets were re-suspended in α-minimum essential medium (α-MEM; Gibco) supplemented with 10 % FBS and 0.16 mg/mL gentamicin (Sanderson Laboratory, Santiago, Chile), plated at a density of 7000 cells/cm^2 and cultured at 37 °C in a 5 % CO_2 atmosphere. When foci reach confluence, cells were detached with 0.25 % trypsin, 2.65 mM EDTA (Sigma-Aldrich, St. Louis, MO, USA), centrifuged and subcultured at 7000 cells/cm^2. After two subcultures, cells were characterized according to their adipogenic and osteogenic differentiation potential. For this, MSCs were incubated with standard adipogenic or osteogenic differentiation media for 14 and 21 days, respectively. To evaluate the adipogenic potential, cultures were stained with Oil Red (Sigma-Aldrich). To evaluate the osteogenic potential, cultures were fixed with 10 % formaldehyde and stained with Alizarin Red (Sigma-Aldrich) as previously described [28]. Immunophenotyping was performed by flow cytometry analysis after immunostaining with monoclonal antibodies against the putative murine MSC markers α-SMA (FITC-conjugated; BD Bioscience, San Jose, CA, USA), Sca-1 (PE-conjugated; eBioscience, San Diego, CA, USA), and CD90 (PECy7-conjugated; eBioscience), or characteristic markers of hematopoietic cell lineages CD45 (AF780-conjugated; BD Bioscience) and CD11b (PECy5-conjugated; eBioscience).

Intravitreal administration of MSCs

Twelve weeks after DM induction, mice were lightly anesthetized and 0.5 % proparacaine (Alcon, Santiago Chile) was topically applied. A cell suspension containing 2×10^5 MSCs, passage 2, in 2 μL saline (DM + MSC mice), or 2 μL saline (DM mice) was slowly injected into the vitreous cavity through the pars plana using a 33-gauge microsyringe (Hamilton, Reno, NV, USA). Eyes showing massive vitreous hemorrhaging after the injection were excluded from the study.

Quantification of RGCs

Animals were euthanatized by cervical dislocation and eyes were enucleated and fixed in 4 % paraformaldehyde (Merck, Darmstadt, Germany). Fixed eyes were orientated to permit radial sectioning and embedded in paraffin. Eyes were sectioned (4 μm), and sections were deparaffinized and incubated with rat anti-mouse beta-3-tubulin antibody (marker of RGCs) (Santa Cruz Biotechnology, Dallas, TX, USA). Afterwards, sections were washed and incubated with anti-rat-Alexa555 secondary antibody (Vector Labs, Burlingame, CA, USA) and counterstained with 4'-6'-diamidino-2-phenylindole (DAPI; Invitrogen, Grand Island, NY, USA). Only sections that included a full length of retina approximately along the horizontal meridian, passing through the ora serrata and the optic nerve in both the temporal and nasal hemispheres were used. The number of RGCs in the ganglion cell layer was quantified by counting labeled cells from the temporal to the nasal ora serrata in five serial sections using the Fluoview FV10i confocal microscope (Olympus, Tokyo, Japan). Samples were blind-analyzed by two independent observers. Data were presented as number of RGCs per 100 µm of retina length [32].

Detection of donor MSCs^{GFP}

For in situ detection of MSCs^{GFP}, eyes were fixed in 4 % paraformaldehyde. One day later, they were embedded in paraffin and radially sectioned. Four-micrometer-thick sections were deparaffinized, incubated with rabbit anti-GFP antibody (eBioscience) at 4 °C overnight, incubated with goat anti-rabbit-FITC antibody (Vector Labs) at room temperature for 1 hour, and counterstained with DAPI. Sections were examined with the Fluoview FV10i confocal microscope.

For the quantification of MSCs^{GFP}, eyes were washed twice with ice-cold PBS, chopped and digested with 1 mg/mL collagenase type II at 37 °C for 30 minutes. Collagenase was inactivated with 10 % FBS and cell suspensions were filtered through a 100-µm strainer and washed twice with ice-cold PBS. To ensure MSC^{GFP} recognition, cells in the suspension were fixed and permeabilized with BD Cytofix/Cytoperm kit (BD Pharmingen, San Jose, CA, USA) and suspended in 1 mL PBS with 2 % FBS plus 1 µL undiluted anti-GFP AlexaFluor647 antibody (Molecular Probes, Grand Island, NY, USA). After incubation at 4 °C for 12 hours, cells were washed, filtered through a 30-µm mesh and acquired in a CyAn ADP flow cytometer (DakoCytomation, Carpinteria, CA, USA) as previously described [33]. Data were analyzed with Summit v4.3 software. Criteria used to consider an event as an MSCs^{GFP} were forward scatter and side scatter similar to ex vivo expanded MSCs and positive fluorescence both in FL1 (GFP) and FL8 (anti-GFP AlexaFluor647) channels. Eyes from untreated diabetic mice were used as autofluorescence controls. Results were presented as number of MSCs^{GFP} per eye.

Evaluation of MSCs^{GFP} differentiation into neural-like or perivascular-like cells

To determine whether donor MSCs differentiated into neural-like cells, eye sections were co-stained with rabbit anti-GFP antibody (eBioscience) and rat anti-mouse beta-III-tubulin antibody (marker of RCGs) (Santa Cruz Biotechnology) or were co-stained with rabbit anti-GFP antibody and rat anti-mouse GFAP

antibody (marker of retinal astrocytes) (Santa Cruz Biotechnology). Afterwards, sections were washed and incubated with anti-rabbit-FITC and anti-rat-Alexa555 secondary antibodies (Vector Labs) and counterstained with DAPI.

To determine whether donor MSCs differentiated into perivascular-like cells, eyes were fixed for 2 hours in 4 % paraformaldehyde. Afterwards, the cornea and lens were removed and the retina was dissected and flattened onto silanized glass slides facilitated by four equidistant radial cuts into the peripheral retina. Wholemounts were permeabilized with 1 % digitonin (Calbiochem) in PBS at room temperature for 1 hour and immunohistochemically stained by overnight incubation with rabbit anti-GFP antibody and rat anti-mouse NG2 antibody (marker of pericytes) (Millipore, Darmstadt, Germany) at 4 °C. Samples were also incubated with Isolectin GS-IB4 from *Griffonia simplicifolia* conjugated to AlexaFluor647 (Life Technology, Grand Island, NY, USA) to allow the detection of retinal capillaries. Afterwards, samples were washed and incubated with anti-rabbit-FITC and anti-rat-Alexa555 secondary antibodies (Vector Labs).

In all cases, retinal tissues without exposure to the primary antibodies were used as controls for immunostaining. Samples were analyzed under confocal microscopy by taking optical sections of 1 µm. Data were analyzed with the Olympus FV10-ASW2.1 software.

Quantification of mRNA levels of neurotrophic, pro-angiogenic and anti-angiogenic factors

Eyes were enucleated and washed two times with ice-cold PBS. Total RNA was purified using Absolutely RNA Miniprep kit (Stratagene, Santa Clara, CA, USA). One microgram of total RNA was used for reverse transcription. RT-PCR reactions were performed in a final volume of 10 µL containing 50 ng cDNA, PCR LightCycler-DNA Master SYBERGreen reaction mix (Roche, Indianapolis, IN, USA), 3 mM MgCl$_2$ and 0.5 µM of the primers for the amplification of NGF, basic fibroblast growth factor (bFGF), GDNF, BDNF, ciliary neurotrophic factor (CNTF), VEGF-α, platelet-derived growth factor (PDGF), angiopoietin 1 (ANG-1) and thrombospondin-1 (TSP-1) (Additional file 1: Table S1), using a Light-Cycler 1.5 thermocycler (Roche). To ensure that amplicons were from mRNA and not for genomic DNA amplifications, controls without reverse transcription were included. Amplicons were characterized according to their size evaluated by agarose gel electrophoresis and to their melting temperature determined in the LightCycler thermocycler. Relative quantifications were performed by the $\Delta\Delta CT$ method. The mRNA level of each target gene was standardized against the mRNA level of

GAPDH, for the same sample. Results were presented as fold-change versus normal mice.

Quantification of protein levels of neurotrophic and anti-angiogenic factors

Eyes were enucleated and washed two times with ice-cold PBS. Samples were mechanically lysed in lysis buffer (RayBiotech, Norcross, GA, USA) containing a protease inhibitor cocktail (Thermo, Waltham, MA, USA) and centrifuged at 12,000 g for 10 minutes. The levels of NGF, bFGF, GDNF and TSP-1 were measured in the supernatant of the lysates using the Mouse beta-NGF ELISA kit (RayBiotech), Mouse FGF basic Quantikine ELISA kit (R&D Systems, Minneapolis MN, USA), Mouse GDNF ELISA kit (MyBioSource, San Diego, CA, USA) and Mouse thrombospondin 1 ELISA kit (MyBioSource), respectively. Data were normalized per mg of protein present in each sample.

Quantification of apoptotic cells

The presence of apoptotic cells in the retina was evaluated by the TUNEL technique in 4-μm thick sections of the eye, using the In Situ Cell Death Detection Kit (Roche) following the manufacturer's instructions. In each experiment, adjacent sections incubated without TdT served as negative control. Nuclei were counterstained with DAPI and samples were observed by confocal microscopy. Each section was scanned systematically from the temporal to the nasal ora serrata looking for fluorescent cells indicative of apoptosis. The number of TUNEL-positive cells per section was determined using the Olympus FV10-ASW2.1 software, and the apoptosis rate was expressed as fold-change versus normal mice.

Quantification of ROS

Retinas were carefully dissected and mechanically lysed in lysis buffer (RayBiotech) containing a protease inhibitor cocktail. For quantification of ROS level, equal volumes of retinal lysates were incubated with 10 μmol/L 2,7-dichloro-dihydro-fluorescein diacetate (H$_2$DCFDA; Invitrogen) for 1 hour at 37 °C. Fluorescence was measured in a fluorimeter (Turner, Sunnyvale, CA, USA) with excitation of 485 nm and emission of 520 nm as previously reported [34]. Data were normalized per mg of protein present in each sample and expressed as fold-change versus normal mice.

Quantification of oxidative damage

Lipid peroxidation was determined by a method that measures the amount of thiobarbituric acid reactivity by the amount of malondialdehyde (MDA) formed during acid hydrolysis of the lipid peroxide compounds using the Lipid Peroxidation MDA Assay kit (Cayman, Ann Arbor, MI, USA). Retinas were mechanically lysed in lysis buffer (RayBiotech) containing a protease inhibitor cocktail and the antioxidant butyl hydroxytoluene (BHT). Lipid peroxidation was quantified sprectophotometrically at 540 nm (Thermo spectronic). MDA level was normalized per mg of protein present in each sample.

Protein nytrosilation was quantified in retinal lysates containing a protease inhibitor cocktail, measuring nitrotyroxine level with the 3-Nitrotyrosine ELISA kit (Abcam, Cambridge, UK) following the manufacturer's instructions. The nitrotyroxine level was normalized per mg of protein present in each sample.

DNA oxidation was measured in genomic DNA isolated from retinal samples by DNAzol (Invitrogen). DNA was digested by incubation with DNAsa RQ1 (Promega, Madison, WI, USA) and 8-OHdG levels were quantified using the Oxidative DNA damage ELISA kit, following the manufacturer's instructions. 8-OHdG levels were normalized per μg of DNA present in each sample.

For all oxidative markers, data were expressed as fold-change versus normal mice.

Evaluation of retinal electrical response

Evaluation of retinal function was carried out by electroretinography (ERG) using the Handheld Multi-species ElectroRetinoGraph Instrument (Ocuscience, Kansas City, MO, USA). In brief, the animals were prepared under red light illumination, anesthesia was provided by the inhaled administration of sevoflurane and pupils were dilated by using tropicamide acetate (Alcon). Body temperature was maintained at 37 °C with a heating plate. All animals recovered from anesthesia after the procedure.

Previous to the measurements, mice were maintained for a period of dark adaptation of 16 hours. The electrical responses were registered by using a corneal electrode with methylcellulose as a coupling agent. Reference and ground electrodes were placed in the ear and in the tail, respectively. Stimulation and parameter recording were carrying out as defined by the international Society for Clinical Electrophysiology of Vision (ISCEV). In brief, flash stimuli were as follows: 0.01 and 3.0 cd · s · m^{-2} in scotopic and photopic conditions. The a-wave amplitude was measured from the baseline to the trough of the a-wave and the b-wave amplitude was measured from the trough of the a-wave to the peak of the b-wave as previously reported [35]. Oscillatory potentials (OPs) were isolated by treating measurements with a digital bandpass filter between 60 and 200 Hz, after dark adaptation using 3.0 cd · s · m^{-2} flash stimuli. The peak amplitudes of the first six waves were recorded from baseline, as previously described [36].

Quantification of retinal vasculature and detection of vascular leakage

Animals were lightly anesthetized and injected in the tail vein with 0.2 mL saline containing 10 mg of 2×10^6 molecular weight fluorescein-dextran (Sigma-Aldrich). Five minutes later, animals were euthanized by cervical dislocation and eyes were enucleated and fixed in 4 % paraformaldehyde for 2 hours. Afterwards, the cornea and lens were removed, and the retina was dissected and flattened onto glass slides facilitated by four equidistant radial cuts into the peripheral retina. The flat-mounted retinas were photographed by confocal microscopy and at least 15 images per animal were analyzed. The fluorescent area, representative of the retinal vascular area, of each image was quantified using the Image J 1.34 software (NIH, Bethesda, USA). Data were presented as retinal vascular area/total retinal area. Retinal vascular leakage was determined by the presence of areas of extravasated FITC-dextran.

Statistical analysis

Data were presented as mean ± standard error of the mean (SEM). Multiple group comparisons were performed by analysis of variance (ANOVA) followed by Bonferroni post-hoc test, while comparisons between two experimental group were performed by Student's t test. $p < 0.05$ were considered statistically significant.

Results

Diabetes induction and retinopathy condition

To induce DM, C57BL6 mice were treated with a single dose of 200 mg/kg STZ, since this protocol causes a rapid and massive destruction of pancreatic beta cells [28]. Diabetic mice were maintained without insulin supplementation to allow for the progression of severe diabetes and the appearance of its complications, including DR. Ten days after STZ administration, blood glucose levels reached their highest concentration, which was three times higher than in normal individuals (Fig. 1a). Twelve weeks after DM induction, elevated blood glucose level was also correlated with a severe reduction in plasma insulin level and a marked increase in glycated hemoglobin level (Fig. 1b and c). At this time, DM mice maintained the same number of RGCs compared to age matched normal mice (Fig. 1d and e); however, retinal electrical response was altered. This could be evidenced by a significant reduction in the amplitudes of the a-wave and b-wave of the ERG of DM mice in comparison to normal mice under stimuli of 0.01 cd·s·m^{-2} and 3.0 cd·s·m^{-2} (Fig. 1f, g and h).

Adipose-derived MSCs were isolated, expanded and characterized according to their adipogenic and osteogenic potential (Fig. 2a–c), and by the presence of the putative murine MSC markers α-SMA, SCA-1 and CD90 and by the non-expression of markers characteristics of hematopoietic cell lineages—CD45 and CD11b (Fig. 2d). Twelve weeks after diabetes induction, DM mice were randomly assigned into two groups: one group that received an intravitreal administration of 2×10^5 MSCs (DM + MSC mice) and another group that received an intravitreal administration of the vehicle (DM mice). Additionally, we also maintained a group of normal non-diabetic mice (Normal mice) as a control.

MSC administration prevents RGCs loss

Since one of the earliest events in the onset of DR is the death of RGCs [13], we tested whether MSC administration was able to prevent the loss of these cells in the retina. As expected, 16 weeks after DM induction (4 weeks after vehicle administration) DM mice presented a significant reduction (~20 %) in the number of RGCs in the retina. However, MSC administration completely prevented this reduction (Fig. 3a and c). The preventive effect was maintained until at least 24 weeks after diabetes induction (12 weeks after MSC administration) when DM + MSC mice presented an equivalent number of RGCs as compared to cells in the normal mice (Fig. 3b and d).

Donor MSCs remains in the eye but do not differentiate into neural-like or perivascular-like cells

One possibility to explain the therapeutic effect observed is that the administered MSCs could differentiate into RGCs, replacing the damaged cells, or into another neural or perivascular cell type that could act as a support cell to prevent RGC death. To assess this possibility, first we evaluated whether administered MSCs could remain in the eye and integrate into the retina of DM mice. Using MSCs isolated from transgenic mice that constitutively express GFP, we observed by flow cytometry and immunohistofluorescence that the number of donor cells in the eyes were decreasing over time, but they were still detectable 12 weeks after their administration (Fig. 4). Most donor cells remained in the vitreous cavity and did not integrate into the retina (Fig. 4c). To evaluate whether MSCs could differentiate into neural or perivascular cells we tried to colocalize GFP with a marker of RGCs (β3-tubulin) (Fig. 5a), with a marker of retinal astrocytes (GFAP) (Fig. 5b) or with a marker of retinal pericytes (NG2) (Fig. 5c). In most cases, GFP -positive cells did not express any of these markers at any of the analyzed times. However, we also detected a small number of MSCs^{GFP} that also expressed the marker of RGC (Additional file 2: Figure S1A), the marker of astrocyte (Additional file 2: Figure S1B) or the marker of pericyte (Additional file 2: Figure S1C).

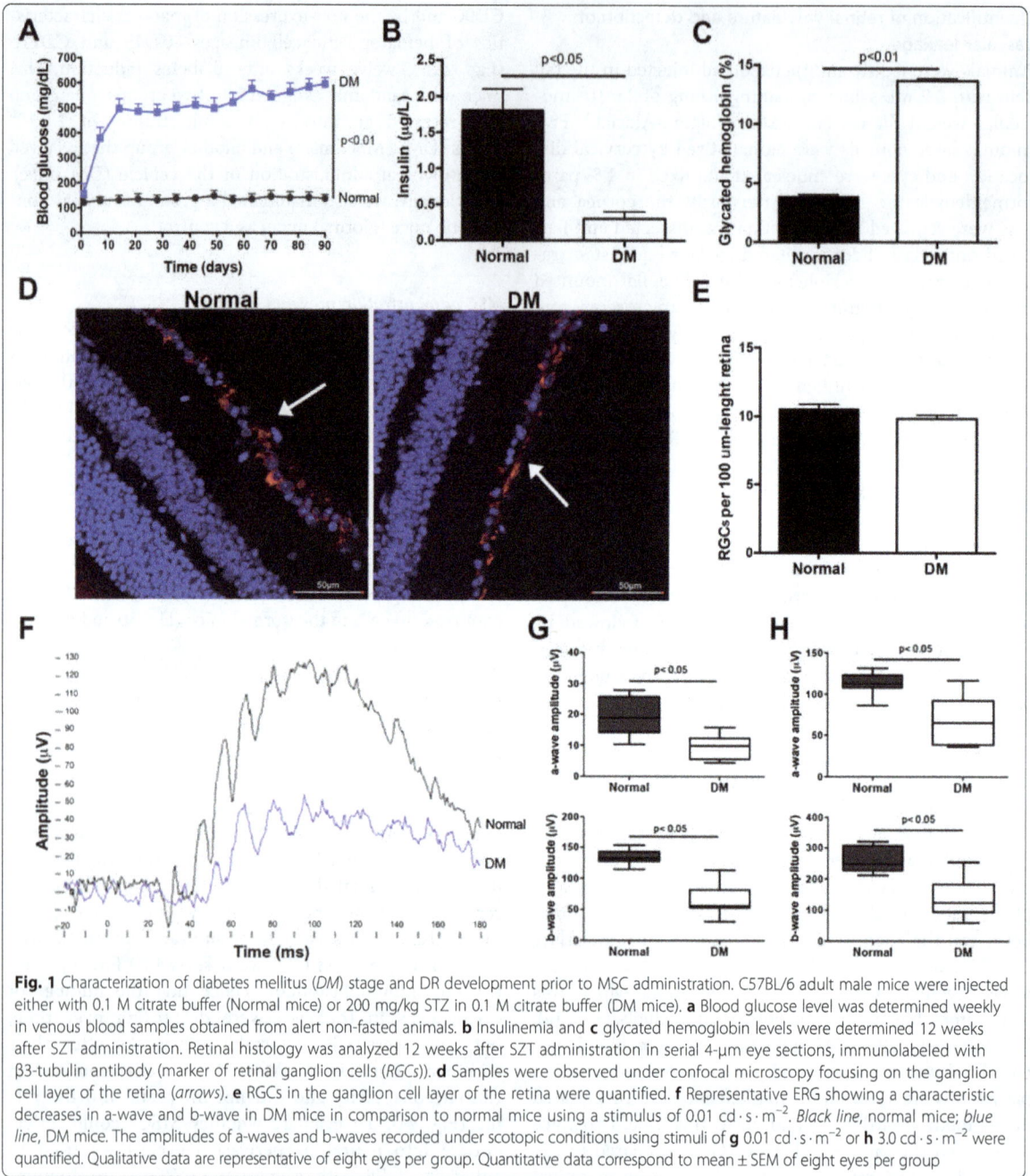

Fig. 1 Characterization of diabetes mellitus (*DM*) stage and DR development prior to MSC administration. C57BL/6 adult male mice were injected either with 0.1 M citrate buffer (Normal mice) or 200 mg/kg STZ in 0.1 M citrate buffer (DM mice). **a** Blood glucose level was determined weekly in venous blood samples obtained from alert non-fasted animals. **b** Insulinemia and **c** glycated hemoglobin levels were determined 12 weeks after SZT administration. Retinal histology was analyzed 12 weeks after SZT administration in serial 4-μm eye sections, immunolabeled with β3-tubulin antibody (marker of retinal ganglion cells (*RGCs*)). **d** Samples were observed under confocal microscopy focusing on the ganglion cell layer of the retina (*arrows*). **e** RGCs in the ganglion cell layer of the retina were quantified. **f** Representative ERG showing a characteristic decreases in a-wave and b-wave in DM mice in comparison to normal mice using a stimulus of 0.01 cd · s · m^{-2}. *Black line*, normal mice; *blue line*, DM mice. The amplitudes of a-waves and b-waves recorded under scotopic conditions using stimuli of **g** 0.01 cd · s · m^{-2} or **h** 3.0 cd · s · m^{-2} were quantified. Qualitative data are representative of eight eyes per group. Quantitative data correspond to mean ± SEM of eight eyes per group

The intravitreal administration of MSCs increases the intraocular levels of neurotrophic factors

MSCs are known to produce, both in vitro and in vivo, a broad range of trophic factors that have been associated with tissue regeneration and neuronal survival [20]. To evaluate whether the therapeutic effects observed after MSC administration could be related to the generation of a pro-regenerative microenvironment, the intraocular levels of several neurotrophic factors were measured.

Four weeks after MSC administration, we observed a significant increase in the mRNA levels of NGF, bFGF and GDNF in the eyes of DM + MSC mice compared to age-matched DM mice and normal mice (Fig. 6a). The mRNA levels of NGF and bFGF remained elevated in the

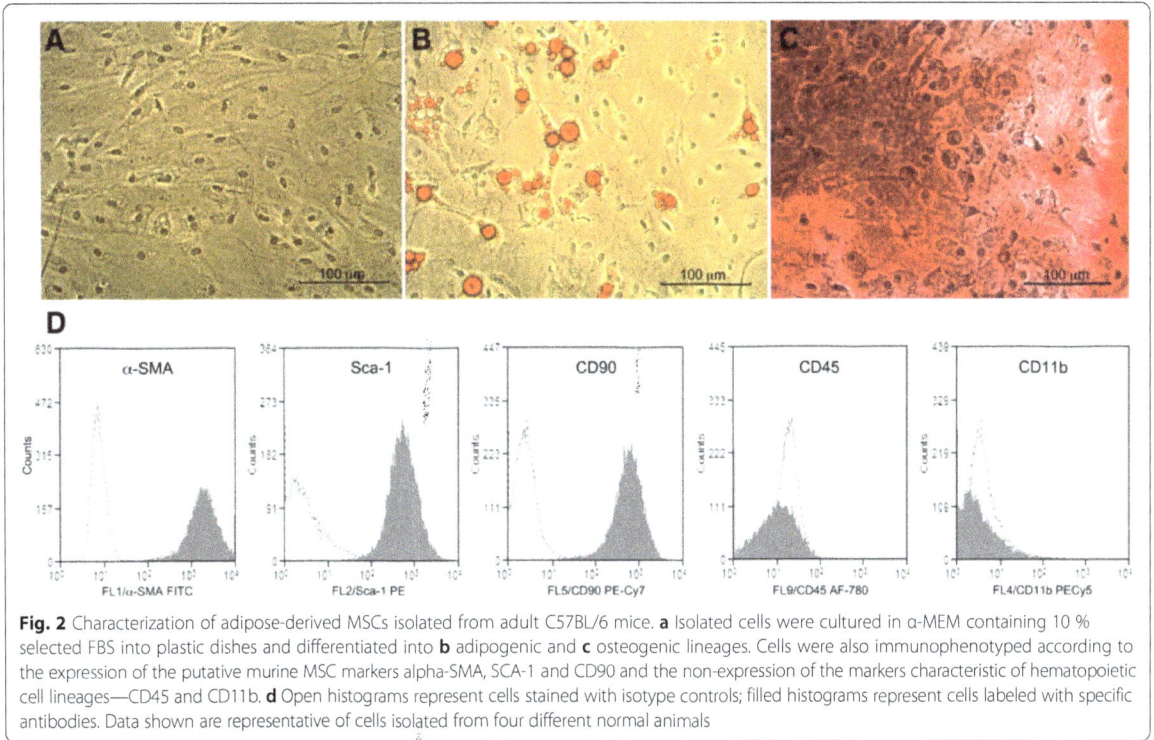

Fig. 2 Characterization of adipose-derived MSCs isolated from adult C57BL/6 mice. **a** Isolated cells were cultured in α-MEM containing 10 % selected FBS into plastic dishes and differentiated into **b** adipogenic and **c** osteogenic lineages. Cells were also immunophenotyped according to the expression of the putative murine MSC markers alpha-SMA, SCA-1 and CD90 and the non-expression of the markers characteristic of hematopoietic cell lineages—CD45 and CD11b. **d** Open histograms represent cells stained with isotype controls; filled histograms represent cells labeled with specific antibodies. Data shown are representative of cells isolated from four different normal animals

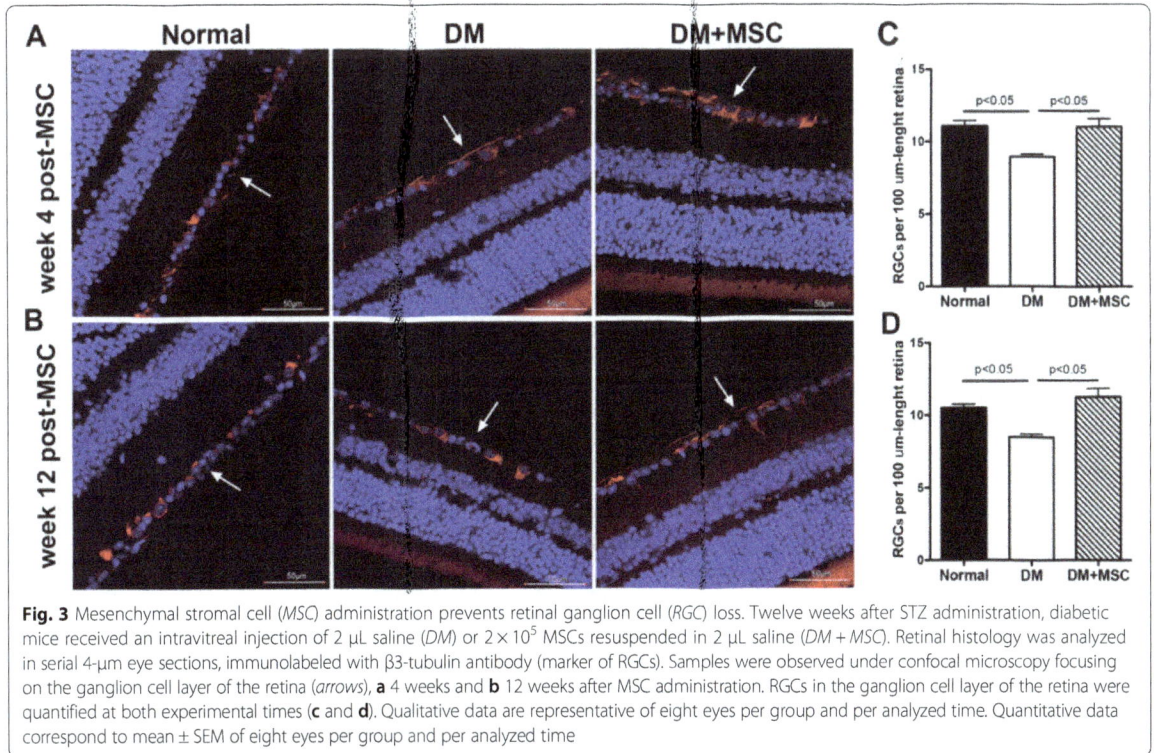

Fig. 3 Mesenchymal stromal cell (*MSC*) administration prevents retinal ganglion cell (*RGC*) loss. Twelve weeks after STZ administration, diabetic mice received an intravitreal injection of 2 μL saline (*DM*) or 2×10^5 MSCs resuspended in 2 μL saline (*DM + MSC*). Retinal histology was analyzed in serial 4-μm eye sections, immunolabeled with β3-tubulin antibody (marker of RGCs). Samples were observed under confocal microscopy focusing on the ganglion cell layer of the retina (*arrows*), **a** 4 weeks and **b** 12 weeks after MSC administration. RGCs in the ganglion cell layer of the retina were quantified at both experimental times (**c** and **d**). Qualitative data are representative of eight eyes per group and per analyzed time. Quantitative data correspond to mean ± SEM of eight eyes per group and per analyzed time

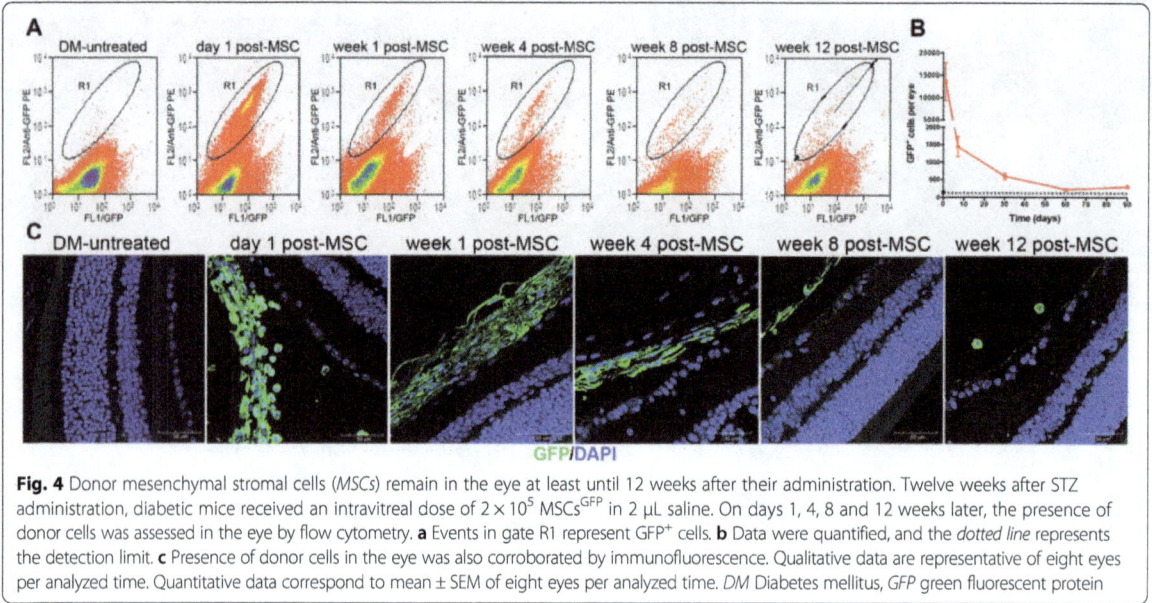

Fig. 4 Donor mesenchymal stromal cells (*MSCs*) remain in the eye at least until 12 weeks after their administration. Twelve weeks after STZ administration, diabetic mice received an intravitreal dose of 2×10^5 MSCs^{GFP} in 2 μL saline. On days 1, 4, 8 and 12 weeks later, the presence of donor cells was assessed in the eye by flow cytometry. **a** Events in gate R1 represent GFP$^+$ cells. **b** Data were quantified, and the *dotted line* represents the detection limit. **c** Presence of donor cells in the eye was also corroborated by immunofluorescence. Qualitative data are representative of eight eyes per analyzed time. Quantitative data correspond to mean ± SEM of eight eyes per analyzed time. *DM* Diabetes mellitus, *GFP* green fluorescent protein

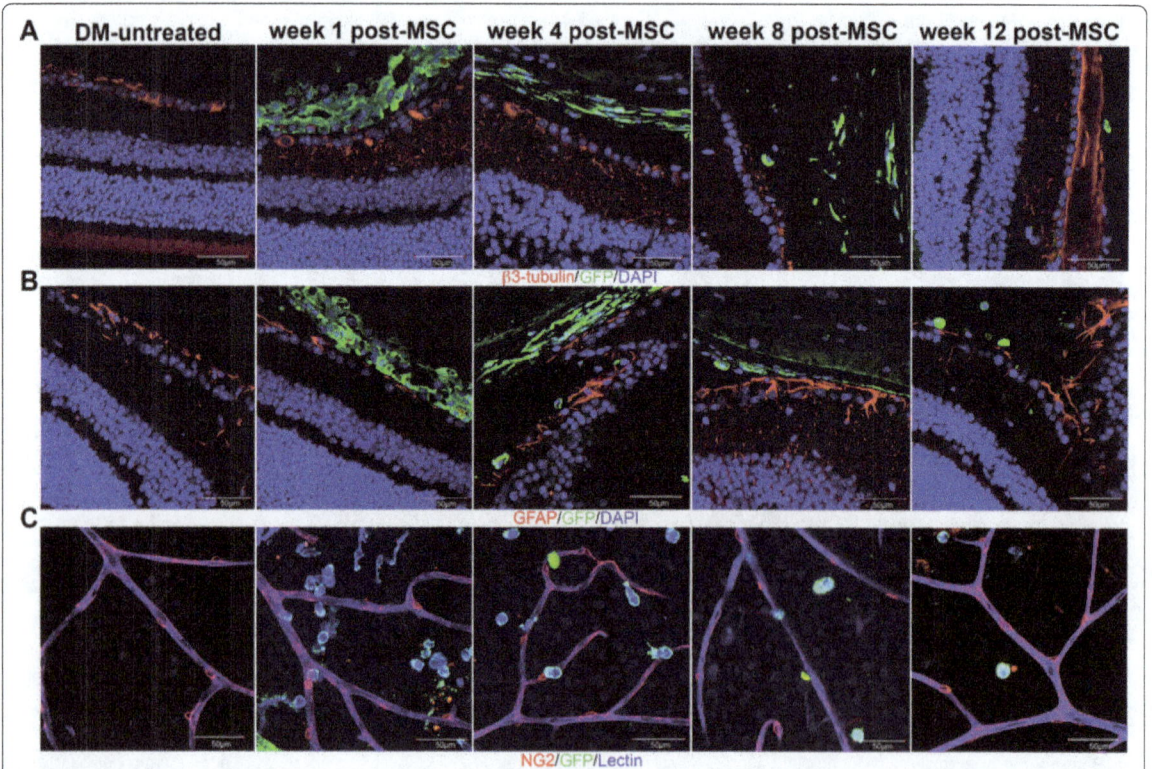

Fig. 5 Most donor mesenchymal stromal cells (*MSCs*) do not differentiate into neural- or perivascular-like cells. On days 1, 4, 8 and 12 weeks after MSC administration, immunofluorescences were performed in eye sections to evaluate differentiation of MSCs into RGCs by **a** colocalization of green fluorescent protein (*GFP*) with the RGC marker β3-tubulin or **b** differentiation into retinal astrocytes by colocalization of GFP with the astrocyte marker GFAP. **c** To evaluate differentiation of MSCs into perivascular cells, immunofluorescences were performed to colocalize GFP with the pericyte marker NG2 in whole mount retinal preparations. Lectin was also added to detect blood vessels. Qualitative data are representative of six eyes per analyzed time

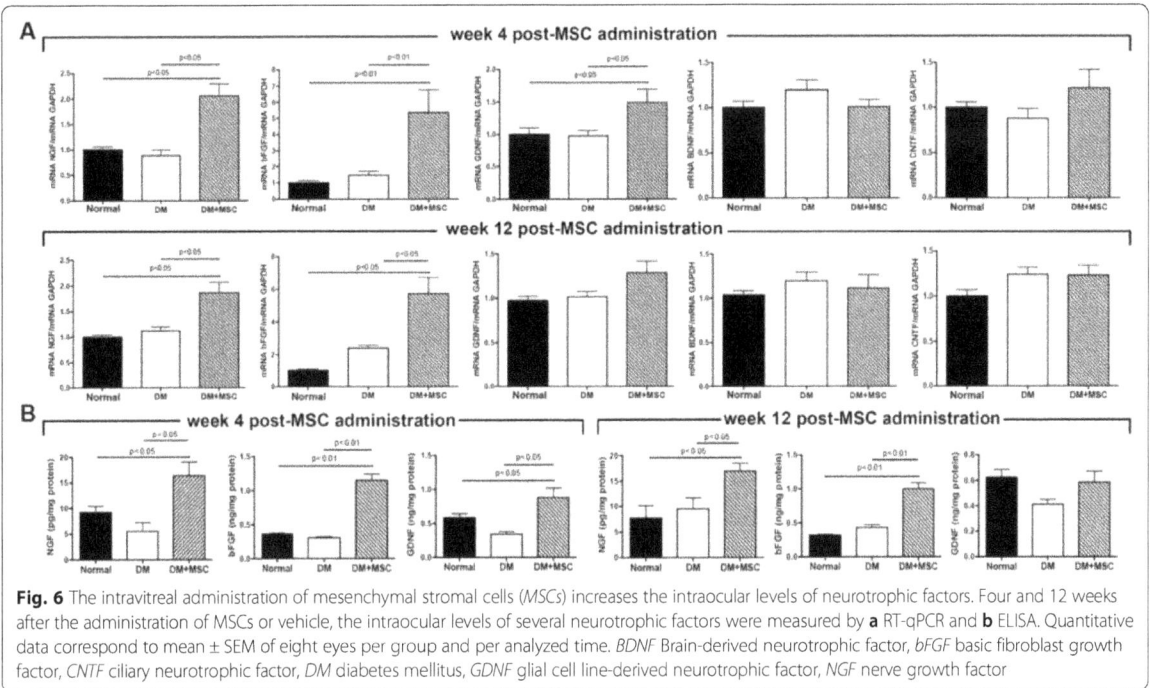

Fig. 6 The intravitreal administration of mesenchymal stromal cells (*MSCs*) increases the intraocular levels of neurotrophic factors. Four and 12 weeks after the administration of MSCs or vehicle, the intraocular levels of several neurotrophic factors were measured by **a** RT-qPCR and **b** ELISA. Quantitative data correspond to mean ± SEM of eight eyes per group and per analyzed time. *BDNF* Brain-derived neurotrophic factor, *bFGF* basic fibroblast growth factor, *CNTF* ciliary neurotrophic factor, *DM* diabetes mellitus, *GDNF* glial cell line-derived neurotrophic factor, *NGF* nerve growth factor

eye of DM + MSC mice 12 weeks post-MSC administration (Fig. 6a). These neurotrophic factors were also elevated at protein levels at both experimental times (Fig. 6b).

One of the main mechanisms associated with the neuroprotective effect of neurotrophic factors is the inhibition of apoptosis. To evaluate whether the microenvironment generated by the secretion of trophic factors could reduce the apoptosis of RGCs, the apoptotic rate was determined by the TUNEL assay. Examination of retinas of diabetic mice showed a significant increase in TUNEL-positive cells both in DM and DM + MSC mice 4 and 12 weeks post-administration of vehicle or MSCs compared to age-matched normal mice (Additional file 3: Figure S2A–D). However, in all cases, apoptotic cells were present in the outer nuclear layer of the retina. At the experimental times evaluated, we could not find any TUNEL-positive cell in the ganglion cell layer of the retina.

The intravitreal administration of MSCs reduces the retinal oxidative damage

Oxidative and nitrative modifications of retinal macromolecules occur promptly in the course of DR and are associated with the early death of retinal neurons [37, 38]. It has been postulated that MSCs could efficiently scavenge reactive species [22]. To evaluate the effect of MSC administration on the oxidative damage of the retina, we measured total ROS level and different markers of oxidative damage 4 and 12 weeks after MSC administration. As expected, diabetes induced a significant increase in the total amount of oxidative species in the retina at both experimental times compared with the oxidative species in age-matched normal mice (Fig. 7a). The elevated ROS levels correlated with a significant increase in the lipid peroxidation level in the retina (Fig. 7b), but were not enough to induce detectable oxidative damage at the protein and DNA levels (Fig. 7c and d). MSC administration induced a small non-significant reduction in ROS levels in the retina (Fig. 7a), but a strong reduction in lipid peroxidation levels at both experiential times, reaching values similar to those observed in normal mice (Fig. 7b).

The intravitreal administration of MSCs has a neutral effect in the electrical response of the retina in DM mice

It is well known that prior to the development of any histologically detectable retinal alteration, diabetic retinas display a significant decrease in neuronal function that could be evidenced by ERG [39]. Thus, we evaluated whether MSC administration could improve the electrical response in the retina of DM mice. For this we performed dark-adapted ERG, 4 and 12 weeks after MSC administration and, using stimuli of 0.01 $cd \cdot s \cdot m^{-2}$ and 3.0 $cd \cdot s \cdot m^{-2}$, we observed that no matter whether diabetic mice were treated with MSCs or only the vehicle, they had a significant reduction in the a-wave and b-wave amplitudes compared to age-matched normal mice (Fig. 8a–e). We also measured Ops, since it has been

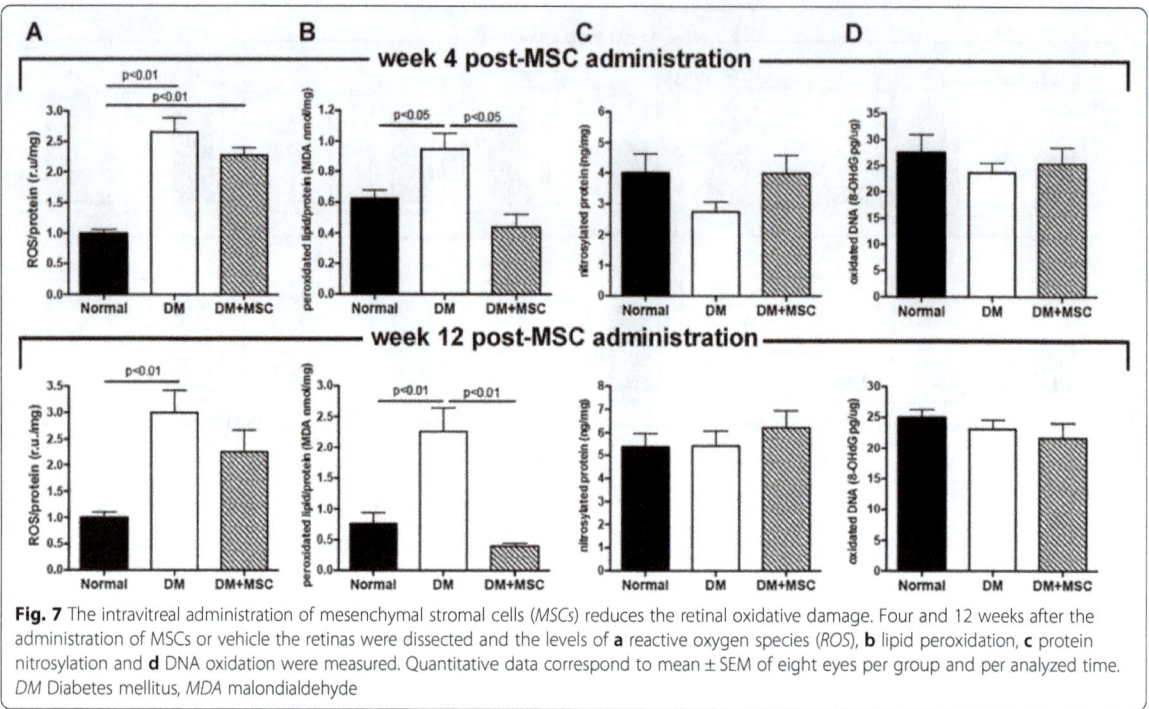

Fig. 7 The intravitreal administration of mesenchymal stromal cells (*MSCs*) reduces the retinal oxidative damage. Four and 12 weeks after the administration of MSCs or vehicle the retinas were dissected and the levels of **a** reactive oxygen species (*ROS*), **b** lipid peroxidation, **c** protein nitrosylation and **d** DNA oxidation were measured. Quantitative data correspond to mean ± SEM of eight eyes per group and per analyzed time. *DM* Diabetes mellitus, *MDA* malondialdehyde

previously reported that OPs are altered in diabetic individuals [40], and OPs are likely due to inner retinal neurotransmission, which represents the synaptic activity between amacrine neurons and RGCs [41]. There were no significant differences in the amplitudes of the OPs between the experimental groups at both experimental times analyzed (Fig. 8f).

The intravitreal administration of MSCs does not induce a provasculogenic microenvironment in the retina

While the growth of new blood vessels is beneficial in some ischemic conditions like ischemic heart disease [42], neovascularization in most ocular diseases, including DR, is devastating to visual function [6]. It is well known that MSCs are able to secrete, in addition to neurotrophic

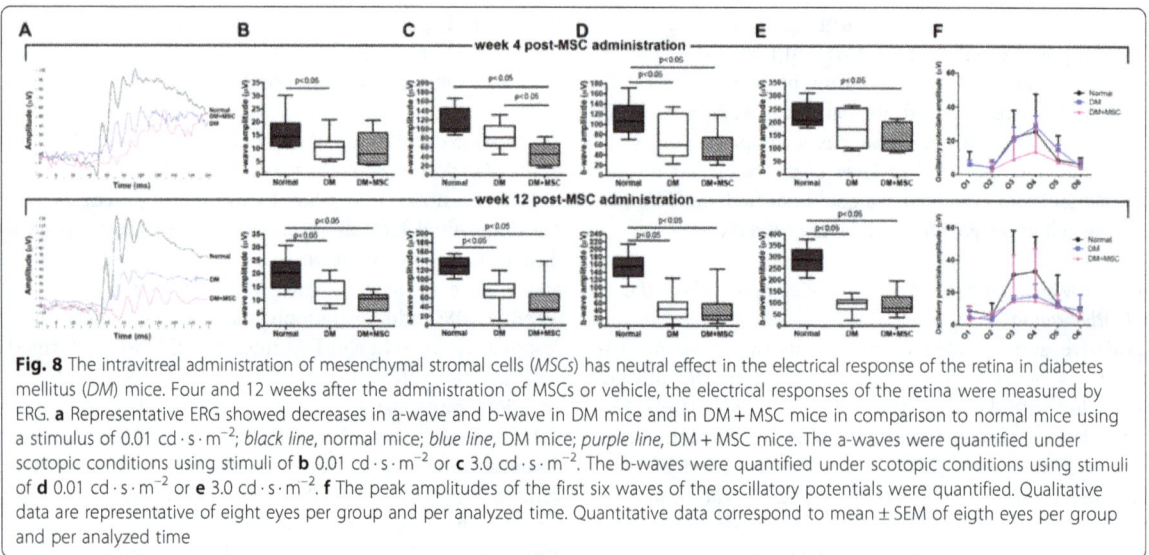

Fig. 8 The intravitreal administration of mesenchymal stromal cells (*MSCs*) has neutral effect in the electrical response of the retina in diabetes mellitus (*DM*) mice. Four and 12 weeks after the administration of MSCs or vehicle, the electrical responses of the retina were measured by ERG. **a** Representative ERG showed decreases in a-wave and b-wave in DM mice and in DM + MSC mice in comparison to normal mice using a stimulus of 0.01 cd·s·m^{-2}; *black line*, normal mice; *blue line*, DM mice; *purple line*, DM + MSC mice. The a-waves were quantified under scotopic conditions using stimuli of **b** 0.01 cd·s·m^{-2} or **c** 3.0 cd·s·m^{-2}. The b-waves were quantified under scotopic conditions using stimuli of **d** 0.01 cd·s·m^{-2} or **e** 3.0 cd·s·m^{-2}. **f** The peak amplitudes of the first six waves of the oscillatory potentials were quantified. Qualitative data are representative of eight eyes per group and per analyzed time. Quantitative data correspond to mean ± SEM of eigth eyes per group and per analyzed time

factors, strong angiogenic factors [43]. This could be a major limitation since MSC administration might worsen DR.

To evaluate if MSC administration induce a provasculogenic microenvironment, the intraocular levels of pro-angiogenic and anti-angiogenic factors were measured. Four and 12 weeks after MSC administration, we did not observe differences in the mRNA levels of the pro-angiogenic factors VEGF-α, PDGF and ANG-1 in the eyes of DM + MSC mice compared to untreated DM mice (Fig. 9a). However, the mRNA level of TSP-1, a potent anti-angiogenic factor, was significantly increased in the eyes of DM + MSC mice compared to untreated DM mice and normal mice at both experimental times (Fig. 9a). The increase in TSP-1 level was also corroborated at protein level by ELISA (Fig. 9b). Additionally, retinal blood vessel density was evaluated in retinal flat-mounts after perfusion with high molecular weight FITC-dextran. At both experimental times, retinas of normal, DM and DM + MSC mice exhibited both superficial and deep vascular layers that extended from the optic nerve to the peripheral areas. The vessels formed a fine radial branching pattern in the superficial retinal layer and a polygonal reticular pattern in the deep retinal

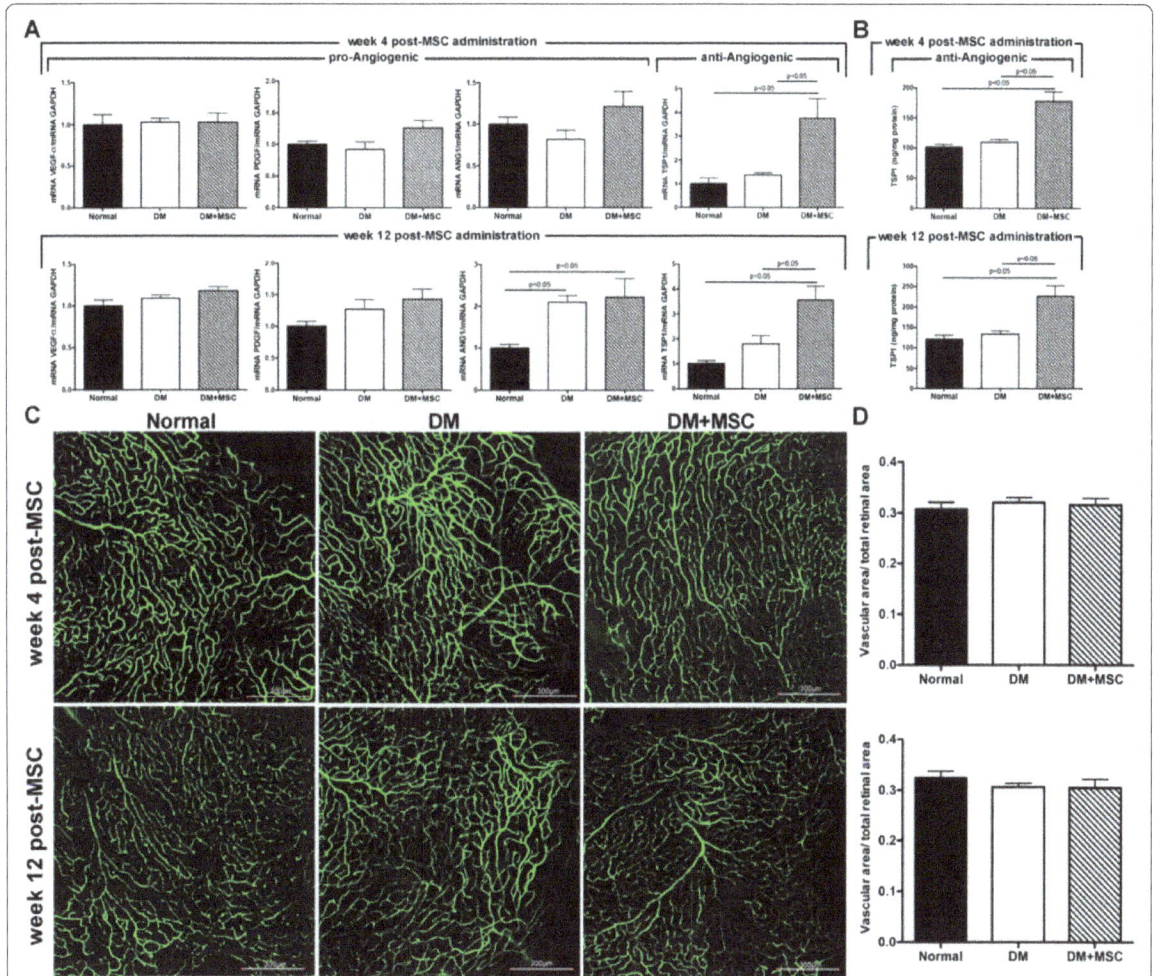

Fig. 9 The intravitreal administration of mesenchymal stromal cells (*MSCs*) does not induce a provasculogenic microenvironment in the retina. Four and 12 weeks after the administration of MSCs or vehicle, the intraocular levels of pro-angiogenic and anti-angiogenic factors were measured by **a** RT-qPCR and **b** ELISA. Retinal blood vessels were visualized at both experimental times by the administration of FITC-dextran in the tail vein. **c** Retinas were dissected, flat mounted and observed by confocal microscopy. **d** Retinal vasculature was quantified and expressed as vascular area/total retinal area. Quantitative data correspond to mean ± SEM of eight eyes per group and per analyzed time. Qualitative data are representative of eight eyes per group and per analyzed time. *ANG1* Angiopoietin 1, *DM* diabetes mellitus, *PDGF* platelet-derived growth factor, *TSP1* thrombospondin 1, *VEGF* vascular endothelial growth factor

layer (Fig. 9c). There were no differences in the amount of blood vessels in the different experimental conditions (Fig. 9d).

An increase in vascular permeability is an early event in the onset of DR [44]. In our model, diabetes induced the appearance of areas of vascular leakage compared to normal mice. However, there was no difference in this parameter when comparing diabetic mice treated with vehicle or with MSCs, since in both cases we could detect the occasional presence of areas of vascular leakage (Additional file 4: Figure S3).

Discussion

Diabetic retinopathy is a major cause of visual impairment and the leading cause of blindness in the Western world [2]. Stem cell-based therapy represents a newly emerging therapeutic approach to treat eye diseases. In this sense, due to the matching of the pathological events that occur at the initial stages of DR, and the cellular and molecular mechanisms associated with the reported MSC therapeutic effects, this approach represents a promising tool for the treatment of DR [44].

One of the first demonstrations that MSCs could play a therapeutic role in DR came from a study in which the intravenous administration of adipose-derived MSCs in an animal model of DM improved the integrity of the blood–retinal barrier [19]. However, in this study it was not clear whether the observed therapeutic effect was secondary to the reduced hyperglycemia, or due to a direct effect of MSCs in the damaged retina. Recently, more direct evidence of the role of MSCs in DR came from studies in which adipose-derived MSCs were intravitreally injected into the Akimba DR model or into diabetic rats treated with STZ [18, 26]. In these studies, the authors showed that MSC administration reduces blood–retinal barrier breakdown and MSCs differentiate into pericytes and integrate into the retinal vasculature [18, 26]. However, it has become more clear that DR affects not only retinal vasculature, but also retinal neuronal and glial cells, and that visual deficits in the early stage of DR are correlated with retinal neurosensory dysfunction that is mainly related to the early loss of RGCs [8, 45]. Thus, in this work we wanted to evaluate whether the intravitreal administration of MSCs was able to prevent the loss of RGCs. For this, we used an animal model of DM induced by the administration of a single high dose of STZ. This animal model is widely used to study the main complications associated with DM since animals develop sustained hyperglycemia and reduced insulinemia due to the massive destruction of pancreatic beta cells, allowing the appearance of the long-term complications associated with DM, including DR [28, 29].

In this animal model, as previously reported [10], diabetes induces a significant reduction in the number of RGCs. This 20 % reduction in the number of RGCs is in agreement with data reported for other animal models of DR [32, 46]. However, MSC administration completely prevents RGC loss, and this effect was maintained at least 12 weeks after the administration.

Both direct differentiation of MSCs into RGCs or indirect support to the neural retina by the secretion of neuroprotective factors or by the reduction of oxidative damage could be related to this therapeutic effect. Here we analyzed the migration of MSCs into the retina and the differentiation potential of these cells into RGCs, but also into astrocytes and pericytes since it has been reported the early damage to these cells could affect neuronal survival and the normal interactions with other retinal cells [47, 48]. We observed that most of the administered cells remained in the vitreous cavity, and they did not express neural or perivascular markers. We could detect only an extremely low number of MSCs that had adopted a putative phenotype of RGCs, astrocytes or pericytes, suggesting that the direct differentiation of MSCs into neural- or perivascular-like cells is not the main mechanism of the therapeutic effect. These results are in agreement with data of Johnson et al., who reported extremely poor retinal integration of MSCs after its intravitreal administration in an animal model of glaucoma [49]. In this sense, several authors have postulated that the pre-treatment of the retina with different drugs is needed to allow the effective incorporation of the transplanted cells [49, 50].

MSCs are known to produce and secrete, both in vitro and in vivo, a broad range of trophic factors, including some potent neurotrophic factors that could potentially prevent retinal neuronal cells from dying [51]. Here we saw that MSC administration increases the intraocular levels of NGF, GDNF and bFGF 4 and 12 weeks after injection. These neurotrophic factors have previously been involved in the inhibition of apoptosis of RGCs in different animal models [52, 53]. Therefore, MSCs remaining in the vitreous cavity could continuously deliver therapeutic molecules locally for a prolonged period of time. Alternatively, these neuroprotective factors could be secreted by the host retinal tissue in response to MSC injection.

Similar results were observed in an animal model of experimental glaucoma induced by the laser photocoagulation of the trabecular meshwork [21]. In this animal model, the intravitreal administration of MSCs provides trophic support to the damaged tissue, and increased RGC survival [21]. Additionally, Yu et al. reported that intravitreal MSC transplantation was neuroprotective after episcleral vein ligature, which can cause moderate ocular hypertension and RGC loss [54], and the same neuroprotective factors were also elevated in the vitreous cavity after the intravitreal administration of human placental stem cells in an animal model of DR [55].

In our model, MSC administration did not reduce the global apoptotic rate in the retina. However, we could not find RGCs undergoing apoptosis at any experimental time analyzed. This could be related to the slow onset and progression of DR, in which the number of cells dying at any given time is very small.

A widespread oxidative damage occurs in the retina of DR patients. Thus, in addition to providing neurotrophic support, MSCs may also give neuroprotection in the retina by modulating the levels of ROS. Previously we have demonstrated that MSCs have a high resistance to ROS due to the robust expression of SOD1, SOD2, CAT and GPX1 enzymes and high levels of glutathione [22]. Furthermore, MSCs constitutively express the enzymes required for the repair of oxidized structures [56]. Therefore, MSCs possess the main enzymatic machinery to detoxify ROS and to correct the oxidative damage. Here we showed that MSC administration induced a small reduction in the ROS level in the retina but a strong reduction in lipid peroxidation levels, one of the main markers of oxidative damage in the retina [57].

Consistent with these observations, the treatment of DM animals with antioxidants or with inhibitors of some of the metabolic pathways that generate ROS reduces oxidative damage of retinal structures and attenuates retinal cell loss [58–61]. Unfortunately, these drugs have low oral bioavailability or are not able to cross the blood–retinal barrier. Therefore large doses must be administrated to maintain therapeutic concentrations inside the retina, limiting their clinical use.

ERG is one of the most accepted techniques to evaluate retinal electrical response in diabetic individuals. In our animal model we observed a significant reduction in the amplitude of the scotopic a-wave and b-wave 3 months after diabetes induction that continued its altered state 4 and 6 months after STZ administration, but we could not detect changes in the amplitude of the OPs. MSC administration has a neutral effect on the electrical response of the retina. In our study we injected MSCs into the vitreous cavity, since it is the nearest location to RGCs, and in human patients intravitreal injection is technically feasible and safe [62]. However, it has been previously reported that cells injected into the vitreous cavity tend to cover the back of the lens and block the passage of light into the eye [21], altering the ERG analysis. Furthermore, we also observed that administered MSCs adopt this location, mainly between 7 and 30 days after its administration (data not shown). Therefore, another administration route, such as transplantation into the subretinal space, should be evaluated in order to avoid this effect.

Additionally, it has been previously shown that axon degeneration of the RGCs can occur with relative sparing of RGC bodies [63]; therefore, the soma protection achieved by the MSC administration in this context could not provide a functional benefit. Further investigation will be required to understand this effect.

One of the main concerns regarding MSC therapy for DR is the secretion of pro-angiogenic growth factors, mainly VEGF and PDGF, that might worsen the course of DR. VEGF is a potent growth factor that enhances vascular permeability, stimulates endothelial cell proliferation and migration, and promotes angiogenesis in the retina [61]. Similarly, PDGF is a pro-angiogenic growth factor that may also promote aberrant neovascularization in DR [64]. Furthermore, PDGF may stimulate the formation and traction of epiretinal membranes in patients with DR [65]. However, it has been reported that, depending on the microenvironment, MSCs could produce trophic factors that may modulate between a pro-angiogenic and anti-angiogenic environment. Here we showed that MSC administration did not increase the intraocular levels of VEGF and PDGF, and did not promote the massive formation of blood vessels or the increase in vascular permeability. Moreover, MSC administration increased the intraocular level of TSP-1, a potent anti-angiogenic factor. In accordance with our results, it has previously been shown that the local administration of MSCs in an animal model of corneal chemical injury, characterized by an accelerated neovascularization process, produced a rapid regression of the new blood vessels [66]. This anti-angiogenic effect was attributed to the secretion of TSP-1 by the administered MSCs, while the ocular levels of the pro-angiogenic factor VEGF were unchanged between the untreated and the MSC-treated group [66].

More studies are required in order to determine the best source and dose of MSCs and the optimal time for their administration. Nevertheless, the ability of MSCs to secrete neurotrophic factors and to reduce oxidative damage of the retina, preventing RGC loss, opens the possibility of using these cells to stop the progressive deterioration of the neural retina in diabetic individuals.

Conclusion

Intravitreal administration of adipose-derived MSCs triggers an effective cytoprotective microenvironment in the retina of diabetic mice. Furthermore, MSC administration does not result in the pathological neovascularization of the retina. Thus, MSCs represent an interesting tool in order to prevent DR.

Additional files

Additional file 1: Table S1. Primer and amplicon characteristics. (PNG 42 kb)

Additional file 2: Figure S1. Donor MSCs differentiate into neural- or perivascular-like cells at very low frequency. One, 4, 8 and 12 weeks after MSC administration immunofluorescences were performed in eye sections to evaluate differentiation of MSCs into RGCs by colocalization of GFP with the ganglion cell marker β3-tubulin (A) or differentiation into astrocytes by colocalization of GFP with the astrocyte marker GFAP (B). To evaluate differentiation of MSCs into perivascular cells, immunofluorescences were performed to colocalize GFP with the pericyte marker NG2 in whole mount retinal preparations. Lectin was also added to detect blood vessels. Arrows indicate cells co-expressing both markers. Qualitative data are representative of six eyes per analyzed time. (PNG 2819 kb)

Additional file 3: Figure S2. MSC administration does not modify the apoptotic rate in the retina of DM mice. Apoptosis in the retina was analyzed by the TUNEL technique in serial 4-μm eye sections 4 weeks after MSC administration (A) and 12 weeks after MSC administration (B). Nuclei were counterstained with DAPI. Arrows indicate TUNEL-positive cells. The number of apoptotic nuclei was quantified at both experimental times and expressed as fold-change versus normal mice (C and D). Qualitative data are representative of eight eyes per analyzed time. Quantitative data correspond to mean ± SEM of eight eyes per group and per analyzed time. (PNG 3862 kb)

Additional file 4: Figure S3. MSC administration does not modify the presence of vascular leakage areas in the retina of DM mice. FITC-dextran was administered in the tail vein 4 and 12 weeks after the administration of MSCs or vehicle. Retinas were dissected, flat mounted and observed by confocal microscopy. Retinal vascular leakage was determined by the presence of areas of extravasated FITC-dextran marked by arrows. Qualitative data are representative of eight eyes per group and per analyzed time. (PNG 2029 kb)

Abbreviations

α-MEM: α-Minimum essential medium; ANG-1: Angiopoietin 1; ANOVA: Analysis of variance; BDNF: Brain-derived neurotrophic factor; bFGF: Basic fibroblast growth factor; BHT: Butyl hydroxytoluene; DAPI: 4′-6′-Diamidino-2-phenylindole; DM: Diabetes mellitus; DR: Diabetic retinopathy; ELISA: Enzyme-linked immunosorbent assay; ERG: Electroretinography; FBS: Fetal bovine serum; GDNF: Glial cell line-derived neurotrophic factor; GFP: Green fluorescent protein; H_2DCFDA: 2,7-Dichloro-dihydro-fluorescein diacetate; MDA: Malondialdehyde; MSC: Mesenchymal stromal cell; NGF: Nerve growth factor; OP: Oscillatory potential; PBS: Phosphate-buffered saline; PDGF: Platelet-derived growth factor; RGC: Retinal ganglion cell; ROS: Reactive oxygen species; RT-qPCR: Real-time quantitative polymerase chain reaction; SEM: Standard error of the mean; STZ: Streptozotocin; TSP-1: Thrombospondin-1; TUNEL: Terminal deoxynucleotidyl transferase (TdT)-mediated dUTP nick end labeling; VEGF: Vascular endothelial growth factor.

Competing interests

The authors declare that they have no competing interests.

Authors' contributions

ME participated in the conception and design of the study, data analysis and manuscript writing. CAU participated in data acquisition, data analysis and manuscript writing. SM participated in data acquisition, data analysis and manuscript writing. KL participated in data acquisition, data analysis and manuscript writing. PC participated in the conception and design of the study and manuscript writing. FE participated in the conception and design of the study, data acquisition, data analysis, financial support and manuscript writing. All authors read and approved the final manuscript.

Acknowledgments

The authors thank Dr. Amelina Albornoz and Dr. Anne Bliss for the English editing of the paper. This work was supported by FONDECYT 1130470 grant to FE.

Author details

[1]Centro de Medicina Regenerativa, Facultad de Medicina Clínica Alemana-Universidad del Desarrollo, Av. Las Condes 12438, Lo Barnechea, Santiago 7710162, Chile. [2]Departamento de Oftalmología, Facultad de Medicina, Universidad de Chile, Av. Independencia 1027, Santiago, Chile.

References

1. Luckie R, Leese G, McAlpine R, MacEwen CJ, Baines PS, Morris AD, et al. Fear of visual loss in patients with diabetes: results of the prevalence of diabetic eye disease in Tayside, Scotland (P-DETS) study. Diabet Med. 2007;24(10):1086–92. doi:10.1111/j.1464-5491.2007.02180.x.
2. Resnikoff S, Pascolini D, Etya'ale D, Kocur I, Pararajasegaram R, Pokharel GP, et al. Global data on visual impairment in the year 2002. Bull World Health Organ. 2004;82(11):844–51.
3. Lee R, Wong TY, Sabanayagam C. Epidemiology of diabetic retinopathy, diabetic macular edema and related vision loss. Eye Vision. 2015;2:17. doi:10.1186/s40662-015-0026-2.
4. Kempen JH, O'Colmain BJ, Leske MC, Haffner SM, Klein R, Moss SE, et al. The prevalence of diabetic retinopathy among adults in the United States. Arch Ophthalmol. 2004;122(4):552–63. doi:10.1001/archopht.122.4.552.
5. Klein R, Knudtson MD, Lee KE, Gangnon R, Klein BE. The Wisconsin Epidemiologic Study of Diabetic Retinopathy XXIII: the twenty-five-year incidence of macular edema in persons with type 1 diabetes. Ophthalmology. 2009;116(3):497–503. doi:10.1016/j.ophtha.2008.10.016.
6. Cheung N, Mitchell P, Wong TY. Diabetic retinopathy. Lancet. 2010;376(9735):124–36. doi:10.1016/S0140-6736(09)62124-3.
7. Ozawa Y, Kurihara T, Sasaki M, Ban N, Yuki K, Kubota S, et al. Neural degeneration in the retina of the streptozotocin-induced type 1 diabetes model. Exp Diabetes Res. 2011;2011:108328. doi:10.1155/2011/108328.
8. Kern TS, Barber AJ. Retinal ganglion cells in diabetes. J Physiol. 2008;586:4401–8. doi:10.1113/jphysiol.2008.156695.
9. Gardner TW, Abcouwer SF, Barber AJ, Jackson GR. An integrated approach to diabetic retinopathy research. Arch Ophthalmol. 2011;129(2):230–5. doi:10.1001/archophthalmol.2010.362.
10. Sasaki M, Ozawa Y, Kurihara T, Kubota S, Yuki K, Noda K, et al. Neurodegenerative influence of oxidative stress in the retina of a murine model of diabetes. Diabetologia. 2010;53(5):971–9. doi:10.1007/s00125-009-1655-6.
11. Brownlee M. The pathobiology of diabetic complications: a unifying mechanism. Diabetes. 2005;54(6):1615–25.
12. Ali TK, Matragoon S, Pillai BA, Liou GI, El-Remessy AB. Peroxynitrite mediates retinal neurodegeneration by inhibiting nerve growth factor survival signaling in experimental and human diabetes. Diabetes. 2008;57(4):889–98. doi:10.2337/db07-1669.
13. Martin PM, Roon P, Van Ells TK, Ganapathy V, Smith SB. Death of retinal neurons in streptozotocin-induced diabetic mice. Invest Ophthalmol Vis Sci. 2004;45(9):3330–6. doi:10.1167/iovs.04-0247.
14. Salam A, Mathew R, Sivaprasad S. Treatment of proliferative diabetic retinopathy with anti-VEGF agents. Acta Ophthalmol (Copenh). 2011;89(5):405–11. doi:10.1111/j.1755-3768.2010.02079.x.
15. Tremolada G, Del Turco C, Lattanzio R, Maestroni S, Maestroni A, Bandello F, et al. The role of angiogenesis in the development of proliferative diabetic retinopathy: impact of intravitreal anti-VEGF treatment. Exp Diabetes Res. 2012;2012:728325. doi:10.1155/2012/728325.
16. Huang Y, Enzmann V, Ildstad ST. Stem cell-based therapeutic applications in retinal degenerative diseases. Stem Cell Rev. 2011;7(2):434–45. doi:10.1007/s12015-010-9192-8.
17. Jeon YJ, Kim J, Cho JH, Chung HM, Chae JI. Comparative analysis of human mesenchymal stem cells derived from bone marrow, placenta, and adipose tissue as sources of cell therapy. J Cell Biochem. 2015. doi:10.1002/jcb.25395.
18. Rajashekhar G, Ramadan A, Abburi C, Callaghan B, Traktuev DO, Evans-Molina C, et al. Regenerative therapeutic potential of adipose stromal cells in early stage diabetic retinopathy. PLoS One. 2014;9(1):e84671. doi:10.1371/journal.pone.0084671.
19. Yang Z, Li K, Yan X, Dong F, Zhao C. Amelioration of diabetic retinopathy by engrafted human adipose-derived mesenchymal stem cells in streptozotocin diabetic rats. Graefes Arch Clin Exp Ophthalmol. 2010;248(10):1415–22. doi:10.1007/s00417-010-1384-z.
20. Caplan AI, Dennis JE. Mesenchymal stem cells as trophic mediators. J Cell Biochem. 2006;98(5):1076–84. doi:10.1002/jcb.20886.
21. Johnson TV, Bull ND, Hunt DP, Marina N, Tomarev SI, Martin KR. Neuroprotective effects of intravitreal mesenchymal stem cell transplantation in experimental glaucoma. Invest Ophthalmol Vis Sci. 2010;51(4):2051–9. doi:10.1167/iovs.09-4509.
22. Valle-Prieto A, Conget PA. Human mesenchymal stem cells efficiently manage oxidative stress. Stem Cells Dev. 2010;19(12):1885–93. doi:10.1089/scd.2010.0093.

23. Lanza C, Morando S, Voci A, Canesi L, Principato MC, Serpero LD, et al. Neuroprotective mesenchymal stem cells are endowed with a potent antioxidant effect in vivo. J Neurochem. 2009;110(5):1674–84. doi:10.1111/j.1471-4159.2009.06268.x.

24. Ma XR, Tang YL, Xuan M, Chang Z, Wang XY, Liang XH. Transplantation of autologous mesenchymal stem cells for end-stage liver cirrhosis: a meta-analysis based on seven controlled trials. Gastroenterol Res Pract. 2015;2015:908275. doi:10.1155/2015/908275.

25. Ho MS, Mei SH, Stewart DJ. The immunomodulatory and therapeutic effects of mesenchymal stromal cells for acute lung injury and sepsis. J Cell Physiol. 2015;230(11):2606–17. doi:10.1002/jcp.25028.

26. Mendel TA, Clabough EB, Kao DS, Demidova-Rice TN, Durham JT, Zotter BC, et al. Pericytes derived from adipose-derived stem cells protect against retinal vasculopathy. PLoS One. 2013;8(5):e65691. doi:10.1371/journal.pone.0065691.

27. Hou HY, Liang HL, Wang YS, Zhang ZX, Wang BR, Shi YY, et al. A therapeutic strategy for choroidal neovascularization based on recruitment of mesenchymal stem cells to the sites of lesions. Mol Ther. 2010;18(10):1837–45. doi:10.1038/mt.2010.144.

28. Ezquer F, Ezquer M, Simon V, Pardo F, Yanez A, Carpio D, et al. Endovenous administration of bone-marrow-derived multipotent mesenchymal stromal cells prevents renal failure in diabetic mice. Biol Blood Marrow Transplant. 2009;15(11):1354–65. doi:10.1016/j.bbmt.2009.07.022.

29. Ezquer F, Giraud-Billoud M, Carpio D, Cabezas F, Conget P, Ezquer M. Proregenerative microenvironment triggered by donor mesenchymal stem cells preserves renal function and structure in mice with severe diabetes mellitus. BioMed Res Int. 2015;2015:164703. doi:10.1155/2015/164703.

30. Dogrul A, Gul H, Yildiz O, Bilgin F, Guzeldemir ME. Cannabinoids blocks tactile allodynia in diabetic mice without attenuation of its antinociceptive effect. Neurosci Lett. 2004;368(1):82–6. doi:10.1016/j.neulet.2004.06.060.

31. Arsie MP, Marchioro L, Lapolla A, Giacchetto GF, Bordin MR, Rizzotti P, et al. Evaluation of diagnostic reliability of DCA 2000 for rapid and simple monitoring of HbA1c. Acta Diabetol. 2000;37(1):1–7.

32. Barber AJ, Antonetti DA, Kern TS, Reiter CE, Soans RS, Krady JK, et al. The Ins2Akita mouse as a model of early retinal complications in diabetes. Invest Ophthalmol Vis Sci. 2005;46(6):2210–8. doi:10.1167/iovs.04-1340.

33. Ezquer F, Ezquer M, Contador D, Ricca M, Simon V, Conget P. The antidiabetic effect of mesenchymal stem cells is unrelated to their transdifferentiation potential but to their capability to restore Th1/Th2 balance and to modify the pancreatic microenvironment. Stem Cells. 2012;30(8):1664–74. doi:10.1002/stem.1132.

34. Al-Shabrawey M, Bartoli M, El-Remessy AB, Platt DH, Matragoon S, Behzadian MA, et al. Inhibition of NAD(P)H oxidase activity blocks vascular endothelial growth factor overexpression and neovascularization during ischemic retinopathy. Am J Pathol. 2005;167(2):599–607. doi:10.1016/S0002-9440(10)63001-5.

35. Krohne TU, Westenskow PD, Kurihara T, Friedlander DF, Lehmann M, Dorsey AL, et al. Generation of retinal pigment epithelial cells from small molecules and OCT4 reprogrammed human induced pluripotent stem cells. Stem Cells Transl Med. 2012;1(2):96–109. doi:10.5966/sctm.2011-0057.

36. Layton CJ, Safa R, Osborne NN. Oscillatory potentials and the b-wave: partial masking and interdependence in dark adaptation and diabetes in the rat. Graefes Arch Clin Exp Ophthalmol. 2007;245(9):1335–45. doi:10.1007/s00417-006-0506-0.

37. Du Y, Miller CM, Kern TS. Hyperglycemia increases mitochondrial superoxide in retina and retinal cells. Free Radic Biol Med. 2003;35(11):1491–9.

38. Abu El-Asrar AM, Meersschaert A, Dralands L, Missotten L, Geboes K. Inducible nitric oxide synthase and vascular endothelial growth factor are colocalized in the retinas of human subjects with diabetes. Eye. 2004;18(3):306–13. doi:10.1038/sj.eye.6700642.

39. Fortune B, Schneck ME, Adams AJ. Multifocal electroretinogram delays reveal local retinal dysfunction in early diabetic retinopathy. Invest Ophthalmol Vis Sci. 1999;40(11):2638–51.

40. Ramsey DJ, Ripps H, Qian H. An electrophysiological study of retinal function in the diabetic female rat. Invest Ophthalmol Vis Sci. 2006;47(11):5116–24. doi:10.1167/iovs.06-0364.

41. Dong CJ, Agey P, Hare WA. Origins of the electroretinogram oscillatory potentials in the rabbit retina. Vis Neurosci. 2004;21(4):533–43. doi:10.1017/S0952523804214043.

42. Sato T, Iso Y, Uyama T, Kawachi K, Wakabayashi K, Omori Y, et al. Coronary vein infusion of multipotent stromal cells from bone marrow preserves cardiac function in swine ischemic cardiomyopathy via enhanced neovascularization. Lab Investig. 2011;91(4):553–64. doi:10.1038/labinvest.2010.202.

43. Dharmasaroja P. Bone marrow-derived mesenchymal stem cells for the treatment of ischemic stroke. J Clin Neurosci. 2009;16(1):12–20. doi:10.1016/j.jocn.2008.05.006.

44. Ezquer F, Ezquer M, Arango-Rodriguez M, Conget P. Could donor multipotent mesenchymal stromal cells prevent or delay the onset of diabetic retinopathy? Acta Ophthalmol (Copenh). 2014;92(2):e86–95. doi:10.1111/aos.12113.

45. Antonetti DA, Barber AJ, Bronson SK, Freeman WM, Gardner TW, Jefferson LS, et al. Diabetic retinopathy: seeing beyond glucose-induced microvascular disease. Diabetes. 2006;55(9):2401–11. doi:10.2337/db05-1635.

46. Zheng L, Howell SJ, Hatala DA, Huang K, Kern TS. Salicylate-based anti-inflammatory drugs inhibit the early lesion of diabetic retinopathy. Diabetes. 2007;56(2):337–45. doi:10.2337/db06-0789.

47. Gardner TW, Antonetti DA, Barber AJ, LaNoue KF, Levison SW. Diabetic retinopathy: more than meets the eye. Surv Ophthalmol. 2002;47 Suppl 2:S253–62.

48. Bringmann A, Pannicke T, Grosche J, Francke M, Wiedemann P, Skatchkov SN, et al. Muller cells in the healthy and diseased retina. Prog Retin Eye Res. 2006;25(4):397–424. doi:10.1016/j.preteyeres.2006.05.003.

49. Johnson TV, Bull ND, Martin KR. Identification of barriers to retinal engraftment of transplanted stem cells. Invest Ophthalmol Vis Sci. 2010;51(2):960–70. doi:10.1167/iovs.09-3884.

50. Minamino K, Adachi Y, Yamada H, Higuchi A, Suzuki Y, Iwasaki M, et al. Long-term survival of bone marrow-derived retinal nerve cells in the retina. Neuroreport. 2005;16(12):1255–9.

51. Chen Q, Long Y, Yuan X, Zou L, Sun J, Chen S, et al. Protective effects of bone marrow stromal cell transplantation in injured rodent brain: synthesis of neurotrophic factors. J Neurosci Res. 2005;80(5):611–9. doi:10.1002/jnr.20494.

52. Sapieha PS, Peltier M, Rendahl KG, Manning WC, Di Polo A. Fibroblast growth factor-2 gene delivery stimulates axon growth by adult retinal ganglion cells after acute optic nerve injury. Mol Cell Neurosci. 2003;24(3):656–72.

53. Schuettauf F, Vorwerk C, Naskar R, Orlin A, Quinto K, Zurakowski D, et al. Adeno-associated viruses containing bFGF or BDNF are neuroprotective against excitotoxicity. Curr Eye Res. 2004;29(6):379–86. doi:10.1080/02713680490517872.

54. Yu S, Tanabe T, Dezawa M, Ishikawa H, Yoshimura N. Effects of bone marrow stromal cell injection in an experimental glaucoma model. Biochem Biophys Res Commun. 2006;344(4):1071–9. doi:10.1016/j.bbrc.2006.03.231.

55. Scalinci SZ, Scorolli L, Corradetti G, Domanico D, Vingolo EM, Meduri A, et al. Potential role of intravitreal human placental stem cell implants in inhibiting progression of diabetic retinopathy in type 2 diabetes: neuroprotective growth factors in the vitreous. Clin Ophthalmol. 2011;5:691–6. doi:10.2147/OPTH.S21161.

56. Silva Jr WA, Covas DT, Panepucci RA, Proto-Siqueira R, Siufi JL, Zanette DL, et al. The profile of gene expression of human mesenchymal stem cells. Stem Cells. 2003;21(6):661–9. doi:10.1634/stemcells.21-6-661.

57. El-Remessy AB, Behzadian MA, Abou-Mohamed G, Franklin T, Caldwell RW, Caldwell RB. Experimental diabetes causes breakdown of the blood-retina barrier by a mechanism involving tyrosine nitration and increases in expression of vascular endothelial growth factor and urokinase plasminogen activator receptor. Am J Pathol. 2003;162(6):1995–2004. doi:10.1016/S0002-9440(10)64332-5.

58. Obrosova IG, Minchenko AG, Vasupuram R, White L, Abatan OI, Kumagai AK, et al. Aldose reductase inhibitor fidarestat prevents retinal oxidative stress and vascular endothelial growth factor overexpression in streptozotocin-diabetic rats. Diabetes. 2003;52(3):864–71.

59. Li J, Wang JJ, Yu Q, Chen K, Mahadev K, Zhang SX. Inhibition of reactive oxygen species by Lovastatin downregulates vascular endothelial growth factor expression and ameliorates blood-retinal barrier breakdown in db/db mice: role of NADPH oxidase 4. Diabetes. 2010;59(6):1528–38. doi:10.2337/db09-1057.

60. Kowluru RA, Koppolu P, Chakrabarti S, Chen S. Diabetes-induced activation of nuclear transcriptional factor in the retina, and its inhibition by antioxidants. Free Radic Res. 2003;37(11):1169–80.

61. Witmer AN, Blaauwgeers HG, Weich HA, Alitalo K, Vrensen GF, Schlingemann RO. Altered expression patterns of VEGF receptors in human diabetic retina and in experimental VEGF-induced retinopathy in monkey. Invest Ophthalmol Vis Sci. 2002;43(3):849–57.

62. Jonas JB, Witzens-Harig M, Arseniev L, Ho AD. Intravitreal autologous bone marrow-derived mononuclear cell transplantation: a feasibility report. Acta Ophthalmol (Copenh). 2008;86(2):225–6. doi:10.1111/j.1600-0420.2007.00987.x.

63. Libby RT, Li Y, Savinova OV, Barter J, Smith RS, Nickells RW, et al. Susceptibility to neurodegeneration in a glaucoma is modified by Bax gene dosage. PLoS Genet. 2005;1(1):17–26. doi:10.1371/journal.pgen.0010004.

64. Vinores SA, Seo MS, Okamoto N, Ash JD, Wawrousek EF, Xiao WH, et al. Experimental models of growth factor-mediated angiogenesis and blood-retinal barrier breakdown. Gen Pharmacol. 2000;35(5):233–9.
65. Mori K, Gehlbach P, Ando A, Dyer G, Lipinsky E, Chaudhry AG, et al. Retina-specific expression of PDGF-B versus PDGF-A: vascular versus nonvascular proliferative retinopathy. Invest Ophthalmol Vis Sci. 2002;43(6):2001–6.
66. Oh JY, Kim MK, Shin MS, Lee HJ, Ko JH, Wee WR, et al. The anti-inflammatory and anti-angiogenic role of mesenchymal stem cells in corneal wound healing following chemical injury. Stem Cells. 2008;26(4): 1047–55. doi:10.1634/stemcells.2007-0737.

4

Mass spectrometry analysis of adipose-derived stem cells reveals a significant effect of hypoxia on pathways regulating extracellular matrix

Simone Riis[1], Allan Stensballe[2], Jeppe Emmersen[1], Cristian Pablo Pennisi[1], Svend Birkelund[2], Vladimir Zachar[1] and Trine Fink[1*]

Abstract

Background: Adipose-derived stem cells (ASCs) are being increasingly recognized for their potential to promote tissue regeneration and wound healing. These effects appear to be partly mediated by paracrine signaling pathways, and are enhanced during hypoxia. Mass spectrometry (MS) is a valuable tool for proteomic profiling of cultured ASCs, which may help to reveal the identity of the factors secreted by the cells under different conditions. However, serum starvation which is essentially required to obtain samples compatible with secretome analysis by MS can have a significant influence on ASCs. Here, we present a novel and optimized culturing approach based on the use of a clinically relevant serum-free formulation, which was used to assess the effects of hypoxia on the ASC proteomic profile.

Methods: Human ASCs from three human donors were expanded in StemPro® MSC SFM XenoFree medium. Cells were cultured for 24 h in serum- and albumin-free supplements in either normoxic (20 %) or hypoxic (1 %) atmospheres, after which the cells and conditioned medium were collected, subfractionated, and analyzed using MS. Prior to analysis, the secreted proteins were further subdivided into a secretome (>30 kDa) and a peptidome (3–30 kDa) fraction.

Results: MS analysis revealed the presence of 342, 98, and 3228 proteins in the normoxic ASC secretome, peptidome, and proteome, respectively. A relatively small fraction of the proteome (9.6 %) was significantly affected by hypoxia, and the most regulated proteins were those involved in extracellular matrix (ECM) synthesis and cell metabolism. No proteins were found to be significantly modulated by hypoxic treatment across all cultures for the secretome and peptidome samples.

Conclusions: This study highlights ECM remodeling as a significant mechanism contributing to the ASC regenerative effect after hypoxic preconditioning, and further underscores considerable inter-individual differences in ASC response to hypoxia. The novel culture paradigm provides a basis for future proteomic studies under conditions that do not induce a stress response, so that the best responders can be accurately identified for prospective therapeutic use. Data are available via ProteomeXchange with identifier PXD003550.

Keywords: ASCs, Proteomics, Hypoxia, Serum-free culture, Secretome, ECM, Mass spectrometry

* Correspondence: trinef@hst.aau.dk
[1]Department of Health Science and Technology, Laboratory for Stem Cell Research, Aalborg University, Fredrik Bajers Vej 3B, Aalborg 9220, Denmark
Full list of author information is available at the end of the article

Background

Human adipose-derived stem cells (ASCs) have come under scrutiny for their putative use in regenerative medicine based on their immunomodulatory, pro-angiogenic, pro-trophic, and anti-apoptotic properties [1–4]. ASCs are of particular interest for the treatment of chronic wounds as they may restart the healing process by reducing inflammation, supporting ingrowth of new vessels into the hypoxic tissue, and promoting fibroblast and keratinocyte proliferation and migration [5]. While the molecular mechanisms are not fully explained, it is apparent that some of these regenerative properties of ASCs are associated with secreted factors in the form of growth factors, cytokines, and extracellular vesicles [6–9]. In addition to the impact on the inflammatory and proliferative phases of wound healing, it also appears that ASCs have a positive effect on the formation and remodeling of the extracellular matrix (ECM) [10]. This is particularly important as the ECM is degraded or otherwise compromised during the chronic wound process [11].

It has been demonstrated that the regenerative potential of ASCs and other mesenchymal stem cells is enhanced by hypoxic conditioning prior to use [12, 13]. Of interest to wound healing, hypoxic treatment induces an increase in ASC secretion of factors relevant to the inflammation and proliferation phases, such as interleukin (IL)-10, basic fibroblast growth factor (bFGF), vascular endothelial growth factor (VEGF), stromal cell-derived factor 1 (SDF-1), insulin-like growth factor-1 (IGF-1), caspase 9 (CASP9), and bcl-2 associated X protein (BAX) [2, 14–18]; 1 % O_2 has especially been shown to increase the secretion of paracrine factors [16]. The identified factors, however, represent only a subset of the entire secretome, and it is probable that as yet unidentified factors play an important role in the wound healing mediated by hypoxia-preconditioned ASCs.

To perform a more global characterization of the effect of hypoxia on ASCs, transcriptional profiling using microarrays has been used to assess the effect on the transcription of a large number of pre-selected genes [19]. High-throughput RNA sequencing that allows for detection of unknown transcripts, and that has been shown suitable for discrimination between distinct ASC subpopulations and between cell populations within the stem cell niche [20, 21], would be another option; however, we have not identified any such study. Both microarrays and RNA sequencing have developed into cost-effective and robust techniques. A serious drawback, however, is the concern regarding a lack of correlation between regulation of transcripts and proteins, particularly when it comes to upregulated genes following perturbation of steady-state conditions [22, 23]. As it is increasingly clear that mRNA abundance does not necessarily reflect the corresponding protein abundance, post-translational modifications, or subcellular location, much effort has

been dedicated to characterize and quantitate the ASC secretome in particular [24], and also to some extent the proteome [25]. Techniques used to characterize the effect of hypoxia on the ASC protein profile range from enzyme-linked immunosorbent assay (ELISA), Western blotting, and antibody-based arrays (allowing for the simultaneous analysis of relatively few pre-selected proteins), to mass spectrometry (MS)-based methods where a much larger gamut of proteins can be identified and quantitated (see [24], and references therein). Generally, the large-scale study of proteins in cells or tissue can be made by two proteomics strategies, either using one- or two-dimensional gel electrophoresis or non-gel-based techniques. In non-gel-based techniques, sample proteins are enzymatically digested in solution phase and separated using liquid chromatography in line with a tandem MS (LC-MS/MS) system [26]. During MS and tandem MS analysis the exact mass and amino acid sequence information for each peptide are obtained. The proteins in the sample can then be identified using database search algorithms and quantified by comparing the peptides to databases containing protein information based on translated genome and mRNA sequences [27].

In previous studies, where the proteome and secretome of ASCs have been analyzed by MS, cells have been expanded in medium supplemented with fetal calf serum (FCS), followed by extensive washes and culture in serum-free conditions for 24–72 h prior to collection of conditioned media/cell lysis [28–30]. This period of serum starvation has been necessary as the presence of high levels of serum proteins, e.g., albumin, found in serum would mask the presence of low-abundant proteins [31]. However, the process of serum starvation in itself frequently leads to unwanted cell responses, including membrane blebbing and growth arrest [32–34], and has been shown to induce a robust response in ASCs, altering the transcription of more than 100 genes [35]. Several serum-free medium formulations that support the clinically compliant expansion of ASCs are commercially available [36]. In particular, studies have shown that ASCs cultured using StemPro® MSC SFM XenoFree (Thermo Fisher) maintain the essential stem cell properties such as immunophenotype and multilineage differentiation potential [37, 38]. However, the applicability of serum-free medium formulations for the culture of cells used for MS protein analysis of ASCs has not previously been reported.

In this paper, we describe a culture protocol that allows the growth of ASCs under clinically relevant conditions for the production of conditioned media compatible with MS. Growth and viability of ASCs have been assessed in media without FCS and excessive bovine protein contamination. Using the optimal medium formulation for culture and preconditioning of the ASCs we used MS-based proteome analysis to compare the secretome and proteome of cells from three different donors cultured under hypoxic (1 % oxygen) and normoxic (20 % oxygen) conditions.

Methods
ASCs
ASCs from three donors were isolated as previously described after written informed consent from the donors and approval by the regional committee on biomedical research ethics of Northern Jutland, Denmark (project no. VN 2005/54) [39–41]. The three different cultures (ASC12, ASC21, and ASC23) have previously been thoroughly characterized by our laboratory both under normoxic and hypoxic conditions, and possess characteristics commensurate with the definition of mesenchymal stem cells [38, 42–47].

Media
Three different media were used: StemPro–, composed of StemPro® MSC SFM XenoFree (Gibco™, Thermo Fisher, www.thermofisher.com) supplemented with 2 mM L-glutamine (Gibco™), and 100 U/mL penicillin and 0.1 mg/mL streptomycin (Gibco™); StemPro+, (StemPro– supplemented with StemPro supplement); and StemPro E8 (StemPro– supplemented with Essential 8™ Medium supplement (Gibco™)).

Cell culture
The cells were cultured in polystyrene tissue flasks (CELLSTAR®, Greiner Bio-One, www.gbo.com) coated with CellStart™ CTS™ (Gibco™), in a standard incubator in an atmosphere of 37 °C, 20 % O_2 and 5 % CO_2. When replating, TrypLE Select (Gibco™) was used for cell detachment. Unless stated otherwise, ASCs were cultured in StemPro+. For all experiments the ASCs were in passage 6–8.

Expansion of ASCs in different media and cell counting
To assess the effect of different media on ASC growth, the cells were seeded at a density of 1500 cells/cm² in 48-well plates (CELLSTAR®) in StemPro+, StemPro–, and StemPro E8, and cultured for up to 72 h. At 24, 48, and 72 h the cell number was determined by measuring the amount of DNA using a Quant-iT™ PicoGreen® dsDNA Assay Kit (Invitrogen™). The fluorescence was measured using a Wallac 1420 Victor Multilabel Counter with excitation and emission at 485 nm and 535, respectively; 6.6 pg DNA/cell was used to calculate the cell number. The media were tested in duplicate on each of the three ASC cultures.

Short-term effect of different media on ASCs
ASCs were seeded in 96-well culture plates at a density 8000 cells/cm². When the culture was 70 % confluent, cells were washed repeatedly with phosphate-buffered saline (PBS). The cells were then incubated for 24 h with either StemPro+, StemPro–, or StemPro E8 media, after which cell morphology and viability were assessed.

For the assessment of cell morphology, phase contrast images were acquired using an Olympus CKX41 microscope (Olympus Life Science) with a PixeLINK PL-A782 camera.

The proportion of viable cells was determined essentially as previously described [48]. In brief, ASCs were stained with 1 µM Yo-PRO-1 (Molecular Probes™), 10 µg/mL PI (Molecular Probes™), and 5 µg/mL Hoechst 33342 (Molecular Probes™), after which images were acquired using an AxioVision software package with a Zeiss Axio Observer.Z1 microscope equipped with an AxioCam MRm camera (Carl Zeiss, Germany). Live, necrotic, and apoptotic cells were counted using Image J 1.47v (NIH, USA), and the proportion of viable cells was calculated as the fraction of Yo-PRO-1 and PI-positive cells relative to the total cell number subtracted from 1. The effect of the media was tested in duplicate on each of the three ASC cultures.

Statistical analysis of cell growth and viability data
The statistical analysis was performed using SigmaPlot 12.0 (Systat Software, Erkrath, Germany). The normal distribution of each group was assessed by means of the Shapiro-Wilk test. Additionally, variance was tested using an Equal Variance Test. Data are represented as mean ± standard error of the mean (SEM). A p value of <0.05 was considered statistically significant. For comparison of more than two groups, a one-way analysis of variance (ANOVA) with Bonferroni's post hoc test was used.

Production and fractionation of conditioned media and cell lysate
For an overview of the steps involved in the production of media and cell lysate for MS, please refer to Fig. 1. For production of conditioned media, ASCs were seeded in T75 tissue culture flasks at a density of 8000 cells/cm², and incubated until approximately 70 % confluence (72 h). The cells were washed thoroughly with PBS to remove any albumin residues and 15 mL fresh StemPro E8 medium was added. Half of the flasks were cultured at 20 % oxygen, the other half at 1 % oxygen. After 24 h, the conditioned medium (CM) was collected, centrifuged, and decanted before protease inhibitors were added (1 tablet per 15 mL medium; Roche Complete Protease inhibitor cocktail, Mini). The resulting CM was first fractionated using spin filters into a high-molecular

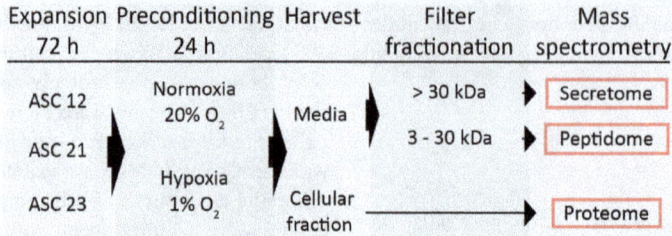

Fig. 1 Preparation of samples for mass spectrometric analysis. Following the expansion of ASCs from three donors for 72 h, cells were cultured under either normoxic or hypoxic conditions for 24 h. The conditioned media were harvested and sequentially fractionated through 30-kDa and 3-kDa spin filters to retain the secretome and peptidome fractions, respectively. The cellular fraction was employed for the analysis of the proteome. *ASC* adipose-derived stem cell

weight secretome fraction (>30 kDa) using a 30-kDa spinfilter (Millipore, Billerica, MA, USA), and, based on the flow-through, a low-molecular weight peptidome fraction (3–30 kDa), where molecules smaller than 3 kDa were removed using a 3-kDa spinfilter (Millipore). After both filtration steps, the retained proteins trapped on the spin filters were washed twice with 4 mL TEAB buffer (50 mM triethylamonium bicarbonate, pH 8.5), and retained in 500 μL TEAB buffer. The protein content was measured spectrophotometrically by protein OD A280 (Nanodrop; Thermo Science, Wilmington, DE), and the samples were stored at −80 °C for further analysis. All experiments were performed for all three cell lines in two separate experiments, each in duplicate.

After harvesting the CM, the ASCs were washed twice in PBS and the cells collected for proteome analysis using a protease and phosphatase inhibited RIPA buffer and subsequently sonicated to ensure complete lysis. Proteome samples were stored at −80 °C until further analysis.

Sample preparation
Secretome
From each sample, a volume corresponding to 25 μg protein was transferred to an Eppendorf tube, and 50 mM TEAB buffer, pH 8.5, was added to a total volume of 100 μL. The proteins were reduced by the addition of 2 μl 0.5 M tris(2-carboxyethyl)phosphine (Thermo Scientific, Waltham, MA, USA) and incubation for 30 min at 37 °C. Next, the proteins were alkylated by the addition of 8 μl 0.5 M chloroacetamide (Sigma-Aldrich, St. Louis, MO, USA) and incubation for 30 min at 37 °C in the dark. Trypsin (0.5 μg) was added to each sample, and the proteins were digested overnight at 37 °C. The enzymatic procedure was stopped by addition of 5 μl 100 % formic acid. Protein digests were dried by vacuum centrifugation and desalted on PorosR3 nanocolumns and resuspended in 30 μL of a solution containing 2 % acetonitrile and 0.1 % formic acid.

Peptidome
The lower molecular weight proteins (3–30 kDa) were concentrated in YM-3 kDa spinfilter (Millipore, Billerica, MA, USA) by centrifugation at 14,000 g at 4 °C. The retained fraction (100 μL) was processed as described for the secretome samples.

Proteome
For the proteome analysis of the cells, a total of 25 μL of protein lysate in modified RIPA buffer was mixed with reducing sample buffer (2× Laemmeli; Bio-Rad, Hercules, CA) and isolated by reducing sodium dodecyl sulfate polyacrylamide gel electrophoresis (SDS-PAGE; (Biorad Any kD Mini-PROTEAN® TGX). The electrophoresis was stopped when the protein had entered 1.5 cm into the running gel according to the applied pre-stained protein marker (SeeBlue Plus2 pre-stained standard; Invitrogen, Carlsbad, CA). The protein load on each gel lane was comparable. The protein was visualized by Coomassie blue (SimplyBlue SafeStain; Invitrogen) and 50 % of each band excised and in-gel digested as described previously [49].

Mass spectrometry analysis
The protein concentrations in the samples were normalized using A280 on a NanoDrop 1000 (Thermo Scientific, Wilmington, DE, USA), and 5 μg total peptide material was analyzed per LC-MS analysis of the low and high molecular weight proteins. A total of 20 % of each gel band was used for each of the biological replicates.

The samples were analyzed using a UPLC-nanoESI MS/MS setup with an UltiMate™ 3000 RSLC nanopump module. The system was coupled online with an emitter for nanospray ionization (New objective picotip 360-20-10) to a QExactive Plus mass spectrometer (Thermo Scientific, Waltham, USA). The peptide material was loaded onto a C18 reversed phase column (Dionex Acclaim PepMap RSLC C18, 2 μm, 100 Å, 100 μm × 2 cm) and separated using a C18 reversed phase column (Dionex Acclaim PepMap RSLC C18, 2 μm, 75 Å, 75 μm × 50 cm) at 40 °C.

The sample was eluted with a gradient of 96 % solvent A (0.1 % FA) and 4 % solvent B (0.1 % FA in ACN), which was increased to 8 % solvent B on a 5-min ramp gradient and subsequently to 30 % solvent B on a 70-min ramp gradient for the proteome samples and 30 min for the secretome samples, at a constant flow rate of 300 nL/min. The mass spectrometer was operated in positive mode, selecting up to 12 precursor ions based on the highest intensity for HCD fragmenting.

Data analysis

The reproducibility and variation of the amounts of peptide per sample were determined using Progenesis QI for Proteomics (NonLinear/Waters, UK). A label-free quantitation by normalized XIC was performed of the trypsin-digested samples by searching the data files using MaxQuant 1.5.2.8 against the *Homo sapiens* Uniprot reference proteome database (UPID5640; April 2015). All standard settings were employed with carbamidomethyl (C) as a static modification and oxidation (M) as a variable modification. All proteins are reported at <1 % false discovery rate (FDR) to ensure only high-confidence protein identifications. The result file from MaxQuant was analyzed in Perseus v1.5.1.6. Initially, all reverse hits were removed from further analysis, and the data was log2-transformed. Two unique peptides or more were required for a protein identification and quantitation to ensure high-quality quantitation. Scatter plots were inspected biological-replicate wise. No replicates were removed due to low reproducibility. To further identify replicate outliers, a principle component analysis (PCA) was performed using all the measured protein abundances in all replicates as an input. For the purpose of conducting PCA, missing values (i.e., proteins where a quantitation value was not obtained for a given replicate analysis) were replaced with values from a normal distribution (width 0.3 and down shift 1.8) to simulate signals from low abundant proteins [50].

MS measurements were grouped preservation-method wise, and two-sided *t* tests were performed to identify statistically significant changing proteins. The statistical tests were corrected for multiple hypothesis testing using permutation-based FDR, and the parameters were chosen to provide sufficient input in the subsequent analysis (FDR = 0.05, s0 = 0.1, and 250 randomizations).

The genes coding for the significantly different expressed proteins found by MS were analyzed for statistical over-representation (enrichment) of Gene Ontology (GO) categories in terms of biological process using Cytoscape (version 3.3.0) with the Biological Networks GO plug-in (BINGO, version 3.0.3) [51, 52]. An adjusted *p* value of 0.05 was used as threshold for significance after correcting for multiple testing by Benjamini and Hochberg FDR correction. Over-representation was computed from hypergeometric test. The dataset for each donor was compared against a reference set of complete *Homo sapiens* GO annotations.

Results

Optimization of culture media

StemPro + has been shown to support ASC growth [53] but is not fully suitable for secretome analysis by MS due to the presence of large amounts of contaminant proteins, e.g., albumin. Low-protein and albumin-free media formulations include StemPro– and StemPro E8, which may offer an alternative for preconditioning of cells for MS. Initially, to compare the performance of the two media with StemPro+, ASCs were expanded in the three different media for up to 72 h. Analysis of the growth rate (Fig. 2a) revealed that StemPro + was superior in supporting a stable rate of cell growth, while in StemPro– cell growth is significantly inhibited over time. Although the growth rate of the cells in StemProE8 was comparable to StemPro + for the first 48 h, it decreased significantly after 72 h. Next, to assess the effect of short-term culture on cell number, viability, and morphology, ASCs were seeded at a higher density and grown in StemPro + for 72 h and then cultured in low-protein and albumin-free media formulations for 24 h. In terms of cell numbers, StemPro + and StemPro E8 were comparable, and StemPro– underperformed (Fig. 2b). Although analysis of cell viability did not suggest that culture in StemPro– caused compromised cell viability (Fig. 2b), cells cultured in StemPro– displayed a slightly more irregular morphology (Fig. 2c). Based on these results, ASCs were expanded in StemPro + and preconditioned in StemPro E8 for 24 h for the production of the CM and proteome samples for MS analysis.

Identification of proteins

For ASC12, ASC21, and ASC23 cultured in normoxia and hypoxia, each of the quadruplicate biological replicates of secretome, peptidome, and proteome was processed as described above, separated and analyzed using UPLC-MS/MS. During this procedure, a list of the generated peptide fragments and information about the retention time, the accurate precursor ion mass and its ion intensity were obtained for each peptide. Using reverse-phase UPLC, the peptides were separated according to their hydrophobicity. When reaching the column, the peptides were ionized by electron spray and the mass of the peptide (precursor ion) was determined followed by fragmentation by collision and mass analysis of the resulting fragments. A good and reproducible separation of the peptides as well as a similar sample loading amount as determined by the total TIC was observed by plotting the m/z against the retention time of each

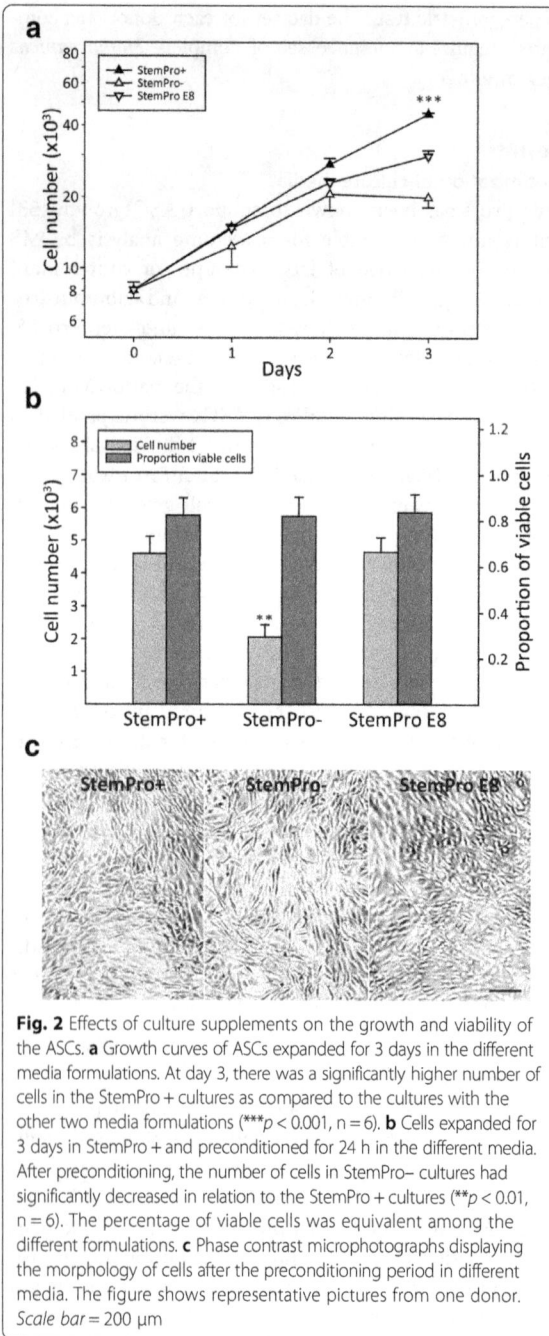

Fig. 2 Effects of culture supplements on the growth and viability of the ASCs. **a** Growth curves of ASCs expanded for 3 days in the different media formulations. At day 3, there was a significantly higher number of cells in the StemPro + cultures as compared to the cultures with the other two media formulations (***$p < 0.001$, n = 6). **b** Cells expanded for 3 days in StemPro + and preconditioned for 24 h in the different media. After preconditioning, the number of cells in StemPro– cultures had significantly decreased in relation to the StemPro + cultures (**$p < 0.01$, n = 6). The percentage of viable cells was equivalent among the different formulations. **c** Phase contrast microphotographs displaying the morphology of cells after the preconditioning period in different media. The figure shows representative pictures from one donor. *Scale bar* = 200 μm

peptide using Progenesis QI for Proteomics (data not shown).

The tandem mass spectra were used to search the Uniprot human reference proteome database with isoforms (UP000005640) to uniquely identify the parent proteins. Combining the MS1 spectra of the peptides and the MS2 spectra of their fragments, quantification of each

protein was obtained. Validation of the UPLC-MS/MS spectra quantification was performed by plotting the intensities of proteins from each replicate against each other (Additional file 1: Figure S1). From this, a good correlation was found, increasing with increasing protein intensity. This is representative for all proteome samples with an average $R = 0.98 \pm 0.01$.

Effect of hypoxic treatment on the ASC secretome

In total, 342 proteins were identified in the secretome of ASC12, ASC21, or ASC23 by label-free quantitative data analysis using MaxQuant and post-processing in Perseus [27]. GO categories, by which at least three proteins were annotated, are shown in Additional file 2 (Table S1). The resulting GO profile showed that many of these proteins had functions that were related to developmental processes including angiogenesis and vasculature development, ECM and cell adhesion/migration, cell survival and cell death, as well as immune regulation. Interestingly, we did not detect any statistically significant hypoxia-induced up- or downregulation of proteins.

Effect of hypoxic treatment on the ASC peptidome

In the peptidome fractions, only a total of 98 proteins were identified and quantified based on strict criteria. GO categories, by which at least three proteins were annotated, are shown in Additional file 3 (Table S2). The GO profile confirmed over-representation proteins with essentially the same functions as the proteins in the secretome. Of the 98 identified proteins, two were found significantly upregulated and three significantly downregulated in the ASC 23 cells (Additional file 4: Table S3). However, for the other ASCs we did not observe any significant effect of hypoxia on the MS peptidome profile.

Effect of hypoxic treatment on the ASC proteome

For cells cultured in normoxia, 94 % of proteins (2741 out of a total of 3228) were common to both ASC 12, ASC21, and ASC 23, with 50–100 proteins uniquely expressed by cells from just one donor (Fig. 3a). For the cells cultured in hypoxia, more than 85 % of proteins were detected in all cells. Despite the high degree of overlap between cells from different donors, the PCA revealed that the variations in protein expression between donors were larger than the variations caused by hypoxic treatment (Fig. 3b). The technical reproducibility of the MS setup used has previously been reported to be very high [54]. A representative volcano plot, depicting an analysis of the effect of hypoxia on protein expression, illustrates that the differences in protein levels between cells cultured in hypoxia and normoxia are less than two-fold for most proteins (Fig. 3c), and statistically significant for a only a relatively small fraction of proteins.

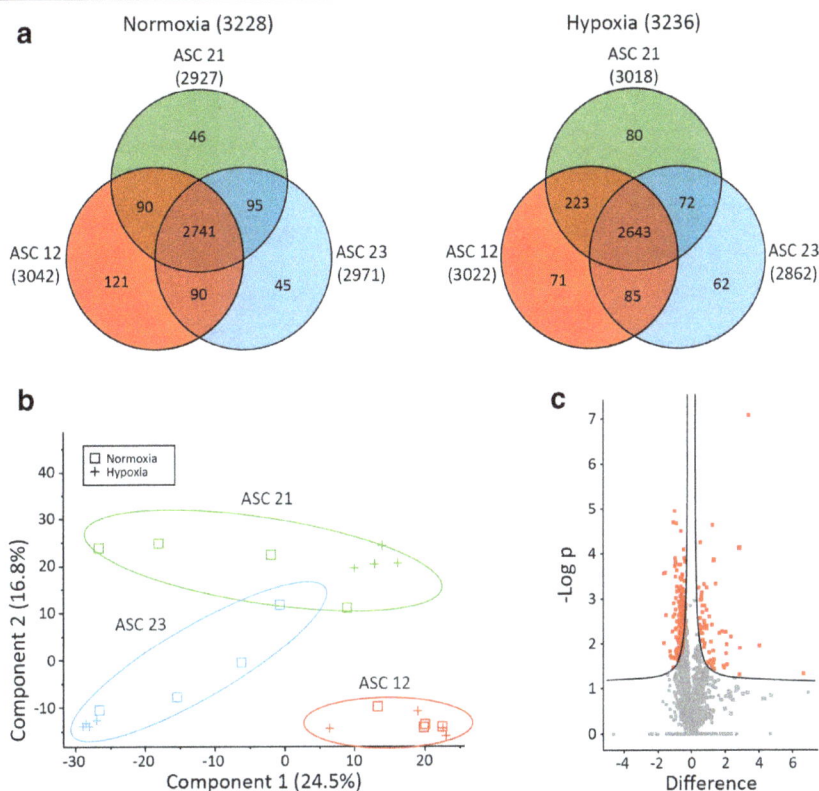

Fig. 3 Analysis of the proteome of the ASCs by mass spectrometry. **a** Venn diagrams showing the concurrence in identified proteins in the proteome from the three donors (ASC 12, 21 and 23) exposed to either normoxic or hypoxic preconditioning. **b** Principal component analysis (PCA) of the proteome fractions in all samples. Shown are PCA score plots of principle components 1 and 2 of the protein abundances as measured in the normoxic and hypoxic samples from all three donors in four biological replicates. *Red*, ASC 12; *green*, ASC 21; *blue*, ASC 23; *square*, normoxic; *cross*, hypoxic. **c** Statistical analysis of the difference between label-free samples of the proteome fraction from hypoxic preconditioned ASC 21 against the proteome fraction from normoxic preconditioned ASC21 with a two-sided t test. The results are visualized by a scatter plot (volcano plot). The −log t test p value is plotted against the t test difference log2 for each protein. The proteins significantly changed between the samples ($p < 0.05$) are in the right and left upper corners (*red*). *ASC* adipose-derived stem cell

Among the proteins regulated by hypoxic preconditioning, 235 (7.1 %) proteins were significantly downregulated, while 82 (2.5 %) were significantly upregulated.

Biological significance of responses to hypoxia

To gain insight into the biological processes in which the differentially regulated proteins are involved, the genes coding for the proteins that were significantly differentially expressed were analyzed in terms of GO.

The GO categories of the proteins identified as significantly upregulated by hypoxic preconditioning comprise several clusters of biological processes, including metabolic processes, cell adhesion and ECM, developmental processes, and responses to stimuli (Fig. 4). When looking into the specific upregulated processes (Table 1), not surprisingly the metabolic processes encompassed mostly those necessary for anaerobic metabolism. Also, processes related to cell adhesion and ECM were represented. Regarding the ECM and cell adhesion/migration, the upregulated proteins included prolyl 4-hydroxylases 1 and 2 (P4HA1 and 2), procollagen-lysine 2-oxoglutarate 5-dioxygenase 1 (PLOD1), and alpha-1 chains of collagen types 1, 3, and 7.

When looking at the GO of proteins that were downregulated by the hypoxic preconditioning, the processes were clustered in terms of metabolism, protein synthesis, cell cycle, and stress response (Fig. 5). A closer look at those clusters (Table 2) revealed that the metabolic processes most downregulated were those related to aerobic metabolism thus complementing the pattern observed for the upregulated metabolic processes. Furthermore, several proteins involved in multiple steps in protein synthesis were downregulated, ranging from mRNA splicing and ribonucleoprotein

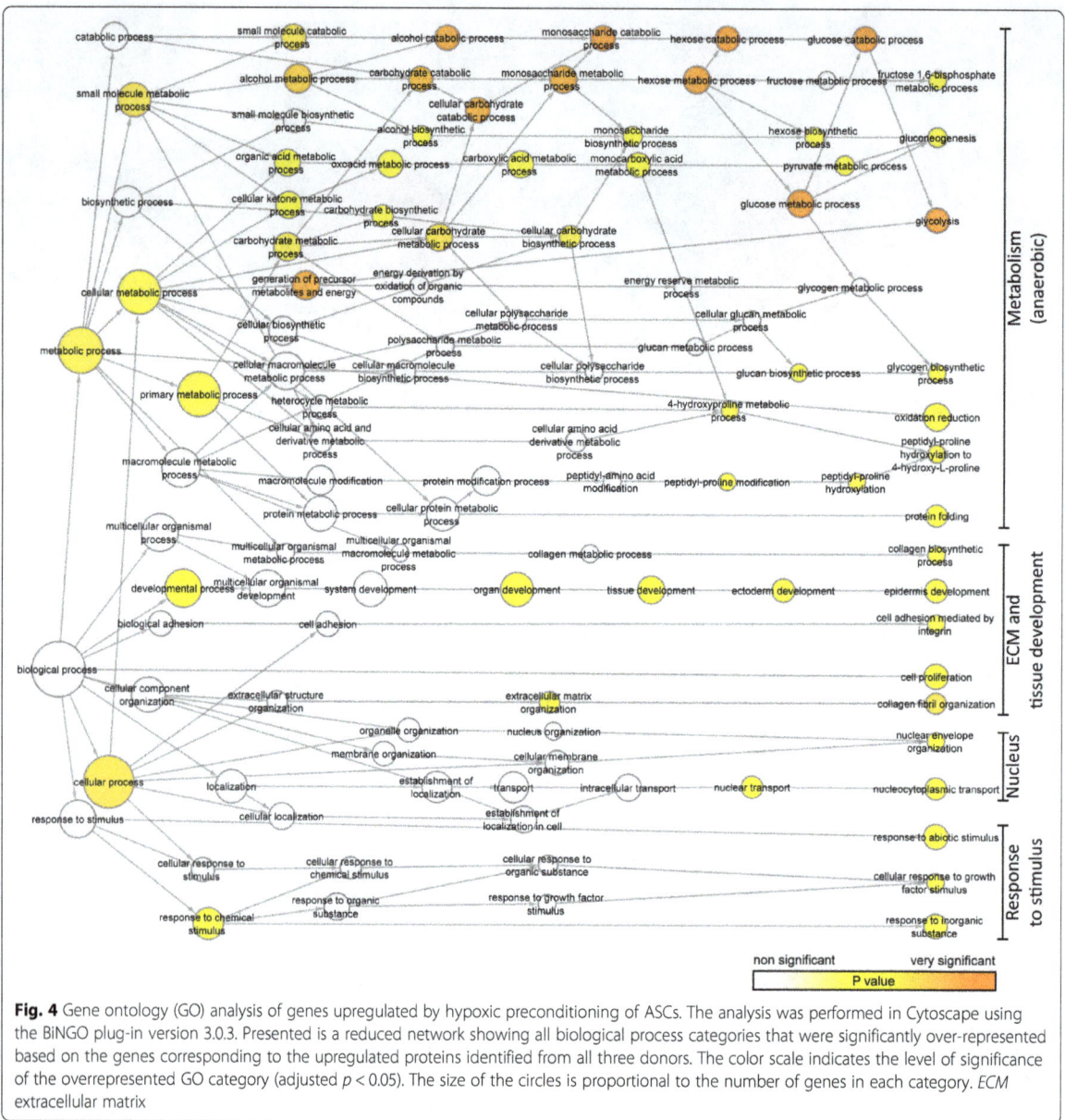

Fig. 4 Gene ontology (GO) analysis of genes upregulated by hypoxic preconditioning of ASCs. The analysis was performed in Cytoscape using the BiNGO plug-in version 3.0.3. Presented is a reduced network showing all biological process categories that were significantly over-represented based on the genes corresponding to the upregulated proteins identified from all three donors. The color scale indicates the level of significance of the overrepresented GO category (adjusted $p < 0.05$). The size of the circles is proportional to the number of genes in each category. *ECM* extracellular matrix

complex assembly to translation elongation. Finally, the downregulated stress response proteins were related to detection of oxygen and DNA damage response.

Discussion

When assessing the secretome of culture cells by MS the usual approach has been to deprive the cells of serum for a given period of time to avoid the presence of interfering proteins. In ASCs, however, serum deprivation has been shown to induce a stress response that might obscure the effect of the variable under experimental assessment, e.g., the oxygen tension [35, 55]. Here, as an alternative, we

assessed a commercial serum-free medium that supports the expansion of ASCs. As high abundant proteins in the medium supplements, such as albumin, represent another source of interference which could dominate the mass spectrum and limit the detection of less abundant secreted proteins and peptides [56], two albumin-free formulations compatible with MS protein analysis were evaluated for the preconditioning phase.

To compare the secretome and proteome of ASCs cultured under hypoxic and normoxic conditions, ASCs were preconditioned in the MS-compatible medium for 24 h at 1 % and 20 % oxygen, after which the conditioned media

Table 1 Enriched biological processes based on upregulated proteins

Description	Genes involved	p value
Metabolism (anaerobic)		
Glycolysis	GPI, LDHA|TPI1, PGK1, ENO2, ALDOA, HK2, PFKP	0.0000
Gluconeogenesis	GPI, TPI1, ENO2	0.0101
Fructose 1,6-bisphosphate metabolic process	ALDOA, PFKP	0.0123
Glycogen biosynthetic process	GYS1, GBE1	0.0285
Oxidation reduction	LDHA, LOX, P4HA1, P4HA2, FTH1, CYP51A1PLOD2, PLOD1, ERO1L, LOXL2, FTL	0.0245
Protein folding	LRPAP1, PFDN4, HSPBP1, PFDN6, ERO1L	0.0469
ECM and tissue development		
Collagen fibril organization	COL1A1, COL3A1, LOX, P4HA1	0.0010
Peptidyl-proline hydroxylation to 4-hydroxy-L-proline	P4HA1, P4HA2	0.0046
Collagen biosynthetic process	COL1A1, COL3A1	0.0123
Epidermis development	COL1A1, COL3A1, CRABP2, COL7A1, TXNIP, PLOD1	0.0181
Cell proliferation	CDV3, LRPAP1, NUMBL, CD81ZAK, FTH1, APOA1, CD276	0.0474
Cell adhesion mediated by integrin	ITGA5, ICAM1	0.0255
Response to stimulus		
Response to inorganic substance	COL1A1, FNTA, BSG, TPM1, TXNIP, ACO1, NDRG1	0.0128
Cellular response to growth factor stimulus	COL1A1, EMD	0.0333
Response to abiotic stimulus	IKBIP, COL1A1, COL3A1, FECHZAK, SLC2A1, TXNIP, ERO1L	0.0333
Nucleus		
Nuclear envelope organization	LMNA, EMD	0.0285
Nucleocytoplasmic transport	LSG1, LMNA, TXNIP, NUTF2, AGFG1	0.0285

ECM extracellular matrix

and the cells were harvested. Subsequently, the conditioned media were fractioned into a secretome and a peptidome fraction and the cells were lysed to explore the proteome.

After analysis of the secreted proteins in both the secretome and the peptidome fraction, we identified a plethora of proteins relevant for stem cell maintenance and tissue regeneration. However, we were not able to detect a significant effect of hypoxic preconditioning on the abundance of these. Similarly, a recent report has shown that hypoxic preconditioning of ASCs does not seem to largely affect the secretion of proteins [57]. To decrease the complexity of the secretome, and thereby increase the chance of detecting low abundance proteins, we chose to fraction the conditioned media into two fractions: the secretome and the peptidome. However, this could be inadequate and further fractionation could have increased the sensitivity towards the low abundant proteins. It has been reported that the abundances of proteins in mammalian cells range from 1 to 10^7 copies per cell [58, 59]. Global LC-MS/MS protein expression studies in cell lines have reported 50 % coverage of all predicted proteins for human cell lines, which is perceived as an excellent proteome coverage with today's technology [58, 59]. To increase this in the future and enable the detection of

quantitative changes for each protein across this range of abundances, extensive fractionation and one or more quantitative strategies are suggested [56]. For targeting of specific very low abundance transcription factors, cytokines, or chemokines, a targeted MS approach will enhance the applicability for detection and quantification [60].

Within the proteome of ASCs a variety of proteins were identified. Among these, a minor proportion was found to be differently expressed in ASCs exposed to hypoxic conditions. Among the most interesting findings, hypoxic conditioning of ASCs increased the presence of proteins involved in regulation of the ECM. Among these proteins were P4HA1 and P4HA2, which are required for collagen synthesis and are essential to the proper three-dimensional folding of newly synthesized procollagen chains [61], and PLOD1, which catalyzes the hydroxylation of lysyl residues in collagen-like peptides on the endoplasmatic reticulum and influences the stability of intermolecular collagen cross-links that provide the tensile strength and mechanical stability of collagen fibrils. Additionally, alpha-1 chains of collagen types 1, 3, and 7 were found to be upregulated. Collagen type 1 alpha chain 1 is a fibril-forming collagen found in most connective tissues and is abundant in the dermis.

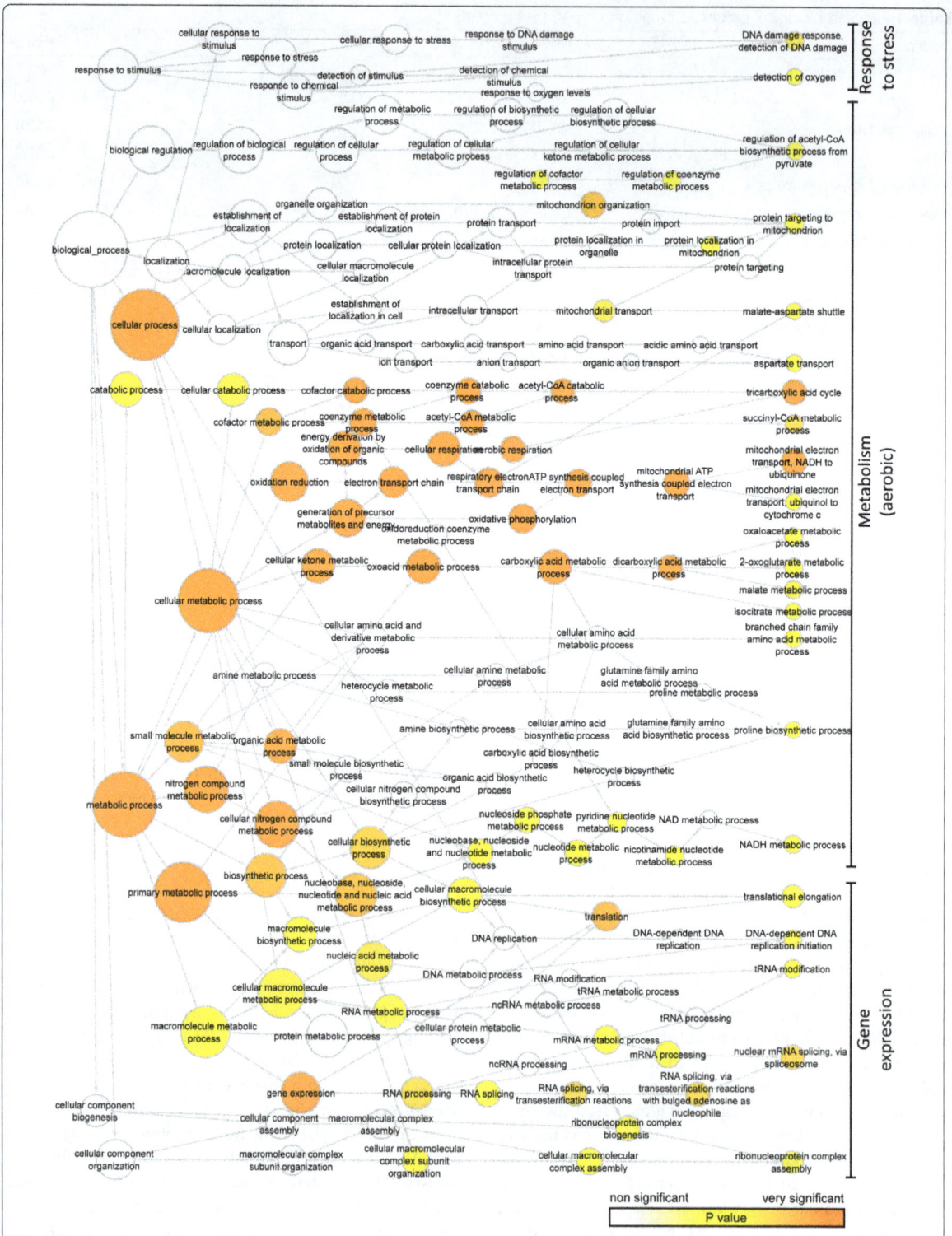

Fig. 5 (See legend on next page.)

Collagen type 3 alpha chain 1 is also a fibril-forming collagen and is found in extensible connective tissues such as the skin and the vascular system, frequently in association with type I collagen. Collagen type 7 alpha chain 1 is restricted to the basement zone beneath stratified squamous epithelia. It functions as an anchoring fibril between the external epithelia and the underlying stroma, ensuring skin integrity and stability. It has been found that intradermal administration of mesenchymal stem cells resulted in production and deposition of collagen type 7 at the dermal-epidermal junction [62]. Among these proteins, P4HA1, P4HA2, PLOD 1, and collagen type 1 alpha chain 1 have been shown to be under transcriptional regulation in fibroblasts by hypoxia-inducible factor 1 (HIF-1) [61, 63]. As our preconditioning

protocol has previously been shown to stabilize HIF-1 in these ASCs [64], the observed upregulation of the ECM-related genes could be mediated via HIF-1. This multifaceted upregulation of proteins involved in ECM production and modulation could contribute to a restoration of an ECM in chronic wounds that is conducive to healing, and thus partly explains the observed wound healing effects of hypoxic ASCs.

Besides the ECM-relevant proteins, we also found that hypoxic preconditioning significantly affected the metabolism of the ASCs. Proteins involved in anaerobic metabolism were found to be upregulated and proteins involved in aerobic metabolism were found to be downregulated. This is well described in the literature as a basic cellular response to hypoxic conditions, as cells

Table 2 Enriched biological processes based on downregulated proteins

Description	Genes involved	p value
Metabolism (aerobic)		
Tricarboxylic acid cycle	CS, FH, SUCLA2, MDH2, IDH3G, SUCLG2SUCLG1, ACO2, PDHB, IDH3A	0.0000
Succinyl-CoA metabolic process	SUCLA2, SUCLG2, SUCLG1	0.0004
mitochondrial electron transport, NADH to ubiquinone	NDUFA9, NDUFB9, NDUFA8, NDUFB8, NDUFS8, NDUFB10 NDUFS5, NDUFA10, NDUFB3, NDUFS3, DLD, NDUFV1	0.0000
Mitochondrial electron transport, ubiquinol to cytochrome c	UQCRC1, UQCR10	0.0335
2-Oxoglutarate metabolic process	IDH3G, DLD, IDH3A	0.0253
Malate metabolic process	FH, MDH2, ME2	0.0042
Isocitrate metabolic process	IDH3G, IDH3A	0.0335
Branched chain family amino acid metabolic process	HIBADH, HIBCH, BCAT2	0.0298
Proline biosynthetic process	PYCR1, PYCR2	0.0479
NADH metabolic process	MDH2, IDH3G, IDH3A	0.0060
Aspartate transport	SLC25A12, SLC25A13	0.0479
Malate-aspartate shuttle	SLC25A12, SLC25A13	0.0137
Regulation of acetyl-CoA biosynthetic process from pyruvate	PDP1, PDHB, DLD	0.0128
Oxaloacetate metabolic process	CS, MDH2, PCK2	0.0106
Protein targeting to mitochondrion	TOMM40, TOMM34, TIMM44, TOMM22	0.0198
Gene expression		
Nuclear mRNA splicing, via spliceosome	SF3B4, PRPF4, SF3A2, SNRPD1, DHX38GEMIN5, USP39, WDR77, LSM2	0.0001
Translational elongation	GFM1, RPL21, RPL22, RPL13, RPL27, EEF2, TUFM	0.0099
DNA-dependent DNA replication initiation	MCM7, MCM3, MCM6	0.0253
tRNA modification	QTRT1, SSB, NSUN2	0.0335
Ribonucleoprotein complex assembly	SF3A2, SNRPD1, CIRBP, GEMIN5, USP39, WDR77	0.0147
Response to stress		
Detection of oxygen	SOD2, ENG	0.0335
DNA damage response, detection of DNA damage	MRPS9, PARP1, MRPS35	0.0017

during hypoxia rapidly enter a state of metabolic crisis which requires a fundamental shift in cellular metabolic strategy to facilitate the entering of an adaptive state which supports tissue survival [65]. This is very effective in ASCs, and hypoxic preconditioning has been shown to enhance the survival and regenerative potential of ASCs in vivo [14, 66–71].

Stress-related proteins were also downregulated by hypoxic conditioning. Among these were the poly(ADP-ribose) polymerase 1, which is involved in the regulation of various important cellular processes such as differentiation and proliferation. Others have found that 1 % O_2 maintains stem cells in an undifferentiated state [72] and we have earlier shown that hypoxia at 1 % O_2 decreases the proliferation of ASCs [16]. Poly(ADP-ribose) polymerase 1 could be involved in this.

Furthermore, proteins involved in protein translation were also found to be downregulated. Among these were proteins involved in mRNA splicing, ribonucleoprotein complex assembly and translation elongation as splicing factors binding to pre-mRNA, ribosomal proteins and the essential translation elongation factor 2 promoting the GTP-dependent translocation of the nascent protein chain from the A-site to the P-site of the ribosomes. It has previously been shown that moderate hypoxia of 1–0.1 % O_2 affects gene expression, transcription, mRNA stability, protein synthesis and post-translational modifications [73]. The downregulation of the broad range of proteins related to translation that we observed in this study could therefore provide an explanation for the general decrease in protein synthesis by hypoxia.

Based on our findings and analysis, the explanation of the regenerative effect of ASCs was not to be found within the proteome. Despite the ECM-relevant proteins and a few others, the approach did not reveal differences in secretome abundance of other low abundant wound healing relevant factors. The majority of the proteins identified by MS were 1.11 ± 2.98-fold regulated and only a very small fraction of these was regulated by hypoxia. Earlier, we have shown hypoxic preconditioning to increase the level of VEGF two-fold as measured by ELISA which was found to be statistically significant compared to the normoxic levels of VEGF [64]. The analytical approach applied in this study allows for the discovery of unknown proteins as the data was compared to complete databases of all known proteins and their isoforms; however, the molecular size of multiple growth factors limits the detection by discovery proteomics. The detection of very low abundance cytokines and chemokines could be improved by targeted MS approaches such as MRM or PRM [60]. By combining our novel growth and fractionation approach, novel wound healing candidates could be discovered.

Conclusions

In this study, we described a new serum-free culturing methodology to obtain conditioned medium from ASCs for MS analysis. Optimal cell growth, viability, and morphology were obtained using StemPro® SFM XenoFree basal medium containing the regular supplements during the expansion phase and, to avoid serum starvation and eliminate albumin interference, Essential 8™ supplements during the conditioning phase. While analysis of the secretome did not reveal any significant hypoxia-induced up- or downregulation of proteins, a relatively small fraction of the proteome was significantly affected by hypoxia. The main effects comprised ECM-relevant proteins ensuring tensile strength, three-dimensional folding, and mechanical stability of collagen fibrils in the dermis ensuring skin integrity and stability. Additionally, we found a switch in metabolism-relevant proteins indicating a change from aerobic to anaerobic metabolism. Although we could not find evidence supporting an enhanced secretion of pro-regenerative proteins by ASCs under hypoxia, this study provides a basis for further studies using proteomic techniques for the characterization of the ASC secretome.

Availability of data

The mass spectrometry proteomics data have been deposited in the ProteomeXchange Consortium [74] via the PRIDE partner repository with the dataset identifier PXD003550.

Additional files

> **Additional file 1: Figure S1.** Scatter plots of all samples from the proteome fraction from all three donors. The log2 transformed protein abundance of all proteins obtained are plotted against each other on the x-axis and y-axis, respectively. Each spot represents the intensity of a protein. Ideally the measurements should yield identical protein abundances, represented by a Pearson's correlation coefficient (R) of 1. (JPG 3314 kb)
>
> **Additional file 2: Table S1.** Over-represented biological processes by gene ontology analysis of proteins identified in the secretome fraction. (DOCX 63 kb)
>
> **Additional file 3: Table S2.** Over-represented biological processes by gene ontology analysis of proteins identified in the peptidome fraction. (DOCX 39 kb)
>
> **Additional file 4: Table S3.** Proteins in the peptidome fraction regulated by hypoxia. (DOCX 16 kb)

Abbreviations

ASC: adipose-derived stem cell; CM: conditioned medium; ECM: extracellular matrix; ELISA: enzyme-linked immunosorbent assay; FCS: fetal calf serum; FDR: false discovery rate; GO: gene ontology; HIF-1: hypoxia-inducible factor 1; LC-MS/MS: liquid chromatography in line with a tandem mass spectrometry; MS: mass spectrometry; P4HA: prolyl 4-hydroxylase; PBS: phosphate-buffered saline; PCA: protein component analysis; PLOD1: procollagen-lysine 2-oxoglutarate 5-dioxygenase 1; TEAB: triethylammonium bicarbonate; VEGF: vascular endothelial growth factor.

Competing interests

The authors declare that they have no competing interests.

Authors' contributions

SR: conception and design, collection and assembly of data, data analysis and interpretation, manuscript writing, final approval of the manuscript. AS: collection and assembly of data, financial support, data analysis and interpretation, manuscript writing, final approval of the manuscript. JE: data analysis and interpretation, final approval of the manuscript. CPP: manuscript writing, final approval of the manuscript. SB: collection and assembly of data, final approval of the manuscript. VZ: data analysis and interpretation, manuscript writing, final approval of the manuscript. TF: Conception and design, financial support, manuscript writing, final approval of the manuscript. All authors read and approved the final manuscript.

Acknowledgements

The authors acknowledge the technical assistance of Hanne Krone Nielsen and Ditte Bech Kristensen. This work was supported in part by funds from Lily Benthine Lunds fond (SR), the Obelske family foundation (TF, AS, and SB), the Svend Andersen foundation (AS and SB) and the SparNord foundation (AS and SB). The funding sources had no influence on study design, collection, analysis, interpretation of data, writing of the report, or the decision to submit the paper for publication.

Author details

[1]Department of Health Science and Technology, Laboratory for Stem Cell Research, Aalborg University, Fredrik Bajers Vej 3B, Aalborg 9220, Denmark. [2]Department of Health Science and Technology, Laboratory for Medical Mass Spectrometry, Aalborg University, Aalborg, Denmark.

References

1. Mattar P, Bieback K. Comparing the immunomodulatory properties of bone marrow, adipose tissue, and birth-associated tissue mesenchymal stromal cells. Front Immunol. 2015;6:560.
2. Rasmussen JG, Frøbert O, Holst-Hansen C, Kastrup J, Baandrup U, Zachar V, et al. Comparison of human adipose-derived stem cells and bone marrow-derived stem cells in a myocardial infarction model. Cell Transplant. 2014;23:195–206.
3. Sadat S, Gehmert S, Song Y-H, Yen Y, Bai X, Gaiser S, et al. The cardioprotective effect of mesenchymal stem cells is mediated by IGF-I and VEGF. Biochem Biophys Res Commun. 2007;363:674–9.
4. Sawada K, Takedachi M, Yamamoto S, Morimoto C, Ozasa M, Iwayama T, et al. Trophic factors from adipose tissue-derived multi-lineage progenitor cells promote cytodifferentiation of periodontal ligament cells. Biochem Biophys Res Commun. 2015;464:299–305.
5. Hassan WU, Greiser U, Wang W. Role of adipose-derived stem cells in wound healing. Wound Repair Regen. 2014;22:313–25.
6. Kapur SK, Katz AJ. Review of the adipose derived stem cell secretome. Biochimie. 2013;95:2222–8.
7. Blazquez R, Sanchez-Margallo FM, de la Rosa O, Dalemans W, Alvarez V, Tarazona R, et al. Immunomodulatory potential of human adipose mesenchymal stem cells derived exosomes on in vitro stimulated T cells. Front Immunol. 2014;5:556.
8. Pascucci L, Alessandri G, Dall'Aglio C, Mercati F, Coliolo P, Bazzucchi C, et al. Membrane vesicles mediate pro-angiogenic activity of equine adipose-derived mesenchymal stromal cells. Vet J. 2014;202:361–6.
9. Lee SH, Jin SY, Song JS, Seo KK, Cho KH. Paracrine effects of adipose-derived stem cells on keratinocytes and dermal fibroblasts. Ann Dermatol. 2012;24:136–43.
10. Song YH, Shon SH, Shan M, Stroock AD, Fischbach C. Adipose-derived stem cells increase angiogenesis through matrix metalloproteinase-dependent collagen remodeling. Integr Biol (Camb). 2016;8(2):205-15. doi:10.1039/c5ib00277j.
11. Demidova-Rice TN, Hamblin MR, Herman IM. Acute and impaired wound healing: pathophysiology and current methods for drug delivery, part 1: normal and chronic wounds: biology, causes, and approaches to care. Adv Skin Wound Care. 2012;25:304–14.
12. Zachar V, Duroux M, Emmersen J, Rasmussen JG, Pennisi CP, Yang S, et al. Hypoxia and adipose-derived stem cell-based tissue regeneration and engineering. Expert Opin Biol Ther. 2011;11:775–86.
13. Madrigal M, Rao KS, Riordan NH. A review of therapeutic effects of mesenchymal stem cell secretions and induction of secretory modification by different culture methods. J Transl Med. 2014;12:260.
14. Lee EY, Xia Y, Kim W-S, Kim MH, Kim TH, Kim KJ, et al. Hypoxia-enhanced wound-healing function of adipose-derived stem cells: increase in stem cell proliferation and up-regulation of VEGF and bFGF. Wound Repair Regen. 2009;17:540–7.
15. Xu L, Wang X, Wang J, Liu D, Wang Y, Huang Z, et al. Hypoxia-induced secretion of IL-10 from adipose-derived mesenchymal stem cell promotes growth and cancer stem cell properties of Burkitt lymphoma. Tumour Biol. 2015. doi:10.1007/s13277-015-4664-8.
16. Rasmussen JG, Frøbert O, Pilgaard L, Kastrup J, Simonsen U, Zachar V, et al. Prolonged hypoxic culture and trypsinization increase the pro-angiogenic potential of human adipose tissue-derived stem cells. Cytotherapy. 2011;13:318–28.
17. He J, Cai Y, Luo L-M, Liu H-B. Hypoxic adipose mesenchymal stem cells derived conditioned medium protects myocardial infarct in rat. Eur Rev Med Pharmacol Sci. 2015;19:4397–406.
18. Qin HH, Filippi C, Sun S, Lehec S, Dhawan A, Hughes RD. Hypoxic preconditioning potentiates the trophic effects of mesenchymal stem cells on co-cultured human primary hepatocytes. Stem Cell Res Ther. 2015;6:237.
19. Pilgaard L, Lund P, Duroux M, Lockstone H, Taylor J, Emmersen J, et al. Transcriptional signature of human adipose tissue-derived stem cells (hASCs) preconditioned for chondrogenesis in hypoxic conditions. Exp Cell Res. 2009;315:1937–52.
20. Dudakovic A, Camilleri E, Riester SM, Lewallen EA, Kvasha S, Chen X, et al. High-resolution molecular validation of self-renewal and spontaneous differentiation in clinical-grade adipose-tissue derived human mesenchymal stem cells. J Cell Biochem. 2014;115:1816–28.
21. Bath C, Muttuvelu D, Emmersen J, Vorum H, Hjortdal J, Zachar V. Transcriptional dissection of human limbal niche compartments by massive parallel sequencing. PLoS One. 2013;8:e64244.
22. Muers M. Gene expression: transcriptome to proteome and back to genome. Nat Rev Genet. 2011;12:518–8.
23. Vogel C, Marcotte EM. Insights into the regulation of protein abundance from proteomic and transcriptomic analyses. Nat Rev Genet. 2012;13:227–32.
24. Kupcova Skalnikova H. Proteomic techniques for characterisation of mesenchymal stem cell secretome. Biochimie. 2013;95:2196–211.
25. Kasap M, Yeğenağa I, Akpinar G, Tuncay M, Aksoy A, Karaoz E. Comparative proteome analysis of hAT-MSCs isolated from chronic renal failure patients with differences in their bone turnover status. PLoS One. 2015;10:e0142934.
26. Wiśniewski JR, Zougman A, Nagaraj N, Mann M. Universal sample preparation method for proteome analysis. Nat Methods. 2009;6:359–62.
27. Cox J, Hein MY, Luber CA, Paron I, Nagaraj N, Mann M. Accurate proteome-wide label-free quantification by delayed normalization and maximal peptide ratio extraction, termed MaxLFQ. Mol Cell Proteomics. 2014;13:2513–26.
28. Frazier TP, Gimble JM, Kheterpal I, Rowan BG. Impact of low oxygen on the secretome of human adipose-derived stromal/stem cell primary cultures. Biochimie. 2013;95(12):2286-96. doi:10.1016/j.biochi.2013.07.011.
29. Zvonic S, Lefevre M, Kilroy G, Floyd ZE, DeLany JP, Kheterpal I, et al. Secretome of primary cultures of human adipose-derived stem cells: modulation of serpins by adipogenesis. Mol Cell Proteomics. 2007;6:18–28.
30. Lee MJ, Kim J, Kim MY, Bae Y, Ryu SH, Lee TG, et al. Proteomic analysis of tumor necrosis factor-alpha-induced secretome of human adipose tissue-derived mesenchymal stem cells. J Proteome Res. 2010;9:1754–62.
31. Eichelbaum K, Krijgsveld J. Exocytosis and endocytosis, vol. 1174. New York, NY: Springer New York; 2014.
32. Kulkarni GV, McCulloch CA. Serum deprivation induces apoptotic cell death in a subset of Balb/c 3 T3 fibroblasts. J Cell Sci. 1994;107(Pt 5):1169–79.
33. Endrich MM, Grossenbacher D, Geistlich A, Gehring H. Apoptosis-induced concomitant release of cytosolic proteins and factors which prevent cell death. Biol Cell. 1996;88:15–22.
34. Urs S, Turner B, Tang Y, Rostama B, Small D, Liaw L. Effect of soluble Jagged1-mediated inhibition of Notch signaling on proliferation and differentiation of an adipocyte progenitor cell model. Adipocyte. 2012;1:46–57.
35. Tratwal J, Mathiasen AB, Juhl M, Brorsen SK, Kastrup J, Ekblond A. Influence of vascular endothelial growth factor stimulation and serum deprivation on gene activation patterns of human adipose tissue-derived stromal cells. Stem Cell Res Ther. 2015;6:62.

36. Riis S, Zachar V, Boucher S, Vemuri MC, Pennisi CP, Fink T. Critical steps in the isolation and expansion of adipose-derived stem cells for translational therapy. Expert Rev Mol Med. 2015;17:e11.

37. Patrikoski M, Juntunen M, Boucher S, Campbell A, Vemuri MC, Mannerström B, et al. Development of fully defined xeno-free culture system for the preparation and propagation of cell therapy-compliant human adipose stem cells. Stem Cell Res Ther. 2013;4:27.

38. Yang S, Pilgaard L, Chase LG, Boucher S, Vemuri MC, Fink T, et al. Defined xenogeneic-free and hypoxic environment provides superior conditions for long-term expansion of human adipose-derived stem cells. Tissue Eng Part C Methods. 2012;18:593–602.

39. Zachar V, Rasmussen JG, Fink T. Isolation and growth of adipose tissue-derived stem cells. In: Methods in molecular biology. Clifton, N.J.: The Royal Society of Chemistry, vol. 698. 2011. p. 37–49.

40. Fink T, Rasmussen J, Lund P, Pilgaard L, Soballe K, Zachar V. Isolation and expansion of adipose-derived stem cells for tissue engineering. Front Biosci (Elite Ed). 2011;3:256–63.

41. Pilgaard L, Lund P, Rasmussen J, Fink T, Zachar V. Comparative analysis of highly defined proteases for the isolation of adipose tissue-derived stem cells. Regen Med. 2008;3:705–15.

42. Pilgaard L, Lund P, Duroux M, Fink T, Ulrich-Vinther M, Søballe K, et al. Effect of oxygen concentration, culture format and donor variability on in vitro chondrogenesis of human adipose tissue-derived stem cells. Regen Med. 2009;4:539–48.

43. Andersen JI, Juhl M, Nielsen T, Emmersen J, Fink T, Zachar V, et al. Uniaxial cyclic strain enhances adipose-derived stem cell fusion with skeletal myocytes. Biochem Biophys Res Commun. 2014;450:1083–8.

44. Prasad M, Zachar V, Fink T, Pennisi CP. Moderate hypoxia influences potassium outward currents in adipose-derived stem cells. PLoS One. 2014; 9:e104912.

45. Fink T, Lund P, Pilgaard L, Rasmussen JG, Duroux M, Zachar V. Instability of standard PCR reference genes in adipose-derived stem cells during propagation, differentiation and hypoxic exposure. BMC Mol Biol. 2008;9:98.

46. Foldberg S, Petersen M, Fojan P, Gurevich L, Fink T, Pennisi CP, et al. Patterned poly(lactic acid) films support growth and spontaneous multilineage gene expression of adipose-derived stem cells. Colloids Surf B Biointerfaces. 2012;93:92–9.

47. Dominici M, Le Blanc K, Mueller I, Slaper-Cortenbach I, Marini F, Krause D, et al. Minimal criteria for defining multipotent mesenchymal stromal cells. The International Society for Cellular Therapy position statement. Cytotherapy. 2006;8:315–7.

48. Pennisi CP, Dolatshahi-Pirouz A, Foss M, Chevallier J, Fink T, Zachar V, et al. Nanoscale topography reduces fibroblast growth, focal adhesion size and migration-related gene expression on platinum surfaces. Colloids Surf B Biointerfaces. 2011;85:189–97.

49. Stensballe A, Andersen S, Jensen ON. Characterization of phosphoproteins from electrophoretic gels by nanoscale Fe(III) affinity chromatography with off-line mass spectrometry analysis. Proteomics. 2001;1:207–22.

50. Deeb SJ, D'Souza RCJ, Cox J, Schmidt-Supprian M, Mann M. Super-SILAC allows classification of diffuse large B-cell lymphoma subtypes by their protein expression profiles. Mol Cell Proteomics. 2012;11:77–89.

51. Maere S, Heymans K, Kuiper M. BiNGO: a Cytoscape plugin to assess overrepresentation of gene ontology categories in biological networks. Bioinformatics. 2005;21:3448–9.

52. Shannon P, Markiel A, Ozier O, Baliga NS, Wang JT, Ramage D, et al. Cytoscape: a software environment for integrated models of biomolecular interaction networks. Genome Res. 2003;13:2498–504.

53. Riis S, Nielsen FM, Pennisi CP, Zachar V, Fink T. Comparative analysis of media and supplements on initiation and expansion of adipose-derived stem cells. Stem Cells Transl Med. 2016;5:314–24.

54. Bennike TB, Kastaniegaard K, Padurariu S, Gaihede M, Birkelund S, Andersen V, et al. Comparing the proteome of snap frozen, RNAlater preserved, and formalin-fixed paraffin-embedded human tissue samples. EuPA Open Proteomics. 2016;10:9–18.

55. Follin B, Tratwal J, Haack-Sørensen M, Elberg JJ, Kastrup J, Ekblond A. Identical effects of VEGF and serum-deprivation on phenotype and function of adipose-derived stromal cells from healthy donors and patients with ischemic heart disease. J Transl Med. 2013;11:219.

56. Hawkridge AM. Practical considerations and current limitations in quantitative mass spectrometry-based proteomics. In: Quantitative proteomics. 2014. p. 1–25.

57. Kalinina N, Kharlampieva D, Loguinova M, Butenko I, Pobeguts O, Efimenko A, et al. Characterization of secretomes provides evidence for adipose-derived mesenchymal stromal cells subtypes. Stem Cell Res Ther. 2015;6:221.

58. Beck M, Schmidt A, Malmstroem J, Claassen M, Ori A, Szymborska A, et al. The quantitative proteome of a human cell line. Mol Syst Biol. 2011;7:549.

59. Nagaraj N, Wisniewski JR, Geiger T, Cox J, Kircher M, Kelso J, et al. Deep proteome and transcriptome mapping of a human cancer cell line. Mol Syst Biol. 2011;7:548.

60. Gallien S, Kim SY, Domon B. Large-scale targeted proteomics using internal standard triggered-parallel reaction monitoring (IS-PRM). Mol Cell Proteomics. 2015;14:1630–44.

61. Gilkes DM, Bajpai S, Chaturvedi P, Wirtz D, Semenza GL. Hypoxia-inducible factor 1 (HIF-1) promotes extracellular matrix remodeling under hypoxic conditions by inducing P4HA1, P4HA2, and PLOD2 expression in fibroblasts. J Biol Chem. 2013;288:10819–29.

62. Kühl T, Mezger M, Hausser I, Handgretinger R, Bruckner-Tuderman L, Nyström A. High local concentrations of intradermal MSCs restore skin integrity and facilitate wound healing in dystrophic epidermolysis bullosa. Mol Ther. 2015;23:1368–79.

63. Deschene K, Céleste C, Boerboom D, Theoret CL. Hypoxia regulates the expression of extracellular matrix associated proteins in equine dermal fibroblasts via HIF1. J Dermatol Sci. 2012;65:12–8.

64. Rasmussen JG, Riis SE, Frøbert O, Yang S, Kastrup J, Zachar V, et al. Activation of protease-activated receptor 2 induces VEGF independently of HIF-1. PLoS One. 2012;7:e46087.

65. Taylor CT. Mitochondria and cellular oxygen sensing in the HIF pathway. Biochem J. 2008;409:19–26.

66. Efimenko A, Starostina E, Kalinina N, Stolzing A. Angiogenic properties of aged adipose derived mesenchymal stem cells after hypoxic conditioning. J Transl Med. 2011;9:10.

67. Stubbs SL, Hsiao ST-F, Peshavariya HM, Lim SY, Dusting GJ, Dilley RJ. Hypoxic preconditioning enhances survival of human adipose-derived stem cells and conditions endothelial cells in vitro. Stem Cells Dev. 2012;21:1887–96.

68. Liu L, Gao J, Yuan Y, Chang Q, Liao Y, Lu F. Hypoxia preconditioned human adipose derived mesenchymal stem cells enhance angiogenic potential via secretion of increased VEGF and bFGF. Cell Biol Int. 2013;37:551–60.

69. Rehman J, Traktuev D, Li J, Merfeld-Clauss S, Temm-Grove CJ, Bovenkerk JE, et al. Secretion of angiogenic and antiapoptotic factors by human adipose stromal cells. Circulation. 2004;109:1292–8.

70. Hollenbeck ST, Senghaas A, Komatsu I, Zhang Y, Erdmann D, Klitzman B. Tissue engraftment of hypoxic-preconditioned adipose-derived stem cells improves flap viability. Wound Repair Regen. 2012;20:872–8.

71. Park B-S, Kim W-S, Choi J-S, Kim H-K, Won J-H, Ohkubo F, et al. Hair growth stimulated by conditioned medium of adipose-derived stem cells is enhanced by hypoxia: evidence of increased growth factor secretion. Biomed Res. 2010;31:27–34.

72. Lin Q, Lee Y-J, Yun Z. Differentiation arrest by hypoxia. J Biol Chem. 2006; 281:30678–83.

73. Ebbesen P, Eckardt K-U, Ciampor F, Pettersen EO. Linking measured intercellular oxygen concentration to human cell functions. Acta Oncol. 2004;43:598–600.

74. Vizcaíno JA, Deutsch EW, Wang R, Csordas A, Reisinger F, Ríos D, et al. ProteomeXchange provides globally coordinated proteomics data submission and dissemination. Nat Biotechnol. 2014;32:223–6.

5

Immunomodulation by mesenchymal stem cells in treating human autoimmune disease-associated lung fibrosis

Ming Liu[1†], Xiansheng Zeng[1,3†], Junli Wang[1†], Zhiping Fu[1†], Jinsong Wang[1,4†], Muyun Liu[4], Dunqiang Ren[1], Baodan Yu[1], Lixia Zheng[1], Xiang Hu[4], Wei Shi[2] and Jun Xu[1*]

Abstract

Background: Interstitial pneumonia in connective tissue diseases (CTD-IP) featuring inflammation and fibrosis is a leading cause of death in CTD-IP patients. The related autoimmune lung injury and disturbed self-healing process make conventional anti-inflammatory drugs ineffective. Equipped with unique immunoregulatory and regenerative properties, mesenchymal stem cells (MSCs) may represent a promising therapeutic agent in CTD-IP. In this study, we aim to define the immunopathology involved in pulmonary exacerbation during autoimmunity and to determine the potential of MSCs in correcting these disorders.

Methods: Lung and blood specimens, bronchoalveolar lavage fluid cells collected from CTD-IP patients, and human primary lung fibroblasts (HLFs) from patients pathologically diagnosed with usual interstitial pneumonia (UIP) and healthy controls were analyzed by histology, flow cytometry and molecular biology. T cell subsets involved in the process of CTD-IP were defined, while the regulatory functions of MSCs isolated from the bone marrow of normal individuals (HBMSCs) on cytotoxic T cells and CTD-UIP HLFs were investigated in vitro.

Results: Higher frequencies of cytotoxic T cells were observed in the lung and peripheral blood of CTD-IP patients, accompanied with a reduced regulatory T cell (Treg) level. CTD-UIP HLFs secreted proinflammatory cytokines in combination with upregulation of α-smooth muscle actin (α-SMA). The addition of HBMSCs in vitro increased Tregs concomitant with reduced cytotoxic T cells in an experimental cell model with dominant cytotoxic T cells, and promoted Tregs expansion in T cell subsets from patients with idiopathic pulmonary fibrosis (IPF). HBMSCs also significantly decreased proinflammatory chemokine/cytokine expression, and blocked α-SMA activation in CTD-UIP HLFs through a TGF-β1-mediated mechanism, which modulates excessive IL-6/STAT3 signaling leading to IP-10 expression. MSCs secreting a higher level of TGF-β1 appear to have an optimal anti-fibrotic efficacy in BLM-induced pulmonary fibrosis in mice.

Conclusions: Impairment of TGF-β signal transduction relevant to a persistent IL-6/STAT3 transcriptional activation contributes to reduction of Treg differentiation in CTD-IP and to myofibroblast differentiation in CTD-UIP HLFs. HBMSCs can sensitize TGF-β1 downstream signal transduction that regulates IL-6/STAT3 activation, thereby stimulating Treg expansion and facilitating anti-fibrotic IP-10 production. This may in turn block progression of lung fibrosis in autoimmunity.

Keywords: Mesenchymal stem cells, Regulatory T cells, Natural killer T cells, Lung fibrosis in autoimmunity, TGF-β1, IL-6, IP-10

* Correspondence: xufeili@vip.163.com
†Equal contributors
[1]State Key Laboratory of Respiratory Diseases, Guangzhou Institute of Respiratory Diseases, The First Affiliated Hospital of Guangzhou Medical University, Guangzhou Medical University, Guangzhou, P. R. China
Full list of author information is available at the end of the article

Background

Interstitial pneumonia (IP) is a heterogeneous group of lung parenchymal disorders, with common pathological features of inflammation and/or fibrosis. Fibrosis in IP patients is often irreversible, resulting in significant morbidity and mortality [1]. IP can be idiopathic (idiopathic pulmonary fibrosis, IPF) or secondary to exposure to a variety of harmful environmental factors. Although the pathogenesis of IP is not yet clear, a subgroup of IP is associated with connective tissue diseases (CTD-IP), including multiple sclerosis, rheumatoid arthritis (RA) and polymyositis/dermatomyositis (PM/DM) [2, 3]. The pathological features of CTD-IP can be nonspecific IP (NSIP), usual IP (UIP), cryptogenic organizing pneumonia (COP), acute interstitial pneumonia and diffuse alveolar damage. The frequency of IP in these CTDs varies, ranging from 20 % to more than 50 % and presenting either before or after these CTDs are diagnosed. More importantly, IP pathologically diagnosed UIP, in particular, is a leading cause of death in these patients. There is no effective treatment currently available, although immunosuppressive and anti-inflammatory drugs, such as corticosteroids, have been widely used.

Recent studies have reported that local and systemic immune activation and impairment of immunological tolerance were detected in CTD-IP patients [4–10]. For example, RA patients had a greater number of CD4-positive T cells in the bronchoalveolar lavage (BAL) fluid than IPF patients [11]. Increased autoantibodies against topoisomerase and Jo-1 were strongly associated with development of IP in multiple sclerosis and PM/DM patients, respectively [12, 13]. Abnormalities in T cells, including T regulatory cells (Tregs) in autoimmunity may play an important role in pulmonary fibrosis in CTD-IP [9]. However, it is still unclear which subsets of immune cells are involved in pulmonary fibrosis and how they affect the development of disease [10, 14], although it is conceivable that dysregulation of the immune system may be an important factor contributing to CTD-IP. Therefore, the characterization of these immunological changes at the molecular and cellular levels in CTD-IP patients and the discovery of novel approaches to correcting these changes will be critical for treating CTD-IP in the future [15, 16].

The immunomodulatory properties of mesenchymal stem cells (MSCs) have recently caused excitement for investigators examining their potential therapeutic application in a variety of immune disorder diseases [17, 18]. MSCs have been tested in rodent models to treat diseases where immunodysregulation is thought to be the main pathogenic mechanism. It has been shown that MSCs can reverse autoimmune response disorder by modulating multiple subsets of immune cells [19]. In addition, their pluripotent nature may also benefit CTD-IP patients by directly or indirectly promoting alveolar repair [20]. Recent studies have demonstrated the capability of MSCs to inhibit bleomycin-induced pneumonitis and fibrosis in a mouse model [21]. However, it has been argued that bleomycin-induced pulmonary fibrosis in a mouse system does not reflect all of the immunological mechanisms involved in human CTD-IP or IPF. Herein, we have characterized the main features of the immune disorder in CTD-IP patients at the active stage of the disease. We found that persistent activation of an innate immune response by high frequency natural killer T cells (NKTs) in the circulation and lung was linked to the promotion of CTD-IP, where the pulmonary fibroblasts obtained a myofibroblast phenotype that persisted in the culture. Using an experimental NKT-peripheral blood mononuclear cell (PBMCs) model in vitro and isolated primary lung fibroblasts from CTD-IP patients pathologically diagnosed with usual interstitial pneumonia (UIP), we demonstrate that MSCs have great potential to inhibit fibrotic development in CTD-IP by sensitization of attenuated TGF-β1 downstream signal, which, in turn, exerts anti-inflammatory and anti-fibrotic effects.

Methods

Subjects

A total of 28 CTD- IP patients (12 patients with rheumatoid arthritis (RA)-IP and 16 patients with polymyositis/dermatomyositis (PM/DM)-IP) who were hospitalized in the affiliated Hospital of Guangzhou Medical University from January 2010 to March 2013 were enrolled in this study, and 23 healthy volunteers were used as control subjects. All patients met the interstitial lung disease and connective tissue disorder-related criteria [22]. The study protocol was approved by the Ethics Committee of the First Affiliated Hospital of Guangzhou Medical University, and informed consent was obtained from all patients and control subjects. Flow cytometry was performed on peripheral blood (PB) and bronchoalveolar lavage fluid (BAL) samples. Baseline characteristics of the studied patients are shown in Tables 1 and 2.

Table 1 Subject characteristics ($n = 51$)[a]

Variable	Subjects	Control (N, %)	CTD-IP (N, %)	P-Value
Total	51	23 (45.1)	28 (54.9)	
Age				0.723
≥50 years	28	12 (42.9)	16 (57.1)	
<50 years	23	11 (47.8)	12 (52.2)	
Gender				0.654
Male	22	10 (45.5)	12 (54.5)	
Female	29	13 (44.8)	16 (55.2)	

[a]Data are expressed as number (percentage)

Table 2 Clinical summary of patients with CTD-IP ($n = 28$)[a]

Characteristics	Data
Age	52 (25–78)
Male gender	12
Female gender	16
Smoking history	
Never	15
Former	6
Current	7
CTD history	28
DM/PM	16
RA	12
Abnormal RF/CRP level	25
Time since CTD diagnosis, yr	4.10 (2–11)
Time since CTD-IP diagnosis, yr	3.48 (0.5–10)

Abbreviations: *CTD* connective tissue disease, *DM/PM* dermatomyositis/
polymyositis, *RA* rheumatoid arthritis, *RF/CRP* rheumatoid
factor/C-reactive protein
[a]Data are expressed as the median (interquartile range) or number

Lung histology and immunohistochemistry
Human lung paraffin sections prepared from lung biopsy specimens of the enrolled patients were stained with hematoxylin and eosin (H&E) for histopathology. Collagen was stained using the Masson trichrome method (Maixin-bio, China). Immunostaining was performed as previously described [23], using antibodies against α smooth muscle actin (α-SMA) (1:400, A2547, Sigma, St Louis, MO, USA) and CD3 (1:100, ab5690, Abcam, Cambridge, UK).

Culture of human lung fibroblasts
Primary human lung fibroblasts (HLFs) were prepared from the lung biopsies of CTD-IP patients ($n = 4$) pathologically diagnosed with usual interstitial pneumonia (UIP). Primary normal human lung fibroblasts (NHLFs) derived from normal tissue areas of surgical lobectomy specimens taken from patients with lung cancer were used as a negative control. Cell culture was performed according to the Primary Lung Fibroblast Culture protocol given in Additional file 1: Methods.

After serum starvation for 24 h, NHLFs were treated with TGF-β1, IL-6 alone or in combination and cytomix (a mixture of TGF-β1, IFN-γ, and IL-1β (all from R&D Systems)) for 48 hours.

Preparation of human bone marrow mesenchymal stem cells
Human bone marrow mesenchymal stem cells (HBMSCs) were isolated from the bone marrow of normal individuals undergoing bone marrow harvest for allogeneic bone marrow transplantation. Informed consent was obtained and the study protocol was approved by the Ethics Committee of the First Affiliated Hospital of Guangzhou Medical University. MSCs derived from umbilical cord (UC) were also isolated. MSCs culture and verification were performed as described in the figure in Additional file 2.

Generation of natural killer T cell-peripheral blood mononuclear cells
Peripheral blood was provided by the Guangzhou Blood Center after approval was given by the Department of Health of Guangdong Province. The generation and identification of natural killer T cell-peripheral blood mononuclear cells (NKT-PBMCs) were performed as described in the figure in Additional file 3.

Cell co-culture
Co-culture of HBMSCs and PBMCs
Cultured HBMSCs or NHLF were added to NKT-PBMCs, PBMC from healthy controls, and IPF patients ($n = 12$) at a 1:20 ratio for 24 or 48 hours. After that, the treated PBMCs were collected for flow cytometry analysis.

Co-culture of HBMSCs and HLFs
HBMSCs were co-cultured with NHLF or CTD-UIP-HLF at a 1:1 ratio using Transwell chambers (Corning, Tewksbury, MA, USA). HBMSCs were plated into the upper chamber, and NHLF or CTD-UIP-HLFs were plated into the lower chamber. CTD-UIP HLF were treated with MSC or TGF-β1 in the absence and presence of neutralizing antibody for either human IP-10 (2 µg/ml) (C) or human TGF-β1(1 µg/ml). The entire culture system was maintained for 48 hours in an incubator containing 5 % CO_2, then NHLFs or CTD-IP-HLFs were lysed for western blot analysis.

Flow cytometry
NKT-PBMCs and whole peripheral blood samples from healthy controls and CTD-IP patients were stained with the following antibodies: CD3-FITC, CD56-PE, CD127-PE, CD45-ECD, CD4-FITC, CD25-PC5, CD4-FITC/CD8-PE/CD3-PC5, FOXP3–PE, and appropriate isotype controls (Beckman Coulter, Indianapolis, IN, USA). Staining was performed according to the manufacturer's instructions.

Western blot
Protein expression and phosphorylation were determined by western blot, as previously described [23]. Briefly, cells were lysed in radioimmunoprecipitation (RIPA) buffer, then subjected to polyacrylamide gel electrophoresis and incubated with primary antibodies at 4 °C overnight, then incubated with secondary antibodies and developed by chemiluminescence reaction (Pierce). Digital chemiluminescent images were obtained and quantified with a Kodak image station 4000R system. Primary antibodies used in

this study were anti-fibronectin (Santa Cruz Biotechnology), anti-vimentin (Santa Cruz Biotechnology), anti-α-SMA antibody (Sigma), anti-STAT3, anti-phosphorylated STAT3, and anti-phosphorylated Smad3 (Cell Signaling).

ELISA and liquid microarray assay

Human TGF-β1 secreted from the cultured cells into medium was measured using an ELISA kit (R&D Systems, Minneapolis, MN, USA). The levels of the cytokines interferon γ (IFN-γ), tumor necrosis factor α (TNF-α), interleukin 8 (IL-8), IL-6, macrophage inflammatory protein-1α (MIP-1α), monocyte chemoattractant protein-1 (MCP-1), MCP-3, IFN-γ-inducible protein 10 (IP-10), and vascular cell adhesion molecule-1 (VCAM-1) were determined by a liquid microarray assay using Luminex technology (Merck Millipore, Billerica, MA, USA).

Animals and experimental groups

C57BL/6 mice aged 8 weeks ($n = 80$) (Guangdong Medical Laboratory Animal Center, China) were randomly divided into four groups: control group (mice treated with saline solution), BLM group (mice challenged with BLM), and MSC treatment groups (treatment of mice with the supernatant from human MSC-BM or MSC-UC). A BLM-induced lung fibrosis mouse model was induced as described [24] by intratracheal addition of 3 U/kg body weight BLM (Nippon Kayaku Co., Ltd. Japan). Supernatants harvested from MSCs (1×10^6) culture were concentrated and intratracheally added to the mouse model 48 hours after BLM administration. Survival rates and lung histological sections were analyzed in mice 21 days after BLM exposure. All animal study protocols were reviewed and approved by the University Committee on Use and Care of Animals at Guangzhou Medical University.

Statistical analysis

All data are expressed as the mean ± SD. Statistical differences between different groups were evaluated using the Student's t test. All analysis was performed using the SPSS 10.0 software package (SPSS, Chicago, IL, USA). A P-value of $P \leq 0.05$ was considered as statistically significant.

Results

Pulmonary interstitial inflammation and fibrosis in CTD-IP patients are accompanied by significantly increased numbers of NKT cells

The histopathology of lung tissue biopsy specimens from healthy controls (Fig. 1a, b) and enrolled CTD-IP patients ($n = 6$) was examined after H&E staining (Fig. 1d, e). Subacute alveolar damage accompanied by patchy alveolar pneumocyte hyperplasia and capillary remodeling was consistently observed (Fig. 1d, e). Moreover, diffuse chronic inflammation and fibrosis were detected in lung parenchyma, resulting in thickened interstitial spaces with accumulation of myofibroblasts and extracellular matrix, especially collagen (Fig. 1d, f and g). By immunostaining, the majority of infiltrated CD3$^+$ T cells were detected in the airway and pulmonary interstitial spaces, as well as lymphoid follicles (Fig. 1h). Furthermore, analyses of inflammatory cells in patients' BAL fluids by flow cytometry showed that more than 85 % of the leucocytes were CD3$^+$ T cells, including CD8$^+$ T cells, CD3$^+$ CD56$^+$ NKT cells and CD4$^+$ T cells (Fig. 1i).

Correlations of the aberrant T subsets and cytokine profiles in the systemic circulation for the impaired pulmonary function

We next determined if the altered lymphocyte profiles also occurred in the systemic circulation of the CTD-IP patients using flow cytometry (Fig. 2). By comparing CTD-IP patients ($n = 28$) with the normal control group ($n = 23$), we found that CD3$^+$ CD56$^+$ NKT-like cells were significantly increased in the peripheral blood of CTD-IP patients (Fig. 2a and d, 6.26 ± 2.74 % in CTD-IP vs. 3.65 ± 1.27 % in controls, $P = 0.003$). Meanwhile, elevation of CD3$^+$ CD8$^+$ cells (29.96 ± 7.62 % in CTD-IP vs. 26.40 ± 4.78 % in control, $P = 0.048$) and reduction of CD3$^+$ CD4$^+$ cells (32.23 ± 6.95 % in CTD-IP vs. 35.71 ± 4.69 % in control, $P = 0.046$) were also detected (Fig. 2a–c). In addition, a reduced number of CD4$^+$ CD25$^+$ FOXP3$^+$ Tregs was observed in the CTD-IP patients compared with normal controls (7.32 ± 2.21 % in CTD-IP vs. 8.36 ± 1.81 % in control, $P = 0.035$), as shown in Fig. 2e.

We then asked if the cytokine profile in the patients' peripheral blood exhibited corresponding changes, which were involved in pulmonary fibrotic development in autoimmunity. As predicted, we detected significantly increased production of pro-inflammatory/fibrotic cytokines, including IL-6, IFN-γ, TNFα, and TGF-β1 in CTD-ILD patients compared with that in normal controls. The augmentation of IL-6 level, rather than TGF-β1, has a negative correlation with a lung function parameter, forced vital capacity (FVC) (Fig. 3a, b), corresponding to a decreased TGF-β1/IL-6 ratio relevant to down-regulation in the Tregs level, which is closely correlated with the declining FVC (Fig. 3c, d). High levels of TNF-α and IFN-γ in circulation associated with an increase in the NKT cell level, was also responsible for reduced FVC (Fig. 3e, f, g).

The autoimmune inflammatory microenvironment induces pulmonary myofibroblast differentiation in CTD-IP

We next tested the impact of a mixture of cytokines (cytomix), which have been demonstrated to be significantly increased in peripheral blood in CTD-IP patients, on myofibroblast development. We detected a myofibroblast differentiation with marked over expression

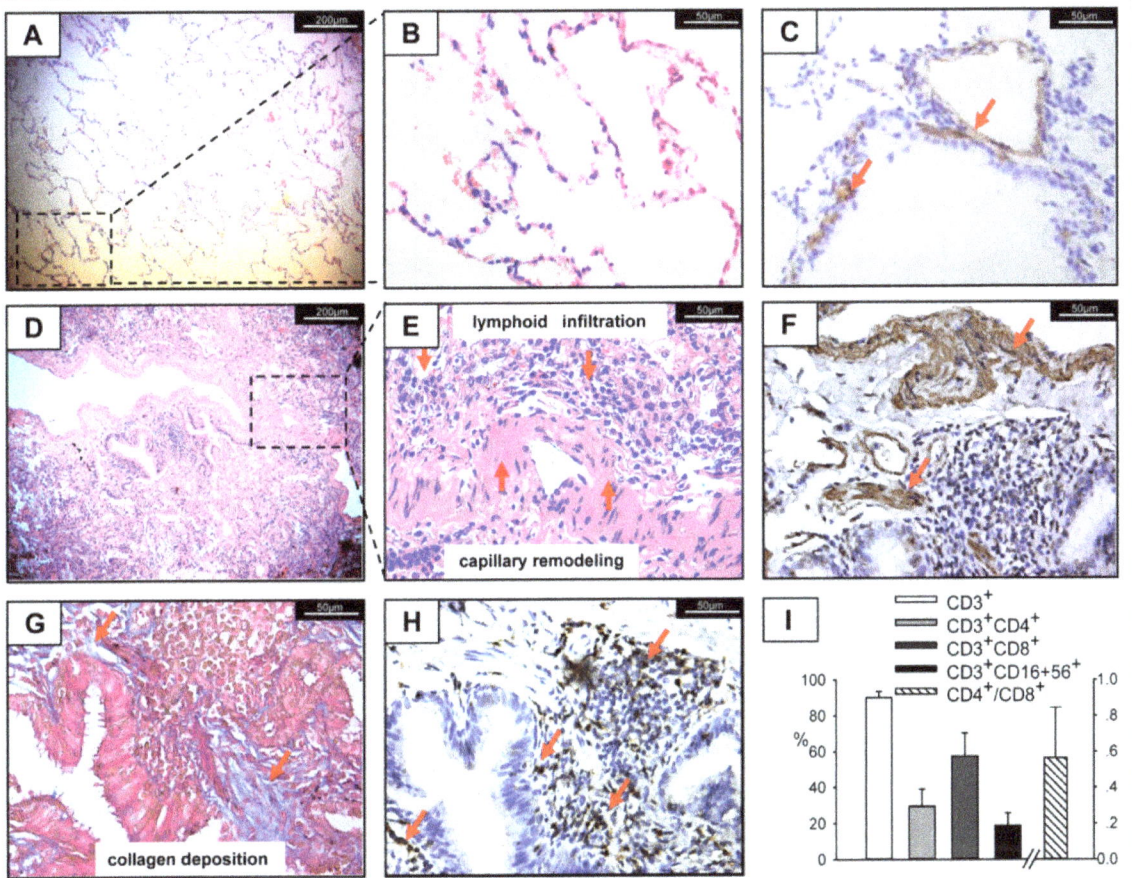

Fig. 1 The frequency of NKT cells is increased in the lung of CTD-IP patients. Representative hematoxylin and eosin (HE) stained lung sections from healthy control (**a**, **b**) and enrolled CTD-IP patients (n = 6) (**d**, **e**) showing areas of sub-acute alveolar damage accompanied by capillary remodeling (**d**, **e**) and lymphoid follicle formation (**d**, **e**) in CTD-IP. Lung sections stained with Masson trichrome (MT) and immunostaining showed increased collagen deposition (blue, **g**), combined with enhanced expression of α-SMA (brown, **f**) in capillaries and interstitial cells compared with healthy control (brown, **c**). Positive CD3 immunostaining was located in the lymphoid follicles (brown, **h**). The *arrows* indicate myofibroblast infiltration with α-SMA-positive staining or T cells with CD3-positive staining. (**a**, **d**) 100× magnification, (**b**), (**c**), (**e**) to (**h**) 400× magnification. **i** Flow cytometric analysis of BALF cells, percentage of CD3$^+$, CD3$^+$ CD4$^+$, CD3$^+$ CD8$^+$, CD3$^+$ CD56$^+$ cells gating on leucocytes and CD8$^+$/CD4$^+$ are presented, and the means ± SD of six cases are shown. *α-SMA* α-smooth muscle actin, *BALF* bronchoalveolar lavage fluid

of αSMA, vimentin, and fibronectin in the normal lung fibroblasts (NHLFs) after exposure to cytomix (Fig. 4a). Low dosage IL-6 addition enhances TGF-β1-induced myofibroblast activation, whereas administration of IL-6 alone can also induce myofibroblast differentiation in a concentration-dependent manner (Fig. 4b).

We observed significantly elevated release of proinflammatory cytokines, including IL-6, IL-8, MIP-1α, MCP-1, MCP-3, VCAM-1 and MIP-1β, from lung fibroblasts (HLFs) derived from CTD-IP patients (n = 4) with pathologically diagnosed usual interstitial pneumonia (UIP) (CTD-UIP HLFs), compared with NHLFs (Fig. 4c, P < 0.05 or P < 0.01). In contrast, production of the anti-fibrotic cytokine IP-10 was significantly reduced in CTD-UIP HLFs (Fig. 4c, P < 0.01). Surprisingly, TGF-β1, which is an

anti-inflammatory, but profibrotic factor, was slightly reduced in the UIP-HLFs. The combined anti-fibrotic effect as measured by the ratio of IP-10 to TGF-β1 was also decreased (4.58 in NHLFs vs. 2.09 in CTD-UIP HLFs).

HBMSCs induce Tregs expansion in either NKT-PBMCs model or PBMCs isolated from IPF patients

Given that human MSCs are emerging as a therapeutic modality in various inflammatory diseases owing to their immunomodulatory properties [25], we examined the regulatory effect of MSCs on cytotoxic NKT cell induction in an established in vitro system, in which high frequency NKT cells can be induced from fresh peripheral blood mononuclear cells (PBMCs) of healthy volunteers by cytokine treatment [26]. In the present

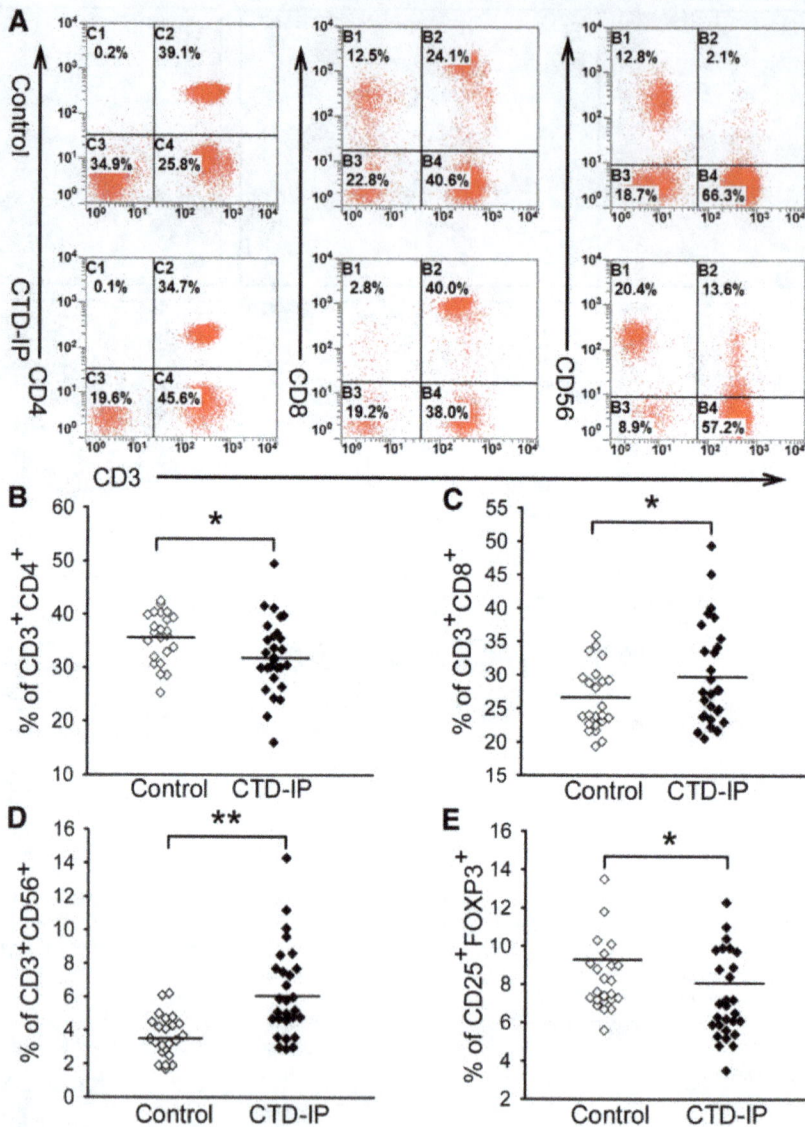

Fig. 2 The frequency of NKT cells in the peripheral blood of CTD-IP patients is increased accompanied by reduction of Tregs. **a** Gating on lymphocytes, flow cytometric analysis of CD3+ CD4+ T cells, CD3+ CD8+ T cells and CD3+ CD56+ cells in the peripheral blood of healthy controls and patients with CTD-IP. **b–e)** scatter plots of the percentage of CD3+ CD4+ T cells, CD3+ CD8+ T cells and CD3+ CD56+ cells gating on lymphocytes, and CD25+ FOXP3+ cells gating on CD4+ cells in the peripheral blood of healthy controls (n = 23) and patients with CTD-IP (n = 28). * P < 0.05, ** P < 0.01 for all comparisons between CTD-IP and control. *NKT* natural killer T cells, *CTD-IP* interstitial pneumonia in connective tissue disease, *Tregs* regulatory T cells

study, CD3+ CD56+ NKT cells were markedly induced (27.3 ± 6.3 %) from PBMCs after cytokine treatment in vitro, compared with less than 5 % of NKT cells in untreated PBMCs. Furthermore, another type of cytotoxic T cell, CD3+ CD8+ T cells, increased 2-fold, while CD3+ CD4+ T cells had a 1-fold reduction in the treated PBMCs compared with untreated PBMCs (Additional file 4: Figure S3). Thus, the alterations of T cell subtypes in cytokine-treated PBMCs in vitro

mimics the changes detected in peripheral blood of CTD-IP patients.

We then investigated the role of human MSCs in modulating T cell subtypes in vitro using the system described above. As shown in Fig. 5b and Additional file 4: Figure S3, co-culture of HBMSCs with NKT-PBMCs in the presence of NKT-inducing agents resulted in a significant reduction in NKT cells from 20.33 ± 1.05 % in the MSC-free control to 15.17 ± 1.75 % with MSC

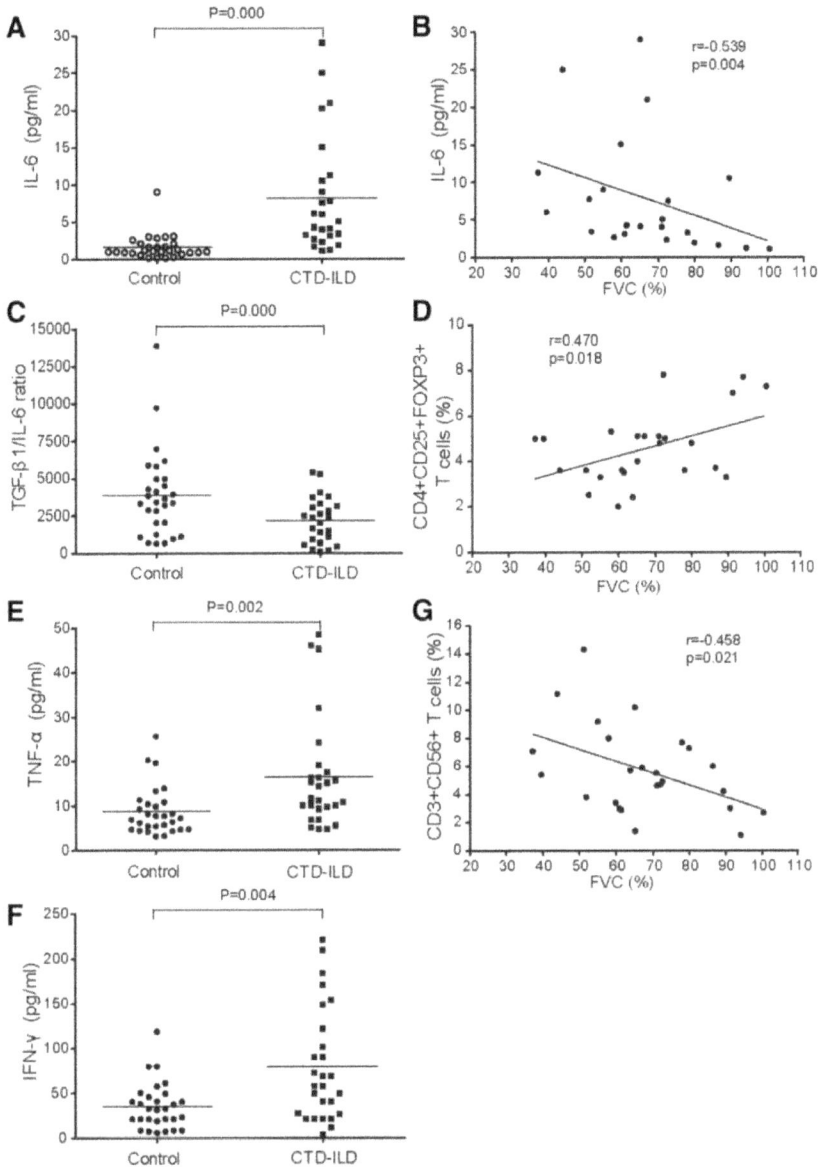

Fig. 3 Correlations of the altered T cell subsets and cytokine profiles with pulmonary functions in the patients with CTD-ILD. **a, c, e, f**) The plasma levels of IL-6, TGF-β/IL-6 ratio, TNF-α, and IFN-γ in the CTD-ILD patients who had not received corticosteroid therapy (n = 27) and healthy control subjects (n = 29). Each point represents one person. The median value for each group is indicated by a *horizontal line*. **b, d, g** Correlations of forced vital capacity (FVC) with the altered T cell subsets and cytokines. **b, d** Correlations of the increased plasma IL-6 level or declining peripheral blood regulatory T cells (Tregs) with worsening FVC. **g** Correlations of the elevation of CD3+CD56+ NKT cells with the reduction of FVC. P values were obtained by Pearson's test. *IL-6* interleukin-6, *TGF-β* transforming growth factor-β, *TNF-α* tumor necrosis factor α, *IFN-γ* interferon γ, *NKT* natural killer T cells, *FVC* forced vital capacity

treatment ($P < 0.05$), and caused a decrease of CD3+ CD8+ T cell induction, but up-regulated CD3+CD4+ and CD4+CD25+ CD127$^{(low/-)}$/foxp3+ T cells, accompanied by significantly diminished IFN-γ and TNF-α, and elevated TGF-β1 and IP-10 in the co-culture supernatants (Fig. 5a $P < 0.01$). A high level of TGF-β1 was also detected in the

culture of HBMSCs alone. The specificity of the MSC's effect was further verified by co-culturing NHLFs with NKT-PBMCs. No effect on NKT cell induction was observed by co-culturing PBMC with NHLFs. Likewise, we confirmed that HBMSCs have the ability to induce Tregs expansion in the IPF patients' PBMCs where there

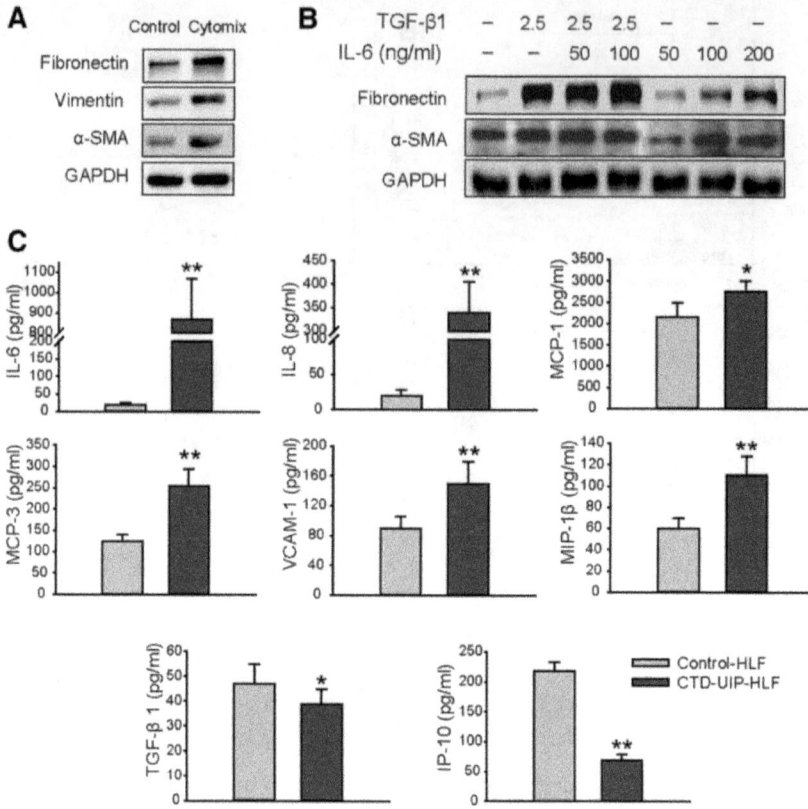

Fig. 4 HLFs differentiation towards myofibroblast after exposure to inflammatory cytomix is linked to the characteristic feature of CTD-UIP HLF's phenotype. **a**, **b** Western blot was performed on normal HLFs treated with cytomix (a mixture of cytokines) (**a**) or TGF-β/IL-6 (**b**) for examination of expression of α-SMA, vimentin, and fibronectin. Data are representative of three independent experiments. **c** Levels of cytokines and chemokines were measured in culture supernatants of human lung fibroblasts (HLF) from patients with CTD-UIP (CTD-UIP HLF) and normal controls (NHLF) using Luminex multiplex technology. Data are representative of two independent experiments. Significance of difference between independent groups of data (mean ± SD) was analyzed by Student's *t* test (two-tailed). * $P < 0.05$, ** $P < 0.01$ for all comparisons between CTD-IP-HLF and NHLF. *CTR-UIP-HLF* HLF isolated from the lung tissues pathologically diagnosed with UIP in CTD-IP patients, *NHLF* normal human lung fibroblasts, *TGF-β* transforming growth factor-β, *IL-6* interleukin-6, *α-SMA* α-smooth muscle actin

was a suppressed Tregs growth compared to normal controls (Fig. 5c).

HBMSCs inhibit the proinflammatory and profibrotic properties of UIP-HLFs through regulation of excessive IL-6 signaling activation

To investigate the role of human MSCs in the modulation of CTD-UIP HLFs, we performed a co-culture of HBMSCs and CTD-UIP HLFs. Similarly, we detected a high level of TGF-β1 in the supernatant either of the co-culture system or HBMSC alone (Fig. 6b, $P < 0.01$), concomitant with marked suppression of IL-6, IL-8, and MCP-1 (Fig. 6a, $P < 0.05$) and a significantly elevated IP-10 secretion in comparison to the co-culture of CTD-UIP HLFs with NHLFs. Co-culture of CTD-UIP HLFs with HBMSCs, but not NHLFs, attenuated α-SMA hyperexpression in the UIP HLFs (Fig. 6c, $P < 0.05$). Furthermore, we found that hyperphosphorylation of

STAT3 attributed to excessive IL-6 secretion in CTD-UIP HLFs was significantly blocked by HBMSC treatment, whereas phosphorylation of Smad3 was slightly upregulated (Fig. 6d).

TGF-β1 hypersecretion in HBMSCs rescues attenuated TGF-β1 downstream signal transduction for induction of expression of anti-fibrotic chemokine IP 10

As TGF-β1 is a profibrotic growth factor that stimulates α-SMA expression and myofibroblast differentiation, we investigated the paradox that TGF-β1 hypersecretion in MSC resulted in an increased level of IP-10 in UIP-HLF and concurrently reduced α-SMA expression. In NHLFs, addition of TGF-β1 elevated α-SMA expression (Fig. 7, $P < 0.05$), accompanied by suppression of IP-10 production (Fig. 7, $P < 0.01$). However, in CTD-UIP HLFs, addition of TGF-β1 significantly increased IP-10 secretion and down-regulated α-SMA

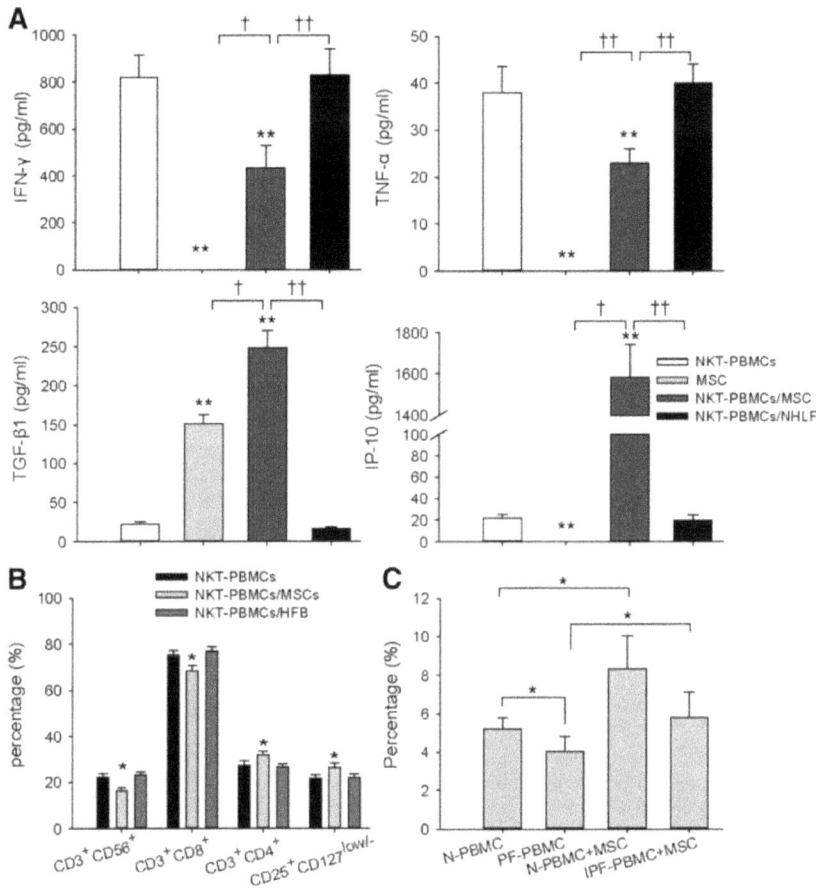

Fig. 5 Immunomodulatory effects of human bone marrow MSCs on aberrant T subsets and cytokines profile. **a, b** NKT-PBMCs were co-cultured with human MSCs or human fibroblasts at a 20:1 ratio of NKT-PBMCs to human MSCs or NHLF prior to cytokines test in the supernatants (**a**) and flow cytometric analysis (**b**) for each group. Triplicate wells were prepared for each group. **a** TNF-α, IFN-γ, TGF-β1, and IP-10 levels in the supernatants of NKT-PBMCs, MSCs, and NKT-PBMCs co-cultured with human bone marrow MSCs or NHLF. ** Significantly different from the NKT-PBMCs group, $P < 0.01$. † $P < 0.05$, †† $P < 0.01$, compared to MSCs or NKT-PBMCs co-cultured with NHLF. Data represent the means ± SD from three independent experiments. **b** Flow cytometric analysis of CD3+ CD56+ cells, CD3+ CD8+ cells, CD3+ CD4+ cells gating on CD45+ cells, and CD25+ CD127(Low/-) Treg cells gating on CD4+ cells, of either NKT-PBMCs (NKT-PBMCs) or NKT- PBMCs co-cultured with human bone MSCs (NKT-PBMCs/MSC, or co-cultured with NHLF (NKT-PBMCs/NHLF). *$P < 0.05$ for comparisons between NKT-PBMCs/MSC and NKT-PBMCs/NHLF or NKT-PBMCs. Data represent the means ± SD from three independent experiments. **c** CD25+ FOXP3+ Treg cells gating on CD4+ cells in the PBMCs of healthy controls and IPF patients ($n = 12$) before and after being co-cultured with MSCs or human fibroblasts. Data represent the means ± SD. *$P < 0.05$. *MSCs* mesenchymal stem cells, *NKT* natural killer T cells, *PBMCs* peripheral blood mononuclear cells, *NHLF* normal human lung fibroblasts, *TNF-α* tumor necrosis factor-α, *IFN-γ* interferon γ, *TGF-β* transforming growth factor-β, *IP-10* interferon γ-induced protein 10

expression (Fig. 7, $P < 0.01$), suggesting that UIP-HLFs have an opposite response to TGF-β1 stimulation compared with NHLFs, and that the negative regulatory effect of IP-10 on α-SMA expression may be downstream of the TGF-β1 pathway.

To elucidate the role of IP-10 elevation induced by TGF-β1-expressing MSCs in modulating UIP-HLFs, a human IP-10-neutralizing antibody (R&D Systems, AF-266-NA) was administrated to HBMSCs, prior to co-culture with CTD-UIP HLFs for 48 hours. The western blot data showed that IP-10 neutralization partly reversed the suppression of α-SMA up-regulation caused by MSC treatment. Similarly, addition of IP-10-neutralizing antibody blocked the efficacy of TGF-β1 administration on CTD-UIP HLFs (Fig. 7c). A consistent result was also observed in HBMSCs treated with TGF-β1-neutralizing antibody, showing that TGF-β1 neutralization in HBMSCs reduced the effect of anti-myofibroblast differentiation on CTD-UIP HLFs (Fig. 7d). This may explain why HBMSCs expressing TGF-β1 have an antifibrotic capability.

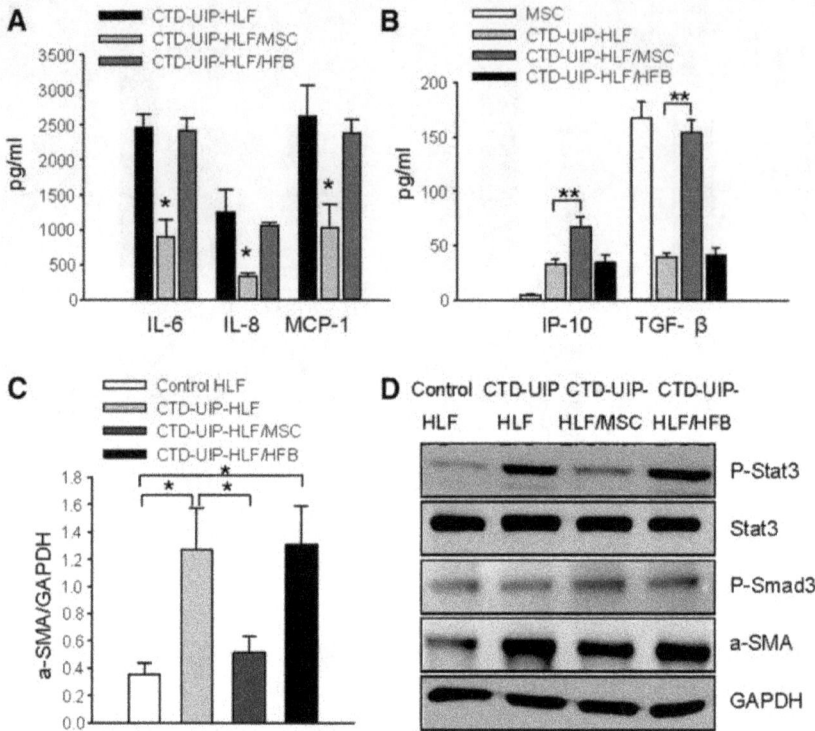

Fig. 6 Immunomodulatory effects of human bone marrow MSCs on CTD-UIP HLFs. **a** IL-6, IL-8, and MCP-1 levels in cultures of CTD-UIP HLF and CTD-UIP HLF pre-treated with either MSCs or NHLF. Triplicate wells were prepared for each group. Data represent the means ± SD from four independent experiments. * Significantly different from CTD-IP-HLF, $P < 0.05$. **b** IP-10 and TGF-β1 levels in cultures of MSCs, CTD-UIP HLF, and CTD-UIP HLF pre-treated with either MSCs or NHLF. Triplicate wells were prepared for each group. Data represent the means ± SD from four independent experiments. * or ** significantly different from the MSC group, $P < 0.05$ or $P < 0.01$ respectively. † $P < 0.05$, †† $P < 0.01$, compared to CTD-UIP HLF pre-treated with NHLF or CTD-UIP HLF without the pretreatment. **c, d** Western blot analysis was performed to assess α-SMA expression and signaling pathways (stat3 and smad3) in NHLF, CTD-UIP HLF, and CTD-UIP HLF pre-treated with either MSCs or NHLF. GAPDH was used as a loading control. Representative blots from three replicates are shown (**d**). Quantification of α-SMA expression (**c**). * Significantly different from the NHLF group with $P < 0.05$. † $P < 0.05$, compared to CTD-UIP HLF pre-treated with NHLF or CTD-UIP HLF without the pretreatment. *MSCs* mesenchymal stem cells, *CTD-UIP-HLF* HLF isolated from lung tissues pathologically diagnosed with UIP in CTD-IP patients, *HLF* human lung fibroblasts, *NHLF* normal human lung fibroblasts, *IP-10* interferon γ-induced protein 10, *TGF-β1* transforming growth factor-β1, α-SMA α-smooth muscle actin

Supernatants harvested from HBMSCs can improve the survival rate in BLM-induced pulmonary fibrosis mice

Finally, we evaluated the antifibrotic efficacy of TGFβ1-hypersecreting HBMSCs in a BLM-induced pulmonary fibrosis mouse model. By making a comparison of antifibrotic capability in supernatants between TGFβ1-high and TGFβ1-low, derived from MSCs originated from different sources, we demonstrate that the supernatant derived from HBMSCs expressing a high level of TGFβ1 has a better therapeutic efficacy on improving the survival rate, as well as reducing pulmonary inflammation and fibrosis than that from MSCs-UC which secrete a lower level of TGFβ1 (Fig. 8).

Discussion

In the present study, we first reported that persistent activation of natural killer T cells (NKTs) is accompanied

by attenuation or deficiency of the regulatory T cell (Treg) response in interstitial pneumonia in connective tissue diseases (CTD-IP). We further disclosed the proinflammatory and profibrotic properties of lung fibroblasts in CTD-IP patients pathologically diagnosed with UIP. To the best of our knowledge, this study is the first to reveal that HBMSCs with a high level of TGF-β1 secretion can redress aberrant TGF-β1 downstream signal transduction for regulation of excessive IL-6/STAT3 signaling, consequent to Treg expansion, and to induction of anti-fibrotic cytokine expression.

NKT cells, a heterogeneous group of T lymphocytes, are known to functionally bridge the innate and adaptive immune system in various immune diseases due to their cytotoxic function and production of the proinflammatory factors IL-4 and IFN-γ [27]. A recent study showed that IFN-γ-producing NKT cells promoted immune

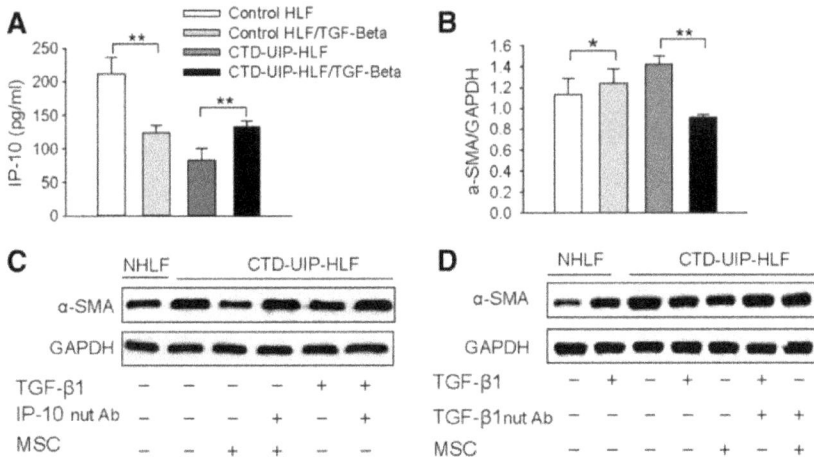

Fig. 7 Suppression of the myofibroblast phenotype in CTD-UIP HLF through the activation of attenuated TGF-β1 signaling and subsequent IP-10 induction. **a, b** IP-10 levels (**a**) and western blot analysis of α-SMA expression (**b**) in NHLF and CTD-UIP HLF in the absence or presence of TGF-β1. Data are representative of three independent experiments. Representative blots from three replicates are shown. Quantification of α-SMA expression by densitometric analysis was performed using Gel-Pro software. * $P < 0.05$, ** $P < 0.01$. **c, d** Representative western blot for α-SMA expression in CTD-UIP HLF treated with MSC or TGF-β1 in the absence and presence of neutralizing antibody for either human IP-10 (2 ug/ml) (**c**), or human TGF-β1(1 ug/ml) (**d**). GAPDH was used as a loading control. Representative blots from three replicates are shown. *CTD-UIP-HLF* HLF isolated from lung tissues pathologically diagnosed with UIP in CTD-IP patients, *HLF* human lung fibroblasts, *TGF-β1* transforming growth factor-β1, *IP-10* interferon γ-induced protein 10, *α-SMA* α-smooth muscle actin, *NHLF* normal human lung fibroblasts

complex (IC)-induced acute lung injury by stimulating production of MIP-1 through both autocrine and paracrine mechanisms, and by enhancing cytokine production from alveolar macrophages and CD11c[+] dendritic cells (DCs) [28]. In the present study, we found that CTD-IP patients with active disease had a higher frequency of NKTs in their peripheral blood and lungs, where the disruption of the normal alveolar architecture was accompanied by patchy alveolar pneumocyte hyperplasia and fibrosing changes. Therefore, uncontrolled activation of a NKT cell-mediated abnormal immune response could contribute to chronic lung injury, inflammation and abnormal repair with diffuse fibrosis in CTD-IP patients. Among T cells subsets, Tregs have a known role in controlling overt inflammation [29]. A systemic defect in Tregs is linked to inferior lung function in the enrolled CTD-IP patients, which is parallel to that observed in patients with idiopathic pulmonary fibrosis (IPF) [9], suggesting that pulmonary fibrotic progression in IPF and CTD-IP patients is associated with the failure of inflammation resolution due to deficiency of Treg manipulation.

A number of investigations have provided convincing evidence showing that interstitial fibroblasts in an inflammatory microenvironment produced by cytotoxic T cell recruitment to lung, are activated and differentiate towards a myofibroblast phenotype [30]. We detected the myofibroblast phenotypes in pulmonary fibroblasts isolated from CTD-UIP lungs, where cytokine/chemokine profiles are characterized with a remarkable increase in IL-6 secretion accompanied by chemokine upregulation, indicating that the abnormal lung interstitial fibroblasts may disturb Treg differentiation whereby cytotoxic immune cells, such as NKT and CD8[+] T cells, maintain activation in lung parenchyma. This may create an uncontrolled positive feedback loop for immune activation and inflammation, which will make conventional anti-inflammatory therapy ineffective in the management of CTD-IP. Breaking this feedback loop so as to restore a normal balance between different subsets of immune cells, rather than using indiscriminate anti-inflammatory agents, may be a promising approach for treating CTD-IP [9, 31].

Many studies reported that MSC-mediated cell therapy is very effective in treating autoimmune diseases [17–19]. We show that HBMSCs induce Treg proliferation in an experimental NKT-PBMC model in vitro, whereas high frequencies of NKT and CD8[+] T cells are reduced. Importantly, we found that the HBMSCs self-secreting a high level of TGF-β1 can facilitate the growth of Tregs in PBMCs isolated from IPF patients as well. These results indicate that MSC-based therapy may allow repair of impaired Tregs through a TGF-β1-dependent regulation, by which cytotoxic T cells are suppressed, rather than by universally inhibiting T cells proliferation.

There is increasing evidence showing that MSCs exert immunosuppressive effects on immune inflammation through the release of many soluble cytokines including

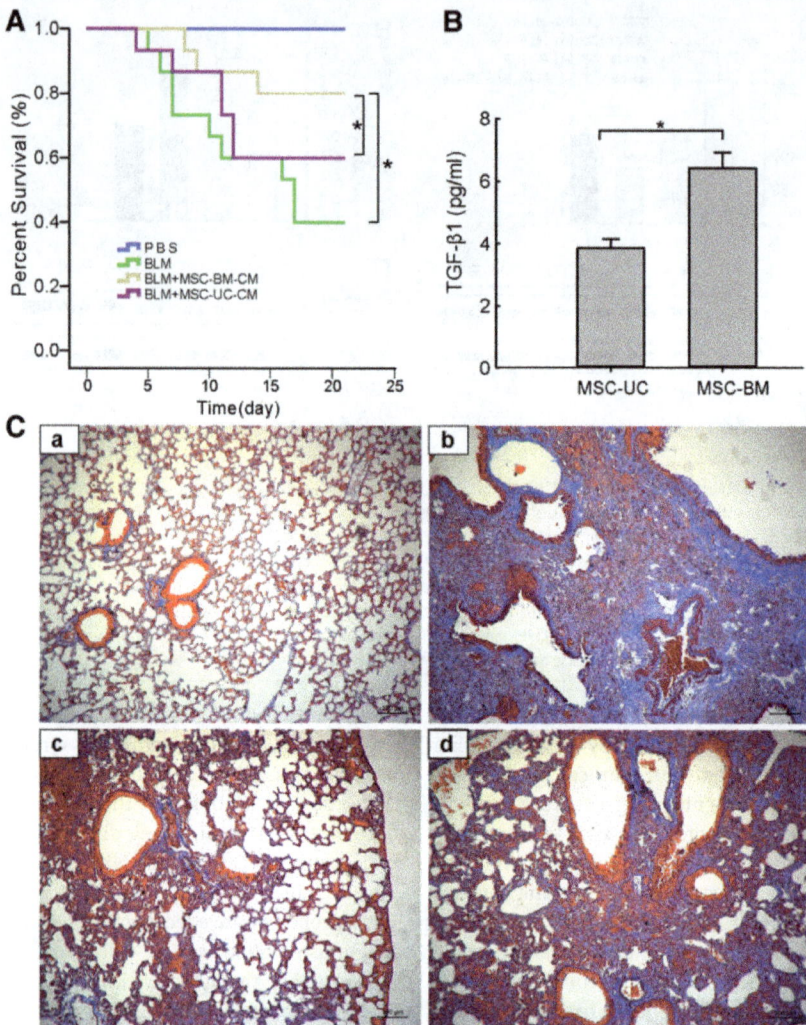

Fig. 8 Mesenchymal stem cells from bone marrow and umbilical cord exert different efficacy in BLM-induced pulmonary fibrosis mouse model. (A) Survival rates of C57BL/6 mice in the control group and BLM-induced group without any treatment or with treatment by supernatant from either MSCs-BM or MSCs-UC. Supernatants harvested from MSC (1×10^6) culture were intratracheally administered to mice 48 hours after BLM treatment. Analysis was conducted by a logrank test based on the Kaplan–Meier method. (B) An enzyme-linked immunosorbent assay demonstrated a significantly higher level of TGF-β1 secreted from HBMSCs than from MSC-UC. (C) Representative Masson staining photomicrographs of the lung tissue sections from mice 21 days after saline exposure (a), BLM exposure (b), BLM exposure with treatment of the supernatant from MSC-BM (c), and BLM exposure with treatment of the supernatant from MSC-UC (d). 200× magnification. *MSCs-BM* mesenchymal stem cells isolated from bone marrow, *MSCs-UC* mesenchymal stem cells isolated from umbilical cord, *TGF-β1* transforming growth factor-β1

TGF-β1, PGE2, indoleamine 2, 3-dioxygenase (IDO), IL-10, and IL-1RA [17, 32–34]. A prominent function of TGF-β1 is regulating immune homeostasis and TGF-β1 deficiency in mice results in excessive inflammation and lethality [35]. Abnormally activated T cells and elevated proinflammatory cytokines, including TNF-α, IFN-γ, and IL-1β, have been detected in TGF-β1 knockout mice [36]. Moreover, endogenous TGF-β1 is essential for the induction of immunosuppressive Treg cells [37, 38]. However, we show a significant up-regulation of the TGF-β1 level accompanied by a reduced Tregs and down-regulation of the ratio of TGF-β1 to IL-6 in the CTD-IP patients, reflecting that the increase of endogenous TGF-β1 released from immunocytes in response to the inflammatory microenvironment could not induce Tregs differentiation owing to IL-6 hypersecretion that causes an imbalance between IL-6 and TGF-β1 in local and systemic modulation of the immune response, thereby disturbing TGF-β1 signaling. A high level of TGF-β1 self-secretion by HBMSCs may therefore be an important mechanism underlying therapeutic effects of MSCs on promoting Tregs expansion in IPF patients [39, 40].

TGF-β signaling is also involved in normal lung development and injury repair [41, 42]. On the contrary, it is able to induce fibroblast proliferation, differentiation, migration, and extracellular matrix production and contraction. In the adult lung, excessive TGF-β-mediated Smad3 signaling, as seen after bleomycin administration, plays a critical role in extensive fibrosis [43]. The current study demonstrates an excessive IL-6 secretion and substantially reduced IP-10 expression, but neither a high level of TGF-β1 nor activated TGF-β-mediated Smad3 signaling in CTD-UIP-HLFs which represent a myofibroblast phenotype. Overproduction of the IL-6 family of cytokines, aberrant activation of their receptors or receptor-associated tyrosine kinases, or epigenetic alterations or mutations in genes encoding negative regulators of STAT3 can bring about persistent STAT3 activation [44–46]. Elevated tyrosine phosphorylation of STAT3 is able to suppress apoptosis and promote angiogenesis and fibrotic proliferation [44]. It has been reported that TGF-β-mediated biological responses are impaired in mice in which STAT3 is excessively activated due to its upstream receptor gp130 mutation. Activated STAT3 in turn elicits the increased expression of the TGF-β signaling inhibitory molecule Smad7, thereby inhibiting the intracellular activity of TGF-β signaling [47].

In general, TGF-β1 can stimulate fibroblast differentiation to the myofibroblast phenotype and suppress myofibroblast apoptosis [48]. However, we show that either HBMSCs self-secreting a high level of TGF-β1 or TGF-β1 added to CTD-UIP-HLFs can induce production of anti-fibrotic chemokine IP-10 [49–52], which may act downstream of TGF-β signaling to negatively regulate activation of myofibroblasts marker [53], leading to attenuation of α-SMA over expression in the treated CTD-UIP-HLFs.

IP-10 is up-regulated after both immune and non-immune mediated tissue injury but is an antifibrotic chemokine involved in tissue repair and remodeling [49, 50, 54]. We and other investigators have found downregulation of IP-10 expression in fibroblasts isolated from CTD-IP (pathologically diagnosed UIP) and IPF lungs, which contributes to the myofibroblast phenotype [55, 56]. Although the ability to inhibit fibroblast migration is thought to be an important mechanism of IP-10 in limiting the development of fibrosis [49, 54], the effect of IP-10 on α-SMA expression in CTD-UIP-HLFs is still unclear. We, for the first time, demonstrate that TGF-β1 released from MSCs may block myofibroblast activation in CTD-UIP HLFs through sensitizing the TGFβ/Smad signaling pathway that is severely attenuated by excessive IL-6/STAT3 signaling, thereby overcoming the proinflammatory phenotype

and relieving the inhibition of IP-10 expression to push against myofibroblast differentiation.

The present study reveals that in patients with CTD-IP, high levels of IL-6 secretion are predominantly associated with pulmonary fibrotic progression. A similar finding reported by Collard and Alhamad has been shown in IPF patients with acute exacerbation [57, 58]. A phase 1b study of placenta-derived mesenchymal stromal cells in IPF patients has recently demonstrated that intravenous MSC administration is feasible and has a good short-term safety profile in patients with moderately severe IPF [59]. Herein we provide, for the first time, clear evidence in vivo showing that MSCs with a higher level of TGFβ1 self-secretion may have an optimal therapeutic efficacy on the counteraction of life-threatening pulmonary fibrotic exacerbation.

Conclusions

Our study provides the first evidence that persistent activation of cytotoxic immune cells, particularly NKTs, accompanied by attenuation or deficiency in Tregs relevant to IL-6 hyper-induction, strongly correlate with fibrotic exacerbation in CTD-IP. MSC-based cell therapy appears to be a promising approach for treating pulmonary fibrotic progression in CTD-IP, the underlying mechanism for which is attributable, at least in part, to the characterization of TGF-β1 hyper-secretion in HBMSCs. This is linked to activation of impaired TGF-β downstream signal pathway, thereby regulating excessive IL-6/STAT3, whereby a relief of suppression in Tregs differentiation and expansion may concomitantly activate anti-fibrotic IP-10 expression. This may in turn block progression of lung fibrosis.

Additional files

Additional file 1: Supplementary material online. Methods. (DOC 79 kb)

Additional file 2: Figure S1. The biological characteristics of human bone marrow MSCs. Flow cytometric analysis at passage 4–6 demonstrated that MSCs were negative for CD14, CD34, CD45, and CD11a, but were positive for CD105, CD90, CD44, and CD29. (TIF 242 kb)

Additional file 3: Figure S2. Cell immunophenotypes of human induced NKT-PBMCs. Cell phenotypes in the peripheral blood of the healthy volunteer and NKT-PBMCs cultures on day 14 were observed. (A), (B), (C), and (D) show the dot plots and summary data of flow cytometric analysis. After the 14-day culture period, a higher frequency of CD3+CD8+ T cells (C), CD3+CD56+ NKT cells (D) and a lower frequency of CD3+CD4+ T cells (B) were observed. The mean ± SD of five cases in each group are shown in (B), (C), and (D). ** P < 0.01. (TIF 481 kb)

Additional file 4: Figure S3. Immunomodulatory effects of human bone marrow MSCs on NKT-PBMCs. NKT-PBMCs were co-cultured with human MSCs or human fibroblasts at a 20:1 ratio of NKT-PBMCs to human MSCs or NHLF prior to flow cytometric analysis. Flow cytometric analysis of CD3+ CD56+ cells, CD3+ CD8+ cells, CD3+ CD4+ cells gating on CD45+ cells, and CD25+ CD127(Low/-) cells gating on CD4+ cells, of either NKT-PBMCs (NKT-PBMCs), or NKT- PBMCs co-cultured with human bone MSCs (NKT-PBMCs/MSC), or co-cultured with NHLF (NKT-PBMCs/NHLF). (TIF 2388 kb)

Abbreviations
BAL: bronchoalveolar lavage; CTD-IP: interstitial pneumonia in connective tissue diseases; CTD-UIP-HLF: HLF isolated from the lung tissues pathologically diagnosed with UIP in CTD-IP patients; FVC: forced vital capacity; H&E: hematoxylin and eosin; HBMSCs: human bone marrow mesenchymal stem cells; HLFs: human primary lung fibroblasts; IDO: indoleamine 2, 3-dioxygenase; IFN-γ: interferon γ; IL-1β: interleukin 1 beta; IP: interstitial pneumonia; IP-10: interferon γ-induced protein 10; IPF: idiopathic pulmonary fibrosis; MSC-BM: MSCs isolated from bone marrow (BM); MSCs: mesenchymal stem cells; MSC-UC: MSCs isolated from umbilical cord (UC); NKTs: natural killer T cells; PBMCs: peripheral blood mononuclear cells; PM/DM: polymyositis/dermatomyositis; RA: rheumatoid arthritis; STAT3: signal transducer and activator of transcription 3; TGF-β: transforming growth factor-β; TNF-α: tumor necrosis factor α; Tregs: regulatory T cells; UIP: usual interstitial pneumonia; VATS: video-assisted thoracoscopic surgery; α-SMA: α-smooth muscle actin.

Competing interests
The authors declare that they have no competing interests.

Authors' contributions
JX and WS were involved in the conception, hypotheses delineation, and design of the study. ML,XSZ,JLW,ZPF, MYL and JSW conducted most laboratory experiments, analyzed the data, interpreted the results; DQR provided technical assistance in the patient primary lung fibroblast culture; BDY, ZPF and LXZ assisted in flow cytometric analysis and experiments of molecular biology. JX, XSZ and JLW discussed analyses, interpretation and presentation and wrote the paper. JX, WS and XH discussed and revised the manuscript prior to submission. All authors have read and approved the final manuscript.

Acknowledgements
We would like to acknowledge Professor S. Holgate (University of Southampton, UK) for providing the human bronchial epithelial cell line. We also are grateful for funding support from the National Key Basic Research Program of China (973 Program, 2009CB522104) to JX, the National Natural Science Foundation of China (Grant No. 81490530, 81490531 H0111 2015) to JX, and the Science and Technology Support Program of the local government in Guangzhou, China (2014, 2015) to JX, Science and Technology Planning Project of Shenzhen City (CXY201107010176A) to JX. The study protocol was also approved by the Ethics Committee of the University of Southern California Keck School of Medicine, XiangYang Central Hospital, and Shenzhen Beike Cell Engineering Research Institute.

Author details
[1]State Key Laboratory of Respiratory Diseases, Guangzhou Institute of Respiratory Diseases, The First Affiliated Hospital of Guangzhou Medical University, Guangzhou Medical University, Guangzhou, P. R. China. [2]Developmental Biology and Regenerative Medicine Program, Department of Surgery, The Saban Research Institute of Children's Hospital Los Angeles, University of Southern California Keck School of Medicine, Los Angeles, CA, USA. [3]Department of Respiratory Medicine, Xiangyang Central Hospital, Xiangyang, Hubei province, P. R. China. [4]Shenzhen Beike Cell Engineering Research Institute, Shenzhen, P. R. China.

References
1. American Thoracic Society/European Respiratory Society International Multidisciplinary Consensus Classification of the Idiopathic Interstitial Pneumonias. This joint statement of the American Thoracic Society (ATS), and the European Respiratory Society (ERS) was adopted by the ATS board of directors, June 2001 and by the ERS Executive Committee, June 2001. Am J Respir Crit Care Med. 2002;165(2):277–304. doi:10.1164/ajrccm.165.2.ats01.
2. Castelino FV, Varga J. Interstitial lung disease in connective tissue diseases: evolving concepts of pathogenesis and management. Arthritis Res Ther. 2010;12(4):213. doi:10.1186/ar3097.
3. Turesson C, O'Fallon WM, Crowson CS, Gabriel SE, Matteson EL. Extra-articular disease manifestations in rheumatoid arthritis: incidence trends and risk factors over 46 years. Ann Rheum Dis. 2003;62(8):722–7.
4. Feghali-Bostwick CA, Tsai CG, Valentine VG, Kantrow S, Stoner MW, Pilewski JM, et al. Cellular and humoral autoreactivity in idiopathic pulmonary fibrosis. J Immunol. 2007;179(4):2592–9.
5. Yang Y, Fujita J, Bandoh S, Ohtsuki Y, Yamadori I, Yoshinouchi T, et al. Detection of antivimentin antibody in sera of patients with idiopathic pulmonary fibrosis and non-specific interstitial pneumonia. Clin Exp Immunol. 2002;128(1):169–74.
6. Dobashi N, Fujita J, Ohtsuki Y, Yamadori I, Yoshinouchi T, Kamei T, et al. Detection of anti-cytokeratin 8 antibody in the serum of patients with cryptogenic fibrosing alveolitis and pulmonary fibrosis associated with collagen vascular disorders. Thorax. 1998;53(11):969–74.
7. Fujita J, Dobashi N, Ohtsuki Y, Yamadori I, Yoshinouchi T, Kamei T, et al. Elevation of anti-cytokeratin 19 antibody in sera of the patients with idiopathic pulmonary fibrosis and pulmonary fibrosis associated with collagen vascular disorders. Lung. 1999;177(5):311–9.
8. Ehrenstein MR, Evans JG, Singh A, Moore S, Warnes G, Isenberg DA, et al. Compromised function of regulatory T cells in rheumatoid arthritis and reversal by anti-TNFalpha therapy. J Exp Med. 2004;200(3):277–85. doi:10.1084/jem.20040165.
9. Kotsianidis I, Nakou E, Bouchliou I, Tzouvelekis A, Spanoudakis E, Steiropoulos P, et al. Global impairment of CD4 + CD25 + FOXP3+ regulatory T cells in idiopathic pulmonary fibrosis. Am J Respir Crit Care Med. 2009;179(12):1121–30. doi:10.1164/rccm.200812-1936OC.
10. Luzina IG, Todd NW, Iacono AT, Atamas SP. Roles of T lymphocytes in pulmonary fibrosis. J Leukoc Biol. 2008;83(2):237–44. doi:10.1189/jlb.0707504.
11. Turesson C, Matteson EL, Colby TV, Vuk-Pavlovic Z, Vassallo R, Weyand CM, et al. Increased CD4+ T cell infiltrates in rheumatoid arthritis-associated interstitial pneumonitis compared with idiopathic interstitial pneumonitis. Arthritis Rheum. 2005;52(1):73–9. doi:10.1002/art.20765.
12. McNearney TA, Reveille JD, Fischbach M, Friedman AW, Lisse JR, Goel N, et al. Pulmonary involvement in systemic sclerosis: associations with genetic, serologic, sociodemographic, and behavioral factors. Arthritis Rheum. 2007;57(2):318–26. doi:10.1002/art.22532.
13. Tillie-Leblond I, Wislez M, Valeyre D, Crestani B, Rabbat A, Israel-Biet D, et al. Interstitial lung disease and anti-Jo-1 antibodies: difference between acute and gradual onset. Thorax. 2008;63(1):53–9. doi:10.1136/thx.2006.069237.
14. Jindal SK, Agarwal R. Autoimmunity and interstitial lung disease. Curr Opin Pulm Med. 2005;11(5):438–46.
15. Skapenko A, Leipe J, Lipsky PE, Schulze-Koops H. The role of the T cell in autoimmune inflammation. Arthritis Res Ther. 2005;7 Suppl 2:S4–S14. doi:10.1186/ar1703.
16. Anderson AE, Isaacs JD. Tregs and rheumatoid arthritis. Acta Reumatol Port. 2008;33(1):17–33.
17. Shi Y, Hu G, Su J, Li W, Chen Q, Shou P, et al. Mesenchymal stem cells: a new strategy for immunosuppression and tissue repair. Cell Res. 2010;20(5):510–8. doi:10.1038/cr.2010.44.
18. Uccelli A, Moretta L, Pistoia V. Mesenchymal stem cells in health and disease. Nat Rev Immunol. 2008;8(9):726–36. doi:10.1038/nri2395.
19. Gerdoni E, Gallo B, Casazza S, Musio S, Bonanni I, Pedemonte E, et al. Mesenchymal stem cells effectively modulate pathogenic immune response in experimental autoimmune encephalomyelitis. Ann Neurol. 2007;61(3):219–27. doi:10.1002/ana.21076.
20. Weiss DJ, Bertoncello I, Borok Z, Kim C, Panoskaltsis-Mortari A, Reynolds S, et al. Stem cells and cell therapies in lung biology and lung diseases. Proc Am Thorac Soc. 2011;8(3):223–72. doi:10.1513/pats.201012-071DW.
21. Murphy S, Lim R, Dickinson H, Acharya R, Rosli S, Jenkin G, et al. Human amnion epithelial cells prevent bleomycin-induced lung injury and preserve lung function. Cell Transplant. 2011;20(6):909–23. doi:10.3727/096368910X543385.
22. Vij R, Strek ME. Diagnosis and treatment of connective tissue disease-associated interstitial lung disease. Chest. 2013;143(3):814–24. doi:10.1378/chest.12-0741.
23. Cherfils-Vicini J, Platonova S, Gillard M, Laurans L, Validire P, Caliandro R, et al. Triggering of TLR7 and TLR8 expressed by human lung cancer cells induces cell survival and chemoresistance. J Clin Invest. 2010;120(4):1285–97. doi:10.1172/JCI36551.
24. Braun RK, Sterner-Kock A, Kilshaw PJ, Ferrick DA, Giri SN. Integrin alpha E beta 7 expression on BAL CD4+, CD8+, and gamma delta T-cells in bleomycin-induced lung fibrosis in mouse. Eur Respir J. 1996;9(4):673–9.
25. Iyer SS, Rojas M. Anti-inflammatory effects of mesenchymal stem cells: novel concept for future therapies. Expert Opin Biol Ther. 2008;8(5):569–81. doi:10.1517/14712598.8.5.569.

26. Lu PH, Negrin RS. A novel population of expanded human CD3 + CD56+ cells derived from T cells with potent in vivo antitumor activity in mice with severe combined immunodeficiency. J Immunol. 1994;153(4):1687–96.

27. Giroux M, Denis F. CD1d-unrestricted human NKT cells release chemokines upon Fas engagement. Blood. 2005;105(2):703–10. doi:10.1182/blood-2004-04-1537.

28. Kim JH, Chung DH. CD1d-restricted IFN-gamma-secreting NKT cells promote immune complex-induced acute lung injury by regulating macrophage-inflammatory protein-1alpha production and activation of macrophages and dendritic cells. J Immunol. 2011;186(3):1432–41. doi:10.4049/jimmunol.1003140.

29. Sakaguchi S. Naturally arising Foxp3-expressing CD25 + CD4+ regulatory T cells in immunological tolerance to self and non-self. Nat Immunol. 2005; 6(4):345–52. doi:10.1038/ni1178.

30. Doucet C, Giron-Michel J, Canonica GW, Azzarone B. Human lung myofibroblasts as effectors of the inflammatory process: the common receptor gamma chain is induced by Th2 cytokines, and CD40 ligand is induced by lipopolysaccharide, thrombin and TNF-alpha. Eur J Immunol. 2002;32(9):2437–49.

31. Strieter RM, Gomperts BN, Keane MP. The role of CXC chemokines in pulmonary fibrosis. J Clin Invest. 2007;117(3):549–56. doi:10.1172/JCI30562.

32. Aggarwal S, Pittenger MF. Human mesenchymal stem cells modulate allogeneic immune cell responses. Blood. 2005;105(4):1815–22. doi:10.1182/blood-2004-04-1559.

33. Glennie S, Soeiro I, Dyson PJ, Lam EW, Dazzi F. Bone marrow mesenchymal stem cells induce division arrest anergy of activated T cells. Blood. 2005; 105(7):2821–7. doi:10.1182/blood-2004-09-3696.

34. Beyth S, Borovsky Z, Mevorach D, Liebergall M, Gazit Z, Aslan H, et al. Human mesenchymal stem cells alter antigen-presenting cell maturation and induce T-cell unresponsiveness. Blood. 2005;105(5):2214–9. doi:10.1182/blood-2004-07-2921.

35. Shull MM, Ormsby I, Kier AB, Pawlowski S, Diebold RJ, Yin M, et al. Targeted disruption of the mouse transforming growth factor-beta 1 gene results in multifocal inflammatory disease. Nature. 1992;359(6397):693–9. doi:10.1038/359693a0.

36. Rubtsov YP, Rudensky AY. TGFbeta signalling in control of T-cell-mediated self-reactivity. Nat Rev Immunol. 2007;7(6):443–53. doi:10.1038/nri2095.

37. Li MO, Sanjabi S, Flavell RA. Transforming growth factor-beta controls development, homeostasis, and tolerance of T cells by regulatory T cell-dependent and -independent mechanisms. Immunity. 2006;25(3):455–71. doi:10.1016/j.immuni.2006.07.011.

38. Huber S, Schramm C, Lehr HA, Mann A, Schmitt S, Becker C, et al. Cutting edge: TGF-beta signaling is required for the in vivo expansion and immunosuppressive capacity of regulatory CD4 + CD25+ T cells. J Immunol. 2004;173(11):6526–31.

39. Kim HY, Kim HJ, Min HS, Kim S, Park WS, Park SH, et al. NKT cells promote antibody-induced joint inflammation by suppressing transforming growth factor beta1 production. J Exp Med. 2005;201(1):41–7. doi:10.1084/jem.20041400.

40. Qi MY, Kai C, Liu HR, Su YH, Yu SQ. Protective effect of Icariin on the early stage of experimental diabetic nephropathy induced by streptozotocin via modulating transforming growth factor beta1 and type IV collagen expression in rats. J Ethnopharmacol. 2011;138(3):731–6. doi:10.1016/j.jep.2011.10.015.

41. Buckley S, Shi W, Barsky L, Warburton D. TGF-beta signaling promotes survival and repair in rat alveolar epithelial type 2 cells during recovery after hyperoxic injury. Am J Physiol Lung Cell Mol Physiol. 2008;294(4):L739–48. doi:10.1152/ajplung.00294.2007.

42. Bartram U, Speer CP. The role of transforming growth factor beta in lung development and disease. Chest. 2004;125(2):754–65.

43. Leask A, Abraham DJ. TGF-beta signaling and the fibrotic response. FASEB J. 2004;18(7):816–27. doi:10.1096/fj.03-1273rev.

44. Jenkins BJ, Roberts AW, Greenhill CJ, Najdovska M, Lundgren-May T, Robb L, et al. Pathologic consequences of STAT3 hyperactivation by IL-6 and IL-11 during hematopoiesis and lymphopoiesis. Blood. 2007;109(6):2380–8. doi:10.1182/blood-2006-08-040352.

45. Heinrich PC, Behrmann I, Haan S, Hermanns HM, Muller-Newen G, Schaper F. Principles of interleukin (IL)-6-type cytokine signalling and its regulation. Biochem J. 2003;374(Pt 1):1–20. doi:10.1042/BJ20030407.

46. Nishihara M, Ogura H, Ueda N, Tsuruoka M, Kitabayashi C, Tsuji F, et al. IL-6-gp130-STAT3 in T cells directs the development of IL-17+ Th with a

47. Jenkins BJ, Grail D, Nheu T, Najdovska M, Wang B, Waring P, et al. Hyperactivation of Stat3 in gp130 mutant mice promotes gastric hyperproliferation and desensitizes TGF-beta signaling. Nat Med. 2005;11(8): 845–52. doi:10.1038/nm1282.

48. Zhang HY, Phan SH. Inhibition of myofibroblast apoptosis by transforming growth factor beta(1). Am J Respir Cell Mol Biol. 1999;21(6):658–65. doi:10.1165/ajrcmb.21.6.3720.

49. Tager AM, Kradin RL, LaCamera P, Bercury SD, Campanella GS, Leary CP, et al. Inhibition of pulmonary fibrosis by the chemokine IP-10/CXCL10. Am J Respir Cell Mol Biol. 2004;31(4):395–404. doi:10.1165/rcmb.2004-0175OC.

50. Bujak M, Dobaczewski M, Gonzalez-Quesada C, Xia Y, Leucker T, Zymek P, et al. Induction of the CXC chemokine interferon-gamma-inducible protein 10 regulates the reparative response following myocardial infarction. Circ Res. 2009;105(10):973–83. doi:10.1161/CIRCRESAHA.109.199471.

51. Pignatti P, Brunetti G, Moretto D, Yacoub MR, Fiori M, Balbi B, et al. Role of the chemokine receptors CXCR3 and CCR4 in human pulmonary fibrosis. Am J Respir Crit Care Med. 2006;173(3):310–7. doi:10.1164/rccm.200502-244OC.

52. Jiang D, Liang J, Hodge J, Lu B, Zhu Z, Yu S, et al. Regulation of pulmonary fibrosis by chemokine receptor CXCR3. J Clin Invest. 2004;114(2):291–9. doi:10.1172/JCI16861.

53. Liang YJ, Luo J, Lu Q, Zhou Y, Wu HW, Zheng D, et al. Gene profile of chemokines on hepatic stellate cells of schistosome-infected mice and antifibrotic roles of CXCL9/10 on liver non-parenchymal cells. PLoS One. 2012;7(8):e42490. doi:10.1371/journal.pone.0042490.

54. Jiang D, Liang J, Campanella GS, Guo R, Yu S, Xie T, et al. Inhibition of pulmonary fibrosis in mice by CXCL10 requires glycosaminoglycan binding and syndecan-4. J Clin Invest. 2010;120(6):2049–57. doi:10.1172/JCI38644.

55. Coward WR, Watts K, Feghali-Bostwick CA, Jenkins G, Pang L. Repression of IP-10 by interactions between histone deacetylation and hypermethylation in idiopathic pulmonary fibrosis. Mol Cell Biol. 2010;30(12):2874–86. doi:10.1128/MCB.01527-09.

56. Lindahl GE, Stock CJ, Shi-Wen X, Leoni P, Sestini P, Howat SL, et al. Microarray profiling reveals suppressed interferon stimulated gene program in fibroblasts from scleroderma-associated interstitial lung disease. Respir Res. 2013;14:80. doi:10.1186/1465-9921-14-80.

57. Alhamad EH, Cal JG, Shakoor Z, Almogren A, AlBoukai AA. Cytokine gene polymorphisms and serum cytokine levels in patients with idiopathic pulmonary fibrosis. BMC Med Genet. 2013;14:66. doi:10.1186/1471-2350-14-66.

58. Collard HR, Calfee CS, Wolters PJ, Song JW, Hong SB, Brady S, et al. Plasma biomarker profiles in acute exacerbation of idiopathic pulmonary fibrosis. Am J Physiol Lung Cell Mol Physiol. 2010;299(1):L3–7. doi:10.1152/ajplung.90637.2008.

59. Chambers DC, Enever D, Ilic N, Sparks L, Whitelaw K, Ayres J, et al. A phase 1b study of placenta-derived mesenchymal stromal cells in patients with idiopathic pulmonary fibrosis. Respirology. 2014;19(7):1013–8. doi:10.1111/resp.12343.

minimum effect on that of Treg in the steady state. Int Immunol. 2007;19(6): 695–702. doi:10.1093/intimm/dxm045.

6

Human bone marrow-derived and umbilical cord-derived mesenchymal stem cells for alleviating neuropathic pain in a spinal cord injury model

Mahmoud Yousefifard[1,2], Farinaz Nasirinezhad[3,4*], Homa Shardi Manaheji[5,6], Atousa Janzadeh[3], Mostafa Hosseini[7,8] and Mansoor Keshavarz[1,2]

Abstract

Background: Stem cell therapy can be used for alleviating the neuropathic pain induced by spinal cord injuries (SCIs). However, survival and differentiation of stem cells following their transplantation vary depending on the host and intrinsic factors of the cell. Therefore, the present study aimed to determine the effect of stem cells derived from bone marrow (BM-MSC) and umbilical cord (UC-MSC) on neuropathic pain relief.

Methods: A compression model was used to induce SCI in a rat model. A week after SCI, about 1 million cells were transplanted into the spinal cord. Behavioral tests, including motor function recovery, mechanical allodynia, cold allodynia, mechanical hyperalgesia, and thermal hyperalgesia, were carried out every week for 8 weeks after SCI induction. A single unit recording and histological evaluation were then performed.

Results: We show that BM-MSC and UC-MSC transplantations led to improving functional recovery, allodynia, and hyperalgesia. No difference was seen between the two cell groups regarding motor recovery and alleviating the allodynia and hyperalgesia. These cells survived in the tissue at least 8 weeks and prevented cavity formation due to SCI. However, survival rate of UC-MSC was significantly higher than BM-MSC. Electrophysiological evaluations showed that transplantation of UC-MSC brings about better results than BM-MSCs in wind up of wide dynamic range neurons.

Conclusions: The results of the present study show that BM-MSC and UC-MSC transplantations alleviated the symptoms of neuropathic pain and resulted in subsequent motor recovery after SCI. However, survival rate and electrophysiological findings of UC-MSC were significantly better than BM-MSC.

Keywords: Spinal cord injuries, Neuropathic pain, Stem cells, Electrophysiologic techniques, Wind up

Background

According to the International Association for the Study of Pain (ISAP), neuropathic pain is a pain caused by damage or diseases affecting the central or peripheral nervous system. Following spinal cord injury (SCI) most patients suffer from long-lasting moderate to severe pain [1–4]. Existing treatments for reducing neuropathic pain have low efficiency in the majority of patients. These treatments include surgical decompression, drug therapy, and palliative care; even new drugs only reduce the pain by 50 % in a quarter of the patients [5]. These therapeutic strategies are purely conservative, and side effects caused by long-term use of the drugs are a great obstacle for applying this method of pain reduction [6, 7]. Neuropathic pain will persist unless the damaged area is healed or pain reduction pathways are amplified. Therefore, researchers are looking to repair the damaged nerve cells.

Intrinsic regeneration of the damaged nerves in the central nervous system is limited, so scientists are trying

* Correspondence: nasirinezhad.f@iums.ac.ir
[3]Physiology Research Center, Iran University of Medical Sciences, Tehran, Iran
[4]Department of Physiology, Iran University of Medical Sciences, Tehran, Iran
Full list of author information is available at the end of the article

to build new nervous contacts at the site of injury to reduce the neuropathic pain [8]. Thus, cell transplantation is thought to be a suitable treatment for SCI. As a result, in recent years ample research has been done in this field, the result of which shows powerful influence of stem cell transplantation in functional recovery after SCI [9, 10]. These studies showed that stem cells are able to proliferate and differentiate into nerve cells such as mature neurons or glial cells under special circumstances [11]. However, survival and differentiation of stem cells following their transplantation varies depending on the host and intrinsic factors of the cell [12, 13]. Based on these findings, it can be stated that the fate of transplanted cells in vivo varies with the intrinsic characteristics of the cells and site of transplantation [14]. However, the optimal source of stem cells is a controversial issue for treating SCI [15, 16].

Mesenchymal stem cells (MSCs) are the main source of cell therapy because of their capability of differentiating into multiple cell types, including blood, adipose tissue, connective tissues, and so forth [17–19]. These cells can easily grow in vitro and exhibit intriguing immunomodulatory properties, non-teratogenicity, and multi-potentiality with high genetic stability. MSCs can maintain regenerative capacity after cryopreservation, improve synaptic transmission, and promote neuronal networks [20–24]. These properties make MSCs prime candidates for various therapeutic applications especially for nervous system repair.

Different sources can be used to isolate MSCs. Some of these resources include umbilical cord [25], placenta [26], bone marrow [27] and adipose tissue [28]. Bone marrow and umbilical cord are rich sources of MSCs. Transplantation of bone marrow-derived mesenchymal stem cells (BM-MSCs) to the injured spinal cord resulted in a significant improvement in sensorimotor of the hindlimb and reduced cavity formation, and show substantial immunosuppressive, anti-proliferative, anti-inflammatory, and anti-apoptotic properties [29]. Human umbilical cord-derived mesenchymal stem cells (UC-MSCs) can differentiate into various neural cells and have beneficial effects on improving functional recovery after SCI. Compared with BM-MSCs, UC-MSCs have higher expansion ability, robust proliferation capacity, and lower risk of bacterial/viral infection [30], while some reports have shown that UC-MSCs evoked an immune response when injected into injured tissues [31].

The difference in the properties of the BM-MSCs and UC-MSCs may have impact on their efficacy in improving SCIs. However, the effectiveness of these cells in reducing neuropathic pain is not fully understood. Therefore, the present study aimed to determine the effect of stem cells derived from bone marrow and umbilical cord on SCI-induced neuropathic pain, and to identify the stem cell population with the highest survival and effectiveness in

transplantation to the site of nerve injury. We selected the rat model because multiple studies of stem cell therapy have been performed in the rat injured spinal cord, and ethical issues do not yet allow us to transplant these cells into human spinal cord.

Materials and methods
Study design
The present experimental study aims to compare the effect of transplanting UCMSCs and BM-MSCs on functional recovery and neuropathic pain caused by SCI. The protocol of the present study was approved by the Tehran University of Medical Sciences Ethics Committee. The researchers adhered to the principles of the Helsinki Declaration and the principles of using laboratory animals as suggested in the National Institutes of Health Guide for Care and Use of Laboratory Animals (Publication No. 85–23, revised 1985) over the course of the study. Most of the materials were obtained from Sigma-Aldrich Company, Germany. For all other cases, the relevant company is given.

Studied animals
Male Wistar rats (n = 72) with a weight range of 140–160 g were used and randomly divided into six groups (12 animals in each group) (Table 1). The animals were obtained from the Laboratory Animal Breeding Center of Iran University of Medical Sciences. All animals were kept in special cages for at least 2 weeks before the initiation of the study for adaptation to the environment. They had free access to water and food (temperature 21 ± 1 °C; 12-hour light/dark cycle). All behavioral tests were performed between 10:00 am and 2:00 pm at room temperature.

Cell culture
The cells were of human source in this study. All samples were obtained with written, informed consent in accordance with the Tehran University of Medical Sciences ethics committee requirements. BM-MSCs were bought from the Royan Institute. The cells were kept in an incubator at 37 °C, 90 % humidity, and 5 % CO_2. They were cultured in cell culture flasks containing DMEM/F_{12} (Dulbecco's modified Eagle's medium/F12; Gibco, Australia), fetal bovine serum 10 % (Gibco) and a combination of penicillin (100 IU/ml), streptomycin sulfate (160 µg/ml), and amphotericin B (10 µg/ml). The medium was changed every 3 days.

UC-MSCs were isolated from Wharton's jelly as follows. After obtaining the mother's consent, the umbilical cord of a healthy infant born by C-section (n = 2) was brought to the cell culture laboratory under sterile conditions and in HBSS (Hank's Balanced Salt Solution) containing penicillin (100 IU/ml), streptomycin sulfate

Table 1 The study groups

Experimental group	Treatment protocol
Control	Healthy animals without treatment
Sham	Laminectomy without SCI induction
SCI	Laminectomy + SCI induction
Vehicle	Laminectomy + SCI + intraspinal injection of cell culture media
BM-MSC	Laminectomy + SCI + intraspinal injection of BM-MSCs
UC-MSC	Laminectomy + SCI + intraspinal injection of UC-MSCs

BM-MSC bone marrow-derived mesenchymal stem cell, *SCI* spinal cord injury, *UC-MSC* umbilical cord-derived mesenchymal stem cell

(160 µg/ml), and amphotericin B (10 µg/ml). UC-MSCs were isolated under sterile conditions. After washing the umbilical cord with 70 % alcohol and phosphate-buffered saline (PBS), amnion and umbilical cord blood vessels were removed accurately and the remaining matrix was chopped into pieces, about 5 mm in diameter. The pieces were moved to 35×10 mm petri dishes, and 1 ml DMEM/F_{12} with 20 % fetal bovine serum (Gibco), penicillin (100 IU/ml), and streptomycin sulfate (150 µg/ml) were added. After 10–15 days of culture and keeping cells in an incubator, cell buds were identified next to the pieces. After seeing cell buds, Wharton's gel pieces were removed from the medium and cell culture continued until the cells reached more than 80 % confluence.

Before transplantation, the surface antigens of the cells were checked using a flow cytometry technique to be sure of their stem cell status. Mesenchymal cells should be negative for CD45 and CD14 but should express CD105, CD29, CD90, and CD44 [32, 33].

SCI induction

A clip compression model was used to induce SCI. This method was introduced in 1978 [34] and validated in subsequent studies [35–37]. Briefly, rats weighting 140–160 g were anesthetized using ketamine (80 mg/kg) and Xylazin (10 mg/kg). After shaving the hair on their back, a 2-cm long incision was made in the T6–T8 area. Muscles were set aside and the spinal cord was exposed with laminectomy. Afterwards, and with much caution, the spinal cord was compressed using a calibrated aneurysm clip providing 20 g/cm^2 pressure. The force of the clip was measured as described previously [38]. The clip was removed after 60 seconds and muscles and skin were sutured separately to close the operation site. Since the animals were incapable of emptying their bladder voluntarily after injury induction, their bladder was emptied at least twice a day until they were able to do so themselves.

Stem cell transplantation

A week after SCI induction, the animals were prepared for transplantation. They were anesthetized using ketamine

(80 mg/kg) and Xylazin (10 mg/kg) and their spinal cord was exposed at the T6–T8 area in the same way as stated above in the SCI induction section. About 1 million cells in a 10-µl volume were then transplanted into the dorsal horn of the spinal cord in two injections, 0.5 mm rostral and caudal of the lesion at a depth of 1 mm below the dorsal surface at a rate of 1 µl/min, using a glass micropipette attached to a stereotaxic injector. Subsequently, the muscles and skin were sutured and the animals were returned to their cages. To confirm the number of cells, the sample was prepared and cell count was performed using trypan blue staining.

Behavioral evaluations

Behavioral tests were carried out every week for 8 weeks after SCI induction. The Basso, Beattie, and Bresnahan (BBB) locomotor scoring scale [39] was used to rate the hind limb motor function. The rats were placed in a container 120 cm in diameter and were studied and rated for 4 minutes. The locomotor behavior of the animals, including hind-limb motor function, weight-bearing, limb coordination, and walking, was assessed and scored.

To evaluate sensory function, four behavioral tests were used, the details of which have been described in a previous study by the authors [40]. In summary, mechanical allodynia was evaluated using the von-Frey test. Eight von-Frey filaments of different diameters were used in an up and down manner to assess the withdrawal threshold of the animal; 50 % withdrawal threshold was then calculated based on the responses. Cold allodynia was evaluated using the acetone test. In this test, about 100 µl acetone was pushed onto the hind paw of the animal. The test was repeated five times for each paw with 1-minute time intervals, and the number of withdrawals was considered as the response and presented as a percentage of the total. Mechanical hyperalgesia was evaluated (Randall-Selitto test) using an analgesia meter. In this test, increasing mechanical tension was applied to both hind limbs with at least 1-minute intervals and their average was recorded. Heat hyperalgesia was assessed by thermal stimulation of the animal's hind paw (Plantar test). The test was repeated three times and the average time was considered as the animal's withdrawal latency. A 25-second cut-off was used for stopping stimulation to avoid tissue damage.

Electrophysiological evaluation

At the end of week 8, a single-unit recording of the dorsal horn of the spinal cord was obtained to evaluate the electrical function of neurons. For this purpose, the animals were deeply sedated (60 mg/kg pentobarbital) and their body temperature was kept at 37 °C over the course of the recording. L1–L2 lamina was then removed for electrophysiological recording. The recording site deviated

0.6 mm from the middle (to the side of the spinal cord) and its depth was 300–700 μm. This depth was selected due to the presence of wide dynamic range (WDR) neurons. These cells receive input from all three types of sensory fibers (Aβ, Aδ, and C), and therefore respond to the full range of stimulation, from light touch to noxious pinch, heat, and chemicals [41]. For each animal, one neuron was evaluated. Electrical stimulation was used to induce responses. Neural response recording started when a stable response from the neuron lasted for at least 1 minute. The spikes of action potential were recorded and their frequency was counted in four delayed window periods after stimulation: 0–20 ms (for Aβ fibers), 20–90 ms (for Aδ fibers), 90–300 ms (for C fibers), and 300–800 ms (for post-discharge). Finally, wind-up phenomenon was calculated based on Jergova et al. [42].

Histological evaluation

The spinal cord was prepared for tissue evaluation in week 8. After transcardial perfusion, the spinal cord was fixated in 4 % paraformaldehyde in 0.1 molar phosphate buffer, pH 7.4, for 24 hours. Then it was kept in 10 %, 20 % and 30 % sucrose solutions for 24 hours each and prepared for serial cross-sectionalizing. Luxol fast blue (LFB) staining was performed to determine the volume of the injury; 20-μm diameter sections were stained [43]. The sections were then observed under a light microscope. The size of the cavity was divided by the total size of the spinal cord section (three sections that had the biggest cavity for each animal; three animals in each group) and were presented as percentages. The data were analyzed with ImageJ software (Wayne Rasband, National Institutes of Health, USA).

Eight weeks after cell transplantation, survival of the cells was evaluated by the aid of immunohistochemistry. The cryostat sections (20 μm) were permeabilized in PBS-T (PBS containing 0.1 % Triton X-100) for 10 minutes and blocked with 10 % fetal bovine serum in PBS-T for 1 hour, and then incubated overnight at 4 °C with the primary antibodies against mouse monoclonal antibody against human nuclei (HuNu; Chemicon Inc., Pittsburgh, PA, USA). Then lamels were washed and goat anti-rabbit IgG conjugated with Alexa-Fluor 594 (Molecular Probes, Eugene, OR, USA) secondary antibodies were added in a 1:100 dilution and incubated at 37 °C for 1 hour. The nucleus of the host cell was also stained using DAPI (4',6-diamidino-2-phenylindole; Molecular Probes, Eugene, OR, USA) and the results were assessed using an Olympus DP72 florescent microscope. In this type of staining, transplanted cells are stained green. All sections were stained (three animals in each groups). The survival rates were calculated based on following formula:

$$Survival\ rate\ (\%) = \frac{total\ number\ of\ survived\ cells\ in\ 8th\ week}{total\ number\ of\ transplanted\ cells} \times 100$$

Statistical analyses

Data were analyzed by SPSS version 21.0 and are presented as mean and standard error. To compare the data gathered from behavioral evaluations of the different groups, two-way analysis of variance with Bonferroni post-hoc test was used, and for assessment of electrophysiological assessment and histological assays, one-way analysis of variance was used. In all analyses, $p < 0.05$ was considered as significant.

Results

Mesenchymal cell characteristics

After isolation, BM-MSCs and UC-MSCs adhered to the bottom of the flask and formed colonies. They became spindle-shaped and suspended cells were removed by medium change. Figure 1 shows the surface antigen profile of these spindle-shaped cells evaluated using flow cytometry. All cells were negative for CD45 and CD14. BM-MSCs were positive for CD44 and CD105 and UC-MSCs expressed CD29 and CD90.

Behavioral evaluations
Motor recovery

After SCI induction, the locomotor score of the animals significantly decreased compared to the sham group (df: 8, 63; F = 79.6; $p < 0.001$). Stem cell transplants did not lead to significant improvement until the fourth week after SCI. However, from weeks 5 to 8 of the study, BM-MSC and UC-MSC transplantations led to progressive improvement of motor recovery in animals compared to the vehicle group (df: 8, 63; F = 366.4; $p < 0.0001$). No difference was seen between the two cell groups regarding motor recovery ($p > 0.99$) (Fig. 2a).

Mechanical allodynia

SCI resulted in a significant decrease in 50 % paw withdrawal threshold of the animals (df: 8, 63; F = 48.6; $p < 0.001$). This decline continued in the vehicle and SCI groups until week 4 and then reached a plateau. In contrast, BM-MSC and UC-MSC transplantations caused improvements in mechanical allodynia (df: 8, 63; F = 1060.2; $p < 0.001$) so that in the seventh and eighth week, this threshold was not significantly different from that of the sham and control groups ($p > 0.05$) (Fig. 2b). There was no significant difference among stem cell treated groups ($p > 0.99$)

Cold allodynia

The percentage of paw withdrawal responses of the animal to cold stimulation significantly increased after SCI

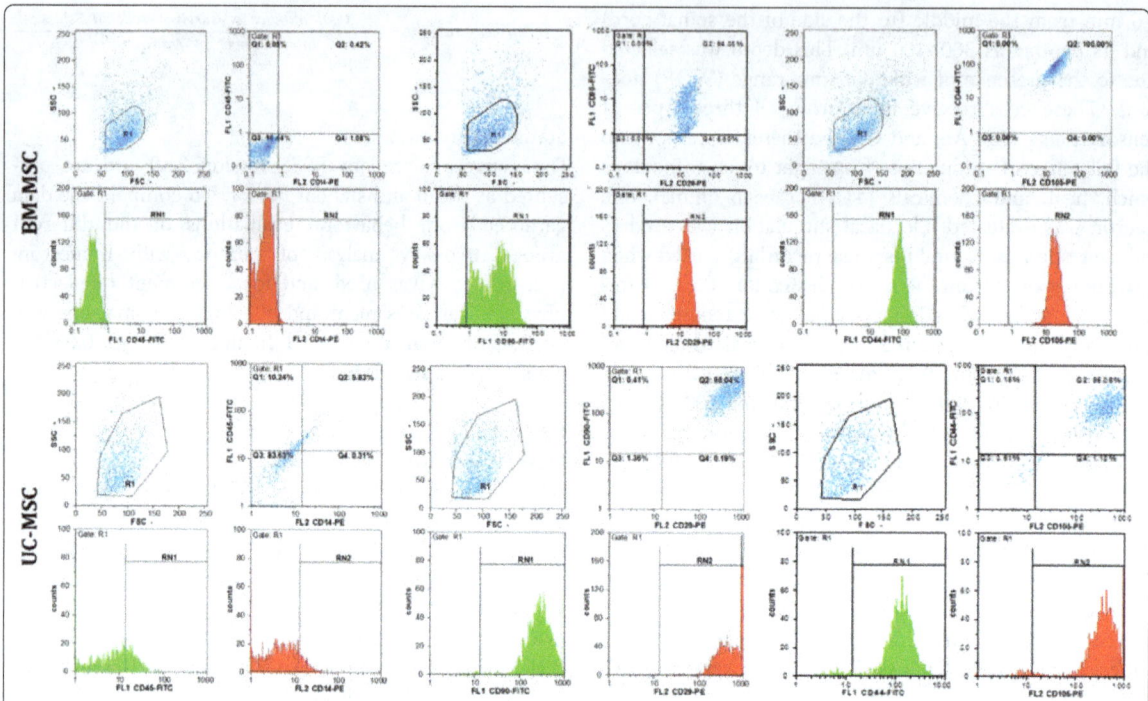

Fig. 1 Immunophenotype results of cells derived from human bone marrow mesenchymal stem cells (*BM-MSCs*) and umbilical cord mesenchymal stem cells (*UC-MSCs*). All cells were positive for CD29, CD44, CD90, and CD105, but negative for CD14, and CD45

induction, compared to the control and sham groups (df: 8, 63; F = 17.7; $p < 0.001$). In SCI and vehicle groups, the percentage of responses continued to increase. On the other hand, BM-MSC and UC-MSC transplantations caused improvements in the rats response to cold stimulation compared to the SCI group (df: 8, 63; F = 426.4; $p < 0.001$), although this threshold did not return to the normal level ($p < 0.05$) (Fig. 2c). There was no significant difference among stem cell treated groups ($p > 0.99$)

Mechanical hyperalgesia

The Randall-Sellito test revealed that SCI caused a decrease in the animal pain threshold under painful mechanical stimulation (df: 8, 63; F = 16.5; $p < 0.001$) (Fig. 2d). Transplantation of BM-MSCs and UC-MSCs led to significant improvement of paw withdrawal threshold compared to the SCI group (df: 8, 63; F = 230.4; $p < 0.001$). There was no significant difference between transplanted animals ($p > 0.99$).

Heat hyperalgesia

As Fig. 2e shows, SCI led to a significant decrease in paw withdrawal threshold due to heat stimulation (df: 8, 63; F = 24.0; $p < 0.001$). BM-MSC and UC-MSC transplantations caused the threshold to rise from the second week onwards and reach the normal level in

week 4 (df: 8, 63; F = 292.0; $p < 0.0001$). There was no significant difference among stem cell treated groups ($p > 0.99$).

Histological evaluation

A big cavity was seen in the SCI and vehicle groups 8 weeks post-SCI. However, BM-MSC and UC-MSC transplantations prevented cavity formation and SCI development (Fig. 3). The size of the cavity was significantly lower in BM-MSC and UC-MSC groups compared to SCI and vehicle groups (df: 16; F = 89.4; $p < 0.001$). However the size of the cavity was not different between transplanted animals ($p > 0.99$).

In addition, immunohistochemistry staining showed that transplanted cells continued to survive in the spinal cord after 8 weeks. In Fig. 4, transplanted cells can be seen in green. The survival rates were 0.36 ± 0.06 % and 0.57 ± 0.06 % in BM-MSC and UC-MSC groups, respectively ($p = 0.01$). The number of surviving cells in the BM-MSC and UC-MSC groups were 2327.0 ± 571.88 and 5728.67 ± 583.15, respectively ($p = 0.002$).

Electrophysiological findings

For electrophysiological investigations, a single-unit recording of WDR neurons in the dorsal horn of the L4 and L5 spinal cord was obtained. The WDR neuron response to electrical stimulation was evaluated 8 weeks

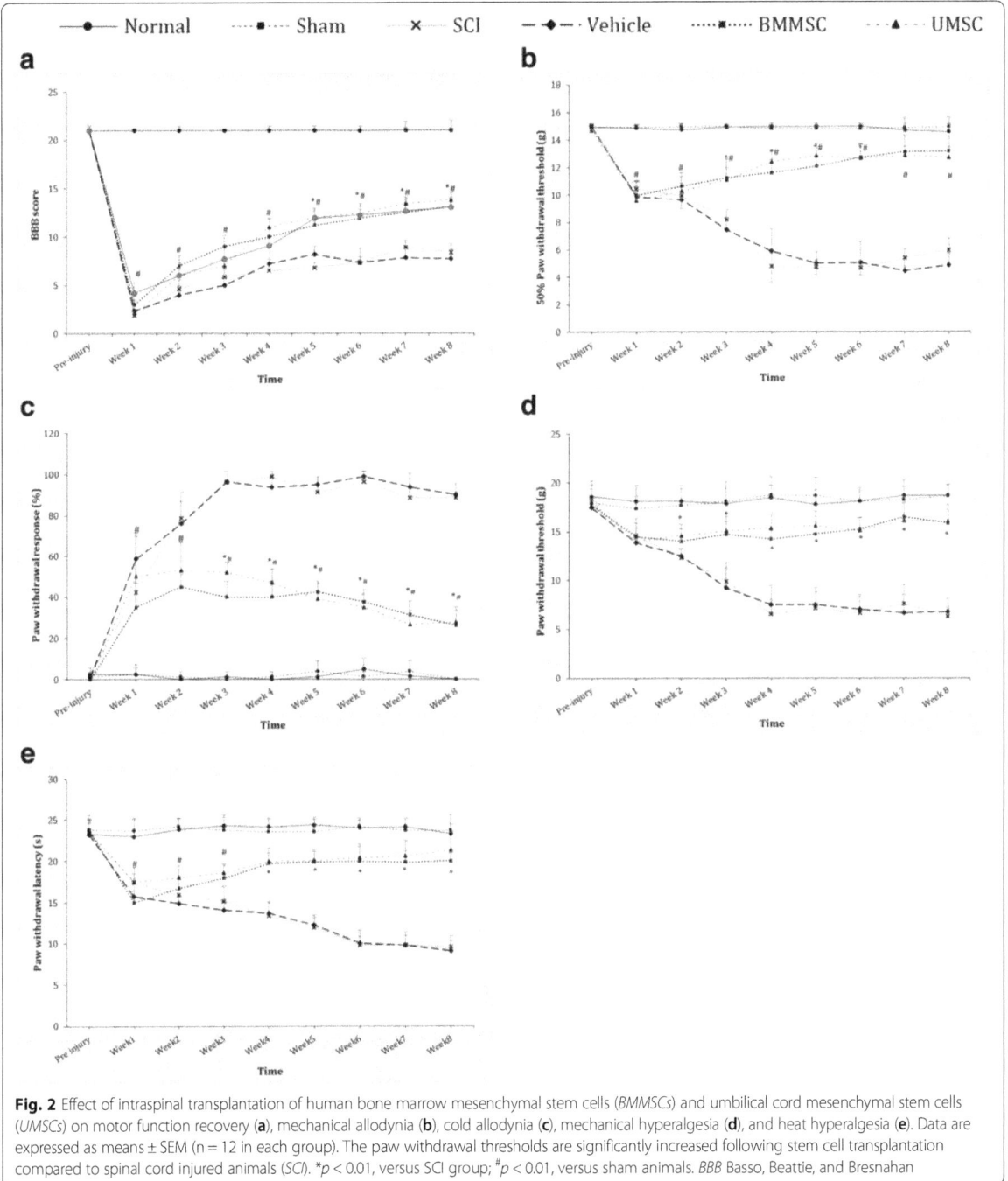

Fig. 2 Effect of intraspinal transplantation of human bone marrow mesenchymal stem cells (*BMMSCs*) and umbilical cord mesenchymal stem cells (*UMSCs*) on motor function recovery (**a**), mechanical allodynia (**b**), cold allodynia (**c**), mechanical hyperalgesia (**d**), and heat hyperalgesia (**e**). Data are expressed as means ± SEM (n = 12 in each group). The paw withdrawal thresholds are significantly increased following stem cell transplantation compared to spinal cord injured animals (*SCI*). *$p < 0.01$, versus SCI group; #$p < 0.01$, versus sham animals. *BBB* Basso, Beattie, and Bresnahan

after transplantation of BM-MSCs and UC-MSCs. Compared to control and sham groups, evoked potentials of the WDR neurons in the SCI and vehicle groups were significantly higher. The response of these neurons to the stimulations received from Aβ (df: 35; F = 28.9; $p < 0.001$), Aδ (df: 35; F = 34.5; $p < 0.001$), and C (df: 35; F = 40.6; $p < 0.001$) fibers, as well as post-discharge response

(df: 35; F = 31.5; $p < 0.001$) and wind up (df: 35; F = 30.6; $p < 0.001$) were higher in the SCI and vehicle groups compared to sham and control groups.

BM-MSC treatment caused the WDR neuron response to stimulations from Aβ ($p = 0.91$), Aδ ($p = 0.87$) and C ($p = 0.99$) fibers to reach that of the control group. Although post-discharge and wind up became significantly

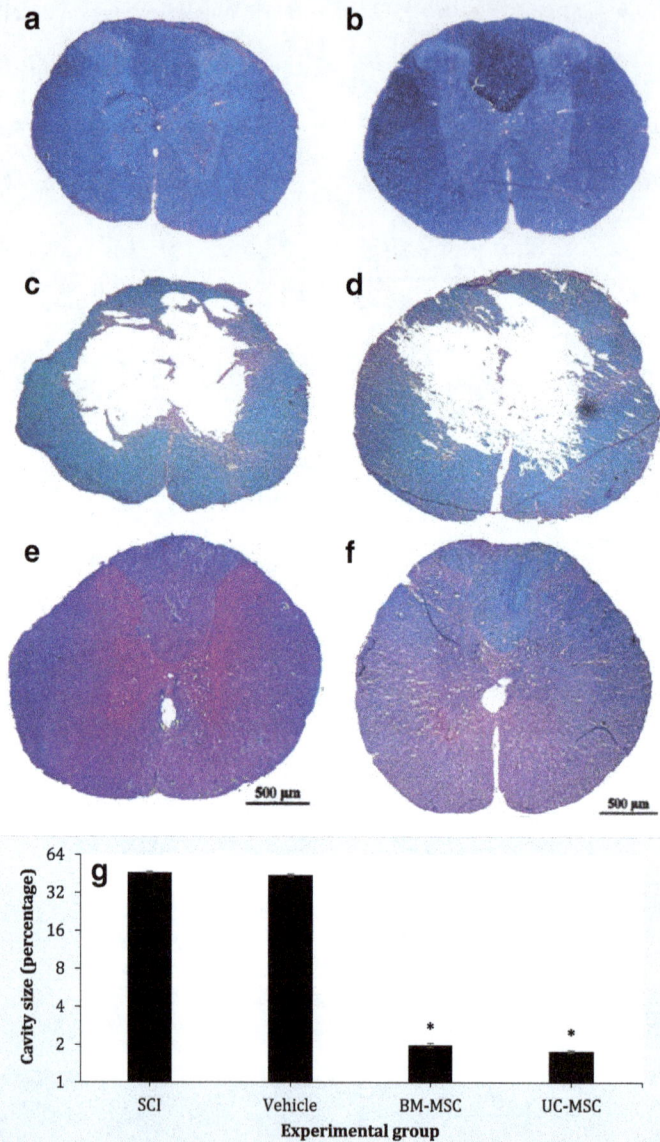

Fig. 3 Luxol fast blue staining for assessment of cavity size in control (**a**), sham (**b**), spinal cord injured (*SCI*) (**c**), vehicle-treated (**d**), human bone marrow-derived mesenchymal stem cell (*BM-MSC*) (**e**) and umbilical cord-derived mesenchymal stem cell (*UC-MSC*) (**f**) animals. Transplantation of human BM-MSCs and UC-MSCs resulted in significantly decreased cavity size (**g**). Original magnification in **a–f**, ×20. Data are expressed as means ± SEM (n = 3 in each group). *$p < 0.001$, versus SCI group

lower than in the SCI group ($p < 0.001$), they did not reach the normal level ($p < 0.05$) (Figs. 5 and 6).

A similar pattern was observed regarding UC-MSC transplants. Transplantation of UC-MSCs led to WDR neuron response to stimulations from Aβ ($p = 0.99$), Aδ ($p = 0.95$) and C ($p = 0.24$) fibers reaching the control group level. However, UC-MSCs did not cause post-discharge and wind up to reach normal levels ($p < 0.05$) although they did significantly decrease compared to the

SCI group ($p < 0.001$). Comparing the two treatments regimes revealed that wind up in the UC-MSC group was significantly lower than the BM-MSC group ($p = 0.008$).

Discussion

The results of the present study showed that transplantation of BM-MSCs and UC-MSCs in the spinal cord alleviate the allodynia and hyperalgesia after SCI. The efficacy of the two types of cells was similar for symptom relief. These

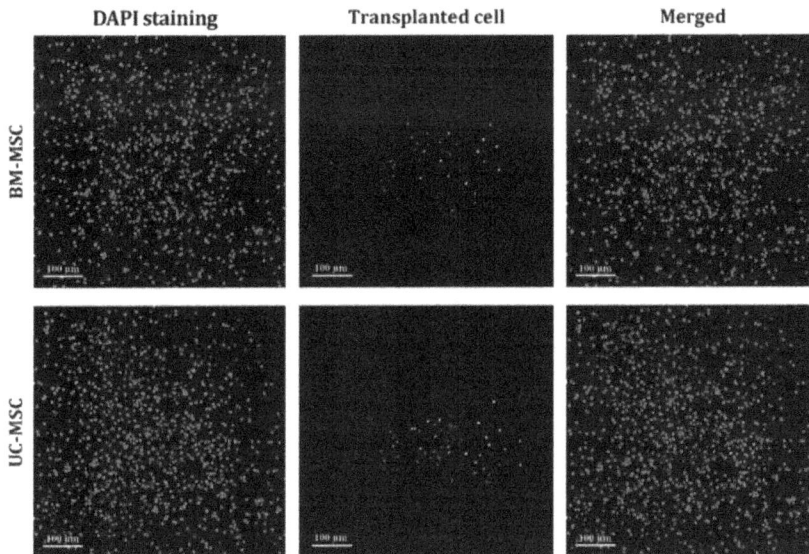

Fig. 4 Immunohistochemistry staining for assessment of human bone marrow-derived mesenchymal stem cell (*BM-MSC*) and umbilical cord-derived mesenchymal stem cell (*UC-MSC*) survival. Host cells are stained by 4',6-diamidino-2-phenylindole (*DAPI*). Mouse monoclonal antibody against human nuclei positive cells (transplanted cells) continue to survive in the spinal cord after 8 weeks (n = 3 in each group). The survival rates were 0.36 ± 0.06 % and 0.57 ± 0.06 % in BM-MSC and UC-MSC groups, respectively (*p* = 0.002)

cells survive in the spinal cord and prevent formation of cavities due to SCI. However, the survival rate of UC-MSCs was significantly higher than BM-MSCs. Electrophysiological evaluation confirmed these findings. Evoked response of the WDR to Aβ, Aδ, and C fiber stimulations, post-discharge and wind up of these second order neurons (WDR) had significantly increased 8 weeks after SCI, while stem cell transplantation decreased the responses to painful stimulation. Statistical analysis showed that animals transplanted with UC-MSCs had better recovery in wind up phenomena. Although the pain threshold in animals transplanted with UC-MSCs and BM-MSCs was not at the level of normal animals, this might be due to an inability of these cells to fully recover post-discharge and wind up phenomena when transplanted into the spinal cord.

Efficacy of MSC transplantation on neuropathic pain depends on numerous factors, such as source of the cells (donor species) [44], number and source of transplanted cells, route of administration (at injury site or intravascular), type of injury (central or peripheral), type of transplant (allogenic or xenogenic), time between injury and cell transplantation, and follow-up duration [45–48]. This explains why MSC transplantation has led to neuropathic pain symptom relief in some studies [44, 45, 47, 49–58] but not others [48, 58–60], or even symptom worsening [46, 48]. The results of the present study revealed that transplantation of 1 million mesenchymal cells derived from bone marrow, umbilical cord, and adipose tissue led to neuropathic pain symptom relief after transplantation.

To our knowledge, the present study is the first in vivo study comparing the efficacy of MSCs derived from different sources on motor recovery and neuropathic pain symptom relief following SCI. Therefore, our results cannot be directly compared to other studies. However, Jin et al. [61] compared the ability of mesenchymal cells derived from bone marrow, umbilical cord, and adipose tissue to differentiate to various tissues in an in-vitro study. They found that mesenchymal cells derived from umbilical cord had stronger proliferation ability and anti-inflammatory effects compared to other cells. They suggest that cells derived from umbilical cord have an advantage over those derived from adult tissues (such as bone marrow and adipose tissue) and can be used as an efficient model in the clinic [61]. In addition, Kern et al. [62] showed that, from morphologic and immune phenotype points of view, BM-MSCs and UC-MSCs are not significantly different. Although the number of colonies was lower in cells isolated from umbilical cord, in the next passages their proliferation and survival was higher than other studied cells [62]. Baksh et al. [63] compared proliferation and differentiation of mesenchymal cells isolated from bone marrow and umbilical cord and revealed that umbilical cord mesenchymal cells keep their mesenchymal characteristics for longer and express signaling pathways similar to those of mesenchymal cells isolated from bone marrow. Panepucci et al. [64] compared gene expression characteristics of

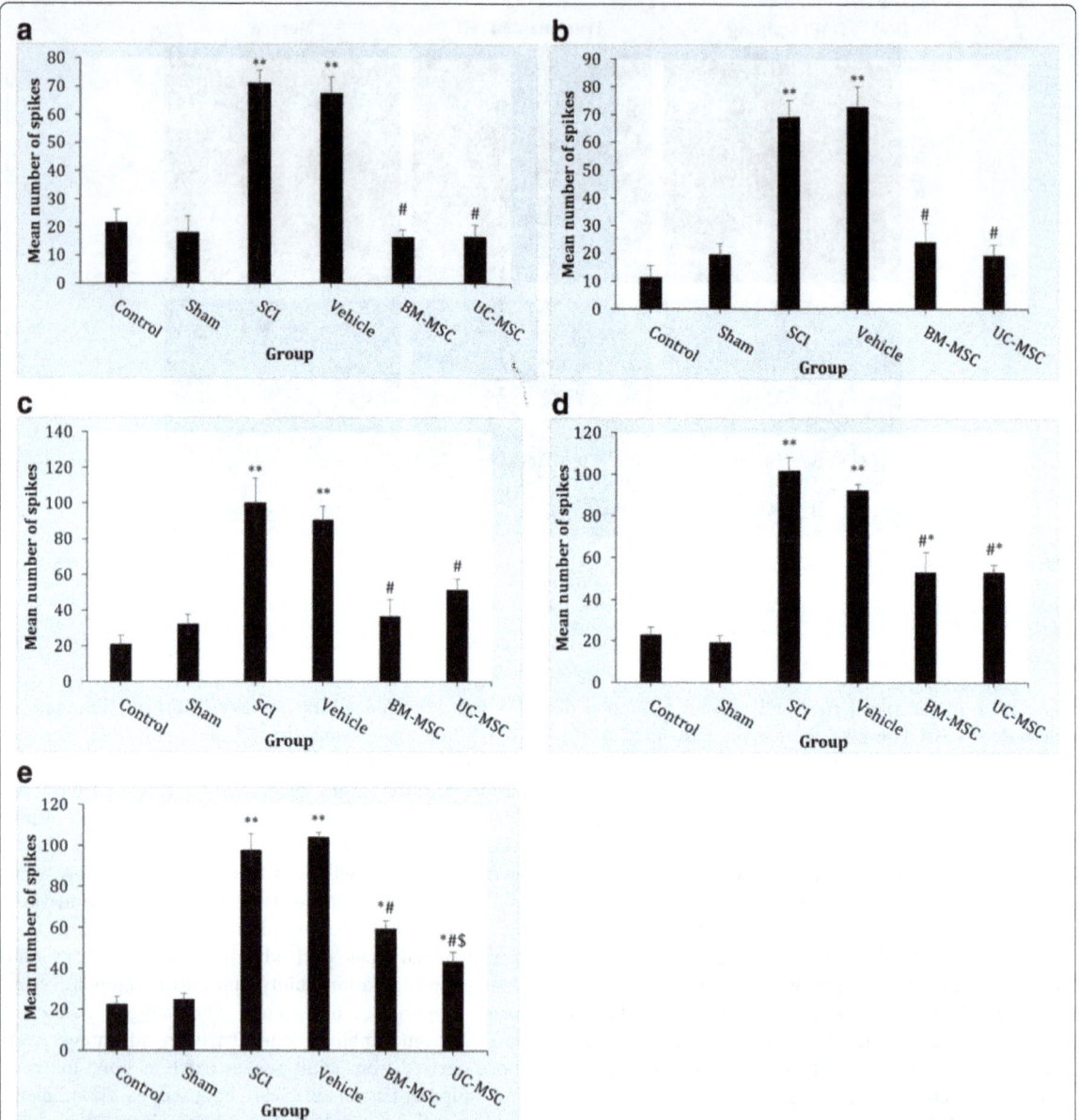

Fig. 5 Single-unit recording of wide dynamic range (WDR) neurons in the dorsal horn of the L4 and L5 spinal cord 8 weeks after transplantation of human bone marrow-derived mesenchymal stem cells (*BM-MSCs*) and umbilical cord-derived mesenchymal stem cells (*UC-MSCs*). Evoked potential of the WDR neurons to stimulations received from Aβ neurons (**a**), Aδ (**b**), and C fibers (**c**), post-discharge response (**d**), and wind up (**e**) are presented as means ± SEM (n = 6 in each group). #$p < 0.001$, versus SCI group; *$p < 0.01$, **$p < 0.001$, versus sham animals; $p < 0.01$, versus BM-MSC treated animals. *SCI* spinal cord injury

mesenchymal cells isolated from bone marrow and umbilical cord and demonstrated that bone marrow-derived cells tend to express genes related to antimicrobial and osteogenesis processes, while umbilical cord-derived cells tend to express genes playing a part in angiogenesis and intracellular matrix renewal. Therefore, it can be suggested that, since mesenchymal cells isolated from umbilical cord have a higher survival rate in the tissue and show more anti-inflammatory and angiogenic effects compared to stem cells derived from bone marrow, they are expected to have higher efficacy in injury healing and symptom relief in neuropathic animals. This might be the reason that wind up recovery in the group treated with UC-MSCs was better than in the group treated with BM-MSCs in this study.

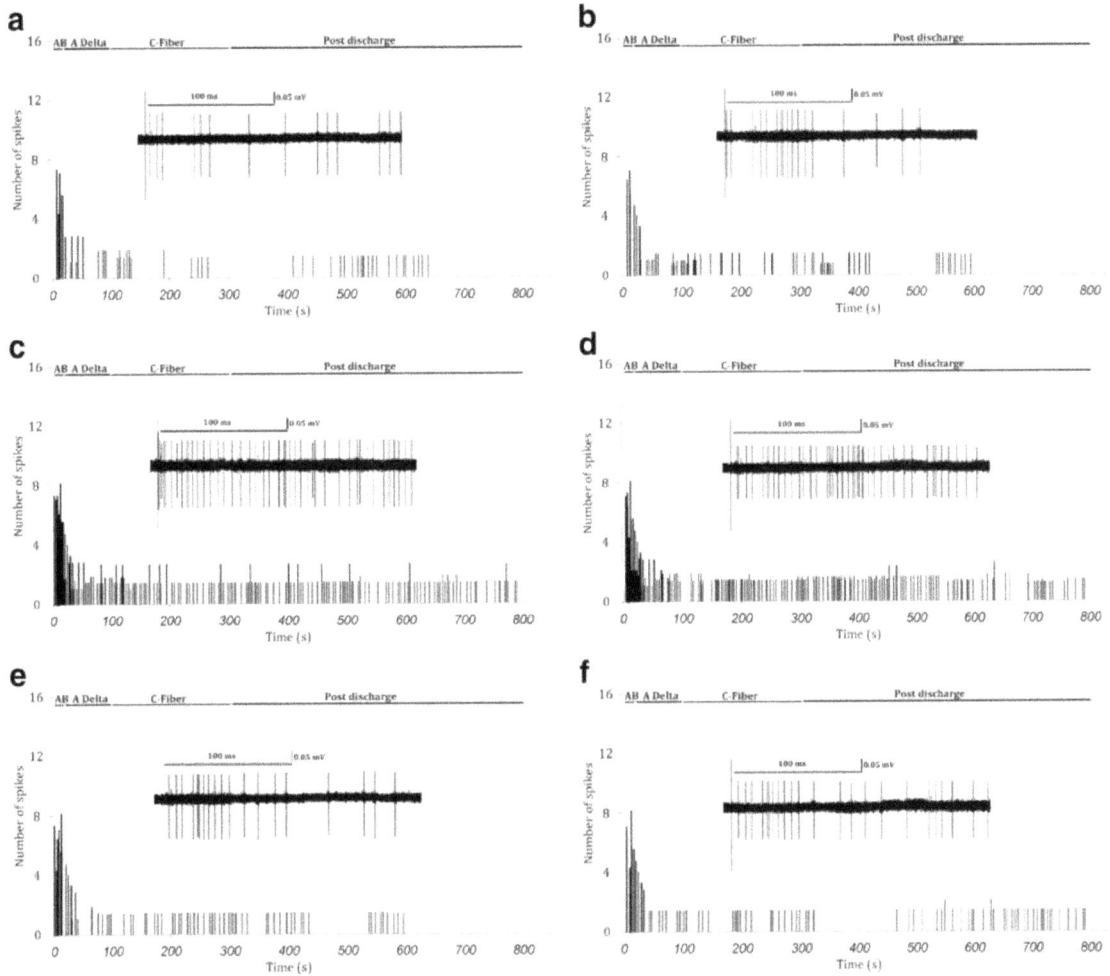

Fig. 6 Examples of the raw evoked responses and post-stimulus time histogram of WDR neurons in control (**a**), sham (**b**), SCI (**c**), vehicle-treated (**d**), human BM-MSC (**e**) and UC-MSC (**f**) animals

The present study showed that BM-MSCs and UC-MSCs are able to survive until 8 weeks after transplantation. Mannoji et al. [65] also reported similar results in this regard. Kim et al. [66] evaluated the effect of bone marrow-derived stem cell transplantation on SCI in rats and determined that these cells are present at the site of injury until 6 weeks after transplantation. They stated that the expression level of neuronal growth factor (NGF) and brain-derived neurotrophic factor (BDNF) in the group treated with mesenchymal cells was higher than their SCI group [66]. In addition, Veeravalli et al. [67] evaluated the effect of umbilical cord-derived stem cell transplantation and revealed that these cells were present at the site of injury at the 3-week follow-up. Transplantation of these cells had induced metalloproteinase 2 expression at the site of injury. Inhibition of this cellular matrix protein resulted in a decrease in the protective effect of mesenchymal cells.

They concluded that transplantation of umbilical cord-derived stem cells prepares the environment for endogenic regeneration by inducing metalloproteinase 2 expression and inhibition of glial scar formation [67].

These studies all indicated that MSCs are able to survive in the tissue for a long time. Most studies show that these cells have a protective role in the tissue and may reduce inflammation caused by SCI. By secretion of cytokines and growth factors, MSCs can also play a role in neural regeneration [68]. Thus, it seems that they provide a favorable environment for endogenic regeneration. In the present study, LFB staining showed that MSC transplantation from two studied sources prevented cavity formation at the site of injury, which might be due to their anti-inflammatory role.

Two inflammation phases are present in SCI. A primary or acute phase at the time of injury causes some cells to die or experience ischemia due to direct compression or a decrease in blood flow. However, the major damage is done in the second phase. This phase lasts weeks or months, and causes spinal cord tissue damage by various mechanisms: apoptosis induction, initiation of astroglial scar formation, central chromatolysis, deficiency in myelin gene expression, degradation of myelin in remaining axons, glutamate over-induction, invasion of immune cells to the site of injury and secretion of inflammatory cytokines, and endothelial damage due to ischemia–reperfusion, and so forth. [69]. Mesenchymal cell transplantation can reverse these damaging processes to some extent. These cells have immunomodulatory characteristics [21, 70–75] and can minimize inflammation and immune system-induced damage if transplanted at the right time [76]. Transplantation of these cells can decrease glial cell hypertrophy and proliferation, and improve recovery with the aid of bioactive molecules, reduction of cytokine secretion, and growth factors. Their angiogenic role can also be used in angiogenesis in the spinal cord [77, 78].

To our knowledge, this study is the first to compare the electrophysiological changes of the spinal cord after BM-MSC and UC-MSC transplantation. The findings of the present study showed that SCI leads to an increase in stimulations from Aβ, Aδ, and C fibers to WDR neurons at below-injury levels, and that MSC treatment recovered these changes to the level of those of intact groups. It has been reported that transplantation of olfactory cells resulted in improvement in touch stimulations and cord dorsum potentials [79]. It was also shown that SCI at the T3 level resulted in complete neural disconnection between the parts on either side of the SCI site. However, after neural stem cell transplantation, an electrical stimulation at the C7 level led to recording of an evoked potential in the T6 area [80]. Erceg et al. [81] showed that transplantation of oligodendrocyte and motoneuron progenitor cells into injured spinal cord could result in restoration of motor pathways that were damaged due to SCI to some extent, and lead to appearance of motor-evoked potential in electrophysiological recordings [81]. Yasuda et al. [82] also revealed that neural stem/progenitor cell transplantation brings about reappearance of motor-evoked potential post-SCI. In addition, Ziegler et al. [83] reported that olfactory ensheathing glial cell transplantation after complete SCI led to reappearance of motor-evoked potentials. In another study, comparison of electrophysiological findings in a rat model showed that somatosensory evoked potential in the BM-MSC treated group was not different from the Schwann cell group [84].

Finally, we should note that, in this study, about 1 million cells were transplanted a week after injury. The reason for the selection of this protocol was its similarity to clinical conditions. In the clinical trials performed recently, mean injected cell count was about 1–3 million cells [85–88]. Additionally, in most clinical conditions, transplantation of cells at the time of injury is not possible and preparation of the patients and cells for transplantation takes at least 1 week to a few months. We cannot determine the mechanism(s) that led to the improvement observed. The cells might have prevented inflammation and stopped the damaging process in the initial days with no further function after that or, on the contrary, they could have continued to exert a protective role. Further study is needed to determine this.

The present study has some limitation. First, the cells are cultured in fetal bovine serum, which is not ideal for application in humans. However, using fetal bovine serum is in accordance with cell culture protocols. Therefore, when translating the result to a clinical trial an alternative for fetal bovine serum will be needed. Second, we compared the efficacy of BM-MSC and UC-MSC xenotransplantation on improvement of SCI 1 week after induction of compression injury. Since we showed the efficacy of BM-MSCs are affected by timing of intervention, the method used for SCI induction, and dosage of cell therapy in a previous meta-analysis [14], generalizing our findings should be done with caution.

Conclusion

The results of the present study showed that BM-MSC and UC-MSC transplantations alleviated the symptoms of neuropathic pain and resulted in subsequent motor recovery after SCI. Efficacy of both sources was similar for symptom relief. These cells survived in the tissue at least 8 weeks and prevented cavity formation due to SCI. However, the survival rate of UC-MSCs was significantly higher than BM-MSCs. Electrophysiological evaluations showed that transplantation of UC-MSCs brings about better results than BM-MSCs in wind up of WDR neurons.

Abbreviations
BBB: Basso, Beattie, and Bresnahan; BM-MSC: Bone marrow-derived mesenchymal stem cell; DMEM: Dulbecco's modified Eagle's medium; LFB: Luxol fast blue; MSC: Mesenchymal stem cell; PBS: Phosphate-buffered saline; SCI: Spinal cord injury; UC-MSC: Umbilical cord-derived mesenchymal stem cell; WDR: Wide dynamic range.

Competing interests
The authors declare that they have no competing interests.

Authors' contributions
FN, MY, and HSM designed the study. MY and AJ participated in data collection. MH and MK participated in analysis and interpretation. MY and FN wrote first draft of the work. All authors revised it critically for important intellectual content. All authors have provided final approval of the version to be published and has agreed to be accountable for all aspects of the work.

Acknowledgment

We kindly appreciate Dr. Zahra Bahari and Mr. Adel Salari for their valuable help. This research has been supported by Tehran University of Medical Sciences & Health Services grant (Number: 92-03-13-22014).

Author details

[1]Electrophysiology Research Center, Tehran University of Medical Sciences, Tehran, Iran. [2]Department of Physiology, School of Medicine, Tehran University of Medical Sciences, Tehran, Iran. [3]Physiology Research Center, Iran University of Medical Sciences, Tehran, Iran. [4]Department of Physiology, Iran University of Medical Sciences, Tehran, Iran. [5]Department of Physiology, Shahid Beheshti University of Medical Sciences, Tehran, Iran. [6]Neuroscience Research Center, Shahid Beheshti University of Medical Sciences, Tehran, Iran. [7]Department of Epidemiology and Biostatistics, School of Public Health, Tehran University of Medical Sciences, Tehran, Iran. [8]Pediatric Chronic Kidney Disease Research Center, Childrens Hospital Medical Center, Tehran University of Medical Sciences, Tehran, Iran.

References

1. Tate DG, Forchheimer MB, Karana-Zebari D, Chiodo AE, Kendall Thomas JY. Depression and pain among inpatients with spinal cord injury and spinal cord disease: differences in symptoms and neurological function. Disabil Rehabil. 2013;35(14):1204–12.
2. Kumru H, Soler D, Vidal J, Navarro X, Tormos J, Pascual-Leone A, et al. The effects of transcranial direct current stimulation with visual illusion in neuropathic pain due to spinal cord injury: an evoked potentials and quantitative thermal testing study. Eur J Pain. 2013;17(1):55–66.
3. Sharp K, Boroujerdi A, Steward O, Luo ZD. A rat chronic pain model of spinal cord contusion injury. Methods Mol Biol (Clifton, NJ). 2012;851:195–203.
4. Finnerup NB. Pain in patients with spinal cord injury. Pain. 2013;154:S71-6.
5. Finnerup NB, Otto M, McQuay HJ. Algorithm for neuropathic pain treatment: an evidence based proposal. Pain. 2005;118:289–305.
6. Backonja MM, Irving GA, Argoff C. Rational multidrug therapy in the treatment of neuropathic pain. Curr Pain Headache Rep. 2006;10:34–8.
7. Marineo G, Iorno V, Gandini C, Moschini V, Smith TJ. Scrambler therapy may relieve chronic neuropathic pain more effectively than guideline-based drug management: results of a pilot, randomized, controlled trial. J Pain Symptom Manag. 2012;43(1):87–95.
8. Hama A, Sagen J. Behavioral characterization and effect of clinical drugs in a rat model of pain following spinal cord compression. Brain Res. 2007;1185:117–28.
9. Sahni V, Kessler JA. Stem cell therapies for spinal cord injury. Nature Rev Neurol. 2010;6(7):363–72.
10. Kabu S, Gao Y, Kwon BK, Labhasetwar V. Drug delivery, cell-based therapies, and tissue engineering approaches for spinal cord injury. J Control Release. 2015;219:141–54.
11. Yazdani SO, Pedram M, Hafizi M, Kabiri M, Soleimani M, Dehghan MM et al. A comparison between neurally induced bone marrow derived mesenchymal stem cells and olfactory ensheathing glial cells to repair spinal cord injuries in rat. Tissue Cell. 2012;44(4):205-13.
12. Sun D, Gugliotta M, Rolfe A, Reid W, McQuiston AR, Hu W, et al. Sustained survival and maturation of adult neural stem/progenitor cells after transplantation into the injured brain. J Neurotrauma. 2011;28(6):961–72.
13. Mark Richardson R, Broaddus WC, Holloway KL, Fillmore HL. Grafts of adult subependymal zone neuronal progenitor cells rescue hemiparkinsonian behavioral decline. Brain Res. 2005;1032(1):11–22.
14. Hosseini M, Yousefifard M, Aziznejad H, Nasirinezhad F. The effect of bone marrow-derived mesenchymal stem cell transplantation on allodynia and hyperalgesia in neuropathic animals: a systematic review with meta-analysis. Biol Blood Marrow Transplant. 2015;21(9):1537–44. doi:10.1016/j.bbmt.2015.05.008.
15. Nakamura M, Okano H. Cell transplantation therapies for spinal cord injury focusing on induced pluripotent stem cells. Cell Res. 2013;23(1):70–80.
16. Volarevic V, Erceg S, Bhattacharya SS, Stojkovic P, Horner P, Stojkovic M. Stem cell-based therapy for spinal cord injury. Cell Transplant. 2013;22(8):1309–23.
17. Klopp AH, Gupta A, Spaeth E, Andreeff M, Marini F. Concise review: Dissecting a discrepancy in the literature: do mesenchymal stem cells support or suppress tumor growth? Stem Cells. 2011;29(1):11–9.
18. Pittenger MF, Mackay AM, Beck SC, Jaiswal RK, Douglas R, Mosca JD, et al. Multilineage potential of adult human mesenchymal stem cells. Science. 1999;284(5411):143–7.
19. Anjos-Afonso F, Siapati EK, Bonnet D. In vivo contribution of murine mesenchymal stem cells into multiple cell-types under minimal damage conditions. J Cell Sci. 2004;117(23):5655–64.
20. Wakitani S, Imoto K, Yamamoto T, Saito M, Murata N, Yoneda M. Human autologous culture expanded bone marrow mesenchymal cell transplantation for repair of cartilage defects in osteoarthritic knees. Osteoarthr Cartil. 2002;10(3):199–206.
21. Nauta AJ, Fibbe WE. Immunomodulatory properties of mesenchymal stromal cells. Blood. 2007;110(10):3499–506.
22. Puissant B, Barreau C, Bourin P, Clavel C, Corre J, Bousquet C, et al. Immunomodulatory effect of human adipose tissue-derived adult stem cells: comparison with bone marrow mesenchymal stem cells. Br J Haematol. 2005;129(1):118–29.
23. Bae JS, Han HS, Youn DH, Carter JE, Modo M, Schuchman EH, et al. Bone marrow-derived mesenchymal stem cells promote neuronal networks with functional synaptic transmission after transplantation into mice with neurodegeneration. Stem Cells. 2007;25(5):1307–16.
24. Cho KJ, Trzaska KA, Greco SJ, McArdle J, Wang FS, Ye JH, et al. Neurons derived from human mesenchymal stem cells show synaptic transmission and can be induced to produce the neurotransmitter substance P by interleukin-1α. Stem Cells. 2005;23(3):383–91.
25. Lu L-L, Liu Y-j, Yang S-G, Zhao Q-J, Wang X, Gong W, et al. Isolation and characterization of human umbilical cord mesenchymal stem cells with hematopoiesis-supportive function and other potentials. Haematologica. 2006;91(8):1017–26.
26. Miao Z, Jin J, Chen L, Zhu J, Huang W, Zhao J, et al. Isolation of mesenchymal stem cells from human placenta: comparison with human bone marrow mesenchymal stem cells. Cell Biol Int. 2006;30(9):681–7.
27. Scherjon SA, Kleijburg-van der Keur C, de Groot-Swings GM, Claas FH, Fibbe WE, Kanhai HH. Isolation of mesenchymal stem cells of fetal or maternal origin from human placenta. Stem Cells. 2004;22(7):1338–45.
28. Cao Y, Sun Z, Liao L, Meng Y, Han Q, Zhao RC. Human adipose tissue-derived stem cells differentiate into endothelial cells in vitro and improve postnatal neovascularization in vivo. Biochem Biophys Res Commun. 2005;332(2):370–9.
29. Dasari VR, Veeravalli KK, Dinh DH. Mesenchymal stem cells in the treatment of spinal cord injuries: a review. World J Stem Cells. 2014;6(2):120–33. doi:10.4252/wjsc.v6.i2.120.
30. Fan C-G, Q-j Z, J-r Z. Therapeutic potentials of mesenchymal stem cells derived from human umbilical cord. Stem Cell Rev Rep. 2010;7(1):195–207. doi:10.1007/s12015-010-9168-8.
31. Cho PS, Messina DJ, Hirsh EL, Chi N, Goldman SN, Lo DP, et al. Immunogenicity of umbilical cord tissue–derived cells. Blood. 2008;111(1):430–8.
32. Sarugaser R, Ennis J, Stanford WL, Davies JE. Isolation, propagation, and characterization of human umbilical cord perivascular cells (HUCPVCs). Methods Mol Biol. 2009;482:269–79. doi:10.1007/978-1-59745-060-7_17.
33. Karaoz E, Aksoy A, Ayhan S, Sarıboyacı AE, Kaymaz F, Kasap M. Characterization of mesenchymal stem cells from rat bone marrow: ultrastructural properties, differentiation potential and immunophenotypic markers. Histochem Cell Biol. 2009;132(5):533–46.
34. Rivlin A, Tator C. Effect of duration of acute spinal cord compression in a new acute cord injury model in the rat. Surg Neurol. 1978;10(1):38–43.
35. Poon PC, Gupta D, Shoichet MS, Tator CH. Clip compression model is useful for thoracic spinal cord injuries: histologic and functional correlates. Spine. 2007;32(25):2853–9. doi:10.1097/BRS.0b013e31815b7e6b.
36. von Euler M, Seiger Å, Sundström E. Clip compression injury in the spinal cord: a correlative study of neurological and morphological alterations. Exp Neurol. 1997;145(2):502–10.
37. Bruce JC, Oatway MA, Weaver LC. Chronic pain after clip-compression injury of the rat spinal cord. Exp Neurol. 2002;178(1):33–48. http://dx.doi.org/10.1006/exnr.2002.8026.
38. Dolan EJ, Tator CH. A new method for testing the force of clips for aneurysms or experimental spinal cord compression. J Neurosurg. 1979;51(2):229–33.
39. Basso DM, Beattie MS, Bresnahan JC. A sensitive and reliable locomotor rating scale for open field testing in rats. J Neurotrauma. 1995;12(1):1–21.
40. Hosseini M, Karami Z, Janzadenh A, Jameie SB, Mashadi ZH, Yousefifard M, et al. The effect of intrathecal administration of muscimol on modulation of

neuropathic pain symptoms resulting from spinal cord injury; an experimental study. Emergency. 2014;2(4):151–7.

41. D'Mello R, Dickenson AH. Spinal cord mechanisms of pain. Br J Anaesth. 2008;101(1):8–16. doi:10.1093/bja/aen088.

42. Jergova S, Hentall ID, Gajavelli S, Varghese MS, Sagen J. Intraspinal transplantation of GABAergic neural progenitors attenuates neuropathic pain in rats: a pharmacologic and neurophysiological evaluation. Exp Neurol. 2012;234:39–49.

43. Goto N. Discriminative staining methods for the nervous system: Luxol fast blue-periodic acid-Schiff-hematoxylin triple stain and subsidiary staining methods. Biotech Histochem. 1987;62(5):305–15.

44. Neuhuber B, Timothy Himes B, Shumsky JS, Gallo G, Fischer I. Axon growth and recovery of function supported by human bone marrow stromal cells in the injured spinal cord exhibit donor variations. Brain Res. 2005;1035(1):73–85.

45. Vaquero J, Zurita M, Oya S, Santos M. Cell therapy using bone marrow stromal cells in chronic paraplegic rats: systemic or local administration? Neurosci Lett. 2006;398(1):129–34.

46. Urdzíková L, Jendelová P, Glogarová K, Burian M, Hájek M, Syková E. Transplantation of bone marrow stem cells as well as mobilization by granulocyte-colony stimulating factor promotes recovery after spinal cord injury in rats. J Neurotrauma. 2006;23(9):1379–91.

47. Klass M, Gavrikov V, Drury D, Stewart B, Hunter S, Denson DD, et al. Intravenous mononuclear marrow cells reverse neuropathic pain from experimental mononeuropathy. Anesth Analg. 2007;104(4):944–8.

48. Amemori T, Jendelová P, Ruzicková K, Arboleda D, Syková E. Co-transplantation of olfactory ensheathing glia and mesenchymal stromal cells does not have synergistic effects after spinal cord injury in the rat. Cytotherapy. 2010;12(2):212–25.

49. Lee B, Kim J, Kim SJ, Lee H, Chang JW. Constitutive GABA expression via a recombinant adeno-associated virus consistently attenuates neuropathic pain. Biochem Biophys Res Commun. 2007;357(4):971–6.

50. Lee KH, Suh-Kim H, Choi JS, Jeun S, Kim EJ, Kim S, et al. Human mesenchymal stem cell transplantation promotes functional recovery following acute spinal cord injury in rats. Acta Neurobiol Exp. 2007;67(1):13–22.

51. Musolino PL, Coronel MF, Hökfelt T, Villar MJ. Bone marrow stromal cells induce changes in pain behavior after sciatic nerve constriction. Neurosci Lett. 2007;418(1):97–101.

52. Guo W, Wang H, Zou S, Gu M, Watanabe M, Wei F, et al. Bone marrow stromal cells produce long-term pain relief in rat models of persistent pain. Stem Cells. 2011;29(8):1294–303.

53. Kumagai G, Tsoulfas P, Toh S, McNiece I, Bramlett HM, Dietrich WD. Genetically modified mesenchymal stem cells (MSCs) promote axonal regeneration and prevent hypersensitivity after spinal cord injury. Exp Neurol. 2013;248:369–80.

54. Siniscalco D, Giordano C, Galderisi U, Luongo L, de Novellis V, Rossi F, et al. Long-lasting effects of human mesenchymal stem cell systemic administration on pain-like behaviors, cellular, and biomolecular modifications in neuropathic mice. Front Integr Neurosci. 2011;5:1–10.

55. Vaysse L, Sol J, Lazorthes Y, Courtade-Saidi M, Eaton M, Jozan S. GABAergic pathway in a rat model of chronic neuropathic pain: modulation after intrathecal transplantation of a human neuronal cell line. Neurosci Res. 2011;69(2):111–20.

56. Zhang J, Wu D, Xie C, Wang H, Wang W, Zhang H, et al. Tramadol and propentofylline coadministration exerted synergistic effects on rat spinal nerve ligation-induced neuropathic pain. PLoS One. 2013;8(8):e72943.

57. Yang CC, Shih YH, Ko MH, Hsu SY, Cheng H, Fu YS. Transplantation of human umbilical mesenchymal stem cells from Wharton's jelly after complete transection of the rat spinal cord. PLoS One. 2008;3(10):e3336.

58. Roh D-H, Seo M-S, Choi H-S, Park S-B, Han H-J, Beitz AJ, et al. Transplantation of human umbilical cord blood or amniotic epithelial stem cells alleviates mechanical allodynia after spinal cord injury in rats. Cell Transplant. 2013;22(9):1577–90.

59. Schäfer S, Berger JV, Deumens R, Goursaud S, Hanisch U-K, Hermans E. Influence of intrathecal delivery of bone marrow-derived mesenchymal stem cells on spinal inflammation and pain hypersensitivity in a rat model of peripheral nerve injury. J Neuroinflammation. 2014;11(1):157.

60. Torres-Espín A, Redondo-Castro E, Hernández J, Navarro X. Bone marrow mesenchymal stromal cells and olfactory ensheathing cells transplantation after spinal cord injury—a morphological and functional comparison in rats. Eur J Neurosci. 2014;39(10):1704–17.

61. Jin HJ, Bae YK, Kim M, Kwon S-J, Jeon HB, Choi SJ, et al. Comparative analysis of human mesenchymal stem cells from bone marrow, adipose tissue, and umbilical cord blood as sources of cell therapy. Int J Mol Sci. 2013;14(9):17986–8001.

62. Kern S, Eichler H, Stoeve J, Klüter H, Bieback K. Comparative analysis of mesenchymal stem cells from bone marrow, umbilical cord blood, or adipose tissue. Stem Cells. 2006;24(5):1294–301. doi:10.1634/stemcells.2005-0342.

63. Baksh D, Yao R, Tuan RS. Comparison of proliferative and multilineage differentiation potential of human mesenchymal stem cells derived from umbilical cord and bone marrow. Stem Cells. 2007;25(6):1384–92. doi:10.1634/stemcells.2006-0709.

64. Panepucci RA, Siufi JLC, Silva WA, Proto-Siquiera R, Neder L, Orellana M, et al. Comparison of gene expression of umbilical cord vein and bone marrow–derived mesenchymal stem cells. Stem Cells. 2004;22(7):1263–78. doi:10.1634/stemcells.2004-0024.

65. Mannoji C, Koda M, Kamiya K, Dezawa M, Hashimoto M, Furuya T, et al. Transplantation of human bone marrow stromal cell-derived neuroregenerative cells promotes functional recovery after spinal cord injury in mice. Acta Neurobiol Exp. 2014;74:479–88.

66. Kim J-W, Ha K-Y, Molon JN, Kim Y-H. Bone marrow–derived mesenchymal stem cell transplantation for chronic spinal cord injury in rats: comparative study between intralesional and intravenous transplantation. Spine. 2013;38(17):E1065–74.

67. Veeravalli KK, Dasari VR, Tsung AJ, Dinh DH, Gujrati M, Fassett D, et al. Human umbilical cord blood stem cells upregulate matrix metalloproteinase-2 in rats after spinal cord injury. Neurobiol Dis. 2009;36(1):200–12.

68. Joyce N, Annett G, Wirthlin L, Olson S, Bauer G, Nolta JA. Mesenchymal stem cells for the treatment of neurodegenerative disease. Regen Med. 2010;5(6):933–46. doi:10.2217/rme.10.72.

69. Oyinbo CA. Secondary injury mechanisms in traumatic spinal cord injury: a nugget of this multiply cascade. Acta Neurobiol Exp (Wars). 2011;71(2):281–99.

70. Montespan F, Deschaseaux F, Sensébé L, Carosella ED, Rouas-Freiss N. Osteodifferentiated mesenchymal stem cells from bone marrow and adipose tissue express HLA-G and display immunomodulatory properties in HLA-mismatched settings: implications in bone repair therapy. J Immunol Res. 2014;2014:1–10.

71. Hou R, Liu R, Niu X, Chang W, Yan X, Wang C, et al. Biological characteristics and gene expression pattern of bone marrow mesenchymal stem cells in patients with psoriasis. Exp Dermatol. 2014;23(7):521–3.

72. Menendez P, Rodriguez R, Delgado M, Rosu-Myles M. Human bone marrow mesenchymal stem cells lose immunosuppressive and anti-inflammatory properties upon oncogenic transformation. Exp Hematol. 2014;42(8):S49.

73. Coulson-Thomas VJ, Gesteira TF, Hascall V, Kao W. Umbilical cord mesenchymal stem cells suppress host rejection the role of the glycocalyx. J Biol Chem. 2014;289(34):23465–81.

74. Alunno A, Montanucci P, Bistoni O, Basta G, Caterbi S, Pescara T, et al. In vitro immunomodulatory effects of microencapsulated umbilical cord Wharton jelly-derived mesenchymal stem cells in primary Sjögren's syndrome. Rheumatology. 2015;54(1):163–8.

75. Wang M, Yang Y, Yang D, Luo F, Liang W, Guo S, et al. The immunomodulatory activity of human umbilical cord blood-derived mesenchymal stem cells in vitro. Immunology. 2009;126(2):220–32.

76. Oudega M, Ritfeld G. Bone marrow-derived mesenchymal stem cell transplant survival in the injured rodent spinal cord. J Bone Marrow Res. 2014;2(146):2–9.

77. Kuchroo P, Dave V, Vijayan A, Viswanathan C, Ghosh D. Paracrine factors secreted by umbilical cord-derived mscs induce angiogenesis in vitro by a VEGF-independent pathway. Stem Cells Dev. 2014;24(4):437–50.

78. Hua J, He Z-G, Qian D-H, Lin S-P, Gong J, Meng H-B, et al. Angiopoietin-1 gene-modified human mesenchymal stem cells promote angiogenesis and reduce acute pancreatitis in rats. Int J Clin Exp Pathol. 2014;7(7):3580.

79. Toft A, Scott DT, Barnett SC, Riddell JS. Electrophysiological evidence that olfactory cell transplants improve function after spinal cord injury. Brain. 2007;130(4):970–84.

80. Lu P, Wang Y, Graham L, McHale K, Gao M, Wu D, et al. Long-distance growth and connectivity of neural stem cells after severe spinal cord injury. Cell. 2012;150(6):1264–73.

81. Erceg S, Ronaghi M, Oria M, Rosello MG, Arago MA, Lopez MG, et al. Transplanted oligodendrocytes and motoneuron progenitors generated from human embryonic stem cells promote locomotor recovery after spinal cord transection. Stem Cells. 2010;28(9):1541–9. doi:10.1002/stem.489.

82. Yasuda A, Tsuji O, Shibata S, Nori S, Takano M, Kobayashi Y, et al. Significance of remyelination by neural stem/progenitor cells transplanted into the injured spinal cord. Stem Cells. 2011;29(12):1983–94. doi:10.1002/stem.767.

83. Ziegler MD, Hsu D, Takeoka A, Zhong H, Ramon-Cueto A, Phelps PE, et al. Further evidence of olfactory ensheathing glia facilitating axonal regeneration after a complete spinal cord transection. Exp Neurol. 2011; 229(1):109–19. doi:10.1016/j.expneurol.2011.01.007.

84. Ban D-X, Ning G-Z, Feng S-Q, Wang Y, Zhou X-H, Liu Y, et al. Combination of activated Schwann cells with bone mesenchymal stem cells: the best cell strategy for repair after spinal cord injury in rats. Regen Med. 2011;6(6):707–20.

85. Ghobrial G, Haas C, Maulucci C, Lepore A, Fischer I. Promising advances in targeted cellular based therapies: treatment update in spinal cord injury. J Stem Cell Res Ther. 2014;4(170):2–7.

86. Yoon SH, Shim YS, Park YH, Chung JK, Nam JH, Kim MO, et al. Complete spinal cord injury treatment using autologous bone marrow cell transplantation and bone marrow stimulation with granulocyte macrophage-colony stimulating factor: phase I/II clinical trial. Stem Cells. 2007;25(8):2066–73.

87. Kumar AA, Kumar SR, Narayanan R, Arul K, Baskaran M. Autologous bone marrow derived mononuclear cell therapy for spinal cord injury: a phase I/II clinical safety and primary efficacy data. Exp Clin Transplant. 2009;7(4):241–8.

88. Jarocha D, Milczarek O, Kawecki Z, Wendrychowicz A, Kwiatkowski S, Majka M. Preliminary study of autologous bone marrow nucleated cells transplantation in children with spinal cord injury. Stem Cells Transl Med. 2014;3(3):395-404.

7

Sphere-forming cells from peripheral cornea demonstrate the ability to repopulate the ocular surface

Jeremy John Mathan, Salim Ismail, Jennifer Jane McGhee, Charles Ninian John McGhee and Trevor Sherwin[*]

Abstract

Background: The limbus forms the outer rim of the cornea at the corneoscleral junction and harbours a population of stem cells for corneal maintenance. Injuries to the limbus, through disease or accidents such as chemical injuries or burns, may lead to significant visual impairment due to depletion of the native stem cells of the tissue.

Methods: Sphere-forming cells were isolated from peripheral cornea for potential use as transplantable elements for limbal stem cell repopulation and limbal reconstruction. Immunocytochemistry, live cell imaging and quantitative PCR were used to characterize spheres and elucidate activity post implantation into human cadaveric corneal tissue.

Results: Spheres stained positively for stem cell markers ΔNP63α, ABCG2 and ABCB5 as well as the basal limbal marker and putative niche marker, notch 1. In addition, spheres also stained positively for markers of corneal cells, vimentin, keratin 3, keratocan and laminin, indicating a heterogeneous mix of stromal and epithelial-origin cells. Upon implantation into decellularized corneoscleral tissue, 3D, polarized and radially orientated cell migration with cell proliferation was observed. Cells migrated out from the spheres and repopulated the entire corneal surface over 14 days. Post-implantation analysis revealed qualitative evidence of stem, stromal and epithelial cell markers while quantitative PCR showed a quantitative reduction in keratocan and laminin expression indicative of an enhanced progenitor cell response. Proliferation, quantified by PCNA expression, significantly increased at 4 days subsequently followed by a decrease at day 7 post implantation.

Conclusion: These observations suggest great promise for the potential of peripheral corneal spheres as transplantable units for corneal repair, targeting ocular surface regeneration and stem cell repopulation.

Keywords: Cell culture, Cornea, Holoclone, Immunocytochemistry, Quantitative PCR, Spheroid

Background

The anterior ocular surface is a continuous sheet of tissue that consists of the transparent cornea and the more peripherally located conjunctiva overlying the opaque, white, sclera. The cornea transitions into the sclera at the zone known as the corneoscleral junction or the limbus. The corneal limbus, the in-vivo location of corneal epithelial stem cells, is the transitional region between the cornea and conjunctiva/sclera. Anatomically, the basal layer of the limbal epithelium appears corrugated because it is arranged in rete pegs (finger-like projections of the epithelium into the stroma below), also named interpalisadal rete ridges, with the upward projections of the stroma termed the palisades of Vogt [1–4]. The longstanding view has been that these rete ridges harbour the cells for corneal maintenance. Indeed, the limbal stem cells divide to give rise to progeny which maintain the structure and function of the cornea [5].

The limbal epithelial crypts (long extensions of the interpalisadal rete ridges) described relatively recently have also been shown to be potential reservoirs of stem cells [6]. Various injurious processes such as intrinsic

* Correspondence: t.sherwin@auckland.ac.nz

Department of Ophthalmology, New Zealand National Eye Centre, Faculty of Medical and Health Sciences, The University of Auckland, Private Bag 92019, Auckland 1010, New Zealand

diseases, chemical injuries and thermal burns [7] can damage the limbal environment and deplete the stem cell population therein, thus impairing the regenerative capacity of the cornea leading to redness and pain, persisting epithelial defects, corneal vascularization, conjunctivalization and ultimately severe visual impairment or blindness. Severe depletion of stem cells within the limbal environment can lead to a condition known as limbal stem cell deficiency (LSCD).

The stem cell niche or microenvironment is responsible for maintaining the stem cell phenotype and directing differentiating cells along a corneal cell differentiation pathway. The importance of the stem cell niche—which includes the cells of the tissue that surround the stem cells along with the signalling molecules they secrete [8–10] as well as the extracellular matrix in directing stem cell differentiation within tissue [11, 12]—is well established, especially within the study of haematopoietic stem cells.

The limbal microenvironment has been shown to be capable of directing the programming of hair follicle stem cells toward a more cornea-like phenotype [13]. In mice, pluripotent stem cells were stimulated to acquire corneal epithelial morphology when exposed to the limbal stromal environment [14]. It is clear that limbal stromal cells are important in maintaining both the limbal and corneal epithelial phenotype. Additionally, the observation that differentiated cells of the central cornea can be stimulated to de-differentiate and develop a dermal phenotype when placed in that environment [15] underscores the need to consider stem cells in context of their immediate microenvironment.

The limbal stroma is an integral part of this microenvironment. Firstly, it is noted that limbal stromal fibroblasts cross the basement membrane and adhere to stem and progenitor cells in the basal epithelium above [16]. In-vitro exposure of limbal epithelial cells to the cells of the limbal stroma in co-culturing experiments also appears to increase certain indicators of stemness. This is manifested as an increase in the expansion capacity of cultured limbal epithelial cells [17] and an increase in putative stem cell markers such as p63 and ABCG2 [18], when compared with cells not cultured in the presence of limbal stromal cells.

Secondly, the limbal stromal environment appears to have the ability to discourage the presence of cells foreign to the niche and promote the presence of cells usually found within the niche. This is seen through the restoration of transparency in previously diseased, conjunctivalized and opacified, corneal epithelia when exposed to limbal stroma conditioned medium [19]. Observations of the loss of the limbal crypt structure being correlated with a decline in stemness of cells [20], the greater success in culturing stem cells in 3D compared with 2D culture systems [21] and the improved clonality when cellular connections are

maintained [22, 23] have lent support for the importance of the 3D extracellular matrix structure in maintaining the stemness of limbal epithelial cells in culture.

The sphere-forming assay is one approach for enriching cell populations with stem cells in vitro. This is a cell culture method that involves incubating cells in a serum-free medium supplemented with growth factors which selectively encourage the survival of stem and progenitor cells [24, 25]. Cells cultured from mice gradually form well-defined spherical entities, ranging from 50 to 150 µm and connected to each other via gap and adherence junctions, over a period of 7 days [24]. The sphere-forming assay has also been successfully used to generate spheres from cells isolated from human ocular tissue [26–28]. These spheres, although enriched with stem cells, are composed of a heterogeneous mix of cells within the continuum from stem to differentiated cells [29]. For peripheral corneal spheres, this is advantageous because this method of stem cell enrichment not only mimics the in-vivo heterogeneity of the natural limbus but also achieves this within a 3D format which allows for intercellular attachment.

Stem cell-enriched sphere-based therapy remains a promising treatment approach for corneal stem cell repopulation. We have previously shown human peripheral corneal spheres to be dynamic entities that demonstrate polarity and directed cell migration and are capable of initiating a wound healing response to injury in vitro [26]. The in-situ behaviour of these peripheral corneal spheres, however, is yet to be characterized. Here we advance on our previous work as we further characterize human peripheral corneal spheres in vitro, implant them into human donor corneoscleral rims and present evidence to show their capacity for in-situ ocular surface repopulation.

Methods
Human tissue
Fresh and frozen human corneoscleral rims and frozen human amniotic membrane were obtained from the New Zealand National Eye Bank with approval from the Northern X Regional Ethics Committee. Consent for human corneal tissue use for the purposes of research is attained prior to eye banking.

Sphere formation and culture
Spheres were generated from human limbal tissue using a cell extraction process and the sphere-forming assay essentially as described previously [30]. A single, entire human donor corneoscleral rim provided by the New Zealand National Eye bank was used to form a sphere batch. The clear cornea component of the rim was used, excising as close to the limbus as possible but excluding the sclera. Briefly, this process initially involved the removal of the corneal endothelium. The corneoscleral region was excised and the tissue was de-epithelialized

using a scalpel blade. The tissue was then incubated in 1.2 U/ml dispase II for 40 min at 37 °C. Subsequently, the tissue was incubated in 2 mg/ml collagenase and 0.5 mg/ml hyaluronidase overnight at 37 °C on a shaker. This extract was strained using a 40-μm strainer to remove undigested material (BD Biosciences, Franklin Lakes, NJ, USA). The filtrate was centrifuged for 7 min at $405 \times g$ and the cell pellet washed with phosphate-buffered saline (PBS). The yield of cells from such an isolation is between 5×10^4 and 1×10^5. Cells were suspended in supplemented Neurobasal-A medium (Neurobasal-A (Life Technologies, Grand Island, NY, USA) with 2 ng/ml epidermal growth factor (Abacus ALS, Auckland, New Zealand), 1 ng/ml fibroblastic growth factor 2 (Abacus ALS), $1 \times$ B27 (50 × stock; Life Technologies), $1 \times$ N2 (100 × stock; Life Technologies), 2 μg/ml heparin (Sigma Aldrich, St Louis, MO, USA), 2 mM GlutaMAX™ Supplement (Life Technologies), $1 \times$ Antibiotic–Antimycotic (Anti-Anti; Life Technologies)) and seeded into wells containing sterile glass coverslips on the well surface. Cells were maintained in culture in humidified incubators at 37 °C in an atmosphere containing 5 % CO_2 to facilitate sphere formation. Fifty per cent of the spent medium was removed and replaced twice weekly. Over the course of 1–2 weeks, cells become adherent to the glass coverslip and aggregate into sphere-like structures. Spheres are maintained in this culture protocol for use in experiments after at least 1 month in sphere culture conditions. This process selects for and concentrates less differentiated cells existing within tissue into sphere-like structures.

Preparation of in-vitro and in-situ sphere attachment surfaces

Poly-L-lysine (Sigma-Aldrich)-coated coverslips were prepared for the immobilization of spheres for immunostaining according to the manufacturer's recommendations. A collagen-coated surface to stimulate sphere cell migration was prepared using Collagen I Rat Protein, Tail (Life Technologies).

Human corneoscleral rims, obtained post surgery and freeze-stored at −80°C for longer than 3 months, were subject to a total of three freeze–thaw cycles to ensure the effective depopulation of the native cells prior to implantation. In a Gelman HLF-120 horizontal laminar flow cabinet and using a Zeiss SV6 Binocular Stereo microscope, frozen and stored human corneoscleral rims were thawed and cut into one-eighth segments using straight scissors. Microsurgical techniques for the implantation of spheres into the epithelial side of the tissue were explored and developed using an ophthalmic surgical microscope (Carl Zeiss, Oberkochen, Germany), a 3.75-mm Short Cut blade (Alcon, Mt Wellington, New Zealand), a Feather MicroScalpel (pfmmedical, Cologne, Germany) and fine forceps.

Spheres implanted onto collagen-coated coverslips and in tissue were incubated with standard culture medium: MEM (1×) GlutaMAX (Life Technologies) supplemented with 10% fetal calf serum and Anti-Anti (Life Technologies). Cell proliferation was identified using Click-iT® EdU Alexa Fluor® 594 Imaging Kit (Life Technologies) by supplementing standard culture medium with 5-ethynyl-2′-deoxyuridine (EDU) at a concentration of 10 μM.

To assess the viability of spheres and implanted cells in tissue, LIVE/DEAD® 2 μM calcein AM and 4 μM ethidium homodimer-1 (Life Technologies) in standard culture medium was used.

Immunocytochemistry

Immobilized spheres and whole-tissue implants were fixed using 4% paraformaldehyde (PFA) (Sigma Aldrich) in PBS and permeabilized in methanol for 10 min at −20 °C. To block non-specific antibody binding, samples were incubated for 2 h on a shaker in 100 mM glycine, 0.1 % Triton X-100 (Serva Electrophoresis GmbH, Heidelberg, Germany), 10 % normal goat serum (NGS; Life Technologies) in PBS. Where relevant, samples were then incubated in the Click-iT® EDU reaction cocktail as per the manufacturer's recommendations for 30 min on a shaker. Samples were then washed in PBS with 3% bovine serum albumin (PBS-B) and incubated overnight at 4 °C with primary antibody prepared in PBS-B + 0.5 % Triton X-100. The primary antibodies used were as follows: anti-ABCB5 at 1:125 (#HPA026975; Sigma Aldrich), anti-ΔNp63 at 1:200 (private order; PickCell Laboratories, Amsterdam, Netherlands), anti-ABCG2 at 1:25 (#14-8888; eBioscience, San Diego, CA, USA), anti-Notch1 at 1:500 (#MS-1339; Thermo Scientific, Waltham, MA, USA), anti-Keratocan at 1:100 (#Sc66941; Santa Cruz, Dallas, TX, USA), anti-Vimentin at 1:200 (#V6630; Novocastra, Newcastle, United Kingdom) and anti-Keratin K3/K76 at 1:50 (#CBL218; Millipore, Billerica, MA, USA). Samples were treated with a 1-h secondary antibody incubation prior to rinsing with PBS and counterstaining with 0.1 μg/ml 4′,6-diamidino-2-phenylindole (DAPI). Secondary antibodies were used at a 1:350 dilution and are as follows: goat anti-mouse Alexa488 (#A11029; Molecular Probes, Eugene, OR, USA) and goat anti-rabbit Alexa488 (#A11034; Molecular Probes).

Tissue sections were fixed with 2.5 % PFA, and then incubated with 2 mg/ml testicular hyaluronidase (Sigma Aldrich) in Tris–HCl for 1 h at 37 °C in a humidity chamber. Samples were permeabilized in methanol at −20 °C for 20 min. Sections were then treated with 20 mM glycine in Tris saline buffer (TSB) for 30 min and blocked in 2 % NGS with 0.1 % Triton X-100 in TSB for 30 min at room temperature. Primary and secondary antibodies were prepared in TSB and incubated and counterstained as already described except that secondary antibodies were incubated for 2 h at room temperature.

For all experiments, controls used included secondary antibody only, primary antibody only and no antibody. For spheres implanted in tissue, non-implanted tissue only stained with both primary and secondary antibodies was used as a control.

Microscopy and imaging

Bright-field images, assessed using the SV6 Binocular Stereo microscope (Carl Zeiss), were captured using a NIKON Digital sight DS-UI camera (NIKON CORPORATION, Tokyo, Japan). Phase-contrast and fluorescence microscopy was performed using the following microscopes: Leica DM IL inverted contrasting microscope (Leica Microsystems, Wetzlar, Germany), 4× magnification 0.1 aperture, C PLAN with Leica Application Suite Version 4.4.0 Build 454; and Leica DM-RA upright fluorescence microscope (Leica Microsystems), 5× magnification 0.15 aperture, HC PL Fluotar and 40× magnification, 1.00 aperture, PL FLUO-TAR Oil PH3 with NIS-Elements Br Microscope Imaging Software version 3.0 and images captured using the NIKON Digital sight DS-UI camera (NIKON). Confocal fluorescence microscopy was performed using the Olympus FV 1000 Confocal laser scanning microscope (Olympus America, Center Valley, PA, USA), 20× magnification, 0.75 aperture U Plan S APO and 60× magnification, 1.35 aperture U Plan S APO oil with the FV10-ASW version 0.4.00 image capture and analysis software.

Quantitative PCR

RNA isolation from pre-implanted and post-implanted sphere cells was performed using the TRIzol® method (Life Technologies) according to the manufacturer's protocol. DNA digestion was performed using DNase I (RNase-free) (#MO303S; New England Biolabs, Ipswich, MA, USA) according to the manufacturer's recommendations although the incubation time was increased to 2 h to ensure complete genomic DNA removal. cDNA synthesis was performed in a Peltier Thermal Cycler PTC-200 (MJ Research, Waltham, MA, USA) using 1× SuperScript® VILO™ cDNA Synthesis Kit (#11754050; Life Technologies) according to the manufacturer's recommendations. Successful cDNA synthesis quality control was performed by PCR using β-actin and GAPDH primers presented in Table 1 and products were analysed by agarose gel electrophoresis detection with gel red dye (Biotium, Hayward, CA, USA). Gels were imaged using the Gel Doc™ EZ Imager (Bio-Rad Laboratories, Hercules, CA, USA) with Image Lab™ software version 5.0 (Bio-Rad Laboratories).

Quantitative PCR was performed using the Lightcycler® 480 SYBR Green I Master mix or the Lightcycler® 480 Probes Master mix (Roche, Auckland New Zealand) as appropriate according to the manufacturer's recommendations using the primer sets (or probe-based assays) presented in Table 1 and purchased from Integrated DNA Technologies (IDT, Singapore). Template cDNA synthesized from an equivalent of 1 ng/µl of RNA was used per 10 µl reaction.

All quantitative PCR experiments were conducted in a Rotor-Gene™ 6000 (Corbett Life Science, Sydney, Australia) and analysed using the Rotor-Gene Q pure detection software version 2.1.0 (Build 9). Cycling conditions for SYBR green detection included an initial activation for 10 min at 95 °C and 40 cycles of 95 °C for 10 sec, 60 °C for 15 sec and 72 °C for 20 sec with detection on the green channel at this third step, while cycling conditions for probe-based assays consisted of initial activation for 10 min at 95 °C and 40 cycles of 95 °C for 10 sec and 58 °C for 45 sec with detection on the green channel at this second step. Quantification of gene expression was performed by measuring 10-fold serial dilutions of purified amplicons with known copy numbers. Two replicates of triplicate measurements for each gene of interest were performed for pre-implantation and post-implantation spheres. The geometric mean for β-actin and GAPDH, the two most stably expressed reference genes across samples, as determined by the statistical algorithm NormFinder, was used for normalization. Following recommendations, non-detects in data were treated as missing values in order to reduce bias [31].

Statistical analysis

Data collection and statistical analysis were performed using Microsoft Excel 2010 version 14.0.7143.5000 (Microsoft Corporation, Washington, DC, USA) and Statistical Package for the Social Sciences (SPSS) v21.0 (IBM, New York, USA). One-way ANOVAs with Tukey post-hoc tests were conducted to analyse significance of inter-group variation in gene expression. $p < 0.05$ was considered significant.

Results

In-vitro sphere characterization

Peripheral corneal spheres were initially characterized in vitro to confirm they were stem-cell enriched and possessed the ability to respond to a collagen matrix with cell migration and division. Spheres immobilized on poly-L-lysine-coated dishes stained positively for putative limbal stem cell markers ΔNp63α, ABCG2 and the recently proposed limbal stem cell marker ABCB5 [32] (Fig. 1a–c) when compared with the background fluorescence emitted by the secondary antibody only (Fig. 1g), primary antibody only and no antibody controls (not shown). Hyper-fluorescent debris were noted at the centre of both test and control spheres. These artefactual signals are commonly observed in the

Table 1 Quantitative PCR primers used in this study

Product of interest	Primer sets	Size of cDNA amplicon (base pairs)	Size of gDNA amplicon (base pairs)
ATP-binding cassette, sub-family G (WHITE), member 2 (ABCG2)	(F): CCTGAGATCCTGAGCCTTTG (R): AAGCCATTGGTGTTTCCTTG	124	184,966
Proliferating cell nuclear antigen (PCNA)	(F): GGCGTGAACCTCACCAGTAT (R): TTCTCCTGGTTTGGTGCTTC	125	0
Vimentin (VIM)	(F): CCAAACTTTTCCTCCCTGAACC (R): GTGATGCTGAGAAGTTTCGTTGA	141	1395
Laminin, alpha 1 (LAMA1)	(F): ACACCGGGAAGTGTCTGAAC (R): GCTTGAGGAGCACCTTTCAC	239	0
Keratocan (KERA)	(F): ATCTGCAGCACCTTCACCTT (R): CATTGGAATTGGTGGTTTGA	167	4043
Actin beta (β-actin)	(F): AACTCCATCATGAAGTGTGACG (R): GATCCACATCTGCTGGAAG	234	345
Glycreraldehyde-3-phosphate dehydrogenase (GAPDH)	(F): CTGACTTCAACAGCGACACC (R): CCCTGTTGCTGTAGCCAAAT	120	224
ABCB5	(F): TACTCTTCCCACTGCCATTG (R): CAATTATCCATCAAGACCATCTATCAAG (Probe): 56-FAM/CCGACCAAG/ZEN/GCGACTGTCTCT/3IABkFQ	106	0
p63alpha	(F): GGGTCGTGAAATAGTCCAGAC (R): CATCCACCTCCCACTGC (Probe): 56-FAM/CACCTCCGT/ZEN/ATCCCACAGATTGCA/3IABkFQ	108	0
alphaSMA	(F): CTGTTGTAGGTGGTTTCATGGA (R): AGAGTTACGAGTTGCCTGATG (Probe): 56-FAM/AGACCCTGT/ZEN/TCCAGCCATCCTTC/3IABkFQ	131	0
Notch1	(F): ACAGATGCCCAGTGAAGC (R): CGAGGTCAACACAGACGAG	112	1289

All primers and probes assays were purchased from Integrated DNA Technologies (Singapore)
F forward, *R* reverse

imaging of spheres and were not considered when analysing the true positive signal.

Spheres also stained positively for the limbal basal epithelial marker notch 1 and the corneal extracellular matrix markers laminin and keratocan (Fig. 1d–f). Laminin staining appeared as hyperfluorescent streaks resembling portions of a basement membrane. Notch 1 and, to a lesser extent, keratocan staining was strongly concentrated in the outer region of the sphere compared with the central sphere, contrasting with the localization of the stem cell markers and laminin (Fig. 1d–f).

Spheres placed on collagen-coated coverslips and incubated in serum-containing medium showed a radial pattern of cell migration outward from the central sphere after 4 days in culture (Fig. 2a–c). Migrating cells stained positively for the differentiated corneal epithelial cell and corneal stromal markers, keratin 3 (Fig. 2a) and vimentin (Fig. 2b) respectively. While keratin 3 staining was stronger in the central sphere in comparison with cells migrating peripherally, vimentin showed a preferential staining pattern in migratory cells. EDU-incorporated cell nuclei (indicating proliferating cells) are detected both within the central sphere and in cells migrating peripherally and are observed to co-localize with both

the differentiation marker keratin 3 as well as the mesenchymal marker vimentin. Vimentin-positive cells immediately migrating out from the sphere had characteristic spindle morphology with long tapering, bidirectional cytoplasmic exstensions (Fig. 2d). Cells are more tightly packed and fibres are radially oriented out from the sphere. In contrast, cells at the leading edge of the migratory wave showed multi-directional cytoplasmic extensions giving the cell a large and spread out appearance (Fig. 2e, f).

Sphere implantation

Frozen-stored corneoscleral tissue was used as a model of limbal stem cell deficiency. It was deemed to possess no viable cells as shown by the absence of DAPI-positive nuclei in sections of the tissue (Fig. 3c). A wedge-shaped 'trough' created by incisions subtending an angle of ~60° with the base of the wedge facing the cornea and the apex facing the sclera exposed the limbus and allowed for the placement of spheres (Fig. 3a). The scleral border of the limbus, for incisions, was visually approximated to be in the region where the tissue was neither completely clear nor completely opaque. Phase-contrast microscopy of implanted spheres showed opaque spheres within the

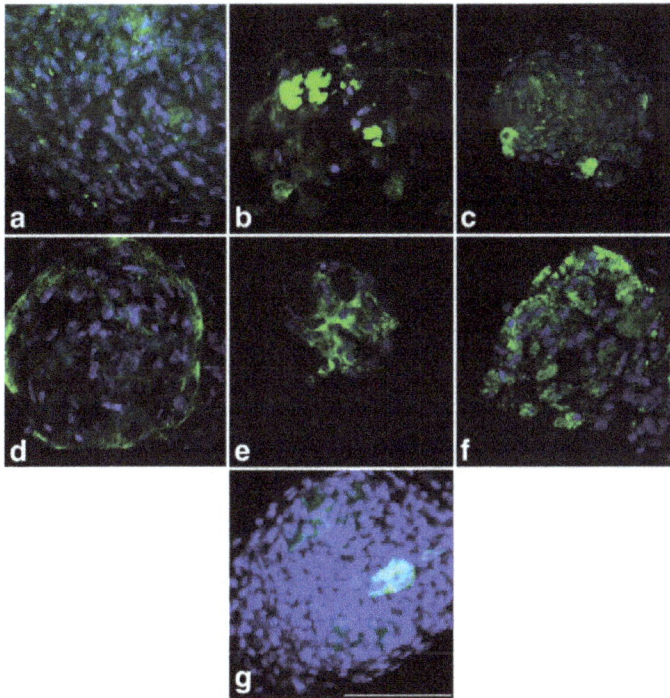

Fig. 1 Immunostaining of peripheral corneal spheres reveals expression of putative stem cell and niche markers. Spheres were imaged at 60× objective magnification by confocal microscopy and labelled with antibodies (*green* signal) for ΔNp63α **a**, ABCG2 **b**, ABCB5 **c**, notch 1 **d**, laminin **e** and keratocan **f**. Representative image of the secondary-antibody-only control **g**. *Blue* signals represent DAPI staining of sphere cell nuclei. Scale bar = 100 μm (Colour figure online)

semi-transparent tissue (Fig. 3d). LIVE/DEAD® staining showed a strong green fluorescent signal (for live cells) confined within implanted spheres in all experiments (Fig. 3b). Implanted spheres remained in place for the duration of each experiment, which was up to 241 h post implantation, despite being submerged in culture medium and subject to physical agitation during handling.

When compared with freshly implanted spheres at 0 h (Fig. 3b), evidence of live cell migration was detected at 25 h post implantation (Fig. 4a). A pronounced increase in cell migration was detected at 72 h post implantation (Fig. 4b) when cells had migrated radially out from the sphere and in multiple focal planes. More extensive cell migration was observed at 217 h post implantation (Fig. 4c). The longest horizontal (visually approximated) diameter of the sphere decreased from 72 to 217 h. In cross-section, sphere cell nuclei appeared dispersed rather than congregating as a single spherical entity (Fig. 4d). Positive EDU labelling was apparent within the sphere and in cells migrating out from the sphere (Fig. 4e), indicating a proliferative response.

Implanted spheres showed active organization of cell migration patterns with 7/11 (63.6 %) spheres exhibiting directed cell migration. This is demonstrated in Fig. 4f by the relative absence of migrating cells in quadrants 2 and 4 and the presence of cells in quadrants 1 and 3 observed in the early stages of cell migration and persisting up to 144 h (not shown).

Spheres placed adjacent to each other showed a cell migration pattern from each sphere towards as well as away from each other. Qualitatively, there appeared to be a disproportionate increase in cell migration from one sphere (left sphere, Fig. 4g and g1) in the direction of the neighbouring sphere (right sphere, Fig. 4g and g1) over time (Fig. 4h).

Ocular surface repopulation by peripheral corneal sphere cells

When spheres implanted in tissue at the limbal region (Fig. 5a) were cultured and imaged at 4 days, live migrating cells appeared outward from the sphere (Fig. 5b). At 7 days a centripetal cell migration pattern from the peripheral cornea out towards the direction of the central cornea (Fig. 5c) was observed. Cells displayed a preferential migration pattern onto clear cornea compared with sclera, seen as cells having migrated further on the corneal side of the implant

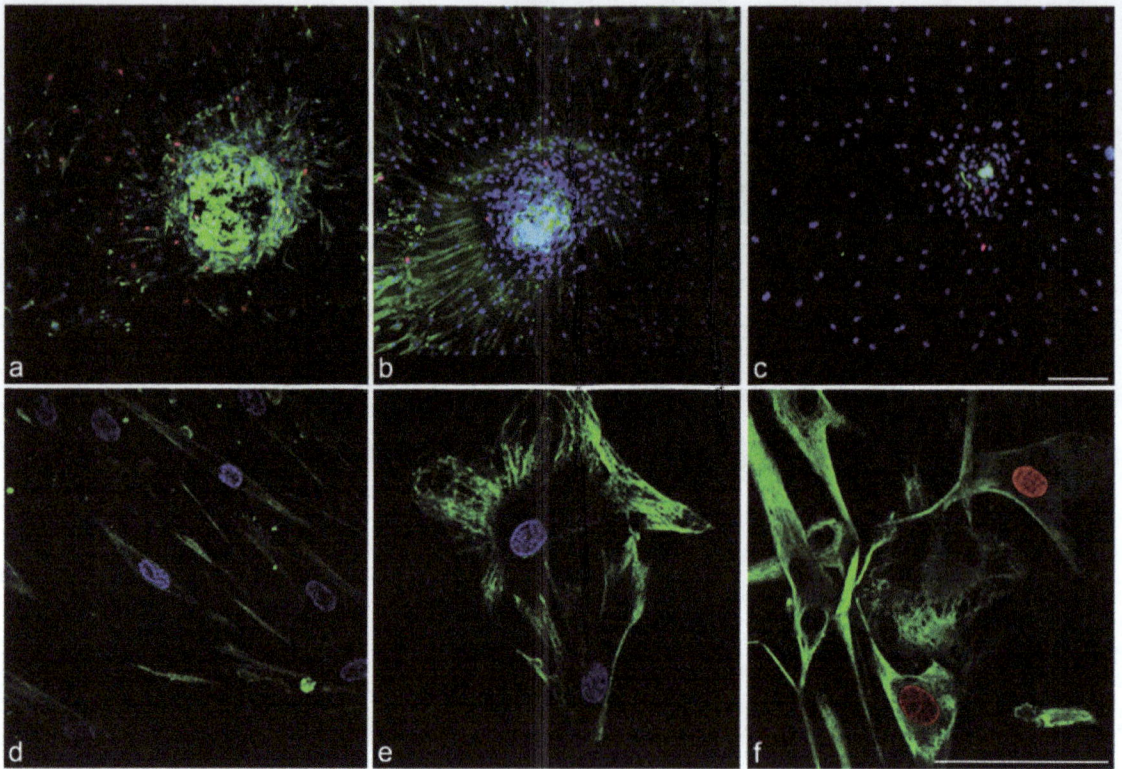

Fig. 2 Immunostaining of peripheral corneal spheres stimulated by collagen I substrate reveals expression of cell proliferation and differentiation markers. Spheres were imaged at 20× objective magnification by confocal microscopy and labelled with antibodies (*green* signal) for keratin 3 **a** and vimentin **b**. The secondary-antibody-only control **c** did not show this signal. *Blue* signals represent DAPI staining of DNA within cell nuclei. Red signals represent EDU-incorporated cell nuclei indicative of proliferating cells. At 60× magnification, vimentin-positive migrated cells proximal to the sphere **d** show a different morphology compared with distal cells **e** and some show positivity for the EDU cell proliferation marker (*red*) **f**. Scale bar =100 μm (Colour figure online)

compared with the scleral side. This was observed on more than 10 separate occasions from spheres derived from four different donors.

Cellular organization differed in cells observed at the cornea, limbus and sclera. In the limbal and scleral region, cells were elongated with a thin spindle appearance in comparison with cells at the corneal region (Fig. 5f, g) where they had a broader appearance. Cells at the leading migratory edge (Fig. 5d) displayed branching cellular processes.

A monolayer of migratory cells, with the longest axis of most cells aligning in the direction of the central cornea, was observed over the corneal surface (Fig. 5e). The alignments of the cellular axes gave the appearance of a relatively uniform migratory column over the cornea. There was an area of circumferentially oriented fibres in the limbal region seen as the longest horizontal axes of most cells aligning perpendicular to the direction of the central cornea (Fig. 5f). Over the sclera, however, cells

occurred in multiple planes of focus whose longest axes did not uniformly align (Fig. 5g).

The number of cells populating the tissue was greater at day 7 post implantation than that which could be provided by sphere cell migration alone, indicating active cell proliferation coupled with the migration. The extent of the migration out towards the cornea was limited only by the size of donor tissue utilized, with cells present at the furthest corneal edge. Notably this extent of cell migration was not observed out towards the sclera (Fig. 5c).

Cross-sectional and whole-mount immunocytochemistry

In cross-section, sphere cells that had migrated over the cornea for 14 days formed a monolayer over the anterior surface (Fig. 6a). Towards the corneal periphery, however, cells had a less organized multilayered appearance (Fig. 6a, dotted region of interest). During the course of migration, cells remained superficial on the anterior

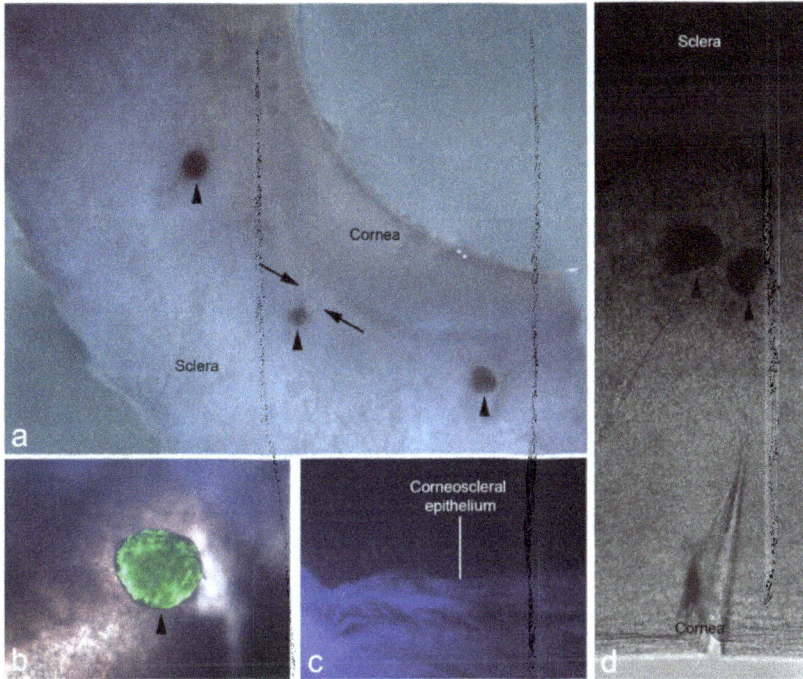

Fig. 3 Implantation of peripheral corneal spheres into donor corneoscleral rims. Spheres (*arrowheads*) were implanted into wedge-shaped incisions made at the limbal region. Under stereomicroscopy, the corneoscleral rim with incisions (*arrows*) and implanted spheres can be clearly visualized **a**. Combined phase-contrast and fluorescence microscopy show an implanted sphere stained positively for live cells with LIVE/DEAD® stain **b**. This signal is confined to the sphere and not detected in the surrounding tissue. A 40× DAPI-stained 10-μm thick cross-section of frozen-stored corneoscleral rim confirmed the absence of DAPI-positive cell nuclei prior to implantation **c**. Through phase-contrast microscopy, the position of the spheres in the semi-transparent region of tissue is shown **d**

surface and did not appear to migrate deep into the tissue substrate.

Immunostaining of cells implanted in tissue for 14 days revealed positive staining for the stem cell marker ΔNp63α (Fig. 6a1). Laminin staining (Fig. 6a2) revealed disorganized clusters of positive signals. Vimentin-positive cells (Fig. 6a3) were identified in cells migrating over the tissue. Additionally, whole-mount staining revealed the presence of the stem cell marker ABCG2 in a few cells (Fig. 6b) and the limbal basal epithelial marker, notch 1 (Fig. 6c). Immunostaining for keratocan using anti-keratin 3 yielded fluorescent signals equivalent to the negative control and therefore were not detected.

Gene expression profile of sphere cells implanted into corneoscleral rims and cultured over time

Expression data were calibrated against the expression value determined for non-implanted spheres. The keratocyte markers keratocan and laminin A1 were significantly reduced in implanted spheres 14 days post implantation compared with non-implanted spheres (Fig. 7). Mean keratocan expression was reduced by 97 % by day 14

(3.10 %) post implantation ($p = 0.000$). Mean laminin A1 expression significantly decreased by 93.3 % in day 7 implants (6.70 %) ($p = 0.014$) and remained at a depressed level at day 14 (21.16 %).

In contrast, mean proliferating cell nuclear antigen (PCNA) expression—a marker of cell proliferation—was increased significantly by 52.15 % at day 4 ($p = 0.007$), then decreased significantly ($p < 0.05$) by 58.70 % from day 4 to day 7 (93.37 %) ($p = 0.004$), returning to the level of the non-implanted sphere and remaining at this level to day 14 (96.67 %).

There was no significant difference in vimentin expression pre and post implantation due to high inter-donor variability in the expression of these markers (data not shown). Additionally, quantitative analyses on expression of the stem cell markers ABCB5, ABCG2 and p63α showed no significant change over time post implantation. Therefore it is unclear whether the stem cell population of the implanted spheres was maintained over the course of culture. Similarly, quantitation of the basal epithelial marker notch 1 and α-SMA, a marker of myofibroblastic transformation, showed no significant change over time post implantation.

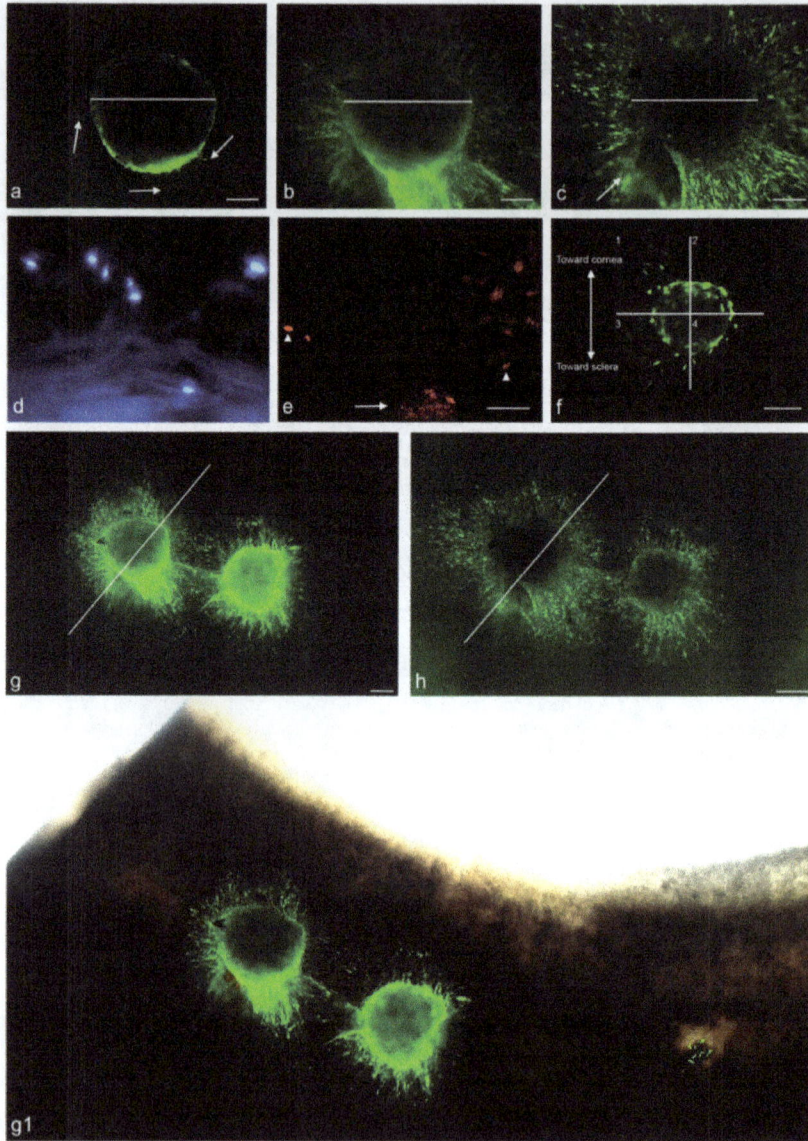

Fig. 4 Implantation of peripheral corneal spheres into corneoscleral tissue results in cell migration, inter-sphere interaction and polarized outgrowth. LIVE/DEAD® staining of sphere-implanted tissue showed a *green* (live cell) fluorescent signal in the outline of the sphere with the beginnings of cell migration and minimal tissue staining (*arrows*) at 25 h **a**. An increase in live cell migration is observed in tissue over time from 72 h **b** to 217 h **c**. *Arrow* indicates a blurred region of tissue staining in a different plane of focus **c**. *White* lines at all three time points are of equal lengths showing a decline in sphere diameter from 72 h **b** to 217 h **c**. In cross-section, cell nuclei labelled using DAPI (*blue*) are dispersed **d**. Confocal imaging of EDU staining shows *red* signals in the sphere (*arrow*) and in migrated cell nuclei (*arrowhead*) **e**. Polarized cell migration is evidenced in implanted spheres **f**. Two spheres implanted adjacent to each other in tissue are imaged first at 72 h **g, g1** and subsequently at 217 h **h**. Here, cells migrate from each sphere in the direction of each other. Migration on the right of the diagonal line appears to have increased from 72 h to 217 h more so than on the left of the diagonal line. Montage imaging of the spheres at 72 h post implantation with light microscope image overlay shows the position of spheres at the limbal region of corneoscleral rims **g1**. Scale bar = 100 µm (Colour figure online)

Discussion

Peripheral corneal sphere characterization

Sphere isolation by the sphere-forming assay in serum-free conditions is well known to promote the survival of stem cells and progenitors, while discouraging the survival of more differentiated cells [24]. Recently, the *ABCB5* gene was convincingly purported as a putative limbal stem cell marker based on its co-expression with

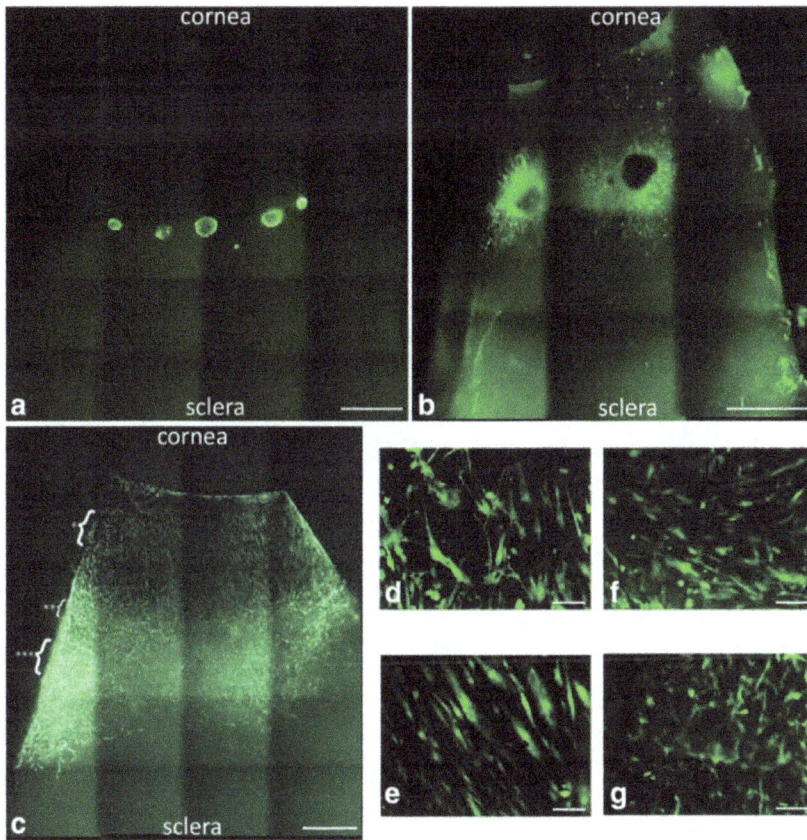

Fig. 5 Peripheral corneal spheres implanted into corneoscral tissue repopulate the ocular surface. LIVE/DEAD® staining (*green*) of implanted spheres, 5× magnification, at 0 h post implantation **a** and 4 days post implantation **b** show cell migration from the spheres appearing as green streaks out from the sphere. At 7 days post implantation **c**, the entire corneal bed appears repopulated with live cells. Representative cells at the leading migratory edge of the corneal surface **d** and cells on the corneal surface (taken from the region indicated by * in c) **e** show differing morphology. Representative cells over the limbal region **f** and sclera **g** (taken from the regions indicated by ** and *** in **c** respectively) show a different cell migration pattern and morphology to that observed in the corneal tissue. Scale bar = 1000 μm for **a, b, e** and 100 μm for **c, d, f, g** (Colour figure online)

the long-standing putative limbal stem cell marker p63α, the reduced ABCB5-expressing cells in limbal stem cell deficiency patients as well as the ability for ABCB5-positive cells to restore an animal model of limbal stem cell deficiency [32]. Our results show that the cells of our peripheral corneal spheres also label positively for ABCB5, which not only provides further circumstantial evidence that ABCB5 may be a marker of human limbal stem cells but also confirms that spheres generated by our protocol are highly likely to be enriched with limbal stem and progenitor cells.

While being stem cell enriched, our peripheral corneal spheres are heterogeneous in nature with respect to cellular origin. We have previously shown exposure of spheres to collagen and differentiation medium produced both vimentin-positive and cytokeratin 3/12-positive cells [30]. We did not find extensive labelling for

either of these markers within the spheres prior to exposure to these stimuli. Thus we propose that whilst the spheres contain cells of both stromal and epithelial origin, they are less differentiated than true epithelial or keratocyte cell types—but these differentiation markers are able to be stimulated, making it difficult to assess the relative proportions of cell populations within the sphere. However, through double immunostaining, Li et al. [33] utilized a similar cell isolation protocol to our laboratory and showed that, after the cell isolation process from tissue, 95 % of cells were vimentin-positive stromal cells while 5 % were pancytokeratin-positive epithelial cells. The corneal epithelial origins of our sphere cells is indicated by detection of basal epithelial markers notch 1 [34] and laminin [35] in immobilized spheres and keratin 3 in spheres given a migratory stimulus. Similarly, the corneal stromal origin of spheres cells is

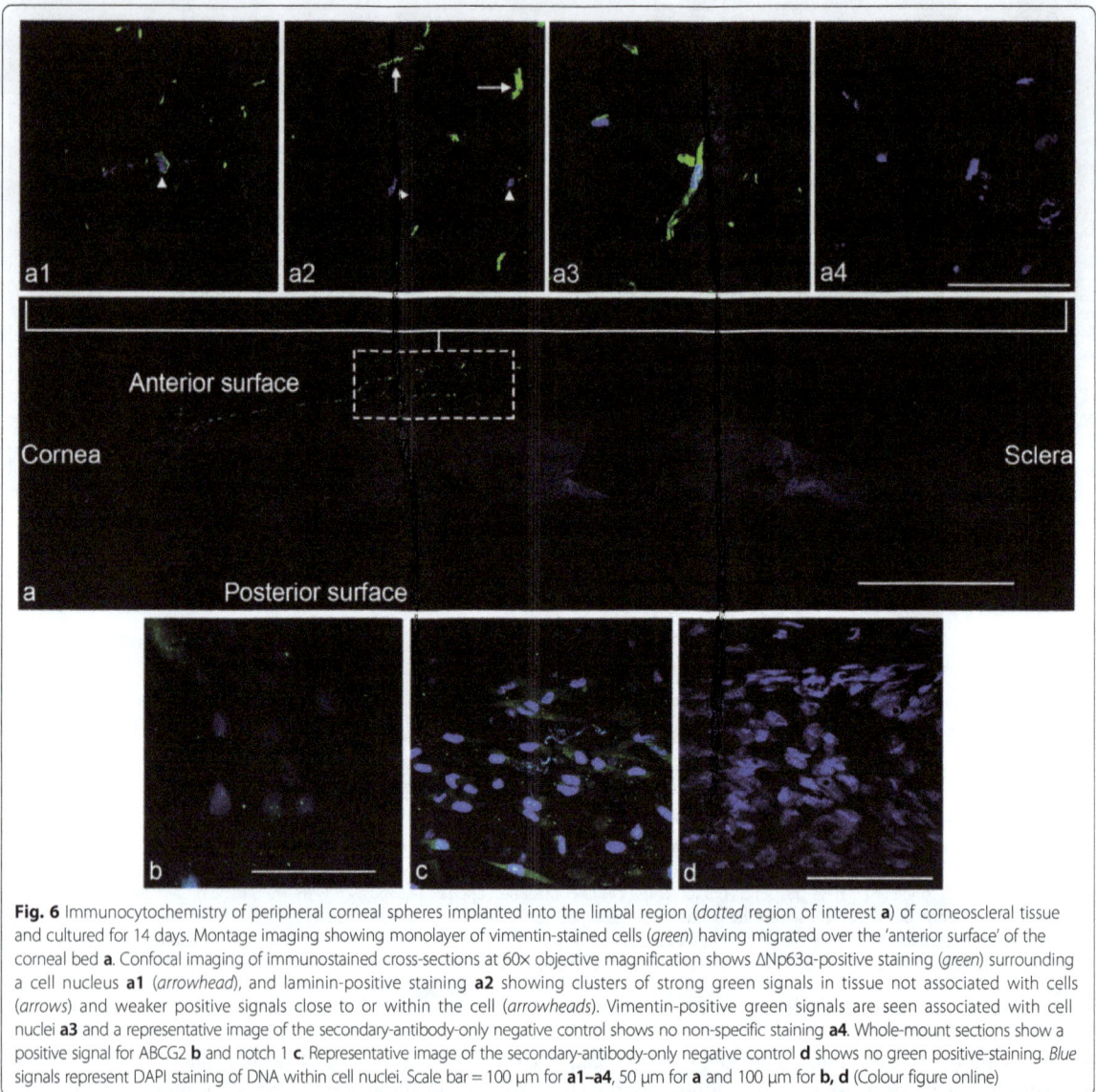

Fig. 6 Immunocytochemistry of peripheral corneal spheres implanted into the limbal region (*dotted* region of interest **a**) of corneoscleral tissue and cultured for 14 days. Montage imaging showing monolayer of vimentin-stained cells (*green*) having migrated over the 'anterior surface' of the corneal bed **a**. Confocal imaging of immunostained cross-sections at 60× objective magnification shows ΔNp63α-positive staining (*green*) surrounding a cell nucleus **a1** (*arrowhead*), and laminin-positive staining **a2** showing clusters of strong green signals in tissue not associated with cells (*arrows*) and weaker positive signals close to or within the cell (*arrowheads*). Vimentin-positive green signals are seen associated with cell nuclei **a3** and a representative image of the secondary-antibody-only negative control shows no non-specific staining **a4**. Whole-mount sections show a positive signal for ABCG2 **b** and notch 1 **c**. Representative image of the secondary-antibody-only negative control **d** shows no green positive-staining. *Blue* signals represent DAPI staining of DNA within cell nuclei. Scale bar = 100 μm for **a1–a4**, 50 μm for **a** and 100 μm for **b, d** (Colour figure online)

evidenced by keratocan [36] in immobilized spheres and vimentin in spheres given a migratory stimulus.

These results align with previous findings which assert that spheres are not simply clusters of functionally isolated cells, but an assemblage of cells within a matrix which mimics the basal corneal epithelium and stroma [37]. Although it may be argued that the extracellular matrix molecules may be from the cell extraction process as opposed to being produced by the cells themselves, laboratory observations showing the initial aggregation of cells followed by the growth of the sphere and our quantitative PCR findings showing a baseline expression of laminin and keratocan in spheres suggest an actively maintained microenvironment within the sphere. Moreover, the persistence of stem cells within spheres in culture over time should attest to the existence of an actively synthesized, functional extracellular matrix given the importance of the limbal niche environment in the maintenance of stem cell character. Collectively, we believe our characterization of these spheres confirmed their identity as stem and progenitor cell-enriched entities of epithelial and stromal origin that would facilitate testing of these as transplantable elements for corneal repair and regeneration.

In-situ corneal repopulation

The implantation of peripheral corneal spheres requires an active process of cellular adherence to the human

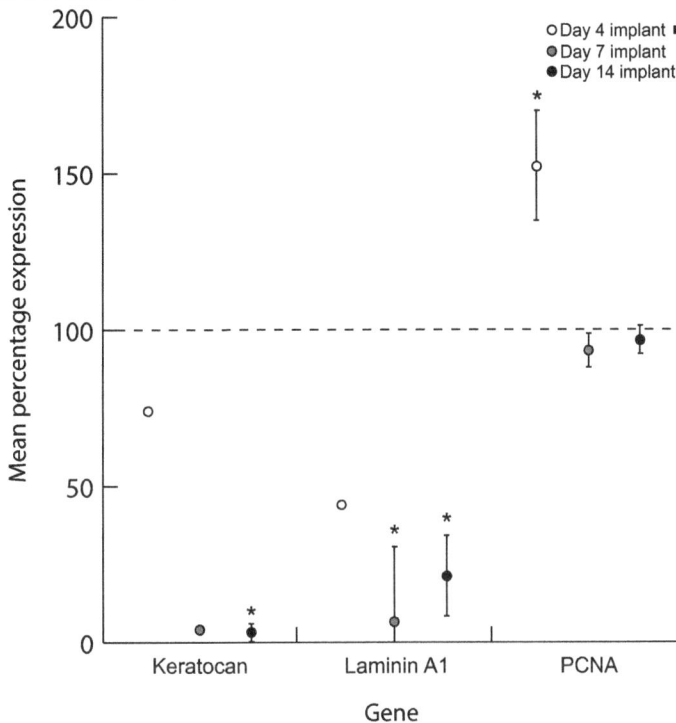

Fig. 7 Expression of keratocan, laminin A1 and PCNA in corneal peripheral spheres implanted into corneoscleral tissue and cultured over 4, 7 and 14 days. All data are calibrated to non-implanted spheres (*dotted line*). Expression data (normalized to β-actin and GAPDH) are expressed as percentages of non-implanted spheres ± 1 standard error of spheres collected from three separate donors. Keratocan and laminin expression declined significantly post implantation. PCNA, however, was significantly elevated 4 days post implantation. There was a significant decrease in PCNA expression from day 4 to day 7. By days 7 and 14, PCNA expression is statistically equivalent to non-implanted sphere PCNA expression. *Black square* for day 4 implant indicates data available from only two donor sets due to tissue processing limitations. *$p < 0.05$. *PCNA* proliferating cell nuclear antigen

limbal substrate provided. Implanted spheres not only survived the implantation and culture process on their new substrate but were able to provide cells which migrated into the foreign limbal environment, thereby demonstrating a capacity for corneal repopulation in tissue. For the first time, we have shown that spheres implanted into the peripheral cornea and cultured over time are able to provide cells which extensively repopulate the entire available area of the corneal bed of the corneoscleral rim segment.

The observed centripetal cell migration pattern from peripheral cornea in the direction of the central cornea supports the well-established theoretical framework for corneal maintenance in vivo where stem cells from the limbus divide and provide centripetally migrating progeny which are responsible for corneal maintenance. Our results align with the longstanding view of corneal maintenance first proposed by Davanger and Evensen [1].

The manner of ocular surface repopulation over the cornea contrasts with the observations over sclera, suggesting that the regional difference in substrate composition exerts an effect on cell migration. Specifically, sphere cells

which completely repopulated corneal tissue demonstrated a preferential migration in the corneal direction in comparison with the scleral direction as evidenced by cells having migrated a greater distance toward the central cornea from the site of implant. Cellular orientation may provide clues to explain this phenomenon. We observe that cells which have repopulated the corneal bed display a capacity to establish themselves in an anatomically appropriate orientation. A close examination of the orientation demonstrated by migrated cells revealed the regular, parallel arrangement of cells aligned with their long axes oriented toward the central cornea, while cells which repopulated the limbal region aligned circumferentially and the limited number of cells within the scleral region appeared in a quasi-random orientation with a lack of cellular alignment in a single direction. We believe that the stimuli for differential cell arrangements after implantation may be two-fold. Firstly, we hypothesize that spheres contain cells which are derived from the limbus and would like to reform a limbus or at least the limbal niche. To this aim, we are not surprised that the cells align differently on the corneal surface to those on the limbal surface and also appear repelled by the scleral surface. Secondly, there may also be residual structural signals for cell arrangement left on the decellularized tissue surface which further aids the seemingly pre-programmed nature of the cellular architecture that appears after implantation. This cellular orientation pattern may reflect the well-established orientation of collagen fibrils distributed within the ocular surface [38]. The limbus sports circumferentially oriented collagen fibrils [39, 40] while the fibrils of the sclera irregularly branch and intersect [38, 41]. The unique arrangement of the collagenous substrate of the cornea probably facilitates a greater extent of cell migration on the corneal side. Further mechanistic studies into the migratory properties of cells from spheres are limited primarily by the availability of human donor tissue for experimentation. Live cell observations in living tissue will be of value to deduce the effect of a biologically active sclera and conjunctiva.

Although the majority of implanted sphere-cell migration was towards the cornea, our results of cellular migration in the scleral direction appear to contribute to the mounting literary evidence surrounding the centrifugal pattern of injury response [26]. Chang et al. [42] demonstrated the centrifugal migration pattern of corneal cells in response to a corneal injury. Majo et al. [43] showed the potential for central corneal cells to participate in the response to injuries of the conjunctival epithelium. Here we show that sphere cells derived from the peripheral cornea and implanted into the limbus are able to elicit a scleral-directed migratory response, suggesting that the structure of the scleral surface, although biologically inactive, is not a barrier to cell migration in our experimental model.

Sphere cell biology post implantation

Characterization of repopulated tissue post sphere implantation showed EDU positivity in cells from implanted spheres agreeing with PCNA quantification showing an initial significant increase in cell proliferation at day 4, followed by a significant decline from day 4 to day 7. These results align with our previous findings that the wound-healing response of spheres in vitro also shows a higher proliferative response by spheres at day 4 in comparison with day 7 and day 14 [44] where, as a result of direct compression injury, sphere cells displayed a more migratory rather than proliferative response at day 7 in comparison with day 4, which agrees with our current in-situ findings. The similarity in the cellular responses observed suggests that the implantation process is akin to a wounding process which generates a similarly reactive biological response by spheres. However, the proliferative response was expected to be higher than we observed. It may be that our initial time window of observation, 4 days post implantation, is

possibly capturing the downward part of the PCNA trend since the wound-healing response post injury in mice was shown to begin as early as 24 h [45].

The downregulation of keratocan and the lack of its immunocytochemical detection reflects the loss of the keratocytic nature in preference for the proliferative/migratory response of implanted cells due to the combined effects of serum exposure as well as the migratory stimulus provided by the ocular surface. Similarly, laminin expression was quantitatively downregulated although there was immunocytochemical evidence of laminin, notch 1 and the stem cell markers ABCG2, ΔNp63α and notch1 post implantation, suggesting the possible maintenance of the basal limbal environment in prolonged culture and despite the predominantly proliferative phenotype post implantation.

Strongly positive vimentin staining of cells with immunochemistry could not be correlated with an increase in vimentin expression in the quantitative PCR data due to high donor variation in the expression of this gene. A larger source of donor tissue may be required to confirm the true trend in vimentin expression. The observed reduction in both stromal and epithelial cell markers post implantation may be indicative of an enhanced stem to progenitor cell response in the corneal repopulation observed over the 14-day time period resulting in the signal from already differentiated stromal and epithelial cells present in spheres being swamped by the increased signal from proliferating and migrating cells.

There appeared to be no significant change in the expression of stem cell, basal epithelial and myofibroblast markers (ABCB5, ABCG2, p63α, notch 1 and α-SMA) over the course of the implantation experiments. This may suggest that the original sphere cell features are being maintained over time but we cannot confirm whether true stem cell repopulation at the limbus has occurred. Further characterization of a larger donor set and longer experimental times may serve to establish the possibility of long-term stem cell repopulation by peripheral corneal spheres.

Conclusion

Peripheral corneal spheres generated by the sphere-forming assay are stem cell enriched, possess properties of the native limbal microenvironment and can be successfully implanted into limbal tissue. This implantation of spheres results in cell migration and proliferation with evidence of cellular differentiation. There is preferential migration towards the cornea, the constraints of which were only the amount of corneal tissue surface provided. Viability of implanted spheres could be maintained beyond 14 days, indicating potential for prolonged restorative capacity. Collectively, these findings give the strongest evidence to date that peripheral corneal spheres could be developed into transplantable units for corneal repair in vivo and play a significant role in therapies targeting ocular surface regeneration and stem cell repopulation.

Abbreviations
ABCB5: ATP-binding cassette sub-family B member 5; ABCG2: ATP-binding cassette sub-family G member 2; DAPI: 4′,6-diamidino-2-phenylindole; EDU: 5-ethynyl-2′-deoxyuridine; PCNA: proliferating cell nuclear antigen.

Acknowledgements
The authors would first and foremost like to honour the tissue donors and their families for an incredibly precious gift. They would also like to acknowledge Associate Professor Dipika Patel for assistance with the surgical methods and Jacqui Ross for imaging and microscopy advice.

Funding
The work presented in this study was funded by grants from the Auckland Medical Research Foundation [1111010], Save Sight Society [3622588] as well as a Faculty of Medical and Health Sciences Summer Scholarship Award and John Hamel McGregor Award both to JM.

Authors' contributions
JJMa and SI participated in the experimental design, data acquisition analysis and manuscript writing. JJMc participated in experimental design. CNJM participated in financial support. TS participated in study design, experimental design and manuscript writing. All authors read, edited and approved the final manuscript

Competing interests
The authors declare that they have no competing interests.

Ethics approval
The University of Auckland Human Participants Ethics Committee (UAHPEC) and the ADHB Research Review Committee (ADHB-RRC) approved this study, which conforms to the Declaration of Helsinki, in 1995 (as revised in Edinburgh 2000).

References
1. Davanger M, Evensen A. Role of the pericorneal papillary structure in renewal of corneal epithelium. Nature. 1971;229:560–1. http://dx.doi.org/10.1038/229560a0.
2. Gipson I, Joyce N, Zieske J. The anatomy and cell biology of the human cornea, limbus, conjunctiva and adnexa. In: Foster C, Azar D, Dohlman C, editors. Smolin and Thoft's the cornea: scientific foundations and clinical practice. Philadelphia: Lippincott Williams & Wilkins; 2005. p. 1.
3. Townsend WM. The limbal palisades of Vogt. Trans Am Ophthalmol Soc. 1991;89:721–56. http://www.ncbi.nlm.nih.gov/pmc/articles/PMC1298638/pdf/taos00010-0740.pdf.
4. Lagali N, Edén U, Utheim TP, Chen X, Riise R, Dellby A, et al. In vivo morphology of the limbal palisades of vogt correlates with progressive stem cell deficiency in aniridia-related keratopathy. Invest Ophthalmol Vis Sci. 2013;54:5333–42. doi:10.1167/iovs.13-11780.
5. Utheim TP. Limbal epithelial cell therapy: past, present, and future. Methods Mol Biol. 2013;1014:3–43.
6. Dua HS, Shanmuganathan VA, Powell-Richards AO, Tighe PJ, Joseph A. Limbal epithelial crypts: a novel anatomical structure and a putative limbal stem cell niche. Br J Ophthalmol. 2005;89:529–32. doi:10.1136/bjo.2004.049742.
7. Dua HS, Saini JS, Azuara-Blanco A, Gupta P. Limbal stem cell deficiency: concept, aetiology, clinical presentation, diagnosis and management. Indian J Ophthalmol. 2000;48(2):83–92. http://www.ijo.in/article.asp?issn=0301-4738.
8. Sugiyama T, Kohara H, Noda M, Nagasawa T. Maintenance of the hematopoietic stem cell pool by CXCL12-CXCR4 chemokine signaling in bone marrow stromal cell niches. Immunity. 2006;25:977–88. http://dx.doi.org/10.1016/j.immuni.2006.10.016.
9. Oatley JM, Oatley MJ, Avarbock MR, Tobias JW, Brinster RL. Colony stimulating factor 1 is an extrinsic stimulator of mouse spermatogonial stem cell self-renewal. Development. 2009;136:1191–9. doi:10.1242/dev.032243.
10. Pinho S, Lacombe J, Hanoun M, Mizoguchi T, Bruns I, Kunisaki Y, et al. PDGFRalpha and CD51 mark human nestin + sphere-forming mesenchymal stem cells capable of hematopoietic progenitor cell expansion. J Exp Med. 2013;210:1351–67. doi:10.1084/jem.20122252;.
11. Tang Y, Rowe R, Botvinick E, Kurup A, Putnam A, Seiki M, et al. MT1-MMP-dependent control of skeletal stem cell commitment via a β1-integrin/YAP/TAZ signaling axis. Dev Cell. 2013;25:402–16. http://dx.doi.org.ezproxy.auckland.ac.nz/10.1016/j.devcel.2013.04.011.
12. Wang CH, Wang TM, Young TH, Lai YK, Yen ML. The critical role of ECM proteins within the human MSC niche in endothelial differentiation. Biomaterials. 2013;34:4223–34. http://dx.doi.org/10.1016/j.biomaterials.2013.02.062.
13. Blazejewska EA, Schlötzer-Schrehardt U, Zenkel M, Bachmann B, Chankiewitz E, Jacobi C, et al. Corneal limbal microenvironment can induce transdifferentiation of hair follicle stem cells into corneal epithelial-like cells. Stem Cells. 2009;27:642–52.
14. Yu D, Chen M, Sun X, Ge J. Differentiation of mouse induced pluripotent stem cells into corneal epithelial-like cells. Cell Biol Int. 2013;37:87–94. doi:10.1002/cbin.10007.
15. Pearton DJ, Yang Y, Dhouailly D. Transdifferentiation of corneal epithelium into epidermis occurs by means of a multistep process triggered by dermal developmental signals. Proc Natl Acad Sci U S A. 2005;102:3714–9. doi:10.1073/pnas.0500344102.
16. Higa K, Kato N, Yoshida S, Ogawa Y, Shimazaki J, Tsubota K, et al. Aquaporin 1-positive stromal niche-like cells directly interact with N-cadherin-positive clusters in the basal limbal epithelium. Stem Cell Res. 2013;10:147–55. http://dx.doi.org/10.1016/j.scr.2012.11.001.
17. Gonzalez S, Deng SX. Presence of native limbal stromal cells increases the expansion efficiency of limbal stem/progenitor cells in culture. Exp Eye Res. 2013;116:169–76. http://dx.doi.org/10.1016/j.exer.2013.08.020.
18. Notara M, Shortt AJ, Galatowicz G, Calder V, Daniels JT. IL6 and the human limbal stem cell niche: a mediator of epithelial-stromal interaction. Stem Cell Res. 2010;5:188–200. http://dx.doi.org/10.1016/j.scr.2010.07.002.
19. Amirjamshidi H, Milani BY, Sagha HM, Movahedan A, Shafiq MA, Lavker RM, et al. Limbal fibroblast conditioned media: a non-invasive treatment for limbal stem cell deficiency. Mol Vis. 2011;17:658–66. http://www.molvis.org/molvis/v17/a75/.
20. Notara M, Shortt AJ, O'Callaghan AR, Daniels JT. The impact of age on the physical and cellular properties of the human limbal stem cell niche. Age. 2013;35:289–300. http://dx.doi.org/10.1007/s11357-011-9359-5.
21. Katikireddy KR, Dana R, Jurkunas UV. Differentiation potential of limbal fibroblasts and bone marrow mesenchymal stem cells to corneal epithelial cells. Stem Cells. 2014;32:717–29. http://dx.doi.org/10.1002/stem.1541.
22. Kawakita T, Shimmura S, Higa K, Espana EM, He H, Shimazaki J, et al. Greater growth potential of p63-positive epithelial cell clusters maintained in human limbal epithelial sheets. Invest Ophthalmol Vis Sci. 2009;50:4611–7. doi:10.1167/iovs.08-2586.
23. Xie H, Chen S, Li G, Tseng SC. Limbal epithelial stem/progenitor cells attract stromal niche cells by SDF-1/CXCR4 signaling to prevent differentiation. Stem Cells. 2011;29:1874–85. doi:10.1002/stem.743.
24. Chen X, Thomson H, Hossain P, Lotery A. Characterisation of mouse limbal neurosphere cells: a potential cell source of functional neurons. Br J Ophthalmol. 2012;96:1431–7. doi:10.1136/bjophthalmol-2012-301546.
25. Reynolds BA, Rietze RL. Neural stem cells and neurospheres—re-evaluating the relationship. Nat Methods. 2005;2:333–6. http://www.nature.com/nmeth/journal/v2/n5/full/nmeth758.html.
26. Yoon JJ, Ismail S, Sherwin T. Limbal stem cells: Central concepts of corneal epithelial homeostasis. World J Stem Cells. 2014;6:391–403. doi:10.4252/wjsc.v6.i4.391.
27. Uchida S, Yokoo S, Yanagi Y, Usui T, Yokota C, Mimura T, et al. Sphere formation and expression of neural proteins by human corneal stromal cells in vitro. Invest Ophthalmol Vis Sci. 2005;46:1620–5.
28. Mimura T, Yamagami S, Usui T, Seiichi, Honda N, Amano S. Necessary prone position time for human corneal endothelial precursor transplantation in a rabbit endothelial deficiency model. Curr Eye Res. 2007;32:617–23. doi:10.1080/02713680701530589.
29. Galli R, Gritti A, Vescovi AL. Adult neural stem cells. Methods Mol Biol. 2008;438:67–84.
30. Yoon JJ, Wang EF, Ismail S, McGhee JJ, Sherwin T. Sphere-forming cells from peripheral cornea demonstrate polarity and directed cell migration. Cell Biol Int. 2013;37:949–60. doi:10.1002/cbin.10119.

31. McCall MN, McMurray HR, Land H, Almudevar A. On non-detects in qPCR data. Bioinformatics. 2014;30:2310–6.

32. Ksander BR, Kolovou PE, Wilson BJ, Saab KR, Guo Q, Ma J, et al. ABCB5 is a limbal stem cell gene required for corneal development and repair. Nature. 2014;511:353–7. http://dx.doi.org/10.1038/nature13426.

33. Li GG, Zhu YT, Xie HT, Chen SY, Tseng SC. Mesenchymal stem cells derived from human limbal niche cells. Invest Ophthalmol Vis Sci. 2012;53:5686–97. doi:10.1167/iovs.12-10300.

34. Thomas PB, Liu YH, Zhuang FF, Selvam S, Song SW, Smith RE, et al. Identification of Notch-1 expression in the limbal basal epithelium. Mol Vis. 2007;13:337–44.

35. Schlotzer-Schrehardt U, Dietrich T, Saito K, Sorokin L, Sasaki T, Paulsson M, et al. Characterization of extracellular matrix components in the limbal epithelial stem cell compartment. Exp Eye Res. 2007;85:845–60. http://dx.doi.org/10.1016/j.exer.2007.08.020.

36. Carlson EC, Liu CY, Chikama T, Hayashi Y, Kao CW, Birk DE, et al. Keratocan, a cornea-specific keratan sulfate proteoglycan, is regulated by lumican. J Biol Chem. 2005;280:25541–7. doi:10.1074/jbc.M500249200.

37. Yoshida S, Shimmura S, Shimazaki J, Shinozaki N, Tsubota K. Serum-free spheroid culture of mouse corneal keratocytes. Invest Ophthalmol Visual Sci. 2005;46:1653–8. http://iovs.arvojournals.org/article.aspx?articleid=2163688.

38. Komai Y, Ushiki T. The three-dimensional organization of collagen fibrils in the human cornea and sclera. Invest Ophthalmol Visual Sci. 1991;32:2244–58. http://iovs.arvojournals.org/article.aspx?articleid=2160522.

39. Aghamohammadzadeh H, Newton RH, Meek KM. X-ray scattering used to map the preferred collagen orientation in the human cornea and limbus. Structure. 2004;12:249–56. http://dx.doi.org/10.1016/S0969-2126(04)00004-8.

40. Newton RH, Meek KM. The integration of the corneal and limbal fibrils in the human eye. Biophys J. 1998;75:2508–12. http://dx.doi.org/10.1016/S0006-3495(98)77695-7.

41. Watson PG, Young RD. Scleral structure, organisation and disease. A review Exp Eye Res. 2004;78:609–23. http://dx.doi.org/10.1016/S0014-4835(03)00212-4.

42. Chang CY, Green CR, McGhee CN, Sherwin T. Acute wound healing in the human central corneal epithelium appears to be independent of limbal stem cell influence. Invest Ophthalmol Vis Sci. 2008;49:5279–86.

43. Majo F, Rochat A, Nicolas M, Jaoudé GA, Barrandon Y. Oligopotent stem cells are distributed throughout the mammalian ocular surface. Nature. 2008;456:250-4.

44. Huang SU, Yoon JJ, Ismail S, McGhee JJ, Sherwin T. Sphere-forming cells from peripheral cornea demonstrate a wound-healing response to injury. Cell Biol Int. 2015;39:1274-87.

45. Gan L, Hamberg-Nyström H, Fagerholm P, Van Setten G. Cellular proliferation and leukocyte infiltration in the rabbit cornea after photorefractive keratectomy. Acta Ophthalmol Scand. 2001;79:488–92. doi:10.1034/j.1600-0420.2001.790512.x.

8

Delineating the effects of 5-fluorouracil and follicle-stimulating hormone on mouse bone marrow stem/progenitor cells

Ambreen Shaikh, Deepa Bhartiya[*], Sona Kapoor and Harshada Nimkar

Abstract

Background: Pluripotent, Lin$^-$/CD45$^-$/Sca-1$^+$ very small embryonic-like stem cells (VSELs) in mouse bone marrow (BM) are resistant to total body radiation because of their quiescent nature, whereas Lin$^-$/CD45$^+$/Sca-1$^+$ hematopoietic stem cells (HSCs) get eliminated. In the present study, we provide further evidence for the existence of VSELs in mouse BM and have also examined the effects of a chemotherapeutic agent (5-fluorouracil (5-FU)) and gonadotropin hormone (follicle-stimulating hormone (FSH)) on BM stem/progenitor cells.

Methods: VSELs and HSCs were characterized in intact BM. Swiss mice were injected with 5-FU (150 mg/kg) and sacrificed on 2, 4, and 10 days (D2, D4, and D10) post treatment to examine changes in BM histology and effects on VSELs and HSCs by a multiparametric approach. The effect of FSH (5 IU) administered 48 h after 5-FU treatment was also studied. Bromodeoxyuridine (BrdU) incorporation, cell cycle analysis, and colony-forming unit (CFU) assay were carried out to understand the functional potential of stem/progenitor cells towards regeneration of chemoablated marrow.

Results: Nuclear OCT-4, SCA-1, and SSEA-1 coexpressing LIN$^-$/CD45$^-$ VSELs and slightly larger LIN$^-$/CD45$^+$ HSCs expressing cytoplasmic OCT-4 were identified and comprised 0.022 ± 0.002 % and 0.081 ± 0.004 % respectively of the total cells in BM. 5-FU treatment resulted in depletion of cells with a 7-fold reduction by D4 and normal hematopoiesis was re-established by D10. Nuclear OCT-4 and PCNA-positive VSELs were detected in chemoablated bone sections near the endosteal region. VSELs remained unaffected by 5-FU on D2 and increased on D4, whereas HSCs showed a marked reduction in numbers on D2 and later increased along with the corresponding increase in BrdU uptake and upregulation of specific transcripts (Oct-4A, Oct-4, Sca-1, Nanog, Stella, Fragilis, Pcna). Cells that survived 5-FU formed colonies in vitro. Both VSELs and HSCs expressed FSH receptors and FSH treatment enhanced hematopoietic recovery by 72 h.

Conclusion: Both VSELs and HSCs were activated in response to the stress created by 5-FU and FSH enhanced hematopoietic recovery by at least 72 h in 5-FU-treated mice. VSELs are the most primitive pluripotent stem cells in BM that self-renew and give rise to HSCs under stress, and HSCs further divide rapidly and differentiate to maintain homeostasis. The study provides a novel insight into basic hematopoiesis and has clinical relevance.

Keywords: Very small embryonic-like stem cells, Hematopoietic stem cells, Follicle-stimulating hormone, 5-Fluorouracil, OCT-4, Bone marrow

* Correspondence: deepa.bhartiya@yahoo.in
Stem Cell Biology Department, National Institute for Research in
Reproductive Health (ICMR), Jehangir Merwanji Street, Parel, Mumbai 400
012, India

Background

In the last decade various groups have reported the presence of non-hematopoietic stem cells in the bone marrow (BM) [1–5] using an array of cell surface markers. In 2006, a rare population of these cells expressing pluripotent markers Oct-4 and Nanog was identified and well characterized in the bone marrow [6]. Over the decade these $LIN^-/CD45^-/SCA-1^+$ in mice and $CD133^+$ in humans, very small embryonic-like stem cells (VSELs) have been described in various murine and human adult tissues including bone marrow [7, 8], cord blood [9–11], testis [12, 13], ovary [14, 15], uterus [16], pancreas [17], and other organs [18, 19]. True to their pluripotent nature, the human and murine VSELs under specific conditions give rise to cells of all the three germ layers in vitro [6, 20, 21]. VSELs are quiescent in nature owing to partial erasure of imprinted genes [22, 23] and thus do not either expand easily in vitro or form teratoma. However, several reports have shown that VSELs enter the cell cycle and mobilize in response to stress or tissue injury [24–30]. In the hematopoietic system, VSELs are considered to be at the top of the hierarchy, because the $CD45^-$ VSELs in vitro give rise to the $CD45^+$ hematopoietic stem cells (HSCs) [31, 32]. Till date, however, the role of VSELs in hematopoiesis in vivo either in steady-state conditions or in response to bone marrow injury has not been studied extensively.

Ratajczak's group [6] was the first to study VSELs $(LIN^-/CD45^-/SCA-1^+)$ and HSCs $(LIN^-/CD45^+/Sca-1^+)$ in mouse BM by flow cytometry. The main distinction between the two cell types is the presence or absence of CD45 expression. CD45 is pan-hematopoietic marker which is expressed by all leukocytes, including HSCs [33]. Others have also reported that cells which express CD45 and CD34 but lack CD38 and lineage antigens $(CD45^+CD34^+CD38^-Lin^-)$ are HSC populations [34] and $LIN^-/CD45^-/SCA-1^+$ are VSELs [20]. Being pluripotent and not committed to the hematopoietic fate, VSELs do not express CD45, are negative for LIN markers, and rather express pluripotent markers. Besides the cell surface marker profile, VSELs and HSCs are also distinguished from each other based on their size and OCT-4 expression pattern. Oct-4 is a nuclear transcription factor crucial to maintain the pluripotent state and its gene is known to be alternatively spliced and has pseudogenes [35–38]. OCT-4 isoform 1 (transcript Oct-4A) is localized in the nucleus, maintains stemness properties, and confers self-renewal function, whereas OCT-4 isoform 2 (transcript Oct-4B) is localized in the cytoplasm and has no assigned biological function as yet [39]. We have reported VSELs with nuclear OCT-4A and slightly bigger progenitors with cytoplasmic OCT-4 in adult testis (spermatogonial stem cells (SSC)) [12], ovary (ovarian stem cells (OSC)) [14], and Wharton's jelly (mesenchymal

cells (MSC)) [40] using a polyclonal OCT-4 antibody which detects both of the isoforms. Based on the OCT-4 staining pattern, we have postulated that immediate descendants or "progenitors" (SSCs, OSCs, MSCs) with cytoplasmic OCT-4 supposedly arise from the pluripotent VSELs. Nuclear form of OCT-4A is no longer required when VSELs initiate differentiation, in the progenitors it is expressed in the cytoplasm, and eventually get degraded as cells differentiate further. Besides OCT-4, VSELs also express several other pluripotent and primordial germ cell markers and thus can easily be distinguished from the HSCs.

Pietras et al. described the heterogeneity that exists among HSCs related to the degree of quiescence. At any given moment, 90–95 % of HSCs exhibit an "activated" phenotype and only 5–10 % are "dormant" [41]. In contrast, all VSELs are quiescent in nature and as a result survive the effects of irradiation and chemotherapy. Ratajczak et al. [31] reported that 1000–1500 cGY total body irradiation in mice resulted in the loss of HSCs whereas VSELs survived and 12 % of them incorporated bromodeoxyuridine (BrdU). Similarly, VSELs survived and their percentage increased in mouse ovaries [15] and testis [13] on treatment with busulphan and cyclophosphamide. Recently we have shown that cord-blood VSELs, which are otherwise quiescent, also enter the cell cycle in response to 24 h of 5-fluorouracil (5-FU) treatment in vitro [11]. The aim of the present study was to understand the differential effects of 5-FU on bone marrow stem/progenitor cells. 5-FU (150 mg/kg) treatment depleted cycling cells, thereby creating stress in the mouse bone marrow and further experiments were undertaken to study whether recovery of bone marrow involved only HSCs or also the VSELs.

Recent reports have shown that, along with cytokines and growth factors, the long-range signaling hormones also affect HSC activity [42]. The hematopoietic progenitors respond in vitro by proliferation to estrogen and androgens [43–45]. Administration of estradiol and an increased level of estrogen as seen during pregnancy also affect survival, self-renewal, and differentiation of HSCs [45]. Accumulating evidence suggests that along with hematopoietic progenitors even VSELs in mouse BM and human cord blood express receptors for sex hormones [46, 47]. Mierzejewska et al. [46] reported that 10-day administration of hormones (follicle-stimulating hormone (FSH), luteinizing hormone, prolactin, androgen, and estrogen) stimulated proliferation of VSELs and HSPCs in vivo as evaluated by the uptake of BrdU. Another study has also recently reported effective mobilization of VSELs and HSCs into circulation in 15 women being treated with FSH to stimulate ovaries in an infertility clinic [48]. Our group had initially reported expression of FSH receptor (FSHR) on ovarian VSELs and OSCs and their modulation by FSH and recently observed similar FSHR expression on testicular VSELs and SSCs and their stimulation by FSH [15, 49–52].

In the present study, different approaches have been used to further characterize and describe the dynamics and functional potential in vitro of VSELs and HSCs in normal and 5-FU-treated adult mouse bone marrow. In addition, the influence of FSH on the recovery of BM after 5-FU treatment and whether FSH stimulates self-renewal, expansion, and/or differentiation of VSELs and HSCs were studied.

Methods

All experimental protocols used in the present study were approved by the Institutional Animal Ethics Committee of NIRRH. Adult Swiss mice 6–8 weeks old maintained in the institute experimental animal facility were used for the study. They were housed in a temperature- and humidity-controlled room on a 12-h light/12-h darkness cycle with free access to food and water.

Experimental design

Characterization of VSELs and HSCs in adult mouse BM

Total nucleated cells obtained from mouse BM were studied for VSELs and HSCs by various methods including flow cytometry (based on size and marker expression: VSELs are 3–5 μm and LIN$^-$/CD45$^-$/SCA-1$^+$, whereas HSCs are ≥6 μm and LIN$^-$/CD45$^+$/SCA-1$^+$), immunolocalization to study pluripotent (OCT-4, SSEA-1, SCA-1), primordial germ cell (STELLA), and proliferation (PCNA) specific markers, and quantitative RT-PCR studies for studying pluripotent (Oct-4A, Nanog, Tert), primordial germ cell (Stella, Fragilis), Sca-1, Oct-4, and proliferation (Pcna) specific transcripts. Due care was taken to design Oct-4 primers which were designed from exon 1 to specifically detect Oct-4A and spanning exons 2–4 to detect total Oct-4 (which majorly includes Oct-4B) as described earlier [15]. Complete lists of the antibodies and primers used in the study are provided in Additional file 1: Tables S1 and S2 respectively.

Effect of 5-FU treatment on mouse BM stem/progenitor cells

Mice were treated with 150 mg/kg body weight of 5-FU (Biochem, Mumbai, India) via the intraperitoneal route. The animals were sacrificed 2, 4, and 10 days (D2, D4, and D10) after the treatment by cervical dislocation and the harvested BM was processed for characterization studies as already described along with histological studies. Besides studying the effect on VSELs and HSCs, flow cytometry analysis was also performed to study the effect of 5-FU on proliferation of various subsets of cells by BrdU uptake and cell-cycle status. The cells that survived 5-FU treatment were also cultured in Methocult medium to study their functional potential.

Effect of FSH treatment on stem/progenitor cells in 5-FU-treated mice

To study the effect of FSH on BM cells in 5-FU-treated mice, recombinant FSH (5 IU, Gonal F, 10 U; Merck Serono, Switzerland) was injected subcutaneously 48 h after animals were treated with 5-FU. BM was harvested after 2 and 5 days of FSH treatment and processed for various studies as already described. Presence of FSH receptors on various cell types was also studied by flow cytometry and immunolocalization studies.

Methods

Isolation of total nucleated cells from mouse BM BM cells from 6–8-week-old adult Swiss mice were isolated. Briefly, BM was flushed from tibias and femurs using DMEM-F12 media (Gibco, Carlsbad, CA, USA) and the cells were passed through 70 μm filter (Falcon, Corning, NY, USA). The filtrate was centrifuged at $1000 \times g$ for 10 min and the pellet obtained was resuspended in 1× RBC lysis buffer (hypotonic ammonium chloride solution) for 10 min. A population of total nucleated cells (TNCs) was obtained after lysis of RBCs and washed twice with DMEM-F12 + 2 % fetal bovine serum (FBS; Gibco). TNCs obtained by this method were used for various studies.

Flow cytometry BM cells from normal, 5-FU-treated, and 5-FU + FSH-treated mice were used for flow cytometry to enumerate Sca-1$^+$/Lin$^-$/CD45$^-$ VSELs and Sca-1$^+$/Lin$^-$/CD45$^+$ HSCs using the gating strategy described by Kucia et al. [6]. A single-cell suspension was prepared and stained with FITC-conjugated rat anti-mouse SCA-1 (BD Biosciences, San Jose, CA, USA), PE rat anti-mouse CD45 (BD Biosciences), and APC mouse Lineage antibody cocktail (BD Pharmingen, San Diego, CA, USA) for 60 min on ice. After washing, the stained cells were run on FACS Aria (BD Biosciences). At least 10^5 events were acquired and results were analyzed by using BD FACS Diva software (BD Biosciences).

BrdU staining Proliferation events in BM cell populations were examined by BrdU incorporation in normal and 5-FU-treated mice by flow cytometry. Briefly, after 5-FU and 5-FU + FSH treatment, the mice were injected with BrdU (1 mg, intraperitoneal; Sigma-Aldrich, St. Louis, MO, USA) daily and a final injection of BrdU was administered 1 h before sacrifice. BM was subsequently isolated and TNCs were immunostained for CD45, LIN markers, SCA-1, and BrdU (FITC BrdU Flow Kit; BD Pharmingen). The manufacturer's protocol was followed and the stained cells were run on FACS Aria. The results obtained were analyzed using FACS Diva software.

Detailed descriptions of the other methods used are presented in Additional file 1.

Statistical analysis Arithmetic means and SDs of our flow cytometry data were calculated, using Graph Pad prism 6 (GraphPad, San Diego, CA, USA) software. Data were analyzed using the Student's t test for unpaired samples and error bars in graphs represent the mean ± SEM. Data from bone marrow HSC and VSEL percentages and numbers are expressed as mean ± SD. Differences were analyzed using ANOVA (one-way or multiple comparisons) as appropriate. The significance level throughout the analyses was chosen to be $p \leq 0.05$.

Results
Mouse BM harbors pluripotent VSELs
Earlier reports [6, 53] and our initial immunofluorescence studies on cell smears of mice bone marrow confirmed the presence of rare, small, spherical cells with high nucleo-cytoplasmic ratio expressing pluripotent stem cell markers including nuclear OCT-4A and SOX-2 and cell-surface SCA-1 and SSEA-1 (Fig. 1a). Interestingly, few larger cells with low nucleo-cytoplasmic ratio and prominent cytoplasm were also observed which expressed cytoplasmic OCT-4 (Fig. 1a; Additional file 1: Figure S1). These cells with cytoplasmic OCT-4 were present in more numbers and were possibly the HSCs. Based on these initial results, the BM cells were investigated further by flow cytometry (Fig. 1b) using the protocol and unique gating strategy reported by Kucia et al. [6]. Small cells (2–8 μm) were gated and analyzed for expression of Lineage (LIN) markers and pluripotent stem cell marker (SCA-1). Next, the LIN$^-$/SCA-1$^+$ cells were gated and analyzed for the expression of hematopoietic marker CD45. The LIN$^-$ and CD45$^-$ non-hematopoietic cells expressing pluripotent marker SCA-1 were the VSELs (LIN$^-$/CD45$^-$/SCA-1$^+$), while the LIN$^-$CD45$^+$ hematopoietic cells with SCA-1 expression were the HSCs (LIN$^-$/CD45$^+$/SCA-1$^+$). As shown in Fig. 1b, the percentage of VSELs (0.022 ± 0.002 %) was almost 3-fold lower than the HSCs (0.081 ± 0.004 %) ($n = 10$). A similar strategy was used to determine the percentage of small cells expressing OCT-4A and SSEA-1 and it was observed that the percentage of LIN$^-$/CD45$^-$ cells co-expressing these markers was almost similar to SCA-1$^+$ VSELs (Additional file 1: Table S3). This co-expression of markers in the small cells suggested that SCA-1$^+$ VSELs also express OCT-4A and SSEA-1. Dual immunostaining of the bone marrow smears confirmed that cells co-expressed SCA-1 with OCT-4A and SCA-1 with SSEA-1 (Fig. 1c).

From confocal microscopy and flow cytometry studies of intact mouse BM, it was thus confirmed that in addition to the abundant LIN$^-$/CD45$^-$/SCA-1$^+$ HSCs with cytoplasmic OCT-4, mouse bone marrow contained very-small-sized LIN$^-$/CD45$^-$/SCA-1$^+$ VSELs expressing pluripotent markers (nuclear OCT-4 and cell-surface SSEA-1).

5-FU treatment spares primitive stem/progenitor cells
The effect of 5-FU was studied on mouse BM cellularity by examining the histological sections of decalcified femur post treatment on D2, D4, and D10 along with untreated controls. 5-FU treatment caused an apparent reduction in cell numbers (Fig. 2b) compared with control (Fig. 2a), with the lowest bone marrow cellularity observed on D4 (Fig. 2c); however by D10, BM sections showed similar histoarchitecture to that of the control sections, confirming endogenous regeneration (Fig. 2d). Also, 5-FU treatment resulted in a marked influx of erythrocytes in the bone marrow on D4 (Fig. 2c). Histological observations were further confirmed by BM total cell count taken on the same days. As expected, the cell count steadily decreased by 3-fold on D2 and by 7-fold on D4 ($p < 0.001$) compared with the control (Fig. 3a). The cell count on D4 ($3.04 \pm 0.79 \times 10^6$) was at least 2-fold lower ($p < 0.001$) than on D2 ($7.65 \pm 0.74 \times 10^6$). On D10 post treatment, the cell numbers ($10.42 \pm 2.3 \times 10^6$) were increased, suggestive of endogenous bone marrow regeneration.

We further studied the effect of 5-FU treatment on cell fractions enriched for VSELs (LIN$^-$/CD45$^-$), HSCs (LIN$^-$/CD45$^+$), and mature hematopoietic cells comprising myeloid and lymphoid progenitors (LIN$^+$/CD45$^+$) by flow cytometry. The percentage of these cells was calculated pre and post 5-FU treatment (Fig. 3b). A 1.3-fold decrease in the LIN$^+$/CD45$^+$ cells was observed compared with control on D4 post treatment (Fig. 3b). However there was an increase in the percentages of both LIN$^-$/CD45$^-$ ($p < 0.01$) and LIN$^-$/CD45$^+$ ($p < 0.001$) compared with control on D4 post treatment (Fig. 3b). The results confirmed that 5-FU affects the cycling myeloid and lymphoid compartments while sparing the primitive stem/progenitor populations that are either slow cycling or quiescent. By D10 an increase in the percentage of LIN$^+$/CD45$^+$ cells was observed with endogenous bone marrow regeneration (Fig. 3b).

Detailed studies were undertaken on the cells that survived in mouse BM on D4 post 5-FU treatment. The quantitative RT-PCR analysis of these cells showed increased expression of pluripotent (Oct-4A, Oct-4, Sca-1 Nanog, and Tert), primordial germ cell (Stella, Fragilis), and proliferation (Pcna) specific transcripts (Fig. 3c). Hematoxylin and eosin (H & E)-stained sections of 5-FU-treated femur on D4 showed the presence of cells with small size, spherical shape, high nucleo-cytoplasmic ratio, and darkly stained nuclei in close proximity of the bone endosteal region (Fig. 3d) These cells expressed nuclear OCT-4 and were observed either singly or in small clusters close to the endosteal region (Fig. 3f). Confocal microscopy studies also demonstrated the expression of

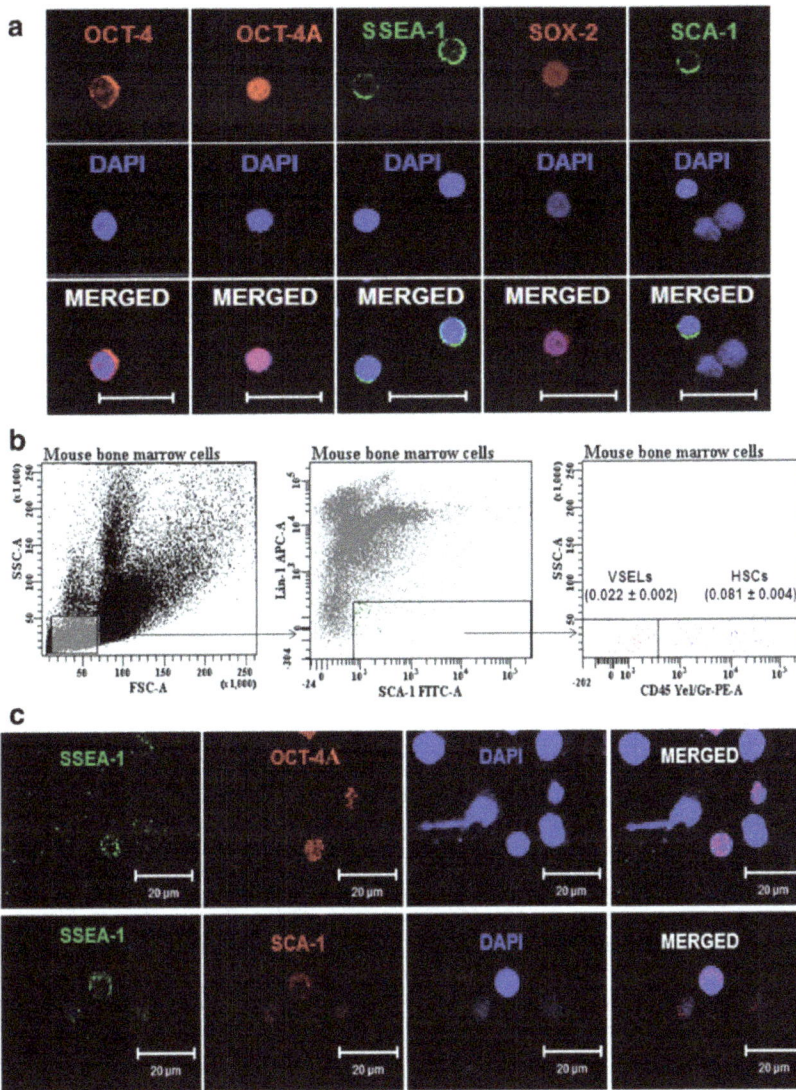

Fig. 1 Pluripotent very small embryonic-like stem cells (*VSELs*) in adult mouse bone marrow. **a** Expression of pluripotent stem cell makers including nuclear OCT-4 (*red*) and SOX-2 (*red*) and cell surface SSEA-1 (*green*) and SCA-1 (*green*) were detected on small-sized cells in bone marrow smears. Also note presence of cytoplasmic OCT-4 in slightly bigger cells. These cells were more abundant and please refer to Additional file 1: Figure S1 to see additional images of cytoplasmic OCT-4-expressing cells. Scale bar = 20 μm. **b** Flow cytometry analysis of VSELs and HSCs in bone marrow. Cells of size 2–8 μm were gated using size calibration beads as reference, followed by sequential selection of LIN⁻/SCA-1⁺ cells. This population was evaluated for expression of CD45. The LIN⁻/SCA-1⁺/CD45⁻ cells were VSELs, while LIN⁻/SCA⁺/CD45⁺ cells were HSCs. The average percentage of VSELs and HSCs with SD from 10 experiments is reported. Representative image is shown. **c** Dual immunofluorescence was performed using anti-SSEA1 antibody (*green*) with nuclear OCT-4 (*red, upper*) and SCA-1 (*red, lower*). Cells co-expressing nuclear OCT-4 and SSEA-1 and SSEA-1 and SCA-1 were observed. Scale bar = 20 μm. In all images, nuclei are counterstained with DAPI. *HSC* Hematopoietic stem cell, *OCT-4* octamer binding transforming factor-4, *SSEA-1* stage-specific embryonic antigen-1, *Sca-1* stem cell antigen-1, *Sox 2* sex-determining region (box 2), *DAPI* 4′,6-diamidino-2-phenylindole (Color figure online)

SSEA-1, SCA-1 (Fig. 3g), and OCT-4 (Fig. 3h) positive cells in 5-FU-treated smears and cryosections respectively.

From these results it is evident that within 48 h of 5-FU treatment all of the cycling cells in the bone marrow were depleted, and endogenous recolonization of the bone marrow occurred by D10. Characterization studies confirmed that VSELs expressing nuclear OCT4-A and SCA-1 survived the effects of 5-FU treatment.

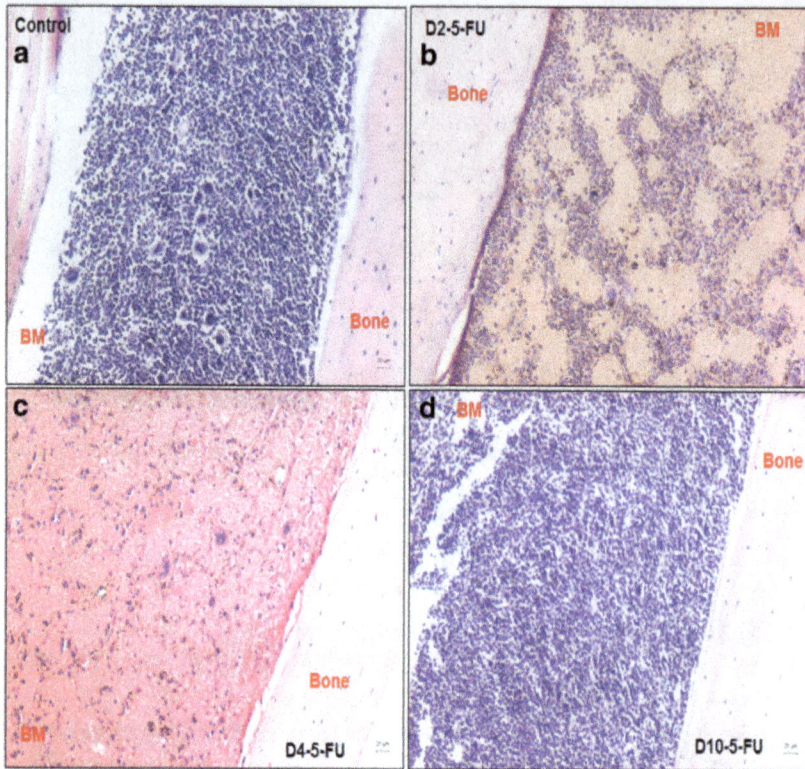

Fig. 2 Effect of 5-FU on bone marrow cellularity. Hematoxylin and eosin-stained sections of decalcified femur (**a**) and affected histology on D2, D4, and D10 (**b–d**) after 5-FU treatment. Gradual decrease in bone marrow cells is observed on D2 (**b**) and D4 (**c**) with lowest cellularity on D4 (**c**) compared with control (**a**). By D10, endogenous recovery of hematopoiesis is seen (**d**). Note an influx of RBCs on D2 and D4. Scale bar = 20 μm. All images are taken on the Nikon 90i microscope. *D2-5-FU, D4-5-FU, D10-5-FU* days post 5-fluorouracil treatment

VSELs survive and proliferate in response to 5-FU treatment

Flow cytometry analysis confirmed survival and an increase in the numbers of LIN$^-$/CD45$^-$/SCA-1$^+$ VSELs and LIN$^-$/CD45$^+$/SCA-1$^+$ HSCs after 5-FU treatment. The VSELs were resistant to 5-FU and the percentage of VSELs increased on D2 after treatment (Fig. 4a). However, to confirm that the increase in percentage was not a result of decreased bone marrow cellularity, absolute numbers were calculated (Table 1). These data showed that the VSELs remained constant after treatment whereas more than 50 % of HSCs were destroyed by 5-FU.

The percentage and absolute numbers of VSELs and HSCs were also calculated on D4 and D10 post treatment (Fig. 4a, Table 1). There was an increase in percentage and absolute numbers of VSELs and HSCs on D4 (VSELs 0.201 ± 0.03 %, HSCs 0.423 ± 0.07 %; Table 1). With the increase in BM cellularity by D10, VSELs reduced slightly but HSC numbers were still high (Table 1). Also the percentage of HSCs was always higher compared with VSELs on all of the days. The

expansion in VSELs and HSCs on D4 suggested that these quiescent cells were activated and entered the cell cycle in response to the stress induced by 5-FU. An increase in numbers of both VSELs and HSCs was observed with higher HSC numbers compared with the VSELs.

The increase in VSEL and HSC number post treatment (Table 1) suggested that they may be proliferating due to the "regenerative pressure" to meet the increased demand of progenitors. This proliferative status of the bone marrow stem/progenitor cells was studied by evaluating the percentage of cycling cells post 5-FU treatment. The percentage of S-phase cells decreased on D2 of 5-FU treatment, further confirming that 5-FU destroyed cycling cells (Fig. 4b). By D4, the percentage of cells in S-phase increased and was associated with a decrease in the G$_0$/G$_1$ cell percentage. This increase suggests self-renewal (activation/involvement) of quiescent cells to restore hematopoiesis by D10.

An in vivo BrdU incorporation assay was then carried out to examine whether both the LIN$^-$/CD45$^+$ (enriched in HSCs) and LIN$^-$/CD45$^-$ (enriched in VSELs) populations

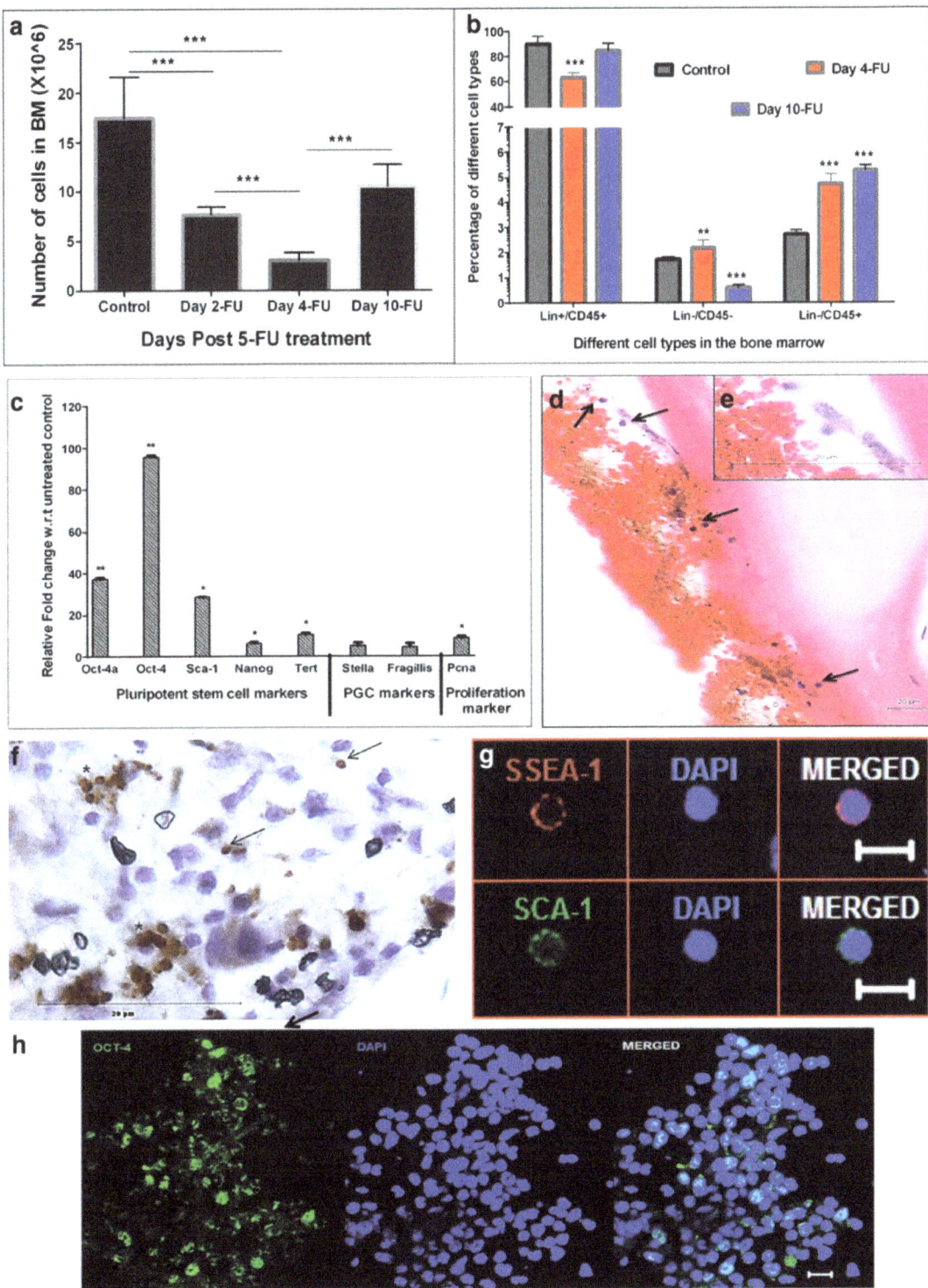

Fig. 3 (See legend on next page.)

Fig. 3 Mature and actively dividing cells in BM are affected by 5-FU whereas primitive and relatively quiescent stem cells are spared. **a** The total BM nucleated cells isolated from two femurs and two tibias of the mice were counted on D2, D4, and D10 after 5-FU treatment and compared with control. Similar to histological data (Fig. 2); the cell count decreased up to D4 (7-fold) and is increased on D10 ($n = 10$, ***$p \leq 0.001$). **b** The percentage of various bone marrow subtypes based on Lin and CD45 expression were evaluated on D4 and D10 after 5-FU treatment. Note that while the percentage of Lin$^+$/CD45$^+$ subtypes decreased on D4, its percentage increased with hematopoietic recovery. Percentage of Lin$^-$/CD45$^-$ (enriched in VSELs) increased on D4 and slightly reduced by D10. The Lin$^-$/CD45$^+$ (enriched in HSCs) percentage also increased by D4, and remained high on D10 ($n = 10$, ***$p \leq 0.001$). **c** Upregulation in the pluripotent, primordial germ cell and proliferation specific transcripts is observed in BM cells that survive 5-FU on D4. The upregulation of these markers confirms the presence and involvement of primitive stem cells in regeneration. The transcript levels on D4 are shown compared with control ($n = 6$, *$p \leq 0.05$; **$p \leq 0.01$). Bars in **a**, **b** represent average ± SD and in **c** average ± SEM. **d–h** Characterization of cells on D4 that survived 5-FU treatment. **d** H & E staining shows VSELs which are spherical in shape and small in size with high nucleo-cytoplasmic ratio located near the endosteal region. *Inset* (**e**) shows cells at higher magnification. **f** Nuclear OCT-4$^+$ VSELs are clearly visualized in bone marrow sections cells in clusters (*) or singly (*arrow*) along the endosteal region; these cells also express (**g**) cell-surface SSEA-1and (**h**) nuclear SCA-1. Presence of nuclear OCT-4, SSEA-1, and SCA-1 confirms that VSELs survive 5-FU treatment. Nuclei are counterstained with DAPI. Scale bar = 20 µm for H & E and immunohistochemistry images, scale bar = 10 µm for immunofluorescence images. *Day 2-FU, Day 4-FU, Day 10-FU* days post 5-FU treatment, *BM* bone marrow, *5-FU* 5-fluorouracil, *H & E* Hematoxylin and Eosin, *OCT-4* octamer binding transforming factor-4, *SSEA-1* stage-specific embryonic antigen-1, *SCA-1*, stem cell antigen-1, *DAPI* 4′,6-diamidino-2-phenylindole

proliferated in response to 5-FU treatment. There was a significant increase in the BrdU uptake by both these cell types ($p < 0.001$) on D4 compared with untreated control, confirming their proliferation (Fig. 4c). BM cryosections on D4 showed the presence of VSELs which co-expressed nuclear OCT-4 and PCNA near the endosteal region of the BM (Fig. 5a). Together, the uptake of BrdU by non-hematopoietic cells and the expression proliferation marker PCNA by the pluripotent OCT4-expressing cells confirms that VSELs underwent self-renewal and thus contributed to restoring hematopoiesis by D10. We further examined the functional potential of cells that survived 5-FU treatment in vitro using a colony-forming unit (CFU) assay. Cells that survived 5-FU responded to the cytokines in the Methocult media and developed 20–25 small and large colonies (Fig. 5b–e).

5-FU treatment mirrored a state of stress in the bone marrow with the need to form blood cells to recolonize the marrow. Several groups have already shown that VSELs remain dormant most of the time, entering the cell cycle only under stress and injury [13, 15, 24–30]. Because of 5-FU-induced stress in the present study, VSELs underwent a short burst of proliferation and then by D10 their numbers became equivalent to the control; in contrast, the HSC numbers initially decreased dramatically, but after D4 showed high proliferative potential. This expansion in HSCs with a small but significant increase in VSELs suggests that VSELs may be proliferating to meet increased demand of their immediate progenitor HSCs. These results are in agreement with the stem cell/progenitor concept wherein the most primitive stem cells undergo rare asymmetric divisions to self-renew and give rise to the progenitors which in turn divide rapidly and undergo clonal expansion followed by differentiation to maintain tissue homeostasis.

FSH enhances hematopoietic recovery from VSELs

To outline the effect of FSH on stem/progenitor cells in the mouse BM, initially the expression of FSHR on the bone marrow cells was studied. Immunofluorescence studies on BM cells that survived 5-FU treatment on D4 showed cell surface expression of FSHR on cells of two distinct sizes (Fig. 6a, b; Additional file 1: Figure S3). Co-expression of SCA-1 and FSHR was also observed, confirming that indeed stem/progenitors express FSHR (Fig. 6c). In addition to FSHR, cells that survived 5-FU treatment also expressed the primordial germ cell (PGC) marker STELLA (Additional file 1: Figure S2) and showed increase in PGC transcripts (Fig. 3c). Immunophenotyping results (Additional file 1: Table S4, Figure S3) showed that only 6 % of LIN$^-$/CD45$^-$ VSELs and 10 % of LIN$^-$/CD45$^+$ HSCs expressed FSHR.

We next determined whether the cells surviving in the BM after 5-FU treatment responded to FSH. Recombinant human FSH (5 IU, subcutaneous) was administered 48 h post 5-FU treatment (5-FU + FSH group) and its effects were examined 2 days later (Fig. 6d) and compared with 5-FU-treated mice without FSH (5-FU – FSH group). Analysis of the cell count (data not shown) and H & E stained sections of bone marrow from both the groups showed a slight increase in bone marrow cellularity in response to FSH treatment. Moreover, on D7 post 5-FU treatment the BM architecture of the 5-FU + FSH group showed restoration of bone marrow hematopoiesis (Fig. 6e). This suggested that FSH enhanced the process of hematopoietic recovery by 72 h in the 5-FU + FSH group because complete recovery was noted in the 5-FU – FSH group only by D10.

The stem/progenitor cells in bone marrow (which survived 5-FU) also responded to FSH treatment. There was a small increase in the absolute numbers of VSELs (5890 untreated vs 6455 FSH treated) and a 1.25-fold

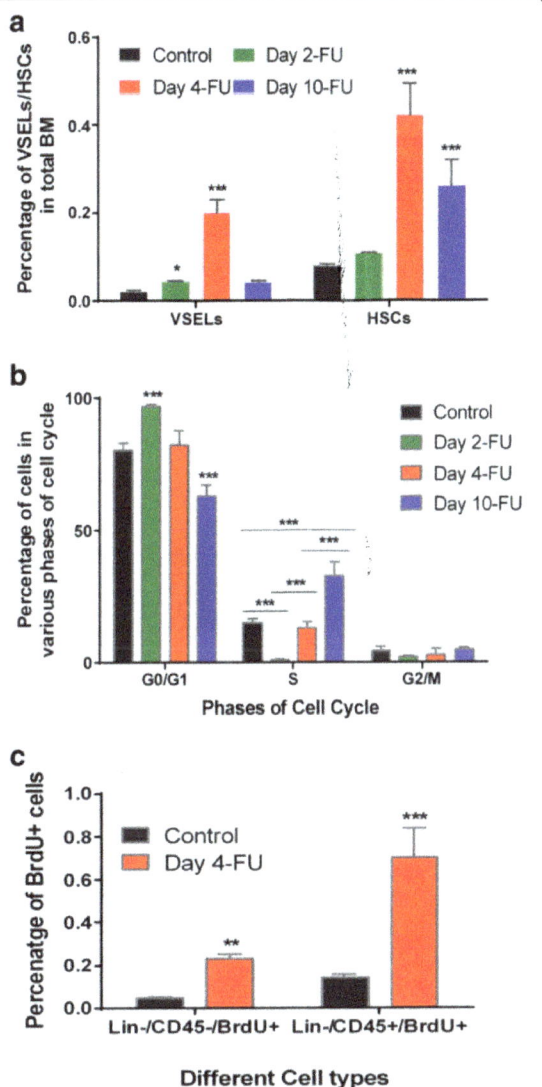

Fig. 4 Mouse BM stem/progenitor cells survive and proliferate in response to 5 FU treatment. **a** Flow cytometry analysis of LIN⁻/CD45⁻ VSELs and LIN⁻/CD45⁺ HSCs showed an increase in percentage post treatment compared with control. Compared with D2, a significant increase was observed on D4. On D10 the percentage of both these cells is reduced ($N = 12$, ***$p \leq 0.001$). **b** Propidium iodide-based cell cycle analysis of BM cells carried out on D2, D4, and D10 after 5-FU compared with the control showed a decrease in S-phase cells on D2 and an increase in the G_0/G_1 cell percentage. By D4 the percentage of S-phase cells increased and continued until D10. The increase in S-phase cells suggests proliferation of cells in response to 5-FU treatment ($N = 6$, ***$p \leq 0.001$). **c** Proliferation of VSELs (LIN⁻/CD45⁻) and HSCs (Lin⁻/CD45⁺) was confirmed by an increase in percentage of BrdU-positive cells on D4 compared with control ($N = 5$, **$p \leq 0.01$,***$p \leq 0.001$). In (**a-c**) Bars represent average ± SD. *Day 2-FU, Day 4-FU, Day 10-FU* days post 5-fluorouracil treatment, *BM* bone marrow, *VSEL* very small embryonic-like stem cell, *HSC* hematopoietic stem cell, *BrdU* bromodeoxyuridine

Table 1 Absolute numbers of VSELs and HSCs in adult mouse bone marrow

Treatment group	VSELs ($\times 10^3$)	HSCs ($\times 10^3$)
Control (no treatment)	3.77 ± 1.01	14.07 ± 3.60
Day 2 post 5-FU	3.49 ± 0.45	8.63 ± 0.70
Day 4 post 5-FU	5.89 ± 0.82	12.29 ± 1.59
Day 10 post 5-FU	4.22 ± 0.80	23.73 ± 3.25

VSEL very small embryonic-like stem cell, *HSC* hematopoietic stem cell, *5-FU* 5-fluorouracil

rise in HSC numbers ($p < 0.001$) after FSH treatment (15,309 FSH treated vs 12,295 untreated) compared with untreated controls (Fig. 6f). To confirm that FSH indeed brought about proliferation of surviving BM stem/progenitor cells and did not just aid in their survival, BrdU incorporation was studied in the LIN⁻/CD45⁻ and LIN⁻/CD45⁺ cells. Similar to our findings in the VSEL and HSC numbers, there was marginal increase (7693 untreated vs 8425 FSH treated) in LIN⁻/CD45⁻/BrdU⁺ cells that are enriched in VSELs along with a remarkable increase (23,699 untreated vs 33,507 FSH treated; 1.41-fold, $p < 0.01$) even in the LIN⁻/CD45⁺ cells (enriched in HSCs, Fig. 6g). The huge expansion in HSCs with minimal increase in VSEL numbers advocates that FSH influences both self-renewal of VSELs and rapid proliferation of progenitors (HSCs) which will then differentiate further to restore hematopoiesis.

The quantitative RT-PCR data of the 5-FU + FSH group showed that treatment also had an effect on the pluripotent stem cell marker expression (Fig. 6h). The transcripts for Oct-4A, Nanog, Sca-1, and Pcna were upregulated compared to the group without FSH, with an almost 21-fold increase ($p < 0.05$) in the expression of Oct-4. FSH treatment also resulted in the upregulation of the transcripts specific for primordial germ cell specific transcripts including a 2-fold increase in Fragilis and a 12-fold increase ($p < 0.05$) in Stella. These results add another level of confirmation to the stimulatory effect of FSH on VSELs and HSCs in the bone marrow post 5-FU.

To conclude, the results of this section show that a fraction of stem/progenitors which survive 5-FU treatment express FSHR and are stimulated by FSH to enhance recolonization by 72 h. A several-fold increase in transcripts specific for Oct-4 (reflecting HSCs) and a marginal increase in Oct-4A (reflecting VSELs) confirmed a larger response of HSCs compared with VSELs which was noted also at the protein level. Further, a 12-fold increase in transcripts specific for Stella vs 2-fold increase in Fragilis suggests that only a fraction of perhaps more committed VSELs express FSHR and respond to FSH. These observations require further investigation.

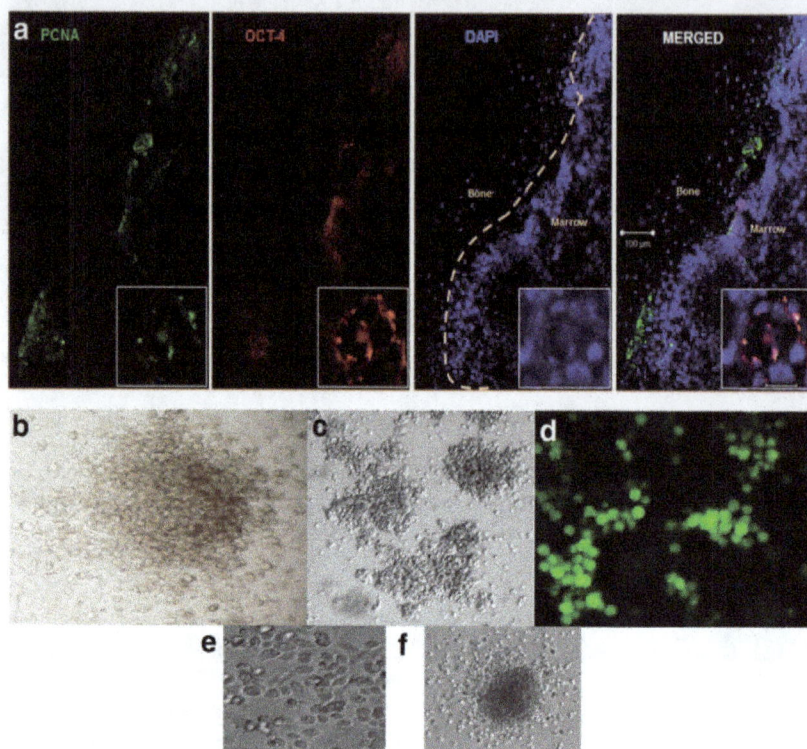

Fig. 5 Proliferation of nuclear OCT-4⁺ VSELs. **a** Dual immunofluorescence on cryosections of D4 femur detected the presence of nuclear OCT-4⁺ cells co-expressing proliferation marker PCNA. Nuclear OCT-4/PCNA co-expressing VSELs were observed close to the endosteal region. The nuclei were counterstained with DAPI. Scale bar = 100 μm. *Inset*: magnified image of cluster of OCT-4/PCNA coexpressing cells. Scale bar = 10 μm. **b, c** Results of the CFU assay: small and large hematopoietic colonies of type CFU-GM or CFU-GEMM were observed on culture of cells that survived 5-FU on D4 in Methocult media. **d** GFP⁺ cell clusters observed after 10 days of culture of 5-FU-treated GFP cells in Methocult media. **e** Typical cobblestone formation representative of hematopoietic colonies seen in 5-FU-treated cell culture. **f** Representative image of colonies observed in CFU assay at 20× magnification, colony formed is of the type CFU-GEMM. *OCT-4* octamer binding transforming factor-4, *PCNA* proliferating cell nuclear antigen, *DAPI* 4′,6-diamidino-2-phenylindole, *CFU-GM* colony-forming unit granulocytes/macrophage, *CFU-GEMM* colony-forming unit-granulocyte, erythroid, macrophage, megakaryocyte

Discussion

VSELs were identified and characterized in various adult organs in mice in 2006 [6] and since then several reports have highlighted their presence in human tissues as well, and even their translational potential is beginning to emerge [54–61]. Results of the present study confirm the presence of LIN⁻/CD45⁻/SCA-1⁺ VSELs (0.022 ± 0.002 %) along with LIN⁻/CD45⁺/SCA-1⁺ HSCs (0.081 ± 0.004 %) in mouse bone marrow. VSELs were small in size and co-expressed nuclear OCT-4, SCA-1, and SSEA-1 and pluripotent (Oct-4A, Sca-1, Nanog, Tert) as well as primordial germ cell (Stella, Fragilis) specific transcripts. HSCs were slightly bigger in size, expressed cytoplasmic OCT-4, and were present in relatively greater numbers. VSELs were found to express both pluripotent and primordial germ cell specific markers since it is postulated that they share a developmental link with primordial germ cells [23]. These results are in agreement with our earlier report on human cord blood [11] and with other studies that have reported high expression of primordial germ cell specific transcripts Vasa, Blimp1, Stella, Prdm14, Nanos3, and Fragilis in VSELs [62].

A precaution taken in the present study for detecting VSELs along with the HSCs was the use of a speed of 1000 × *g* to spin-down cells at various steps during processing since VSELs are invariably lost during processing at lower speeds. The inability to detect VSELs in mouse BM in a recent study [63] is most probably because a low speed of 400 xg was used while processing, which probably resulted in loss of VSELs in the supernatant. Similarly VSELs have been invariably and unknowingly discarded along with the RBCs [10] on density gradient centrifugation of cord blood samples. VSELs are endogenous pluripotent stem cells and a possible, novel candidate for regenerative medicine in addition to human embryonic and induced pluripotent stem cells.

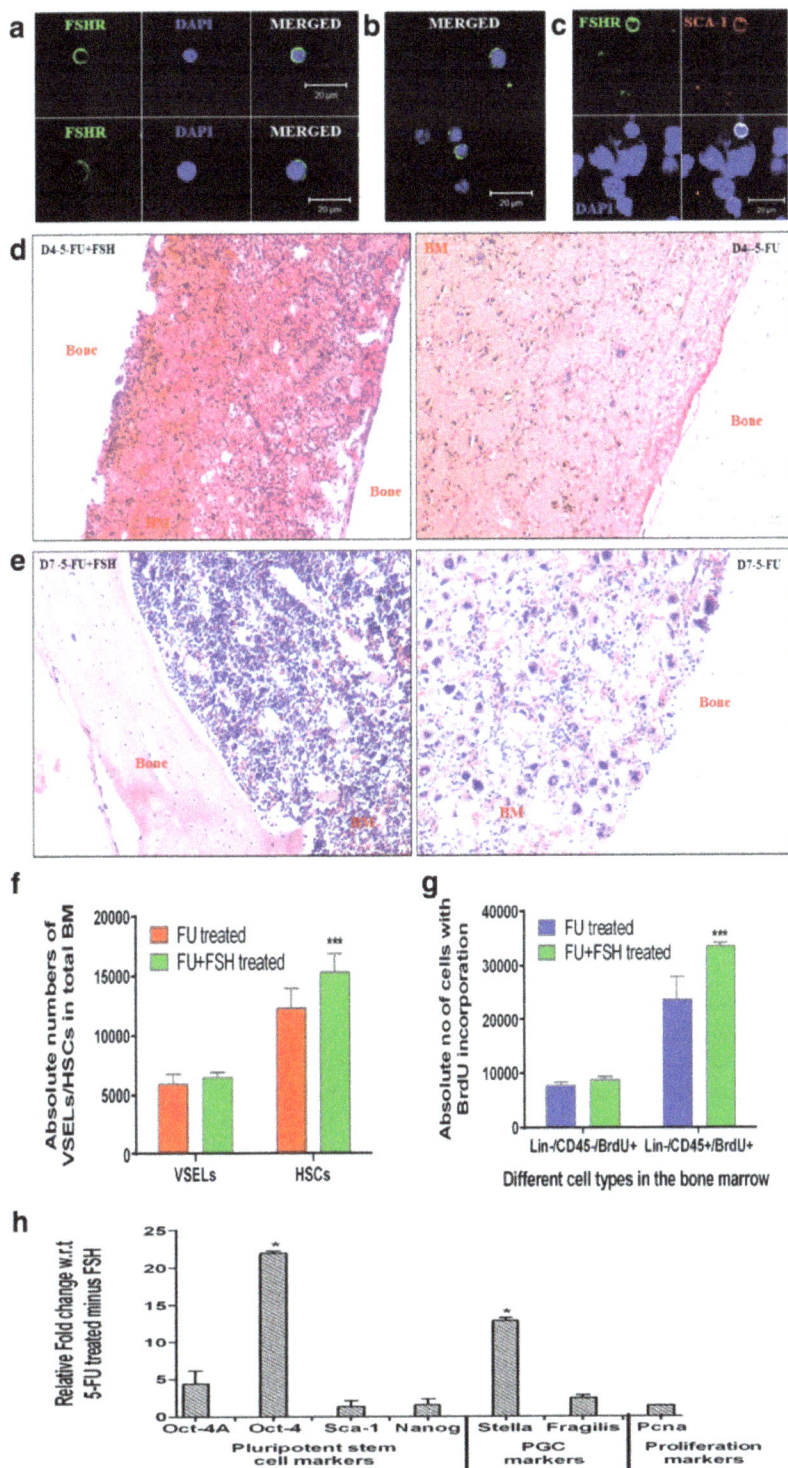

Fig. 6 (See legend on next page.)

(See figure on previous page.)

Fig. 6 Hematopoiesis recovery is augmented after FSH treatment. **a**, **b** Cell surface expression of FSHR (*green*) on both small VSELs (*top panel*) and slightly bigger HSCs (*bottom panel*) which survive 5-FU treatment in mouse BM on D4. **c** Co-expression of SCA-1 (*red*) and FSHR (*green*) on the same cells. The nuclei are counterstained with DAPI. Scale bar = 20 μm. **d**, **e** H & E sections on D4 showed increased BM cellularity after FSH treatment on D4 (**d**) compared with FSH minus control. **e** In 5-FU + FSH-treated sections, the endogenous bone marrow regeneration was almost complete by D7 after 5-FU treatment compared with untreated controls. **f** Flow cytometry data showed an increase in number of VSELs and HSCs on FSH treatment. **g** BrdU uptake also increased on FSH treatment slightly in the primitive Lin⁻/CD45⁻ (enriched in VSELs) while a significant increase ($p < 0.001$) was seen in Lin⁻/CD45⁺ (HSC-enriched) cells. **h** Upregulation of transcripts specific for pluripotent, primordial germ cells, and proliferation markers on FSH treatment compared with untreated controls. Significant increase in Oct-4 and Stella transcripts was observed. Please note that Oct-4 transcripts (which reflect HSCs) were more than Oct-4A (which reflects VSELs). Bars in **f**, **g** represent average ± SD and in **h** average ± SEM. Representative of five experiments in **f**, **g** and in **h** data obtained from three experiments. *$p \leq 0.05$. ***$p \leq 0.001$. *D4-5-FU*, *D7-5-FU* days post 5-FU treatment, *FSH* follicle-stimulating hormone, *BM* bone marrow, *5-FU* 5-fluorouracil, *H & E* hematoxylin and eosin, *VSEL* very small embryonic-like stem cell, *FSHR* follicle stimulating hormone receptor, *SCA-1* stem cell antigen-1, *HSC* hematopoietic stem cell, *OCT-4* octamer binding transforming factor-4, *BrdU* bromodeoxyuridine (Color figure online)

One of the main objectives of the present study was to examine the effects of 5-FU on VSELs and to investigate whether VSELs contribute to hematopoiesis in vivo. Under steady-state conditions, VSELs remain quiescent but in response to tissue injury or stress they are activated and mobilize to the affected area [24–30]. A condition of bone marrow stress was thus used in the present study to understand the role of VSELs in bone marrow regeneration. 5-FU, a chemotherapeutic agent, rapidly depletes the cycling myeloid and lymphoid progenitors [64–67]. This creates a dearth of progenitors in the bone marrow and cytotoxic stress is established. Additionally, treatment with 5-FU spares the slow cycling and dormant cells, thereby resulting in bone marrow enriched in primitive stem cells [68–71]. Even in the present study, as expected 5-FU caused 7-fold depletion in bone marrow cellularity by D4 and spontaneous endogenous recovery of the bone marrow occurred by D10, which established a basal timeline for the later experiments.

The kinetics of VSELs and HSCs was studied further within this period of 2–10 days post treatment. VSELs, being the most primitive cells in the bone marrow and quiescent under normal conditions, were resistant to the effects of 5-FU confirmed by no change in their numbers on D2 ($3.49 \pm 0.45 \times 10^3$) compared with untreated control ($3.77 \pm 1.01 \times 10^3$). On the other hand, more than 50 % of HSCs were depleted in response to 5-FU ($8.63 \pm 0.70 \times 10^3$) on D2 compared with untreated control ($14.07 \pm 3.60 \times 10^3$) in agreement with earlier reports [67, 69, 72]. The bias in the effect of 5-FU on HSCs was possibly due to the presence of two separate subpopulations within the HSCs, comprising of the quiescent cells that survive 5-FU cytotoxicity and the actively dividing cells which get depleted. These two subpopulations have been described as long- and short-term repopulating subsets of HSCs which show varying degree of radioprotection [73].

The increase in numbers of both VSELs and HSCs in response to 5-FU confirmed at both protein and mRNA levels and by BrdU and cell cycle analysis in the present study suggests that the stem/progenitor cells get activated in response to stress, and both proliferate but the extent of proliferation of HSCs is much higher compared with VSELs. This compensatory expansion of stem/progenitors initiates regeneration of chemoablated BM. Our results are in agreement with earlier reports by Ratajczak et al. [31], who reported BrdU incorporation in 12 % of the VSELs post irradiation in mouse bone marrow. Similarly, Baldrige et al. [74] also reported activation of both intermediate blood progenitors and primitive long term-HSCs (LT-HSCs) after chronic bacterial infection. Conclusive evidence that VSELs survive and proliferate in BM after 5-FU treatment is provided by us for the first time in the literature by dual immunofluorescence studies on BM sections. Nuclear OCT-4A and PCNA-expressing VSELs are clearly evident on D4 along the endosteal region of the BM which is considered to be the niche for stem cells [75, 76]. The niche maintains stem cells pool in vivo and regulates the hematopoietic recovery after myelo-suppression [77–79]. The regenerative ability of cells that survive chemoablation in BM was also confirmed by CFU assay in vitro that resulted in colony formation in Methocult media, thus implying that surviving 5-FU cells are functional and can form mature blood lineages to recolonize the BM.

Furthermore, we report two distinct sizes of cells (possibly VSELs and HSCs) expressing FSHR in ablated BM. Similar expression of FSHR on stem/progenitors has been reported in mouse BM and cord blood [46, 47] and on ovarian and testicular stem/progenitors cells by our group [49–52]. However further studies are required to examine this in a more quantitative manner because we found only 6 % of LIN⁻/CD45⁻ VSELs and 10 % of LIN⁻/CD45⁺ HSCs expressing FSHR by immunophenotyping studies. Our group has earlier shown that FSH treatment stimulates ovarian and testicular VSELs to self-renew in chemoablated gonads and also stimulates clonal expansion of progenitors [15, 50, 52]. Recently, Zbucka-Kretowska et al. [48] reported that FSH mobilized BM stem/progenitor cells into circulation but not endothelial progenitor cells in 15 female patients on FSH therapy for ovarian stimulation in an infertility clinic. The augmented hematopoietic recovery on D7 in response to FSH treatment noted in the present study after compared with D10 in untreated mice provides further evidence that indeed FSH influences stem/progenitor cells in BM. The enhanced recovery on FSH treatment was found to be accompanied with a marginal increase in the number of VSELs and a dramatic increase in HSCs and differentiated cells.

True stem cells are expected to divide asymmetrically (ACD) and give rise to two distinct cell types, whereas the progenitors that arise by ACD of stem cells undergo rapid symmetric divisions (SCD) and differentiation to maintain tissue homeostasis. Although the HSCs are the best investigated human somatic stem cells, which cells undergo ACD is still not clearly understood [80]. Emerging evidence suggests that (1) VSELs are the most primitive small-sized stem cells in the BM in addition to actively dividing, larger sized HSCs, (2) VSELs give rise to HSCs based on in-vitro studies [31, 32] and the OCT-4 expression pattern shown in the present study wherein VSELs express nuclear OCT-4A and HSCs express cytoplasmic OCT-4, and (3) both VSELs and HSCs are stimulated and take part in regeneration of BM under stress conditions shown in the present study and also reported earlier [13, 15, 31]. Based on this we suggest that VSELs may correspond to the LT-HSCs. A recent report of a similar imprinting pattern of erasure at the IgF2-H19 locus to maintain quiescence in LT-HSCs [81] and VSELs [82] provides further evidence of a link between these cells. VSELs are probably the true stem cells in the hematopoietic system that undergo ACD to self-renew and give rise to HSCs which are actively dividing progenitors that further differentiate into various blood cell types. This was recently discussed in detail in various adult organs [83].

Certain changes in the niche may push the stem cells to undergo uncontrolled proliferation resulting in cancer/leukemia. If this was true the stem cells and cancer cells should share similar markers. Recently Zhao et al. [84] found a significantly increased nuclear OCT-4A expression in patients with acute leukemia. This observation combined with current understanding that VSELs are the most primitive cells expressing nuclear OCT-4A, it can be hypothesized that in the case of leukemia, changes in the niche push the relatively quiescent VSELs to undergo uncontrolled proliferation. This link between VSELs and HSCs and leukemia stem cells warrants further investigation.

Conclusions

To summarize, bone marrow houses pluripotent VSELs which give rise to the HSCs. VSELs and a subset of HSCs

survive 5-FU-induced stress and are activated to recolonize the chemoablated BM. VSELs/HSCs also express FSHR and thus FSH treatment expedites BM regeneration by almost 72 h. This opens up a newer avenue for use of FSH in the clinical setting while treating bone marrow failure and irradiation patients. Our results provide better understanding towards the stem/progenitor cell hierarchy in the hematopoietic system. Studying the factors released by the niche cells which activate stem/progenitor cells to regenerate BM cells under stress conditions and also which result in uncontrolled proliferation and expansion of Oct-4A$^+$ VSELs in leukemia patients is warranted.

Additional file

Additional file 1: Description of methods, primers and antibodies used in the study. (PDF 340 kb)

Competing interests
The authors declare that they have no competing interests.

Authors' contributions
AS was responsible for the conception and design of the study, data collection, analysis and interpretation, and manuscript writing. DB was responsible for the conception and design of the study, data analysis and interpretation, and manuscript writing. SK was responsible for the data collection, analysis and interpretation, and manuscript writing. HN was responsible for the data collection and interpretation. All authors read and approved the final version of the manuscript.

Acknowledgments
The authors thank Dr Mukherjee, Gayatri Shinde, Sushma Khavale, Nivedita Dhavale, and Puja Soni for their help in carrying out flow cytometry studies and thank Shobha Sonawane and Reshma Gaonkar for their help with confocal studies. They are also thankful to Sandhya Anand, Hiren Patel, Varsha Pursani, and Jarnail Singh for their help with animal studies. The authors acknowledge University Grants Commission (UGC), Government of India, New Delhi for support towards the doctoral program of Ambreen Shaikh. Financial support for the study was provided by Indian Council of Medical Research, Department of Health Research, Government of India, New Delhi, India [NIRRH:RA/303/09-2015].

References
1. Shi Q, Rafii S, Wu MH, Wijelath ES, Yu C, Ishida A, et al. Evidence for circulating bone marrow-derived endothelial cells. Blood. 1998;92:362–7.
2. Jiang Y, Jahagirdar BN, Reinhardt RL, Schwartz RE, Keene CD, Ortiz-Gonzalez XR, et al. Pluripotency of mesenchymal stem cells derived from adult marrow. Nature. 2002;418:41–9.
3. D'Ippolito G, Diabira S, Howard GA, Menei P, Roos BA, Schiller PC. Marrow-isolated adult multilineage inducible (MIAMI) cells, a unique population of postnatal young and old human cells with extensive expansion and differentiation potential. J Cell Sci. 2004;117:2971–81.
4. Peister A, Mellad JA, Larson BL, Hall BM, Gibson LF, Prockop DJ. Adult stem cells from bone marrow (MSCs) isolated from different strains of inbred mice vary in surface epitopes, rates of proliferation, and differentiation potential. Blood. 2004;103:1662–8.
5. Beltrami AP, Cesselli D, Bergamin N, Marcon P, Rigo S, Puppato E, et al. Multipotent cells can be generated in vitro from several adult human organs (heart, liver and bone marrow). Blood. 2007;110:3438–46.
6. Kucia M, Reca R, Campbell FR, Zuba-Surma E, Majka M, Ratajczak J, et al. A population of very small embryonic-like (VSEL) CXCR4(+) SSEA-1(+)Oct-4+ stem cells identified in adult bone marrow. Leukemia. 2006;20:857–69.

7. Zuba-Surma EK, Kucia M, Wu W, Klich I, Lillard Jr JW, Ratajczak J, et al. Very small embryonic-like stem cells are present in adult murine organs: ImageStream-based morphological analysis and distribution studies. Cytometry A. 2008;73A:1116–27.

8. Sovalat H, Scrofani M, Eidenschenk A, Pasquet S, Rimelen V, Hénon P. Identification and isolation from either adult human bone marrow or G-CSF-mobilized peripheral blood of CD34(+)/CD133(+)/CXCR4(+)/Lin(–)CD45(–) cells, featuring morphological, molecular, and phenotypic characteristics of very small embryonic-like (VSEL) stem cells. Exp Hematol. 2011;39:495–505.

9. Kucia M, Halasa M, Wysoczynski M, Baskiewicz-Masiuk M, Moldenhawer S, Zuba-Surma E, et al. Morphological and molecular characterization of novel population of CXCR4+ SSEA-4+ Oct-4+ very small embryonic-like cells purified from human cord blood: preliminary report. Leukemia. 2007; 21:297–303.

10. Bhartiya D, Shaikh A, Nagvenkar P, Kasiviswanathan S, Pethe P, Pawani H, et al. Very small embryonic-like stem cells with maximum regenerative potential get discarded during cord blood banking and bone marrow processing for autologous stem cell therapy. Stem Cells Dev. 2012;21:1–6.

11. Shaikh A, Nagvenkar P, Pethe P, Hinduja I, Bhartiya D. Molecular and phenotypic characterization of CD133 and SSEA4 enriched very small embryonic-like stem cells in human cord blood. Leukemia. 2015;9:1909–17.

12. Bhartiya D, Kasiviswanathan S, Unni SK, Pethe P, Dhabalia JV, Patwardhan S, et al. Newer insights into premeiotic development of germ cells in adult human testis using Oct-4 as a stem cell marker. J Histochem Cytochem. 2010;58:1093–106.

13. Anand S, Bhartiya D, Sriraman K, Patel H, Manjramkar DD. Very small embryonic-like stem cells survive and restore spermatogenesis after busulphan treatment in mouse testis. J Stem Cell Res Ther. 2014;4:216.

14. Parte S, Bhartiya D, Telang J, Daithankar V, Salvi V, Zaveri K, et al. Detection, characterization and spontaneous differentiation in vitro of very small embryonic-like putative stem cells in adult mammalian ovary. Stem Cells Dev. 2011;20:1451–64.

15. Sriraman K, Bhartiya D, Anand S, Bhutda S. Mouse ovarian very small embryonic-like stem cells resist chemotherapy and retain ability to initiate oocyte-specific differentiation. Reprod Sci. 2015;7:884–903.

16. Gunjal P, Bhartiya D, Metkari S, Manjramkar D, Patel H. Very small embryonic-like stem cells are the elusive mouse endometrial stem cells-a pilot study. J Ovarian Res. 2015;8:9.

17. Bhartiya D, Mundekar A, Mahale V, Patel H. Very small embryonic-like stem cells are involved in regeneration of mouse pancreas post-pancreatectomy. Stem Cell Res Ther. 2014;5:106.

18. Liu Y, Gao L, Zuba-Surma EK, Peng X, Kucia M, Ratajczak MZ, et al. Identification of small Sca-1(+), Lin(–), CD45(–) multipotential cells in the neonatal murine retina. Exp Hematol. 2009;37:1096–107.

19. Kassmer SH, Bruscia EM, Zhang PX, Krause DS. Nonhematopoietic cells are the primary source of bone marrow-derived lung epithelial cells. Stem Cells. 2012;30:491–9.

20. Taichman RS, Wang Z, Shiozawa Y, Jung Y, Song J, Balduino A, et al. Prospective identification and skeletal localization of cells capable of multilineage differentiation in vivo. Stem Cells Dev. 2010;19:1557–70.

21. Havens AM, Shiozawa Y, Jung Y, Sun H, Wang J, McGee S, et al. Human very small embryonic-like cells generate skeletal structures, in vivo. Stem Cells Dev. 2012;22:622–30.

22. Shin DM, Zuba-Surma EK, Wu W, Ratajczak J, Wysoczynski M, Ratajczak MZ, et al. Novel epigenetic mechanisms that control pluripotency and quiescence of adult bone marrow-derived Oct4 (+) very small embryonic-like stem cells. Leukemia. 2009;23:2042–51.

23. Mierzejewska K, Heo J, Kang JW, Kang H, Ratajczak J, Ratajczak MZ, et al. Genome-wide analysis of murine bone marrowderived very small embryonic-like stem cells reveals that mitogenic growth factor signaling pathways play a crucial role in the quiescence and ageing of these cells. Int J Mol Med. 2013;32:281–90.

24. Kucia M, Wysoczynski M, Wu W, Zuba-Surma EK, Ratajczak J, Ratajczak MZ. Evidence that very small embryonic like (VSEL) stem cells are mobilized into peripheral blood. Stem Cells. 2008;26:2083–92.

25. Zuba-Surma EK, Kucia M, Dawn B, Guo Y, Ratajczak MZ, Bolli R. Bone marrow derived pluripotent very small embryonic-like stem cells (VSELs) are mobilized after acute myocardial infarction. J Mol Cell Cardiol. 2008; 44:865–73.

26. Paczkowska E, Kucia M, Koziarska D, Halasa M, Safranow K, Masiuk M, et al. Clinical evidence that very small embryonic-like stem cells are mobilized into peripheral blood in patients after stroke. Stroke. 2009;40:1237–44.

27. Abdel-Latif A, Zuba-Surma EK, Ziada KM, Kucia M, Cohen DA, Kaplan AM, et al. Evidence of mobilization of pluripotent stem cells into peripheral blood of patients with myocardial ischemia. Exp Hematol. 2010;38:1131–42.

28. Gharib SA, Dayyat EA, Khalyfa A, Kim J, Clair HB, Kucia M, et al. Intermittent hypoxia mobilizes bone marrow derived very small embryonic-like stem cells and activates developmental transcriptional programs in mice. Sleep. 2010;33:1439–46.

29. Drukała J, Paczkowska E, Kucia M, Młyńska E, Krajewski A, Machaliński B, et al. Stem cells, including a population of very small embryonic-like stem cells, are mobilized into peripheral blood in patients after skin burn injury. Stem Cell Rev. 2012;8:184–94.

30. Marlicz W, Zuba-Surma E, Kucia M, Blogowski W, Starzynska T, Ratajczak MZ. Various types of stem cells, including population of very small embryonic-like stem cells, are mobilized into peripheral blood in patients with Crohn's disease. Inflamm Bowel Dis. 2012;18:1711–22.

31. Ratajczak J, Wysoczynski M, Zuba-Surma E, Wan W, Kucia M, Yoder MC, et al. Adult murine bone marrow-derived very small embryonic-like stem cells differentiate into the hematopoietic lineage after coculture over OP9 stromal cells. Exp Hematol. 2011;39:225–37.

32. Ratajczak J, Zuba-Surma E, Klich I, Liu R, Wysoczynski M, Greco N, et al. Hematopoietic differentiation of umbilical cord blood-derived very small embryonic/epiblast-like stem cells. Leukemia. 2011;25:1278–85.

33. Shivtiel S, Kollet O, Lapid K, Schajnovitz A, Goichberg P, Kalinkovich A, et al. CD45 regulates retention, motility, and numbers of hematopoietic progenitors, and affects osteoclast remodeling of metaphyseal trabecules. J Exp Med. 2008;205:2381–95.

34. Ogata K, Satoh C, Tachibana M, Hyodo H, Tamura H, Dan K, et al. Identification and hematopoietic potential of CD45- clonal cells with very immature phenotype (CD45-CD34-CD38-Lin-) in patients with myelodysplastic syndromes. Stem Cells. 2005;23:619–30.

35. Xu G, Yang L, Zhang W, Wei X. All the tested human somatic cells express both Oct4A and its pseudogenes but express Oct4A at much lower levels compared with its pseudogenes and human embryonic stem cells. Stem Cells Dev. 2015;24:1546–57.

36. Jez M, Ambady S, Kashpur O, Grella A, Malcuit C, Vilner L, et al. Expression and differentiation between OCT4A and its pseudogenes in human ESCs and differentiated adult somatic cells. PLoS One. 2014;9:e89546.

37. Samardzija C, Quinn M, Findlay JK, Ahmed N. Attributes of Oct4 in stem cell biology: perspectives on cancer stem cells of the ovary. J Ovarian Res. 2012;5:37.

38. Liedtke S, Enczmann J, Waclawczyk S, Wernet P, Kögler G. Oct4 and its pseudogenes confuse stem cell research. Cell Stem Cell. 2007;1:364–6.

39. Atlasi Y, Mowla SJ, Ziaee SA, Gokhale PJ, Andrews PW. OCT4 spliced variants are differentially expressed in human pluripotent and nonpluripotent cells. Stem Cells. 2008;26:3068–74.

40. Bhartiya D. Are mesenchymal cells indeed pluripotent stem cells or just stromal cells? OCT-4 and VSELs biology has led to better understanding. Stem Cells Int. 2013;2013:547501.

41. Pietras EM, Warr MR, Passegué E. Cell cycle regulation in hematopoietic stem cells. J Cell Biol. 2011;195:709–20.

42. Nakada D, Levi BP, Morrison SJ. Integrating physiological regulation with stem cell and tissue homeostasis. Neuron. 2011;70:703–18.

43. Selleri C, Catalano L, De Rosa G, Fontana R, Notaro R, Rotoli B. Danazol: in vitro effects on human hemopoiesis and in vivo activity in hypoplastic and myelodysplastic disorders. Eur J Haematol. 1991;47:197–203.

44. Carreras E, Turner S, Paharkova-Vatchkova V, Mao A, Dascher C, Kovats S. Estradiol acts directly on bone marrow myeloid progenitors to differentially regulate GM-CSF or Flt3 ligand-mediated dendritic cell differentiation. J Immunol. 2008;180:727–38.

45. Nakada D, Oguro H, Levi BP, Ryan N, Kitano A, Saitoh Y, et al. Oestrogen increases haematopoietic stem-cell self-renewal in females and during pregnancy. Nature. 2014;505:555–8.

46. Mierzejewska K, Borkowska S, Suszynska E, Adamiak M, Ratajczak J, Kucia M, et al. Hematopoietic stem/progenitor cells express several functional sex hormone receptors-novel evidence for a potential developmental link between hematopoiesis and primordial germ cells. Stem Cells Dev. 2015;24:927–37.

47. Abdelbaset-Ismail A, Suszynska M, Borkowska S, Adamiak M, Ratajczak J, Kucia M, et al. Human hematopoietic stem/progenitor cells express several functional sex hormone receptors. J Cell Mol Med. 2016;20:134–46.

48. Zbucka-Kretowska M, Eljaszewicz A, Lipinska D, Grubczak K, Rusak M, Mrugacz G, et al. Effective Mobilization of very small embryonic-like stem cells and hematopoietic stem/progenitor cells but not endothelial progenitor cells by follicle-stimulating hormone therapy. Stem Cells Int. 2016;2016:8530207.

49. Bhartiya D, Sriraman K, Gunjal P, Modak H. Gonadotropin treatment augments postnatal oogenesis and primordial follicle assembly in adult mouse ovaries? J Ovarian Res. 2012;5:32.

50. Patel H, Bhartiya D, Parte S, Gunjal P, Yedurkar S, Bhatt M. Follicle stimulating hormone modulates ovarian stem cells through alternately spliced receptor variant FSH-R3. J Ovarian Res. 2013;6:52.

51. Parte S, Bhartiya D, Manjramkar DD, Chauhan A, Joshi A. Stimulation of ovarian stem cells by follicle stimulating hormone and basic fibroblast growth factor during cortical tissue culture. J Ovarian Res. 2013;6:20.

52. Patel H, Bhartiya D. Testicular stem cells express follicle-stimulating hormone receptors and are directly modulated by FSH. Reproductive sciences (Article in press)

53. Ratajczak MZ, Zuba-Surma EK, Wojakowski W, Ratajczak J, Kucia M. Bone marrow—home of versatile stem cells. Transfus Med Hemother. 2008;35:248–59.

54. Bhartiya D, Patel H. Very small embryonic-like stem cells are involved in pancreatic regeneration and their dysfunction with age may lead to diabetes and cancer. Stem Cell Res Ther. 2015;6:96.

55. Anand S, Patel H, Bhartiya D. Chemoablated mouse seminiferous tubular cells enriched for very small embryonic-like stem cells undergo spontaneous spermatogenesis in vitro. Reprod Biol Endocrinol. 2015;13:33.

56. Guerin CL, Loyer X, Vilar J, Cras A, Mirault T, Gaussem P, et al. Bone-marrow-derived very small embryonic-like stem cells in patients with critical leg ischaemia: evidence of vasculogenic potential. Thromb Haemost. 2015;113:1084–94.

57. Abouzaripour M, RagerdiKashani I, Pasbakhsh P, Atlasy N. Intravenous transplantation of very small embryonic like stem cells in treatment of diabetes mellitus. J Med Biotechnol. 2015;7:22–31.

58. Chen ZH, Lv X, Dai H, Lou D, Chen R, Zou GM. Hepatic regenerative potential of mouse bone marrow very small embryonic-like stem cells. J Cell Physiol. 2015;230:1852–61.

59. Bhartiya D, Hinduja I, Patel H, Bhilawadikar R. Making gametes from pluripotent stem cells—a promising role for very small embryonic-like stem cells. Reprod Biol Endocrinol. 2014;12:114.

60. Lee SJ, Park SH, Kim YI, Hwang S, Kwon PM, Han IS, et al. Adult stem cells from the hyaluronic acid-rich node and duct system differentiate into neuronal cells and repair brain injury. Stem Cells Dev. 2014;23:2831–40.

61. Dawn B, Tiwari S, Kucia MJ, Zuba-Surma EK, Guo Y, Sanganalmath SK, et al. Transplantation of bone marrow-derived very small embryonic-like stem cells attenuates left ventricular dysfunction and remodeling after myocardial infarction. Stem Cells. 2008;26:1646–55.

62. Shin DM, Liu R, Klich I, Wu W, Ratajczak J, Kucia M, et al. Molecular signature of adult bone marrow-purified very small embryonic-like stem cells supports their developmental epiblast/germ line origin. Leukemia. 2010;24:1450–61.

63. Nakatsuka R, Iwaki R, Matsuoka Y, Sumide K, Kawamura H, Fujioka T, et al. Identification and characterization of Lineage-CD45-Sca-1+ VSEL phenotypic cells residing in adult mouse bone tissue. Stem Cells Dev. 2016;25:27–42.

64. Hodgson GS, Bradley TR. Properties of hematopoietic stem cells surviving 5-fluorouracil treatment: evidence for a pre-CFU-S cell? Nature. 1979;281:381–2.

65. Lerner C, Harrison DE. 5-fluorouracil spares hemopoietic stem cells responsible for long-term repopulation. Exp Hematol. 1990;18:114–8.

66. Van Zant G. Studies of hematopoietic stem cells spared by 5-fluorouracil. J Exp Med. 1984;159:679–90.

67. Randall TD, Weissman IL. Phenotypic and functional changes induced at the clonal level in hematopoietic stem cells after 5-fluorouracil treatment. Blood. 1997;89:3596–606.

68. Harrison DE, Lerner CP. Most primitive hematopoietic stem cells are stimulated to cycle rapidly after treatment with 5-fluorouracil. Blood. 1991;78:1237–40.

69. Szilvassy SJ, Cory S. Phenotypic and functional characterization of competitive long-term repopulating hematopoietic stem cells enriched from 5-fluorouracil-treated murine marrow. Blood. 1993;81:2310–20.

70. Ogawa M, Shih JP, Katayama N. Enrichment for primitive hemopoietic progenitors of marrow cells from 5-fluorouracil treated mice and normal mice. Blood Cells. 1994;20:7–11.

71. Randall TD, Weissman IL. Characterization of a population of cells in the bone marrow that phenotypically mimics hematopoietic stem cells: resting stem cells or mystery population? Stem Cells. 1998;16:38–48.

72. Radley JM, Scurfield G. Effects of 5-fluorouracil on mouse bone marrow. Br J Haematol. 1979;43:341.

73. Morrison SJ, Weissman IL. The long-term repopulating subset of hematopoietic stem cells is deterministic and isolated by phenotype. Immunity. 1994;1:661–73.

74. Baldridge MT, King KY, Boles NC, Weksberg DC, Goodell MA. Quiescent hematopoietic stem cells are activated by IFN-gamma in response to chronic infection. Nature. 2010;465:793–7.

75. Lord BI, Testa NG, Hendry JH. The relative spatial distributions of CFUs and CFUc in the normal mouse femur. Blood. 1975;46:65–72.

76. Calvi LM, Adams GB, Weibrecht KW, Weber JM, Olson DP, Knight MC, et al. Osteoblastic cells regulate the hematopoietic stem cell niche. Nature. 2003;425:841–6.

77. Salter AB, Meadows SK, Muramoto GG, Himburg H, Doan P, Daher P, et al. Endothelial progenitor cell infusion induces hematopoietic stem cell reconstitution in vivo. Blood. 2009;113:2104–7.

78. Ding L, Saunders TL, Enikolopov G, Morrison SJ. Endothelial and perivascular cells maintain hematopoietic stem cells. Nature. 2012;481:457–62.

79. Doan PL, Russell JL, Himburg HA, Helms K, Harris JR, Lucas J, et al. Tie2+ bone marrow endothelial cells regulate hematopoietic stem cell regeneration following radiation injury. Stem Cells. 2013;31:327–37.

80. Murke F, Castro S, Giebel B, Görgens A. Concise Review: Asymmetric cell divisions in stem cell biology. Symmetry. 2015;7:2025–37.

81. Venkatraman A, He XC, Thorvaldsen JL, Sugimura R, Perry JM, Tao F, et al. Maternal imprinting at the H19-Igf2 locus maintains adult hematopoietic stem cell quiescence. Nature. 2013;500:345–9.

82. Shin DM, Liu R, Wu W, Waigel SJ, Zacharias W, Ratajczak MZ, et al. Global gene expression analysis of very small embryonic-like stem cells reveals that the Ezh2-dependent bivalent domain mechanism contributes to their pluripotent state. Stem Cells Dev. 2012;21:1639–52.

83. Bhartiya D. Stem cells, progenitors & regenerative medicine: a retrospection. Indian J Med Res. 2015;141:154–61.

84. Zhao Q, Ren H, Feng S, Chi Y, He Y, Yang D, et al. Aberrant expression and significance of OCT-4A transcription factor in leukemia cells. Blood Cells Mol Dis. 2015;5:90–6.

9

Single-stranded DNA binding protein Ssbp3 induces differentiation of mouse embryonic stem cells into trophoblast-like cells

Jifeng Liu[1], Xinlong Luo[1,3], Yanli Xu[1], Junjie Gu[1,2], Fan Tang[1,2], Ying Jin[1,2]* and Hui Li[1,2]*

Abstract

Background: Intrinsic factors and extrinsic signals which control unlimited self-renewal and developmental pluripotency in embryonic stem cells (ESCs) have been extensively investigated. However, a much smaller number of factors involved in extra-embryonic trophoblast differentiation from ESCs have been studied. In this study, we investigated the role of the single-stranded DNA binding protein, Ssbp3, for the induction of trophoblast-like differentiation from mouse ESCs.

Methods: Gain- and loss-of-function experiments were carried out through overexpression or knockdown of Ssbp3 in mouse ESCs under self-renewal culture conditions. Expression levels of pluripotency and lineage markers were detected by real-time quantitative reverse-transcription polymerase chain reaction (qRT-PCR) analyses. The global gene expression profile in Ssbp3-overexpressing cells was determined by affymetrix microarray. Gene ontology and pathway terms were analyzed and further validated by qRT-PCR and Western blotting. The methylation status of the Elf5 promoter in Ssbp3-overexpressing cells was detected by bisulfite sequencing. The trophoblast-like phenotype induced by Ssbp3 was also evaluated by teratoma formation and early embryo injection assays.

Results: Forced expression of Ssbp3 in mouse ESCs upregulated expression levels of lineage-associated genes, with trophoblast cell markers being the highest. In contrast, depletion of Ssbp3 attenuated the expression of trophoblast lineage marker genes induced by downregulation of Oct4 or treatment with BMP4 and bFGF in ESCs. Interestingly, global gene expression profiling analysis indicated that Ssbp3 overexpression did not significantly alter the transcript levels of pluripotency-associated transcription factors. Instead, Ssbp3 promoted the expression of early trophectoderm transcription factors such as Cdx2 and activated MAPK/Erk1/2 and TGF-β pathways. Furthermore, overexpression of Ssbp3 reduced the methylation level of the Elf5 promoter and promoted the generation of teratomas with internal hemorrhage, indicative of the presence of trophoblast cells.

Conclusions: This study identifies Ssbp3, a single-stranded DNA binding protein, as a regulator for mouse ESCs to differentiate into trophoblast-like cells. This finding is helpful to understand the regulatory networks for ESC differentiation into extra-embryonic lineages.

Keywords: Ssbp3, Mouse embryonic stem cells, Trophoblast, Differentiation

* Correspondence: yjin@sibs.ac.cn; lihuilh@shsmu.edu.cn
[1]Laboratory of Molecular Developmental Biology, Shanghai Jiao Tong University School of Medicine, Shanghai 200025, China
Full list of author information is available at the end of the article

Single-stranded DNA binding protein Ssbp3 induces differentiation of mouse embryonic stem...

121

Background

The formation of the early blastocyst represents the first lineage specification of mammalian embryos. At this stage, the mammalian embryo forms an internal cavity, and contains two types of cells: the inner cell mass (ICM) growing on the interior and the monolayer of the trophectoderm (TE) growing on the exterior [1, 2]. The ICM possesses developmental pluripotency, and develops into the three germ layers as well as the extra-embryonic mesoderm portion of the placenta, while the TE gives rise to the extra-embryonic ectoderm portion of the placenta [3].

Embryonic stem cells (ESCs) and trophoblast stem cells (TSCs) are cell culture derivatives of the ICM and TE, respectively [4, 5]. These two stem cell types possess distinctly different abilities to self-renew and generate progenies, which are under the control of both extrinsic signals and intrinsic regulatory networks. Although derived from the ICM, ESCs can undergo *trans*-differentiation towards extra-embryonic trophoblast lineage via overexpression of TE master regulators, such as Cdx2 and Gata3 [6–8]. In contrast to the extensive studies on the molecular regulation of ESC self-renewal and pluripotency, only a few of the factors involved in extra-embryonic lineage differentiation of ESCs have been defined.

Ssbp3 (single-stranded DNA binding protein 3) belongs to a family of single-stranded DNA binding proteins which recognizes pyrimidine-rich single-stranded DNA and regulates transcription. It has two homologues, Ssbp2 and Ssbp4 [9]. In 1998, Bayarsaihan et al. discovered that Ssbp3 could bind a conserved sequence, containing pyrimidines almost exclusively, on one strand in chicken collagen α2 (I) promoter regions to regulate its expression [10]. Further studies revealed the function of Ssbp3 in regulating mouse head morphogenesis by protecting Ldb1 from Rlim-mediated ubiquitination in development. Ssbp3 null mice lacked the head structure from the anterior to the ear, accompanied by a small size and spine deformity [11–16]. However, the function of Ssbp3 in early embryonic and extra-embryonic development remains unknown.

In the present study, we report that Ssbp3 plays an important role in regulating mouse ESC differentiation to trophoblast-like cells. We show that the expression of Ssbp3 was increased in ESC differentiation models towards trophoblast lineages induced by either Oct4 downregulation or supplementation of bone morphogenetic protein (BMP)4 and basic fibroblast growth factor (bFGF). Forced expression of Ssbp3 substantially promoted differentiation. The cells overexpressing Ssbp3 had gene expression patterns resembling those of trophoblast cells and formed teratomas containing hemorrhages in vivo. In contrast, depletion of Ssbp3

compromised the induction of trophoblast differentiation from ESCs. Genome-wide gene expression analysis indicated that Ssbp3 overexpression in ESCs activated mitogen-activated protein kinase (MAPK)/extracellular signal-regulated kinase (Erk)1/2 and transforming growth factor (TGF)-β pathways, which are known to be critical for mouse trophoblast development [17–20], and induced the expression of trophoblast lineage-associated markers. Taken together, our results revealed that Ssbp3 played an important role in the control of ESCs to differentiate into trophoblast-like cells.

Methods
Cell culture

Mouse E14T and ZHBTc4 ESCs (kind gifts from Dr. Austin Smith) were cultured under feeder-free conditions in GMEM (Glasgow's minimum essential medium; Invitrogen) supplemented with 10 % fetal bovine serum (FBS; Hyclone), 1 mM sodium pyruvate (Sigma), 0.1 mM non-essential amino acids (NEAA; Invitrogen), 2 mM L-glutamine (Invitrogen), 100 μM β-mercaptoethanol (Invitrogen), 100 U/mL penicillin and 100 μg/mL streptomycin (Hyclone), and 1000 U/mL recombinant leukemia inhibitory factor (LIF; Chemicon).

Mouse TSCs (a kind gift from Dr. Shaorong Gao) were maintained in the medium containing 30 % TS medium and 70 % mouse embryonic fibroblast (MEF)-conditioned TS medium with 25 ng/mL human recombinant fibroblast growth factor 4 (FGF4; Invitrogen) and 1 μg/mL heparin (Sigma) in 0.1 % gelatin-coated dishes. In detail, the TSC medium based on RPMI 1640 (Invitrogen) was supplemented with 20 % FBS (Hyclone), 1 mM sodium pyruvate (Sigma), 2 mM L-glutamine (Invitrogen), 100 μM β-mercaptoethanol (Invitrogen), and 100 U/mL penicillin and 100 μg/mL streptomycin (Hyclone) [21].

MEF and 293T cells were cultured in Dulbecco's modified Eagle's medium (DMEM; Invitrogen) with 10 % FBS (Hyclone), and 100 U/mL penicillin and 100 μg/mL streptomycin (Hyclone) [22].

The trophoblast induction medium was modified based on the TX medium [23], and contained the KnockOut DMEM/F-12 (Invitrogen), 64 μg/mL L-ascorbic acid-2-phosphate magnesium (ACC), insulin-transferrin-selenium (Life Technologies), 543 μg/mL NaHCO$_3$ (Sigma), 1 μg/mL heparin (Sigma), 2 mM L-glutamine (Life Technologies), 100 U/mL penicillin and 100 μg/mL streptomycin (Hyclone), 10 ng/mL BMP4 (HUMANZYME), and 25 ng/mL bFGF (HUMANZYME).

Plasmids and antibodies

The coding sequences of full length Ssbp3 [NCBI Reference Sequence: NM_023672.2], its truncations, and full length Cdx2 [NCBI Reference Sequence: NM_007673.3] were amplified by polymerase chain reaction (PCR) from complementary DNA (cDNA) of mouse ESCs or TSCs, respectively. PCR products were inserted into the pPy-CAGIP expression vector (a kind gift from Dr. Ian Chambers). Primers used for gene cloning are listed in Additional file 1 (Table S1). Lentiviral vector pLKO.1-hygro (Addgene) was used to express short-hairpin RNAs (shRNAs) that targeted mouse Ssbp3, Cdx2, or Gata3 mRNA, respectively [24]. For lentiviral packaging, psPAX2 and pMD2.G vectors (kind gifts from Dr. Didier Trono) were used. shRNA targeting sequences are also listed in Additional file 1 (Table S1). The inserted sequences and ligation sites of all constructs were validated by DNA sequencing. Western blotting was carried out with primary antibodies against Ssbp3 (Genetex), Cdx2 (Biogenex), Oct4 (produced and affinity-purified in our laboratory), phospho-Erk1/2 (Cell Signaling Technology), total Erk1/2 (Cell Signaling Technology), and β-Tubulin (Sigma). For immunofluorescence staining, cells were incubated with primary antibody against Cdx2 (Biogenex) and then Cy3 conjugated goat anti-mouse secondary antibody (ProteinTech Group).

Transfection and infection
Mouse ESCs were plated in six-well plates at a density of 1.0×10^5 cells per well. Twenty-four hours later, cells were transfected with plasmids via Lipofectamine™2000 (Invitrogen), or infected with viral particles.

Western blotting
Proteins extracted from the indicated cells were subjected to Western blotting with the amount of 100 μg proteins per lane, as described previously [25].

Immunofluorescence staining
Mouse ESCs were plated in four-well plates containing coverslips at a density of 1.0×10^5 cells per well and were transfected with an empty vector or an Ssbp3 expression plasmid on the second day, respectively. The cells were cultured at 37 °C for an additional 96 h. They were then fixed with 4 % paraformaldehyde for 15 min, permeabilized with 0.2 % Triton X-100 for 10 min, and blocked with 3 % bovine serum albumin (BSA) for 30 min. Next, cells were incubated with the primary anti-Cdx2 antibody (1:500 diluted) overnight at 4 °C. The next day, cells were incubated with the fluorescent Cy3-goat anti-mouse antibody (1:200 diluted) for 1 h at room temperature in the dark. The nuclei were further stained with 4',6-diamidino-2-phenylindole (DAPI; Invitrogen) for 3 min. Finally, coverslips were dried and affixed to slides using a fluorescent mounting medium [26].

qRT-PCR analyses
Total RNA was extracted and transcribed into cDNA as previously described [27]. GAPDH was used to normalize gene expression levels in cells, and 28S rRNA was used as an internal control in teratoma analysis. Real-time quantitative reverse transcription PCR (qRT-PCR) analyses were performed on an ABI7900 using the FastStart Universal SYBR Green Master (Roche) as previously described [27]. Primers used for qRT-PCR analyses are listed in Additional file 1 (Table S1).

Microarray assay and data analysis
Mouse ESC samples transfected with an empty vector or an Ssbp3 expression plasmid for 96 h were collected for gene expression detection using the Affymetrix GeneChip® Mouse Genome 430 2.0 Array. Duplicate samples were subjected to array analysis. The procedures including RNA extraction, cDNA synthesis, labeling, hybridization, washing, and scanning were performed by the Shanghai Biotechnology Corporation. Datasets were submitted to the database for annotation, visualization, and integrated discovery (DAVID), and analyzed by gene ontology (GO) analysis and Kyoto encyclopedia of genes and genomes (KEGG) pathway mapping. Data are publically available at the National Center for Biotechnology Information with Gene Expression Omnibus accession number GSE67562.

Bisulfite sequencing
The bisulfite treatment of DNA was performed using the EZ-DNA methylation direct kit (Zymo Research) according to the manufacturer's instructions. Elf5 primers were the same as described previously [28]. Amplified products were purified using gel filtration columns, cloned into the pGEM-T Easy Vector (Promega), and sequenced using SP6 primers. For each promoter sequence, ten randomly selected clones were sequenced.

Teratomas
Mouse ESCs transfected with an empty vector or an Ssbp3 expression plasmid were selected with 1 μg/mL puromycin (Sigma) for 48 h. Then 1.5×10^6 cells were intramuscularly injected into NOD/SCID mice [29]. Teratomas were collected 6 weeks later for qRT-PCR detection of gene expression and histological analysis.

Chimera
Mouse ESCs constitutively expressing the H2B-GFP fusion gene were transfected with a vector or an Ssbp3 expression plasmid. Twenty-four hours later, cells were selected with 1 μg/mL puromycin for 2 days. Two cells were injected into each 8-cell-stage mouse embryo. The embryos were subsequently transferred into the uterus of pseudo-pregnant female mice following standard procedures. Injected embryos at embryonic day 6.5 (E6.5) and E14.5 were collected and imaged, respectively. Fluorescence images were taken by a laser confocal

scanning microscope (Zeiss, LSM710) or a stereo zoom microscope (Zeiss, Axio Zoom.V16).

Statistical analysis

All data were from three biological replicates and are presented as means ± SD. Pairwise statistical significance was determined by the Student's t test. $p < 0.05$ was considered statistically significant.

Results

Forced expression of Ssbp3 induces differentiation of mouse ESCs with a trophoblast-like gene expression pattern

To study the function of Ssbp3 in mouse ESCs, we overexpressed Ssbp3 in E14T ESCs cultured under a self-renewal condition, in parallel with an empty vector as a negative control (Fig. 1a). Obvious differentiation was observed 96 h after transfection with the Ssbp3 plasmid, as evidenced by changes in the cell morphology and the intensity of alkaline phosphatase (AKP) staining (Fig. 1b). Our qRT-PCR results showed that the expression levels of key pluripotency-associated markers such as Oct4, Sox2, Nanog, and Rex1 were not altered significantly in ESCs overexpressing Ssbp3 compared with control ESCs at day 4 (Fig. 1c). However, most trophoblast marker genes tested, including early TSC markers (Cdx2, Gata3, Elf5) and relative mature trophoblast markers (Hand1, Dlx3, Esx1, Psx1, Krt8) [30], were significantly upregulated with greater than 10-fold increases (Fig. 1d), while most extra-embryonic endoderm marker genes detected such as Gata6, Gata4, Sox7, and Sox17 were modestly upregulated about five-fold (Fig. 1e). Meanwhile, expression levels of germ layer markers (the endoderm, mesoderm, and ectoderm) displayed mild alterations with less than two-fold changes (Fig. 1e, f and g). Consistently, immunofluorescence staining and Western blotting results revealed the enhanced expression of Cdx2 at the protein level 96 h after overexpression of Ssbp3 (Fig. 1h and i). We then evaluated the effect of Ssbp3 on ESCs under a differentiation condition induced by LIF withdrawal. Consistent with the results obtained under the self-renewal condition, trophoblast marker genes were dramatically upregulated (Figure S1A in Additional file 2). Overexpression of Ssbp3 also elevated extra-embryonic endoderm markers modestly (Figure S1B in Additional file 2) with slight alterations in the expression levels of germ layer-specific and pluripotency markers (Figure S1B, C, D and E in Additional file 2). These results implied that overexpression of Ssbp3 could induce mouse ESC differentiation biased towards trophoblast-like lineages.

Ssbp3 protein contains three different regions responsible for different functions: a well-conserved FORWARD/LUFS domain at the N-terminal end, through which Ssbp3 interacts with other proteins; a highly conserved proline-rich sequence in the middle critical for embryonic head development; and a C-terminal end possessing transcriptional activity [14, 31, 32]. To determine which region conferred Ssbp3 the ability to induce ESC differentiation, truncation mutants lacking the C-terminal, or middle, or N-terminal region were constructed (Fig. 1j) and transfected into ESCs, respectively. Unexpectedly, none of the truncation mutants displayed the same ESC differentiation function as did the full length Ssbp3 (Fig. 1k). Therefore, it is likely that the effect of Ssbp3 for inducing ESC differentiation requires its intact structure.

We next compared gene expression changes induced by Ssbp3 and Cdx2 overexpression, as Cdx2 is known as a key regulator for the trophoblast development, and overexpression of Cdx2 in mouse ESCs has been shown to efficiently induce trophoblast differentiation [7]. Our qRT-PCR results showed that expression patterns of various lineage markers in ESCs overexpressing Ssbp3 resembled those in ESCs overexpressing Cdx2 (Fig. 1l), suggesting that Ssbp3 might have a role similar to Cdx2 for induction of ESC differentiation. Moreover, we found that both mRNA and protein levels of Ssbp3 were substantially higher in TSCs than in ESCs, further supporting the association of Ssbp3 with trophoblast lineages at both mRNA and protein levels (Fig. 1m, n).

Ssbp3 depletion attenuates the activation of trophoblast gene expression induced by downregulation of Oct4 in mouse ESCs

Mouse ESCs are usually considered to have a weak ability, if any, to generate trophoblast cell types by spontaneous differentiation [33]. However, genetic manipulation such as reduction of Oct4 or Tet1 [29, 34], or induction of Cdx2, Gata3, Arid3a, Brog5, or other key trophoblast-associated factors, can convert mouse ESCs to TS-like cells [6, 7, 35–38]. Here, we used the ZHBTc4 mouse ESC line as an in vitro differentiation model for trophoblast induction as previously reported [34]. In this cell line, both alleles of endogenous Oct4 loci were deleted and Oct4 expression was controlled by a tetracycline (Tc)-regulated Oct4 transgene.

In line with published results, Tc treatment reduced Oct4 expression rapidly at both the mRNA and protein levels, and robustly induced trophoblast differentiation (Fig. 2a, b, c). We found that the expression of Ssbp3 at the mRNA and protein levels increased gradually with Tc treatment (Fig. 2b, c), adding more evidence for the potential association of Ssbp3 expression with trophoblast differentiation.

To determine whether Ssbp3 plays a role in the trophoblast differentiation induced by Oct4 downregulation, two shRNA sequences (shSsbp3-976 and shSsbp3-

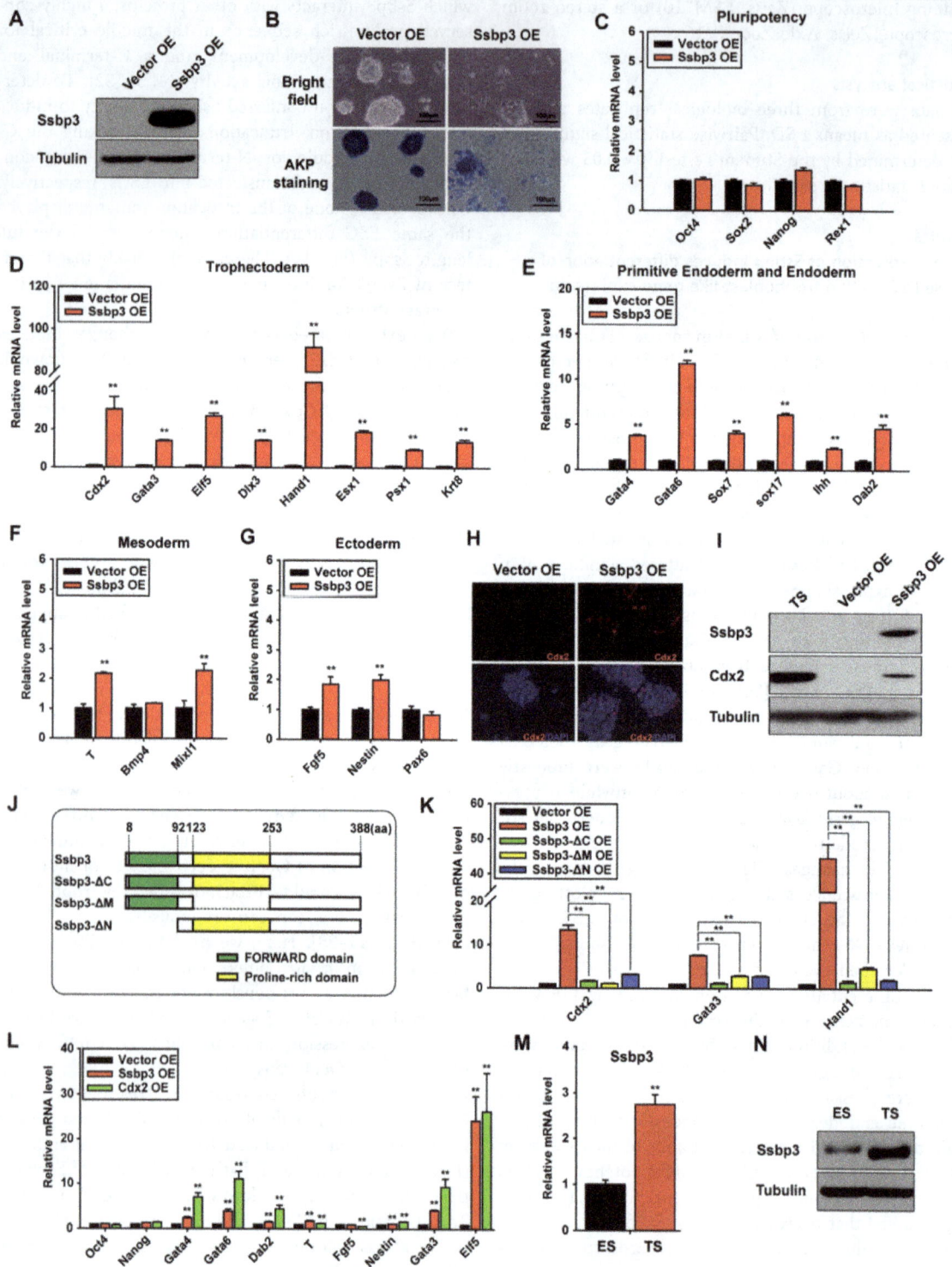

Fig. 1 (See legend on next page.)

(See figure on previous page.)
Fig. 1 Forced expression of Ssbp3 induces differentiation of mouse ESCs with a trophoblast-like gene expression pattern. **a** Western blotting of Ssbp3 protein levels in ESCs transfected with a vector, or an Ssbp3 plasmid. Twenty-four hours after transfection, ESCs were selected by puromycin for an additional 72 h. **b** Morphology changes and AKP staining of ESCs overexpressing Ssbp3. **c–g** Expression levels of pluripotency and lineage markers in ESCs overexpressing Ssbp3 determined by qRT-PCR analyses. Pluripotency markers (**c**), trophoblast markers (**d**), primitive endoderm and endoderm markers (**e**), mesoderm markers (**f**), and ectoderm markers (**g**). The average mRNA level in cells transfected with the control vector was set at 1.0. Data are shown as mean ± SD (n = 3). *p < 0.05, **p < 0.01. **h** Immunofluorescence staining of ESCs 96 h after transfection. Samples were stained with anti-Cdx2 (*red*) antibody, and DAPI staining highlighted the nuclei (*blue*). **i** Western blotting of Cdx2 protein levels in ESCs 96 h after transfection. β-Tubulin was used as an internal control. **j** The schematic diagram of the Ssbp3 coding sequence and various truncation mutants. **k** qRT-PCR analysis for the expression levels of trophoblast markers in ESCs 96 h after transfection of plasmids as indicated. The average mRNA level in cells transfected with the control vector was set at 1.0. Data are shown as mean ± SD (n = 3). *p < 0.05, **p < 0.01. **l** qRT-PCR analysis for mRNA levels of pluripotency and lineage markers 96 h after transfection of indicated plasmids. The average mRNA level in cells transfected with the control vector was set at 1.0. Data are shown as mean ± SD (n = 3). *p < 0.05, **p < 0.01. **m, n** The mRNA and protein levels of Ssbp3 in ESCs and TSCs determined by qRT-PCR (**m**) and Western blotting (**n**). *ESC* embryonic stem cell, *OE* overexpression, *TSC* trophoblast stem cell

1304) targeting different Ssbp3 coding sequences were used to establish two stable Ssbp3 knockdown ZHBTc4 ESC lines. A control cell line (shNT) was generated using a non-targeting shRNA sequence (Fig. 2d). qRT-PCR results showed that Tc treatment for 96 h induced expression of trophoblast lineage markers in shNT cells as reported previously [34]. However, levels of these trophoblast markers were markedly lower in shSsbp3-976 and shSsbp3-1304 cells than in control cells, although they were still higher than those in ESCs in the absence of Tc treatment (Fig. 2e). Hence, Ssbp3 was partially required for activation of trophoblast marker genes induced by Oct4 depletion.

Ssbp3 depletion weakens the trophoblast gene expression induced by BMP4 and bFGF treatment in ESCs

To determine the role of Ssbp3 in an additional trophoblast differentiation model, we developed a new cytokine-induced trophoblast differentiation protocol based on two published protocols. One protocol was developed in Hubert Schorle's laboratory using a chemically defined TX medium containing bFGF and TGF-β [23]; the other protocol used BMP4 to enable mouse ESCs to differentiate into trophoblasts in defined culture conditions [39]. In this study, we used BMP4 to replace TGF-β in the TX medium and named the new medium the Bb medium, as we found that the combination of

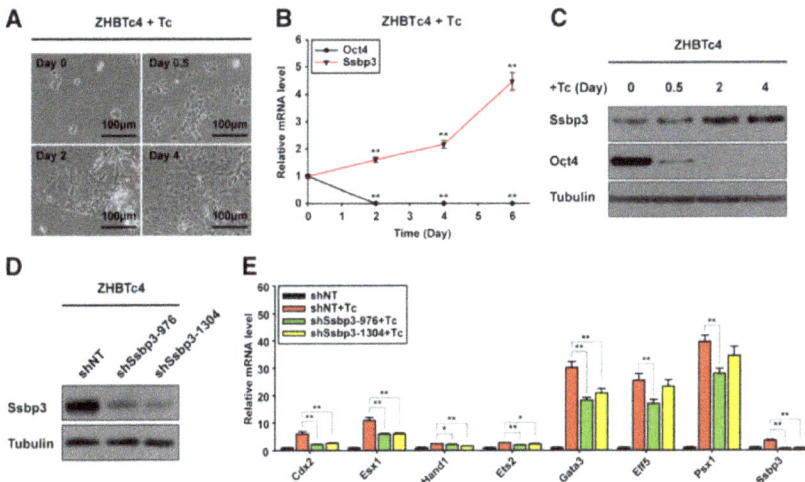

Fig. 2 Ssbp3 depletion attenuates the activation of trophoblast gene expression induced by downregulation of Oct4 in mouse ESCs. **a** The morphology of ZHBTc4 cells after treatment with Tc. Differentiation was triggered by Tc-mediated downregulation of Oct4. **b** Expression levels of Ssbp3 during differentiation of the ZHBTc4 cell line were determined by qRT-PCR analysis. The average mRNA level in ZHBTc4 cells cultured without Tc was set at 1.0. Data are shown as mean ± SD (n = 3). *p < 0.05, **p < 0.01. **c** Protein levels of Ssbp3 and Oct4 during ZHBTc4 cell differentiation were determined by Western blotting. **d** Western blot analysis of the silencing efficiency of shRNAs targeting the Ssbp3 coding sequence in ZHBTc4 cells. **e** The expression levels of trophoblast-specific markers in cells expressing NT control shRNA or specific Ssbp3-targeting shRNA were determined by qRT-PCR analysis. Expression of Oct4 was downregulated by treatment with Tc for 96 h in ZHBTc4 ESCs. The average mRNA level in shNT-expressing ZHBTc4 cells cultured in the absence of Tc was set at 1.0. Data are shown as mean ± SD (n = 3). *p < 0.05, **p < 0.01. *NT* non-targeting, *shRNA* short-hairpin RNA, *Tc* tetracycline

BMP4 and bFGF in the TX medium could model ESC differentiation into trophoblast lineages more efficiently. Cells lost typical ESC morphology and became flat 2 days after they were cultured in the Bb medium. The majority of the cells became cobblestone-shaped at day 6 (Fig. 3a). Transcript levels of Cdx2 and Ssbp3 increased gradually during the differentiation (Fig. 3b).

To investigate the role of Ssbp3 in this triggering agent-induced trophoblast differentiation model, we established two Ssbp3 stable knockdown E14T lines using the same shRNA sequences mentioned above (Fig. 3c); a control line was generated with the shNT

sequence. At day 4 of the Bb medium-induced differentiation, the expression of trophoblast transcription factors (Dlx3, Gata3, Psx1, Esx1, and Hand1) were evidently up-regulated, with elevated levels ranging from 15-fold to several hundred-fold compared with levels in undifferentiated ESCs (Fig. 3e–i), further validating that the newly developed protocol could robustly induce trophoblast differentiation in ESCs. In accordance with the results obtained in the Oct4 reduction-induced trophoblast differentiation model, depletion of Ssbp3 attenuated the induction of most trophoblast markers significantly (Fig. 3d–i), supporting the hypothesis that Ssbp3 plays

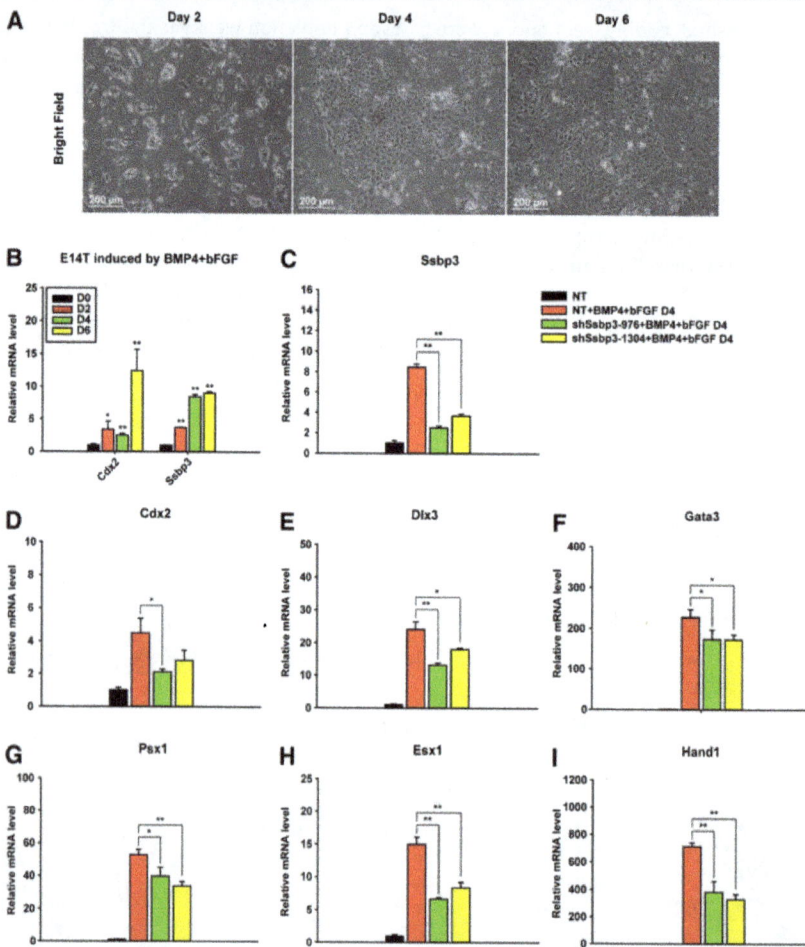

Fig. 3 Ssbp3 depletion weakens the trophoblast gene expression induced by BMP4 and bFGF treatment in ESCs. **a** The morphology of E14T cells after treatment with bone BMP4 and bFGF at the indicated time points. **b** Expression levels of Ssbp3 gradually increased in E14T cells treated with BMP4 and bFGF. The expression levels of Ssbp3 were determined by qRT-PCR analysis. The average mRNA level in E14T cells cultured without treatment was set at 1.0. Data are shown as mean ± SD ($n = 3$). *$p < 0.05$, **$p < 0.01$. **c** The silencing efficiency of shRNAs targeting Ssbp3 in E14T cells was detected by qRT-PCR analysis at day 4. **d–i** The expression levels of trophoblast-specific markers in cells expressing NT control shRNA or specific Ssbp3-targeting shRNAs were determined by qRT-PCR analysis. E14T cells were induced to trophoblast differentiation by treatment with BMP4 and bFGF for 96 h. The average mRNA level in shNT-expressing E14T cells without treatment was set at 1.0. Data are shown as mean ± SD ($n = 3$). *$p < 0.05$, **$p < 0.01$. *bFGF* basic fibroblast growth factor, *BMP4* bone morphogenetic protein 4, *NT* non-targeting, *shRNA* short-hairpin RNA

an important role for the full induction of trophoblast lineage markers during ESC differentiation into trophoblast-like cells.

Overexpression of Ssbp3 induces a trophoblast-like transcriptional program

To investigate the transcriptional effect of ectopic expression of Ssbp3 on a genome-wide scale, we compared the global transcriptional profile of Ssbp3-overexpressing cells with that of control cells using affymetrix microarray. Biological duplicate samples were used in this assay. One thousand eight hundred and eighty differentially expressed genes (DEG) were identified with a cutoff threshold of two-fold, including 980 upregulated and 900 downregulated genes, respectively (Fig. 4a, left panel; Table S2 in Additional file 3). The top 30 DEGs induced were listed in the heatmap shown in the right panel of Fig. 4a. As anticipated, the majority of the listed top genes are specifically related to the trophoblast development. Notably, critical TSC master regulators such as Cdx2, Elf5 and Gata3 were included in the list. The findings suggested that ectopic expression of Ssbp3 evoked an overall startup of a trophoblast-like transcriptional program in mouse ESCs.

To further define the ability of Ssbp3 to induce ESC differentiation to a trophoblast-like phenotype, we compared the global expression profile of Ssbp3-overexpressing ESCs with previously published profiles of Cdx2- and Gata3-overexpressing ESCs [6]. As shown, 60.1 % (1141 out of the 1880) of the DEGs induced by Ssbp3 were shared with the DEGs induced by Cdx2 or Gata3 (Fig. 4b; Table S3 in Additional file 4). GO analysis illustrated that these overlapped genes were strongly enriched in terms associated with transcription regulation and placenta development (Fig. 4c). Moreover, we analyzed the upregulated genes in Ssbp3-overexpressing cells using a batch query tool at the website of the Mouse Genome Informatics (MGI). Recovered mammalian phenotype (MP) terms contained multiple trophoblast subtypes and developmental stages (Table S4 in Additional file 5), similar to the MP terms recovered in Cdx2- and Gata3-overexpressing cells [6]. These results further validated the role of Ssbp3 for inducing a trophoblast-like transcriptional program in mouse ESCs. In addition, the remaining 39.9 % (739 out of the 1880) of the DEGs specifically induced by Ssbp3 (Fig. 4b; Table S3 in Additional file 4) were also analyzed by GO analysis, and they were most enriched in terms related to skeletal system development and morphogenesis (Figure S2 in Additional file 6), well in accordance with the previously reported phenotype of truncation of anterior skull bones and mild skeletal defects in other body parts in Ssbp3 knockout mice [13, 14].

Since TS-specific master genes Cdx2 and Elf5 were listed among the top 20 upregulated genes induced by overexpression of Ssbp3, we were interested to know whether Cdx2 or Elf5 acted as the downstream factors of Ssbp3, and were therefore involved in the trophoblast-like differentiation program from ESCs driven by exogenous expression of Ssbp3. To address this question, shRNAs based Cdx2 or Elf5 stable knockdown ESC lines were established. The results of qRT-PCR analyses revealed that knockdown of either Cdx2 or Elf5 obviously compromised the ability of Ssbp3 to induce the expression of trophoblast marker genes (Fig. 4d, e), indicating that Cdx2 and Elf5 probably play important roles for Ssbp3 to drive the trophoblast-like transcription program from ESCs.

Overexpression of Ssbp3 decreases the methylation level at the Elf5 promoter in mouse ESCs

The methylation status of the Elf5 promoter was reported to have a robust difference between ESCs and TSCs [40, 41], with a methylated and repressive status in ESCs and a hypo-methylated and active status in TSCs. We compared the methylation status at four separate regions of the Elf5 promoter positioning from −1000 bp to +400 bp with respect to the transcription start site (TSS) among ESCs overexpressing Ssbp3 at day 6, undifferentiated ESCs, and TSCs. Bisulfite sequencing analysis showed that all regions were extensively methylated in undifferentiated ESCs but hypo-methylated in TSCs. However, in Ssbp3-overexpressing ESCs, the percentage of methylated CpG dinucleotides was higher in all four regions compared to TSCs, but lower in regions 2, 3 and 4 compared to undifferentiated ESCs (Fig. 5). Thus, it seems that overexpression of Ssbp3 in ESCs under a self-renewal culture condition generated a specific Elf5 promoter methylation pattern in which the methylation level was between undifferentiated ESCs and TSCs, in line with the elevated Elf5 mRNA level in Ssbp3-overexpressing ESCs.

Ssbp3 overexpression activates MAPK/Erk1/2 and TGF-β pathways

Trophoblast differentiation involves multiple signaling pathways. We examined which pathway was activated upon Ssbp3 overexpression. KEGG pathway analysis of all 1880 DEGs induced by Ssbp3 overexpression showed that MAPK/Erk1/2 signaling and TGF-β signaling pathways were markedly enriched (Fig. 6a). These two pathways had been previously reported to be critical for establishment of the trophectoderm in vivo or maintenance of TSC proliferation in vitro [18, 20]. Our qRT-PCR analysis for these signaling-related components verified their activation in Ssbp3-overexpressing cells (Fig. 6b, c). Besides, Western blotting showed enhanced phosphorylation levels of Erk1/2 in Ssbp3-overexpressing cells, suggesting a possible interplay between Ssbp3 and the MAPK/Erk1/2 pathway (Fig. 6d). Functionally,

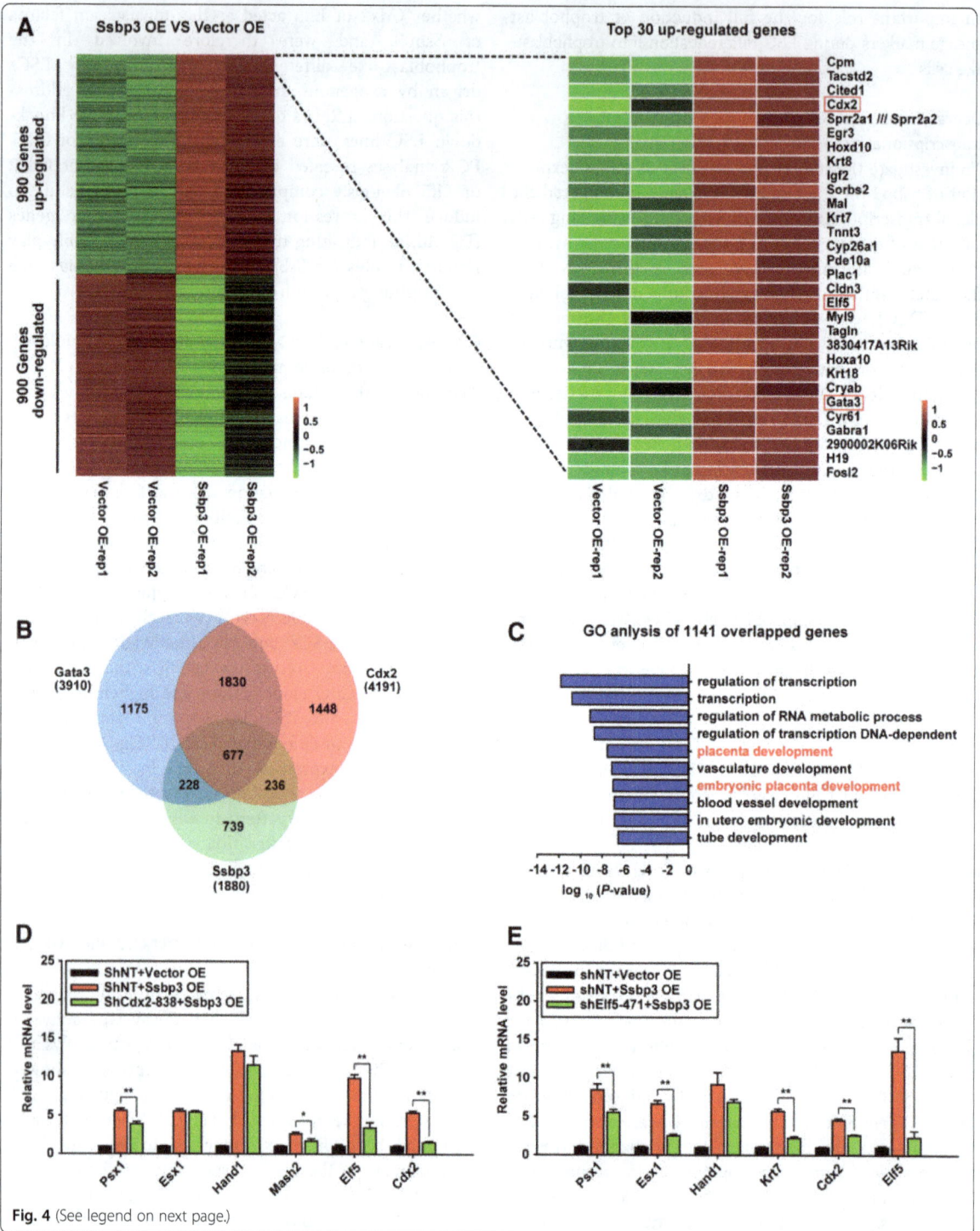

Fig. 4 (See legend on next page.)

(See figure on previous page.)
Fig. 4 Overexpression of Ssbp3 induces a trophoblast-like transcriptional program. **a** Heatmap of the DEGs induced by Ssbp3 overexpression in ESCs (fold change >2). *Green* and *red* values represent fold changes for down- and upregulation, respectively. Heatmap in the *right panel* shows the top 30 upregulated genes in detail. **b** Venn diagram showing the overlap of the DEGs induced by Ssbp3 (*green*), Gata3 (*blue*), or Cdx2 (*orange*) overexpression, with the number of genes indicated. Out of 1880 DEGs induced by Ssbp3, 1141 DEGs were shared with Gata3 or Cdx2. **c** Significantly enriched GO terms of the 1141 DEGs shared between Ssbp3 and Cdx2 or between Ssbp3 and Gata3. **d, e** qRT-PCR analysis for expression levels of trophoblast-specific markers in Cdx2 and Elf5 stable knockdown cell lines 96 h after Ssbp3 overexpression. The average mRNA level in stable cell line expressing shNT was set at 1.0. Data are shown as mean ± SD (*n* = 3). **p* < 0.05, ***p* < 0.01. *GO* gene ontology, *NT* non-targeting, *OE* overexpression, *shRNA* short-hairpin RNA

treatment of Ssbp3-overexpressing ESCs with a MAP kinase-ERK kinase (MEK) inhibitor, PD0325901 [42], diminished the upregulation of Cdx2 induced by Ssbp3 overexpression (Fig. 6e). The result was consistent with the report that PD0325901 treatment impaired Cdx2 expression and function in early embryos [18]. Taken together, these data indicated that Ssbp3 could activate MAPK/Erk1/2 and TGF-β pathways, which might be, at least partially, responsible for the activation of the trophoblast lineage marker genes induced by Ssbp3 overexpression.

Teratomas derived from Ssbp3-overexpressing ESCs contain hemorrhage
To test the developmental potential of Ssbp3-overexpressing ESCs in vivo, ESCs transfected with the Ssbp3 expression plasmid were intramuscularly injected into NOD/SCID mice. ESCs transfected with an empty vector were injected as a negative control. Teratomas were obtained from these mice 6 weeks later. Three teratomas were generated from vector-transfected ESCs and 10 teratomas were generated from Ssbp3-overexpressing ESCs (Fig. 7a). Teratomas generated from Ssbp3-overexpressing ESCs were larger

and heavier compared with teratomas derived from control ESCs (Fig. 7a and b). Of note, Ssbp3-overexpressing ESC-produced teratomas contained obvious hemorrhage. Histologically, both types of teratomas contained tissues and cells representing all three embryonic germ layers (Fig. 7c and d). However, clusters of trophoblast giant cells possessing large cytoplasm and large nuclei could be apparently observed in the internal hemorrhagic regions of Ssbp3-overexpressing ESC-produced teratomas, but not in control ESC-generated teratomas (Fig. 7e). Furthermore, expression levels of trophoblast markers were much higher in teratomas derived from Ssbp3-overexpressing ESCs compared with control teratomas (Fig. 7f). Therefore, overexpression of Ssbp3 aroused the tendency of ESCs to differentiate into trophoblast-like cell types in vivo.

ESCs overexpressing Ssbp3 mainly contribute to the placenta part in chimeric embryos
To study the in vivo differentiation ability of ESCs overexpressing Ssbp3, we injected mouse ESCs genetically labeled with green fluorescent protein (GFP) and overexpressing Ssbp3 into 8-cell-stage embryos (E2.5) and

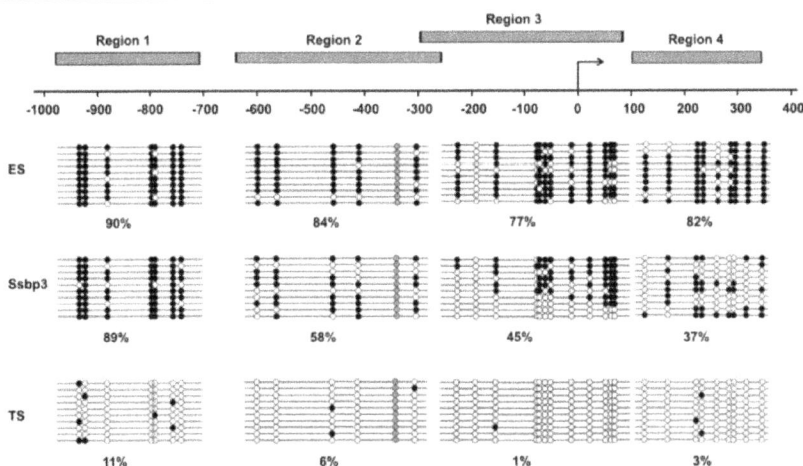

Fig. 5 Overexpression of Ssbp3 decreases the methylation level at the Elf5 promoter in mouse ESCs. Clonal bisulfite sequencing of the 1 kb promoter region of Elf5 in ESCs, TSCs, and ESCs overexpressing Ssbp3 were shown. Each line represents an individual clone with *open circles* depicting unmethylated sites and *filled circles* indicating methylated sites. The CpG dinucleotide at −355 bp is polymorphic and absent where shaded *gray*. *ESC* embryonic stem cell, *TSC* trophoblast stem cell

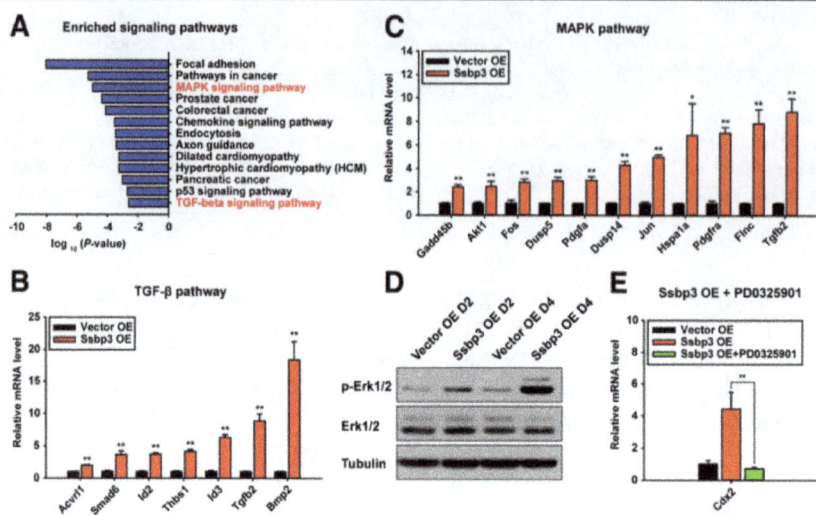

Fig. 6 Ssbp3 overexpression activates MAPK/Erk1/2 and TGF-β pathways. **a** Significantly enriched signaling pathways of all DEGs upon overexpression of Ssbp3 by KEGG pathway analysis. **b**, **c** qRT-PCR analysis for expression levels of upregulated genes related to MAPK/Erk/1/2 and TGF-β pathways in Ssbp3-overexpressing ESCs. The average mRNA level in cells transfected with the control vector was set at 1.0. Data are shown as mean ± SD ($n = 3$). *$p < 0.05$, **$p < 0.01$. **d** Western blotting of p-Erk1/2 and total Erk1/2 protein levels in Ssbp3-overexpressing cells. Cells transfected with an Ssbp3 plasmid or a vector were collected at 48 h and 96 h after transfection, respectively. **e** qRT-PCR analysis for expression levels of Cdx2 in Ssbp3-overexpressing ESCs with or without PD0325901 treatment. The average mRNA level in cells transfected with the control vector in the absence of PD0325901 was set at 1.0. Data are shown as mean ± SD ($n = 3$). *$p < 0.05$, **$p < 0.01$. *Erk* extracellular signal-regulated kinase, *MAPK* mitogen-activated protein kinase, *OE* overexpression, *TGF* transforming growth factor

traced their distribution at E6.5 and E14.5, respectively (Figure S3A in Additional file 7). We carried out two batches of experiments. In the first batch of experiment, we injected 30 embryos with control cells overexpressing a vector and 50 embryos with ESCs overexpressing Ssbp3. At E6.5, one embryo injected with the control ESCs and nine embryos injected with ESCs overexpressing Ssbp3 were obtained. In contrast to the whole embryo distribution of vector-transfected ESCs, in six out of nine embryos, ESCs overexpressing Ssbp3 were located predominantly in the trophoblast giant cell region and ectoplacental cone, as well as in extra-embryonic endoderm, but not in the epiblast of E6.5 embryos (Figure S3B in Additional file 7). In the second batch of experiments, we injected 50 embryos with control cells and 50 embryos with Ssbp3-overexpressing ESCs. At E14.5, out of 100 injected embryos, only one embryo injected with Ssbp3-overexpressing ESCs was obtained. Consistent with the result obtained in the first batch of experiments, the Ssbp3-overexpressing ESCs mainly contributed to the placenta part (Figure S3C in Additional file 7). These data suggest an active role of Ssbp3 for promoting trophoblast-like lineage differentiation in vivo.

Discussion

The present study shows that Ssbp3 might be an important regulator of trophoblast lineage differentiation. The

following lines of experimental evidence support this proposal: i) Ssbp3 was highly expressed in TSCs, and its expression gradually increased during ESC trophoblast differentiation induced by Oct4 knockdown or treatment with BMP4 and bFGF; ii) overexpression of Ssbp3 dramatically induced the expression of trophoblast marker genes; iii) depletion of Ssbp3 attenuated the induction of trophoblast-associated genes induced by downregulation of Oct4 or treatment with BMP4 and bFGF in ESCs; iv) Ssbp3-overexpressing ESCs generated large aggressive teratomas with massive internal hemorrhage in vivo; and v) when injected into 8-cell-stage embryos, Ssbp3-overexpressing ESCs mainly contributed to the placenta part. Interestingly, downregulation of Ssbp3 decreased the induction of Cdx2 and Esx1 more dramatically than other trophoblast marker genes when Oct4 was downregulated or when BMP4 and bFGF were included in the ESC culture medium. It is possible that Ssbp3, as a DNA binding protein, regulates Cdx2 and Esx1 directly, although we do not have any experimental evidence for this.

Our global gene expression analysis revealed that Ssbp3 could drive a trophoblast-like transcriptional program in ESCs. About 60 % of the 1880 DEGs induced by Ssbp3 overexpression overlapped with the previously reported DEGs induced by Cdx2 or Gata3 [6], indicating that these three regulators may share functional mechanisms to drive trophoblast differentiation. GO analysis

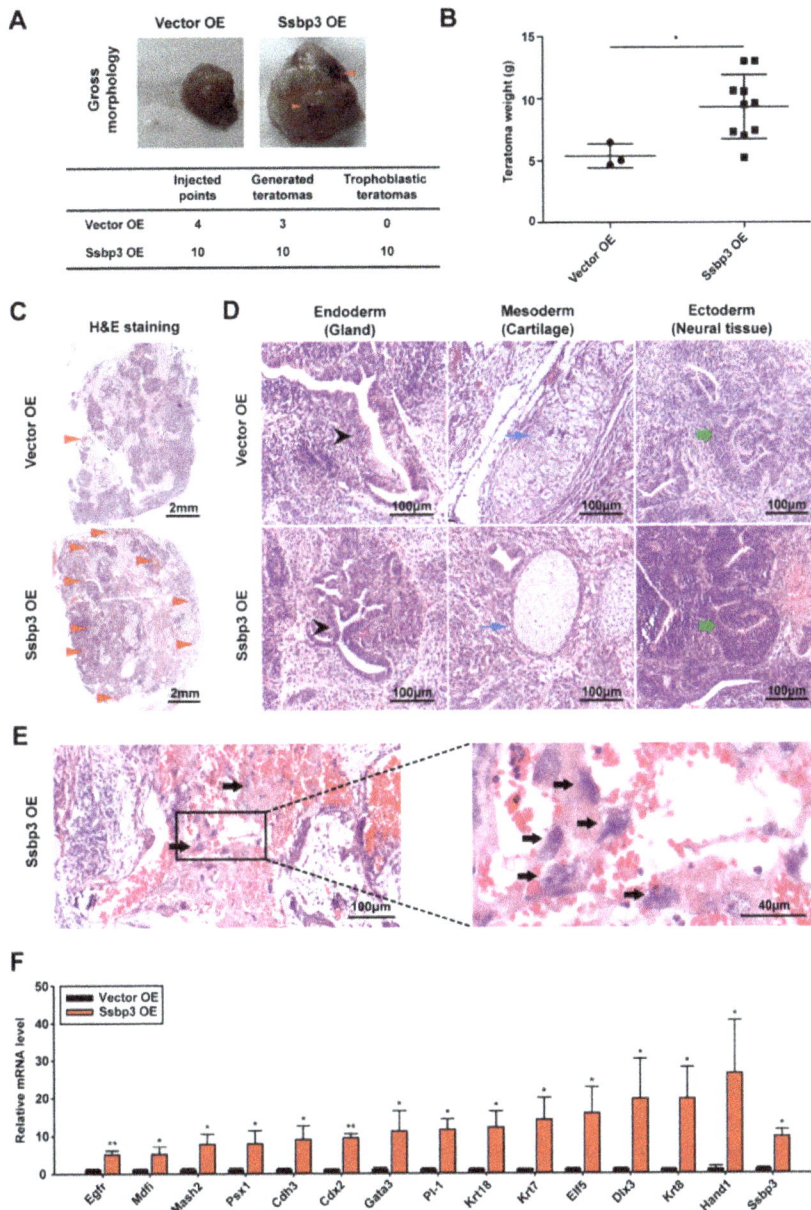

Fig. 7 Teratomas derived from Ssbp3-overexpressing ESCs contain hemorrhage. **a** Gross appearance of teratomas derived from control cells or Ssbp3-overexpressing ESCs (*upper panel*). The number of teratomas examined is presented in the table (*lower panel*). **b** The net weight of teratomas derived from control cells and Ssbp3-overexpressing ESCs. **c** Cross-section of teratomas derived from control cells (*upper panel*) or Ssbp3-overexpressing ESCs (*lower panel*). **d** Histology of teratomas derived from control cells (*upper panel*) or Ssbp3-overexpressing ESCs (*lower panel*) showing tissue complexity (hematoxylin and eosin staining). *Arrowheads* mark the endoderm (*black*), mesoderm (*blue*), and ectoderm (*green*) cells. **e** Hematoxylin and eosin-stained images for sections of a teratoma derived from Ssbp3-overexpressing ESCs. The trophoblast cluster (*arrow heads, left panel*) and trophoblast giant cells with the enlarged nuclei (*arrow heads, right panel*) are indicated. **f** qRT-PCR analysis for expression levels of trophoblast-specific markers in teratomas derived from control cells or Ssbp3-overexpressing ESCs. The average mRNA level in teratomas derived from control cells was set at 1.0. Data are shown as mean ± SD ($n = 3$). *$p < 0.05$, **$p < 0.01$. *OE* overexpression

illustrated that the DEGs shared by Ssbp3 and Cdx2 or Ssbp3 and Gata3 were highly enriched in terms associated with transcription regulation and placenta development. Cdx2 and Elf5, two key regulators of TSC self-renewal and maintenance, were robustly induced by ectopic expression of Ssbp3. Notably, Cdx2 or Elf5 depletion impaired activation of trophoblast lineage marker genes induced by Ssbp3 overexpression. These findings imply that Ssbp3 may act as an upstream regulator for the expression of Cdx2 and Elf5 during trophoblast differentiation, although it is not clear whether Ssbp3 executes this function directly or not. Although Ssbp3, Cdx2, and Gata3 can all activate trophoblast lineage marker genes, each has its distinct targets. In fact, GO analysis demonstrates that the specific DEGs regulated by Ssbp3 are strongly related to embryonic skeletal system development and the pattern specification process, indicating that Ssbp3 plays a role in embryogenesis as previously reported [13, 14]. Furthermore, KEGG analysis showed that MAPK/Erk1/2 and TGF-β pathways were activated in Ssbp3-overexpressing ESCs. All of the bioinformatic analyses, including GO, KEGG, and MP, support the hypothesis that Ssbp3 is closely associated with the transcriptional regulatory circuitry of trophoblast cell differentiation. As Ssbp3 was highly expressed in TSCs and activated transcription factors and signaling pathways required for TSC self-renewal, we propose that Ssbp3 may function at, or from, an early stage of trophoblast lineage specification.

The detailed mechanism by which Ssbp3 executed its function in regulation of trophoblast gene expression is not clear yet. Based on the facts that the overlapped DEGs between Ssbp3 and Cdx2 or Ssbp3 and Gata3 were strongly enriched in terms associated with transcriptional regulation and that Ssbp3 was reported to have transcriptional activity in its C-terminus [32], we anticipate that it may function through regulating gene expression as a transcription factor. However, where and how it binds the genome remains unclear. In fact, Ssbp3 binding sites, including both single- and double-stranded DNAs, are still rarely identified. So far only a pyrimidine-rich element in the promoter region of the chicken α2 (I) collagen gene was identified as a single-stranded DNA binding site by Ssbp3 [10]. Further investigations with chromatin immunoprecipitation coupled with DNA-sequencing (ChIP-seq) assays may help to address this question.

Our study shows that Ssbp3 plays an important role during trophoblast lineage specification in vitro. However, previous studies reported that Ssbp3 null mice showed skeletal abnormalities, but nothing was mentioned about placental abnormalities or embryonic lethality [11, 12]. The lack of early embryonic lethality (at or before implantation) in Ssbp3 knockout mice might be explained by functional redundancy from other family members. Supporting this hypothesis, we found that Ssbp2, another member of the Ssbp3 family, had a similar role to Ssbp3. Overexpression of Ssbp2 changed ESC morphology and led to increased expression of trophoblast-associated markers to an extent more dramatic than other lineage markers (data not shown). Therefore, double-knockout of Ssbp3 and Ssbp2 in mice could be helpful to verify the assumption.

It is known that Ssbp3 plays a critical role in the head morphogenesis during mouse embryonic development through activating the Lim1-Ldb1 transcriptional complex via interaction with Ldb1 [13]. To explore the effect of Ldb1 on the induction of trophoblast marker gene expression induced by Ssbp3, we up- and downregulated Ldb1 in ESCs together with Ssbp3 overexpression, respectively. No obvious effects were found (data not shown), suggesting that the trophoblastic gene induction activity of Ssbp3 may be Ldb1-independent. Further exploration of proteins interacting with Ssbp3 will facilitate our understanding of how Ssbp3 controls the trophoblast-like transcriptional program and activates related pathways.

Conclusions

Ectopic expression of Ssbp3 robustly induced trophoblast lineage marker gene expression in mouse ESCs. Conversely, depletion of Ssbp3 attenuated the trophoblast gene activation. Ssbp3 controlled a trophoblast-like transcriptional program by inducing the expression of early trophectoderm master transcription factors such as Cdx2, Gata3, and Elf5, and by activating MAPK/Erk1/2 and TGF-β pathways, two signaling pathways essential for trophoblast development. Both gain- and loss-of-function experiments support the notion that Ssbp3 might be an important regulator involved in trophoblast differentiation from mouse ESCs. The study is significant for better understanding the regulatory networks controlling ESC differentiation into extra-embryonic lineages and should shed light on the study of trophoblast development of mouse early embryos.

Additional files

Additional file 1: Table S1. Primers used in this study for gene cloning, qRT-PCR, and shRNA targeting sequences. (XLSX 16 kb)

Additional file 2: Figure S1. Forced expression of Ssbp3 induces mouse ESC differentiation with a bias to trophoblast lineages under the LIF withdrawal condition. (A–E) Expression levels of pluripotency and lineage specific markers in ESCs overexpressing Ssbp3 were determined by qRT-PCR analyses, including markers for the trophoblast (A), primitive and definitive endoderm (B), mesoderm (C), ectoderm (D), and pluripotency (E). The average mRNA level in cells transfected with the control vector was set at 1.0. Data are shown as mean ± SD ($n = 3$). *$p < 0.05$, **$p < 0.01$. (TIF 306 kb)

Single-stranded DNA binding protein Ssbp3 induces differentiation of mouse embryonic stem...

133

Additional file 3: Table S2. Differentially expressed genes induced by Ssbp3. (XLSX 36 kb)

Additional file 4: Table S3. Shared or specific DEGs induced by Ssbp3, Cdx2, or Gata3 overexpression. (XLSX 110 kb)

Additional file 5: Table S4. MGI Batch Report—mammalian phenotype (MP) ontology of the upregulated genes induced by Ssbp3 overexpression. (XLSX 15 kb)

Additional file 6: Figure S2. GO analysis of 739 Ssbp3 specific genes compared with DEGs induced by Gata3 or Cdx2 overexpression. The significantly enriched GO terms are strongly related to embryonic skeletal system development and pattern specification process. (TIF 78 kb)

Additional file 7: Figure S3. Ssbp3-overexpressing ESCs mainly contribute to the placenta part in chimeric embryos. (A) GFP-labeled E14T cells overexpressing a vector or an Ssbp3 expression plasmid were injected into wild-type 8-cell-stage embryos. For each embryo, two GFP-labeled cells were injected. (B) Strong contribution of injected E14T cells overexpressing Ssbp3 to the extra-embryonic lineage in E6.5 chimeric embryos. Gross and fluorescence images of E6.5 chimeric embryos developed from embryos described in A. (C) E14T cells overexpressing Ssbp3 predominantly contributed to the placenta part of the E14.5 chimeric embryo. (TIF 7678 kb)

Abbreviations
AKP, alkaline phosphatase; bFGF, basic fibroblast growth factor; BMP, bone morphogenetic protein; BSA, bovine serum albumin; cDNA, complementary DNA; DAPI, 4',6-diamidino-2-phenylindole; DAVID, database for annotation, visualization, and integrated discovery; DEG, differentially expressed genes; DMEM, Dulbecco's modified Eagle's medium; E, embryonic day; ERK, extracellular signal-regulated kinase; ESC, embryonic stem cell; FBS, fetal bovine serum; FGF4, fibroblast growth factor 4; GO, gene ontology; ICM, inner cell mass; KEGG, Kyoto encyclopedia of genes and genomes; LIF, leukemia inhibitory factor; MAPK, mitogen-activated protein kinase; MEF, mouse embryonic fibroblast; MEK, MAP kinase-ERK kinase; MGI, Mouse Genome Informatics; MP, mammalian phenotype; PCR, polymerase chain reaction; qRT-PCR, real-time quantitative reverse transcription polymerase chain reaction; shRNA, short-hairpin RNA; Tc, tetracycline; TE, trophectoderm; TGF, transforming growth factor; TSC, trophoblast stem cell

Acknowledgments
We thank Dr. Austin Smith for providing mouse E14T and ZHBTc4 ESC lines, and Dr. Shaorong Gao for the TSC line. We thank Dr. Ian Chambers for the pPyCAGIP vector and Dr. Didier Trono for the psPAX2 and pMD2.G vectors. This study was supported by a grant from the National Natural Science Foundation of China (31271456), a grant from the Ministry of Science and Technology of China (2010CB965101), and the Chinese Academy of Science Grant (XDA01010102). The study was also supported by a grant from the Ministry of Science and Technology of China (2013CB966801) and the major research projects, integrated projects (91419309) funded by National Nature Science Foundation of China.

Authors' contributions
JL and HL collected data, performed data analysis, and prepared the manuscript. XL, YX, JG, and FT collected data and prepared the manuscript. YJ and HL conceived the study, and prepared and revised the manuscript. All authors read and approved the final manuscript.

Competing interests
The authors declare that they have no competing interests.

Author details
[1]Laboratory of Molecular Developmental Biology, Shanghai Jiao Tong University School of Medicine, Shanghai 200025, China. [2]Key Laboratory of Stem Cell Biology, Institute of Health Sciences, Shanghai Institutes for Biological Sciences, Chinese Academy of Sciences, Shanghai Jiao Tong University School of Medicine, New Life Science Building A, Room 1328, 320 Yue Yang Road, Shanghai 200032, China. [3]Present address: KU Leuven Department of Development and Regeneration, Stem Cell Institute Leuven, Herestraat 49, 3000, Leuven, Belgium.

References
1. Lanner F. Lineage specification in the early mouse embryo. Exp Cell Res. 2014;321:32–9.
2. Cockburn K, Rossant J. Making the blastocyst: lessons from the mouse. J Clin Invest. 2010;120:995–1003.
3. Fleming TP. A quantitative analysis of cell allocation to trophectoderm and inner cell mass in the mouse blastocyst. Dev Biol. 1987;119:520–31.
4. Evans MJ, Kaufman MH. Establishment in culture of pluripotential cells from mouse embryos. Nature. 1981;292:154–6.
5. Tanaka S, Kunath T, Hadjantonakis AK, Nagy A, Rossant J. Promotion of trophoblast stem cell proliferation by FGF4. Science. 1998;282:2072–5.
6. Ralston A, Cox BJ, Nishioka N, Sasaki H, Chea E, Rugg-Gunn P, Guo G, Robson P, Draper JS, Rossant J. Gata3 regulates trophoblast development downstream of Tead4 and in parallel to Cdx2. Development. 2010;137:395–403.
7. Tolkunova E, Cavaleri F, Eckardt S, Reinbold R, Christenson LK, Scholer HR, Tomilin A. The caudal-related protein cdx2 promotes trophoblast differentiation of mouse embryonic stem cells. Stem Cells. 2006;24:139–44.
8. Home P, Ray S, Dutta D, Bronshteyn I, Larson M, Paul S. GATA3 is selectively expressed in the trophectoderm of peri-implantation embryo and directly regulates Cdx2 gene expression. J Biol Chem. 2009;284:28729–37.
9. Castro P, Liang H, Liang JC, Nagarajan L. A novel, evolutionarily conserved gene family with putative sequence-specific single-stranded DNA-binding activity. Genomics. 2002;80:78–85.
10. Bayarsaihan D, Soto RJ, Lukens LN. Cloning and characterization of a novel sequence-specific single-stranded-DNA-binding protein. Biochem J. 1998;331(Pt 2):447–52.
11. Chen L, Segal D, Hukriede NA, Podtelejnikov AV, Bayarsaihan D, Kennison JA, Ogryzko VV, Dawid IB, Westphal H. Ssdp proteins interact with the LIM-domain-binding protein Ldb1 to regulate development. Proc Natl Acad Sci U S A. 2002;99:14320–5.
12. van Meyel DJ, Thomas JB, Agulnick AD. Ssdp proteins bind to LIM-interacting co-factors and regulate the activity of LIM-homeodomain protein complexes in vivo. Development. 2003;130:1915–25.
13. Nishioka N, Nagano S, Nakayama R, Kiyonari H, Ijiri T, Taniguchi K, Shawlot W, Hayashizaki Y, Westphal H, Behringer RR, Matsuda Y, Sakoda S, Kondoh H, Sasaki H. Ssdp1 regulates head morphogenesis of mouse embryos by activating the Lim1-Ldb1 complex. Development. 2005;132:2535–46.
14. Enkhmandakh B, Makeyev AV, Bayarsaihan D. The role of the proline-rich domain of Ssdp1 in the modular architecture of the vertebrate head organizer. Proc Natl Acad Sci U S A. 2006;103:11631–6.
15. Gungor C, Taniguchi-Ishigaki N, Ma H, Drung A, Tursun B, Ostendorff HP, Bossenz M, Becker CG, Becker T, Bach I. Proteasomal selection of multiprotein complexes recruited by LIM homeodomain transcription factors. Proc Natl Acad Sci U S A. 2007;104:15000–5.
16. Xu Z, Meng X, Cai Y, Liang H, Nagarajan L, Brandt SJ. Single-stranded DNA-binding proteins regulate the abundance of LIM domain and LIM domain-binding proteins. Genes Dev. 2007;21:942–55.
17. Saba-El-Leil MK, Vella FD, Vernay B, Voisin L, Chen L, Labrecque N, Ang SL, Meloche S. An essential function of the mitogen-activated protein kinase Erk2 in mouse trophoblast development. EMBO Rep. 2003;4:964–8.
18. Lu CW, Yabuuchi A, Chen L, Viswanathan S, Kim K, Daley GQ. Ras-MAPK signaling promotes trophectoderm formation from embryonic stem cells and mouse embryos. Nat Genet. 2008;40:921–6.
19. Slager HG, Lawson KA, van den Eijnden-van Raaij AJ, de Laat SW, Mummery CL. Differential localization of TGF-beta 2 in mouse preimplantation and early postimplantation development. Dev Biol. 1991;145:205–18.
20. Erlebacher A, Price KA, Glimcher LH. Maintenance of mouse trophoblast stem cell proliferation by TGF-beta/activin. Dev Biol. 2004;275:158–69.
21. Quinn J, Kunath T, Rossant J. Mouse trophoblast stem cells. Methods Mol Med. 2006;121:125–48.
22. Lu R, Yang A, Jin Y. Dual functions of T-box 3 (Tbx3) in the control of self-renewal and extraembryonic endoderm differentiation in mouse embryonic stem cells. J Biol Chem. 2011;286:8425–36.
23. Kubaczka C, Senner C, Arauzo-Bravo MJ, Sharma N, Kuckenberg P, Becker A, Zimmer A, Brustle O, Peitz M, Hemberger M, Schorle H. Derivation and maintenance of murine trophoblast stem cells under defined conditions. Stem Cell Rep. 2014;2:232–42.

24. Stewart SA, Dykxhoorn DM, Palliser D, Mizuno H, Yu EY, An DS, Sabatini DM, Chen IS, Hahn WC, Sharp PA, Weinberg RA, Novina CD. Lentivirus-delivered stable gene silencing by RNAi in primary cells. RNA. 2003;9:493–501.
25. Li L, Sun L, Gao F, Jiang J, Yang Y, Li C, Gu J, Wei Z, Yang A, Lu R, Ma Y, Tang F, Kwon SW, Zhao Y, Li J, Jin Y. Stk40 links the pluripotency factor Oct4 to the Erk/MAPK pathway and controls extraembryonic endoderm differentiation. Proc Natl Acad Sci U S A. 2010;107:1402–7.
26. Jin Y, Xu XL, Yang MC, Wei F, Ayi TC, Bowcock AM, Baer R. Cell cycle-dependent colocalization of BARD1 and BRCA1 proteins in discrete nuclear domains. Proc Natl Acad Sci U S A. 1997;94:12075–80.
27. Zhang Z, Liao B, Xu M, Jin Y. Post-translational modification of POU domain transcription factor Oct-4 by SUMO-1. FASEB J. 2007;21:3042–51.
28. Lee HJ, Hinshelwood RA, Bouras T, Gallego-Ortega D, Valdes-Mora F, Blazek K, Visvader JE, Clark SJ, Ormandy CJ. Lineage specific methylation of the Elf5 promoter in mammary epithelial cells. Stem Cells. 2011;29:1611–9.
29. Koh KP, Yabuuchi A, Rao S, Huang Y, Cunniff K, Nardone J, Laiho A, Tahiliani M, Sommer CA, Mostoslavsky G, Lahesmaa R, Orkin SH, Rodig SJ, Daley GQ, Rao A. Tet1 and Tet2 regulate 5-hydroxymethylcytosine production and cell lineage specification in mouse embryonic stem cells. Cell Stem Cell. 2011;8:200–13.
30. Giakoumopoulos M, Golos TG. Embryonic stem cell-derived trophoblast differentiation: a comparative review of the biology, function, and signaling mechanisms. J Endocrinol. 2013;216:R33–45.
31. Bayarsaihan D. SSDP1 gene encodes a protein with a conserved N-terminal FORWARD domain. Biochim Biophys Acta. 2002;1599:152–5.
32. Wu L. Structure and functional characterization of single-strand DNA binding protein SSDP1: carboxyl-terminal of SSDP1 has transcription activity. Biochem Biophys Res Commun. 2006;339:977–84.
33. Beddington RS, Robertson EJ. An assessment of the developmental potential of embryonic stem cells in the midgestation mouse embryo. Development. 1989;105:733–7.
34. Niwa H, Miyazaki J, Smith AG. Quantitative expression of Oct-3/4 defines differentiation, dedifferentiation or self-renewal of ES cells. Nat Genet. 2000; 24:372–6.
35. Niwa H, Toyooka Y, Shimosato D, Strumpf D, Takahashi K, Yagi R, Rossant J. Interaction between Oct3/4 and Cdx2 determines trophectoderm differentiation. Cell. 2005;123:917–29.
36. Rhee C, Lee BK, Beck S, Anjum A, Cook KR, Popowski M, Tucker HO, Kim J. Arid3a is essential to execution of the first cell fate decision via direct embryonic and extraembryonic transcriptional regulation. Genes Dev. 2014;28:2219–32.
37. Vong QP, Liu Z, Yoo JG, Chen R, Xie W, Sharov AA, Fan CM, Liu C, Ko MS, Zheng Y. A role for borg5 during trophectoderm differentiation. Stem Cells. 2010;28:1030–8.
38. Kuckenberg P, Buhl S, Woynecki T, van Furden B, Tolkunova E, Seiffe F, Moser M, Tomilin A, Winterhager E, Schorle H. The transcription factor TCFAP2C/AP-2gamma cooperates with CDX2 to maintain trophectoderm formation. Mol Cell Biol. 2010;30:3310–20.
39. Hayashi Y, Furue MK, Tanaka S, Hirose M, Wakisaka N, Danno H, Ohnuma K, Oeda S, Aihara Y, Shiota K, Ogura A, Ishiura S, Asashima M. BMP4 induction of trophoblast from mouse embryonic stem cells in defined culture conditions on laminin. In Vitro Cell Dev Biol Anim. 2010;46:416–30.
40. Ng RK, Dean W, Dawson C, Lucifero D, Madeja Z, Reik W, Hemberger M. Epigenetic restriction of embryonic cell lineage fate by methylation of Elf5. Nat Cell Biol. 2008;10:1280–90.
41. Kuckenberg P, Peitz M, Kubaczka C, Becker A, Egert A, Wardelmann E, Zimmer A, Brustle O, Schorle H. Lineage conversion of murine extraembryonic trophoblast stem cells to pluripotent stem cells. Mol Cell Biol. 2011;31:1748–56.
42. Rinehart J, Adjei AA, Lorusso PM, Waterhouse D, Hecht JR, Natale RB, Hamid O, Varterasian M, Asbury P, Kaldjian EP, Gulyas S, Mitchell DY, Herrera R, Sebolt-Leopold JS, Meyer MB. Multicenter phase II study of the oral MEK inhibitor, CI-1040, in patients with advanced non-small-cell lung, breast, colon, and pancreatic cancer. J Clin Oncol. 2004;22:4456–62.

10

Restrained Th17 response and myeloid cell infiltration into the central nervous system by human decidua-derived mesenchymal stem cells during experimental autoimmune encephalomyelitis

Beatriz Bravo[1], Marta I. Gallego[2], Ana I. Flores[3], Rafael Bornstein[4], Alba Puente-Bedia[1], Javier Hernández[1], Paz de la Torre[3], Elena García-Zaragoza[2], Raquel Perez-Tavarez[5], Jesús Grande[3], Alicia Ballester[1] and Sara Ballester[1*]

Abstract

Background: Multiple sclerosis is a widespread inflammatory demyelinating disease. Several immunomodulatory therapies are available, including interferon-β, glatiramer acetate, natalizumab, fingolimod, and mitoxantrone. Although useful to delay disease progression, they do not provide a definitive cure and are associated with some undesirable side-effects. Accordingly, the search for new therapeutic methods constitutes an active investigation field. The use of mesenchymal stem cells (MSCs) to modify the disease course is currently the subject of intense interest. Decidua-derived MSCs (DMSCs) are a cell population obtained from human placental extraembryonic membranes able to differentiate into the three germ layers. This study explores the therapeutic potential of DMSCs.

Methods: We used the experimental autoimmune encephalomyelitis (EAE) animal model to evaluate the effect of DMSCs on clinical signs of the disease and on the presence of inflammatory infiltrates in the central nervous system. We also compared the inflammatory profile of spleen T cells from DMSC-treated mice with that of EAE control animals, and the influence of DMSCs on the in vitro definition of the Th17 phenotype. Furthermore, we analyzed the effects on the presence of some critical cell types in central nervous system infiltrates.

Results: Preventive intraperitoneal injection of DMSCs resulted in a significant delay of external signs of EAE. In addition, treatment of animals already presenting with moderate symptoms resulted in mild EAE with reduced disease scores. Besides decreased inflammatory infiltration, diminished percentages of CD4$^+$IL17$^+$, CD11b$^+$Ly6G$^+$ and CD11b$^+$Ly6C$^+$ cells were found in infiltrates of treated animals. Early immune response was mitigated, with spleen cells of DMSC-treated mice displaying low proliferative response to antigen, decreased production of interleukin (IL)-17, and increased production of the anti-inflammatory cytokines IL-4 and IL-10. Moreover, lower RORγT and higher GATA-3 expression levels were detected in DMSC-treated mice. DMSCs also showed a detrimental influence on the in vitro definition of the Th17 phenotype.

(Continued on next page)

* Correspondence: sballes@isciii.es
Alicia Ballester and Sara Ballester were co-principal investigators.
[1]Instituto de Salud Carlos III, Unidad Funcional de Investigación en Enfermedades Crónicas, Laboratory of Gene Regulation, Carretera de Majadahonda-Pozuelo Km 2, 28220 Madrid, Spain
Full list of author information is available at the end of the article

(Continued from previous page)

Conclusions: DMSCs modulated the clinical course of EAE, modified the frequency and cell composition of the central nervous system infiltrates during the disease, and mediated an impairment of Th17 phenotype establishment in favor of the Th2 subtype. These results suggest that DMSCs might provide a new cell-based therapy for the control of multiple sclerosis.

Keywords: Placental mesenchymal stem cells, Th17, EAE, Immunomodulation

Background

Multiple sclerosis (MS) is a progressive inflammatory disorder of the central nervous system (CNS) elicited by an immune reaction against self-neuroantigens. Compelling support for the autoimmune etiology of MS is provided by the experimental autoimmune encephalomyelitis (EAE) animal model of CNS inflammation. In this model, illness is triggered by the immune response to experimentally supplied auto-antigens [1]. This reaction is driven by auto-responsive T cells in lymph nodes and spleen able to migrate to the CNS and cross the blood–brain barrier, where they find their cognate antigen in the context of resident antigen-presenting cells, such as microglia or astrocytes, or of immigrant macrophages or dendritic cells. These events favor an inflammatory environment with CNS injury, characterized by loss of the insulating myelin sheath of neuronal axons resulting in motor disability. EAE can also be induced by transplanting T helper (Th) cells delivering interleukin (IL)-17 (Th17) or interferon (IFN)-γ (Th1) [2]. Since IFN-γ has, however, demonstrated a dual role in disease pathogenesis, the IL-17 pathway is considered a more appropriate therapeutic target [3, 4]. Two other subsets of CD4$^+$ cells, namely Th2 and T regulatory (Treg) cells, are able to control or ameliorate EAE disease evolution [5–7] by secreting cytokines such as IL-4 [8], IL-10 [9], and transforming growth factor (TGF)-β [10]. Several other cytokines are able to modify EAE course, such as IL-27—a main negative regulator of Th17 development [11–14].

Although significant progress has been made in MS therapy, none of the available treatments achieves a halt or reversion of disability progression. Hence, development of new therapeutic strategies is a crucial challenge. To this end, stem cells have been introduced into the MS scenario in recent years. While some reports support possible advantages of embryonic over adult stem cells [15], ethical concerns about the use of the former promote the study of adult stem cells [16, 17]. Several phase I/II clinical trials underway in MS patients are evaluating the therapeutic potential of mesenchymal stem cells (MSCs) derived from different tissues, such as bone marrow (BM), adipose tissue, placental or umbilical cord blood [16–23]. Preliminary results indicate that administration of MSCs to patients with MS is feasible and safe. In addition, some studies have reported a degree of structural, functional,

and physiological improvement after treatment, consistent with the immunomodulatory and neuroprotective effects of MSCs. Despite early clinical stabilization or improvement in some of these patients, further controlled trials are warranted to evaluate alternative cell sources and administration schedules which might affect MS disease course more consistently. Some preclinical data from experimental models would appear to supply grounds for postulating neuroprotective and immunomodulatory properties for MSCs. Neuroprotection by MSCs is suggested by reports showing stimulation of oligodendrogenesis, oligodendrocyte progenitor migration, remyelination, and reduction of axonal loss [24–27]. Some MSC subpopulations deliver active neurotrophins [28], and human neural progenitors obtained in vitro from MSCs improved neurological function in EAE [29]. Immunomodulatory effects ascribed to MSCs throughout EAE treatment include hepatocyte growth factor production [30], prostaglandin E2 secretion [31], promotion of IL-27 [14], inhibition of IL-17 and tumor necrosis factor-alpha (TNFα) production [24, 32], downregulation of IFNγ T-cell expression [33] and T-cell anergy [34].

Most of the studies focused on adult MSC therapy for EAE have used BM as a cell source. However, MSCs derived from other tissues, such as adipose [35, 36], endometrial [37], umbilical cord [38, 39], or placenta [40, 41], have also been shown to influence EAE development. MSC content in adult BM is limited, invasive procedures are required for MSC procurement, and the number and differentiation capacity of such cells decrease with the age of the donor. Placenta-derived MSCs present several advantages over other sources, since the cells can be isolated without any donor injury and a large amount of cells with high differentiation ability can be obtained [42]. We previously characterized a subset of human decidua-derived MSCs (DMSCs) with capacity to differentiate at the clonal level into the three embryonic germ layers [42–45]. DMSCs display some properties of embryonic cells and others of adult stem cells, as they express transcription factors involved in pluripotency (Oct-4 and Rex-1) and organogenesis (GATA-4), though not embryonic markers (SSEA-1, -3, -4 and TRA-1-81) [42] expressed by other placental-derived MSCs [46]. DMSCs are of maternal origin and show higher proliferation rates and differentiation capacity than do BM-derived

MSCs [47, 48], thereby making them biologically different to MSCs derived from other adult sources.

In the present study we have evaluated the therapeutic potential of DMSCs on EAE. Results showed that a prophylactic treatment with DMSCs was able to delay the onset and reduce the severity of the disease substantially for as long as treatment was maintained. Furthermore, the therapeutic utility of DMSCs was also demonstrated in animals which were initially treated when they presented with moderate symptoms, with this resulting in a mild course of EAE. DMSC treatment reduced CNS injury areas and modulated the peripheral immune response, leading to an anti-inflammatory profile of spleen T cells. The frequency and cell composition of CNS infiltrates were also modified, with the percentages of $CD4^+IL-17^+$, $CD11b^+Ly6G^+$, and $CD11b^+Ly6C^+$ cells being reduced by DMSC treatment.

Methods

Animals, EAE induction, clinical evaluation and treatments

All experiments were conducted under institutional ethical and safety guidelines, with approval number 017/15 of the Madrid Regional Authority's Ethics Committee, in accordance with European Union legislation. C57BL/6 mice were bred and maintained at the Institution's animal facility. Mice were housed in groups of 4–5. To induce EAE, 10- to 14-week-old female mice were anesthetized by intraperitoneal administration of ketamine and xylazine and immunized by subcutaneous injection in flanks with 200 μg myelin oligodendrocyte glycoprotein (MOG)$_{35-55}$ peptide (Peptide 2.0, Chantilly, VA, USA) in complete Freund adjuvant containing 2.5 mg/ml *Mycobacterium tuberculosis* H37RA (Difco) in a total volume of 100 μl. *Bordetella pertussis* toxin (300 ng in 100 μl) was administered intraperitoneally on the day of antigen inoculation and 48 hours later (D0 and D2 post-immunization (p.i.), respectively). Groups of 7–10 animals were used for each experiment. Clinical signs were scored on a 0–5 scale as follows: no clinical signs, 0; loss of tail tonicity, 1; rear limb weakness, 2; paralysis of one rear limb, 3; paralysis of two rear limbs, 4; full paralysis of four limbs, 5. At value 4, animals were sacrificed to avoid further progress of the disease. Score values were calculated as the average of the evaluations assigned to each mouse by three independent observers in blind inspection. For DMSC treatments, cells at passage 6–8 with 95–98 % viability were used. At this passage number, the cells still preserve a high proliferation and multilineage differentiation capacity [42]. One million cells were administered in 100 μl phosphate-buffered saline (PBS) by intraperitoneal injection to every treated animal on the days indicated for each experiment.

Isolation of human DMSCs and culture

Human placentas from healthy mothers were supplied by the Department of Obstetrics and Gynecology under written consent previously approved by the Ethics Committee at the Hospital Universitario 12 de Octubre. DMSC isolation and culture was performed as previously described [42]. Briefly, placental membranes were digested with trypsin-versene (Lonza, Spain), and the cells were seeded at 1.2×10^5 cells/cm^2 and cultured at 37 °C, 5 % CO_2 and 95 % humidity in Dulbecco's modified Eagle medium (DMEM; Lonza) supplemented with 2 mM L-glutamine, 0.1 mM sodium pyruvate, 55 μM B-mercaptoethanol, 1 % nonessential amino acids, 1 % penicillin/streptomycin, 10 % fetal bovine serum and 10 ng/ml epidermal growth factor 1 (EGF-1; Sigma-Aldrich Química, Spain). The morphology, phenotype and MSC characteristics of DMSCs have been previously reported [42]. Cells were cryopreserved and, before use, were thawed and passaged at a density of around 5×10^4 cells/cm^2 until passage 6–8.

Mouse cell isolation and culture

Mouse spleen cells were obtained as previously described [49]. $CD4^+$ cells were magnetically sorted (Miltenyi Biotech) to 90–95 % purity, and tested by flow cytometry with anti-CD4 antibody (L3T4; Miltenyi Biotech). Total spleen population or purified $CD4^+$ cells from each group of animals were pooled, washed and suspended in Click's medium [50] before in vitro culture. For anti-CD3/anti-CD28 stimulation, cells were cultured in microwell plates coated with anti-CD3 (Y-CD3-1, 10 μg/ml) [51] and soluble anti-CD28 (clone 37.51, 1 ng/ml; eBioscience, Hatfield, UK). For antigenic stimulation, 25 μM MOG$_{35-55}$ was used in cell cultures. Th17 phenotype skewing conditions were achieved by IL-6 and TGFβ treatment as previously described [52]. Briefly, anti-CD3/anti-CD28 stimulation was supplemented with 20 ng/ml IL-6 (eBioscience), 5 ng/ml TGFβ (eBioscience), 25 μg/ml anti-IL-4 (11B11; ATCC HB188) and 25 μg/ml anti-IFN-γ (R46A2; ATCC HB170). Cocultures of DMSC-murine spleen cells were performed at a ratio of 1:7. First, plates were seeded with DMSCs in DMEM supplemented with EGF-1 (10 ng/ml; Sigma-Aldrich Química). After 12 hours this medium was removed and spleen cells were added in Click's medium with soluble anti-CD3 (25 μg/ml) and anti-CD28 (1 μg/ml). For isolation of CNS inflammatory infiltrates, animals were sacrificed and perfused through the left ventricle with 200 ml PBS to wash out leukocytes present within the blood vessels. Spinal cords and brains were removed, and tissue from each mouse was homogenized through a 100-μm pore strainer. After centrifugation, the pellet was dissolved in 30 % Percoll (Amersham) and the homogenate mix was layered over 80 % Percoll. Infiltrating cells were collected from the 30–80 % interface, after centrifugation at 3,000 rpm for 30 minutes at room temperature without brake. Spleen cells for early immune response analysis were obtained from animals at day 7–10 p.i., while CNS studies were performed on mice at day 20 p.i.

T-cell proliferation and cytokine expression measurements

Total spleen cells or purified CD4$^+$ cells (2×10^5) were split into p-96 well microtiter and subjected to T lymphocyte-specific stimuli (anti-CD3/anti-CD28 antibodies or MOG$_{35-55}$, as specified above) through 72 hours. Colorimetric assay based on MTT was used as a T-cell proliferation measurement according to that described in [53]. Cytokines released to the medium were quantified by enzyme-linked immunosorbent assay (ELISA), with each sample assayed in quintuplicate. Capture and biotin-conjugated antibodies were, respectively: eBio17CK15A5 and eBio17B7 (e-Bioscience) for IL-17A; Jes5-2A5 and Jes5-16E3 (BD-Bioscience) for IL-10; and 11B11 and BVD6-24G2 (Becton-Dickinson) for IL-4. Obtention of total RNA for retrotranscription and quantitative real-time polymerase chain reaction (RT-qPCR) were performed as previously reported [54]. The primer pairs used for each gene were as follows:

IL-17 forward: 5' -GAAGCTCAGTGCCGCCA-3' ;
IL-17 reverse: 5' -TTCATGTGGTGGTCCAGCTTT-3' ;
IL-4 forward: 5' -ATCCTGCTCTTCTTTCTCG-3' ;
IL-4 reverse: 5' -GATGCTCTTTAGGCTTTCC-3' ;
IL-10 forward: 5' -TGCTATGCTGCCTGCTCTTA-3' ;
IL-10 reverse: 5' -GCTCCACTGCCTTGCTCTTA-3' ;
RORγT forward: 5' -CCGCTGAGAGGGCTTCAC-3' ;
RORγT reverse: 5' -TGCAGGAGTAGGCCACATTA CA-3' ;
GATA-3 forward: 5' -AGAACCGGCCCCTTATCAA-3' ;
GATA-3 reverse: 5' -AGTTCGCGCAGGATGTCC-3' ;
Foxp3 forward: 5' -ACCACCTTCTGCTGCCACTG-3' ;
Foxp3 reverse: 5' -TGCTGTCTTTCCTGGGTGTACC-3' ;
β-actin forward: 5' -TGTTACCAACTGGGACGACA-3' ; and,
β-actin reverse: 5' -GGGGTGTTGAAGGTCTCAAA-3' .

PCR product quality was checked by a melting curve analysis for each sample and the reaction efficiencies were checked to be near 2. Each result was normalized by the housekeeping β-actin gene expression. Relative quantification of gene expression analysis was performed using the Pfaffl method [55].

Flow cytometry cell staining

For all experiments, cells were incubated in 0.5 µg FcBlock (BD Bioscience) for 10 minutes at room temperature. Surface molecule staining was performed in the dark for 30 minutes at 4 °C. Cells were then washed twice with staining buffer followed by fixation in 1 % paraformaldehyde. Antibodies for surface markers were: anti-CD4 clone RM4-5 biotin (eBioscience); anti-CD8α clone 53–6.7 biotin (eBioscience); anti-CD19 clone 1D3 biotin (eBioscience); anti-NK1.1 clone PK136 biotin (eBioscience); anti-CD11c clone HL3 biotin (BD-Pharmingen); anti-CD11b clone M1/70 APC (eBioscience); anti-Ly6G clone AL-21 APC-Cy7 (BD-Pharmingen); and anti-Ly6C clone 1A8 PE (BD-Pharmingen). For biotin antibodies, streptavidin PE-Cy7 (eBioscience) was used to detect positive cells. For intracellular staining, anti-IL17 (clone TC11-18H10-PE; BD-Pharmingen), anti-RORγT (clone AFKJS-9-APC; eBioscience) and anti-Foxp3 (clone FJK-16S; eBioscience) were used on cells previously permeabilized and fixed by Cytofix/CytopermTM (Becton Dickinson) and Staining Set Kit (eBioscience), respectively, for cytoplasmic IL-17 and nuclear RORγT detection. Cells were acquired on a BD FACSCantoTM II. Data were collected by BD FACS Diva software and analyzed by FlowJo software (Tree Star Inc.). Fluorescence minus one (FMO) controls were used for gating analysis to distinguish positive and negative cell populations. Propidium iodide staining was used for live/dead discrimination. Compensation was carried out using single color controls, and compensation matrices were calculated and applied by FlowJo software.

Histopathology

Mice were anesthetized by intraperitoneal administration of ketamine-xylazine and transcardially perfused with 4 % paraformaldehyde. Spinal cords and brains were fixed in 4 % paraformaldehyde. Vibratome free-floating slices (15–30 µm) were preserved in 0.1 M phosphate buffer. For detection of demyelinating and inflammatory lesions, slices were subjected to luxol fast blue (LFB)–periodic acid-Schiff (PAS)–hematoxylin triple staining according to Goto [56] and hematoxylin–eosin staining, performed as previously described [52]. Perivascular infiltrates were quantified by examining hematoxylin–eosin serial sections along the brain and spinal cord of each animal. Slices were classified as positive or negative for infiltrate quantification. For immunodetection of CD4$^+$ and GFAP$^+$ cells, free-floating spinal cord sections were boiled in a microwave oven in 10 mM sodium citrate buffer for antigen retrieval. Prior to incubation with antibodies, endogenous peroxidase activity was inhibited with 2 % hydrogen peroxide in CD4 immunohistochemistry samples, and tissue autofluorescence was minimized by 2 % sodium borohydride treatment of the immunofluorescence samples. As primary antibodies, L3T4 (Sino Biological Inc), EPR1034Y (Millipore) and PC10 (Abcam) were, respectively, used for detection of CD4, GFAP and PCNA (as a marker of cell proliferation). Overnight incubation with primary antibody was followed by 1-hour incubation with Biotin-conjugated goat anti-rabbit (Jackson Inmunoresearch Lab Inc.) for CD4 detection, or Alexa Fluor 594 Donkey anti-rabbit IgG (Invitrogen) and Alexa Fluor 488 Donkey anti-mouse IgG (Invitrogen) for immunofluorescence of GFAP and PCNA, respectively. Thereafter, samples for CD4 immunohistochemistry were exposed for 30 minutes to Vectastain ABC reagent (Vector Laboratories) and to DAB developing solution (Vector Laboratories), and counterstained with hematoxylin for visualization using a Leica DM2000 microscope. Fluorescent images were captured by confocal microscopy using a Leica TCS SP5 AOBS Confocal Microscope (Leica Microsystems GmbH, Wetzlar, Germany), and analysis was performed with Image J software designed by the NIH (MD, USA). In all cases, specificity of staining was confirmed by controls omitting the primary antibody.

Statistics

Statistical analyses were performed with Graph Pad Prism version 5.02 (Graph Pad software, Inc). The t-test was used for unpaired data; in cases where n ≤ 10 (with a minimum of 5), Welch's correction was introduced in order not to assume equal variances. Contingency table analysis for comparison of perivascular infiltrate quantification and disease incidence was performed using Chi-square test (n > 30) or Fisher's test (n < 30, with a minimum of 14). The area under the curve (AUC) was calculated from EAE clinical course for each mouse, and differences between groups were analyzed by the Mann–Whitney test. Statistical significance is indicated as $*p < 0.05$, $**p < 0.01$ or $***p < 0.001$.

Results

DMSC treatment delays the development of EAE and restrains early Th17 response

To determine whether DMSC administration had protective effects on EAE, we designed a preventive approach comprising three intraperitoneal injections of cells at days −1, 3 and 6, with the day of MOG inoculation being established as day 0. Daily monitoring of score values showed that DMSC treatment resulted in a significant delay in the onset of EAE symptoms. The first clinical signs were apparent in the EAE control group at day 10–13 p.i., whereas most of the DMSC-treated mice did not show symptoms of established disease (score values higher than 1) before day 25–30 p.i. (Fig. 1a). Although DMSC-treated animals ultimately attained score values near those of the EAE control group, significant differences between both groups of mice remained until at least day 30 p.i. (Fig. 1b). We also evaluated the AUC as a measure of disease severity. When such analysis was restricted to the late phase of the

clinical course (from day 30 to day 55 p.i.), no difference was found between groups. However, overall clinical course examination showed significantly lower values for the DMSC group, in line with the beneficial effect of DMSCs until at least day 30 p.i. (Fig. 1c). Furthermore, data from individual mice demonstrated notably delayed disease onset (Fig. 1d) and decreased disease incidence evaluated at day 20 p.i. (Fig. 1e) for DMSC-treated animals with respect to the EAE group.

For a more in-depth examination of differences between DMSC-treated and untreated animals, we chose the disease phase with the most striking divergences in clinical signs (days 10–20 after MOG inoculation). Histopathological images of cerebellum and spinal cord sections from animals sacrificed at day 20 p.i. supported that DMSCs attenuated CNS pathology in EAE. LFB–PAS–hematoxylin staining showed important areas of myelin disruption in EAE mice whereas DMSC administration contributed to preservation of myelin integrity (Fig. 2a and b). In addition, DMSC-treated mice showed a smaller degree of infiltration, as both the number of analyzed hematoxylin–eosin stained sections showing perivascular infiltration (Fig. 2c) and the number of perivascular infiltrates per section (Fig. 2d) were strongly reduced in DMSC-treated mice. Moreover, CD4$^+$ cells were frequent in the infiltration areas of EAE animals, while CD4 immunoreactivity was rather limited in DMSC-treated mice (Fig. 2e). Likewise, GFAP staining of spinal cord from EAE animals showed swollen astrocytic processes, indicative of astrocytic reactivity, with less severity in DMSC-treated mice (Fig. 2f). Furthermore, the use of anti-PCNA as a marker of cell proliferation revealed a lessening of astrocyte-division activity by DMSC treatment.

We next studied the early immune response in both groups of animals by comparing T-cell reactivity in total spleen cell populations from DMSC-treated and untreated EAE mice shortly after MOG$_{35-55}$ inoculation (day 7–10 p.i.). Cells were stimulated by anti-CD3 and anti-CD28 antibodies to induce broad T-lymphocyte activation, and by MOG$_{35-55}$ to induce antigen-specific T-cell reactivity. In both cases, cells from DMSC-treated mice showed reduced proliferation and IL-17 release as compared to EAE control mice, while expression of the anti-inflammatory cytokines IL-4 and IL-10 were markedly increased in cultures of cells from DMSC-treated animals as against control (Fig. 3a and b). These results were also observed when purified CD4$^+$ cell fractions were evaluated, suggesting a direct effect on this cell population without the need for other intermediate cell types (Fig. 3c). Moreover, analysis of IL-17, IL-4 and IL-10 mRNA levels from DMSC-treated versus untreated animals showed marked differences between the two groups which correlated with the results of soluble cytokine release (Fig. 3d). RORγT and GATA-3 transcription

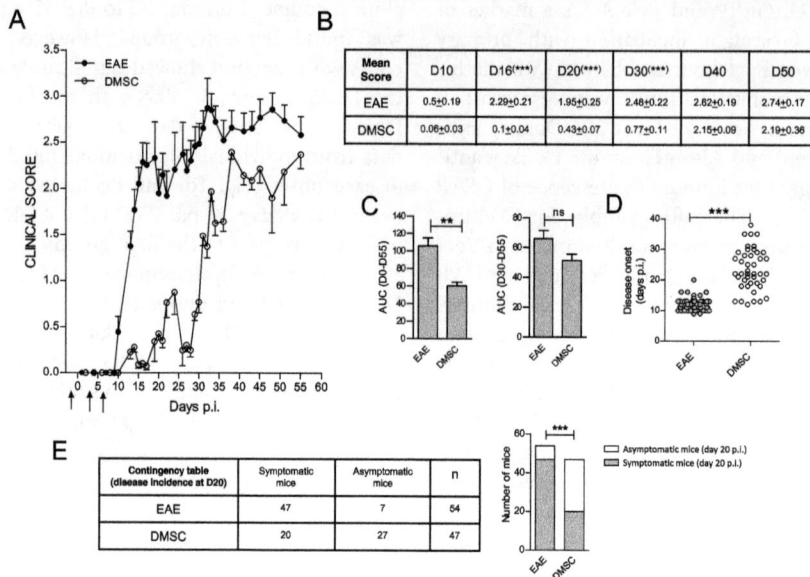

Fig. 1 Decidua-derived mesenchymal stem cell (*DMSC*) treatment delays onset of MOG-experimental autoimmune encephalomyelitis (*EAE*). Groups of 7–10 C57BL/6 mice were established by MOG$_{35-55}$ peptide inoculation. **a** DMSC group animals received 1×10^6 DMSCs at days −1, 3 and 6 post-immunization (*p.i.*) (*arrows*). Daily mean clinical scores along EAE course are shown from one representative (n = 10 mice/group) out of five independent experiments. **b** Statistical significance by unpaired *t*-test for mean scores at different days after immunization. **c** Area under the curve (*AUC*) was calculated from EAE clinical course for each mouse between days 0 and 55 (D0–55) or between days 30 and 55 p.i. (D30–D50), and differences between groups were analyzed using the Mann–Whitney test. Standard error of the means are shown. Data from individual mice included in five independent experiments were used to examine **d** individual disease onset and **e** disease incidence at day 20 p.i.; the differences between groups were analyzed by *t*-test and by Chi-square test, respectively (n for each group is indicated). The bar graphs for representation of the disease incidence contingency table show the numbers of symptomatic and asymptomatic mice at day 20 p.i. for each group. **p < 0.01, ***p < 0.001. *ns* Not significant

factors are known to be the master transcription factors that control the definition of Th17 and Th2 phenotypes, respectively [57]. We also analyzed their mRNA levels and found that, while RORγT expression was downregulated, GATA-3 mRNA levels were upregulated in CD4$^+$ cells from DMSC-treated mice as compared to EAE untreated mice (Fig. 3e), which suggests that the Th17 phenotype is restrained while the Th2 subset is favored by DMSC treatment. Conversely, quantification of the mRNA levels of Foxp3, the transcription factor controlling Treg cell development, did not show differences between CD4$^+$ cells from untreated and DMSC-treated EAE mice (Figure S1 in Additional file 1).

Impaired establishment of Th17 phenotype in DMSC-treated animals is independent of experimentally induced inflammatory response

To ascertain whether the effects of DMSC treatment on T-cell activity were dependent on the inflammatory response triggered by MOG inoculation, healthy mice were subjected to three successive DMSC doses every 3 days to emulate the treatment procedure used for EAE animals. As previously observed for EAE mice, T-cell proliferation and IL-17 expression were lower in

DMSC-primed mice than in naïve animals (Fig. 4), suggesting that DMSCs do not require an inflammatory environment to downregulate the inflammatory potential of T cells.

We next examined the influence of DMSCs on Th17 phenotype establishment in vitro. When spleen T cells and DMSCs were cocultured in vitro under nonpolarizing effector T cell condition (Fig. 5a) or under Th17-skewing condition in the presence of IL-6 and TGFβ (Fig. 5b), soluble IL-17 release was reduced as compared to cultures of exclusively spleen cells. The same results were obtained when the CD4$^+$ cell population was cocultured with DMSCs (Figure S2 in Additional file 2), implying that, as in the case of in vivo treatments, there is a direct effect of DMSCs on CD4$^+$.

This deficiency in IL-17 production seems to be acquired in perpetuity, since removal of DMSCs from the culture before a second round of T-cell stimulation was unable to restore the levels of IL-17 released by the T cells that had been previously cocultured with DMSCs (Fig. 5c and d). Accordingly, flow cytometry analysis of IL-17$^+$ and RORγT$^+$ cells showed a decrease in both Th17 cell markers expressed by the T cells cocultured with DMSCs (Fig. 5e and f).

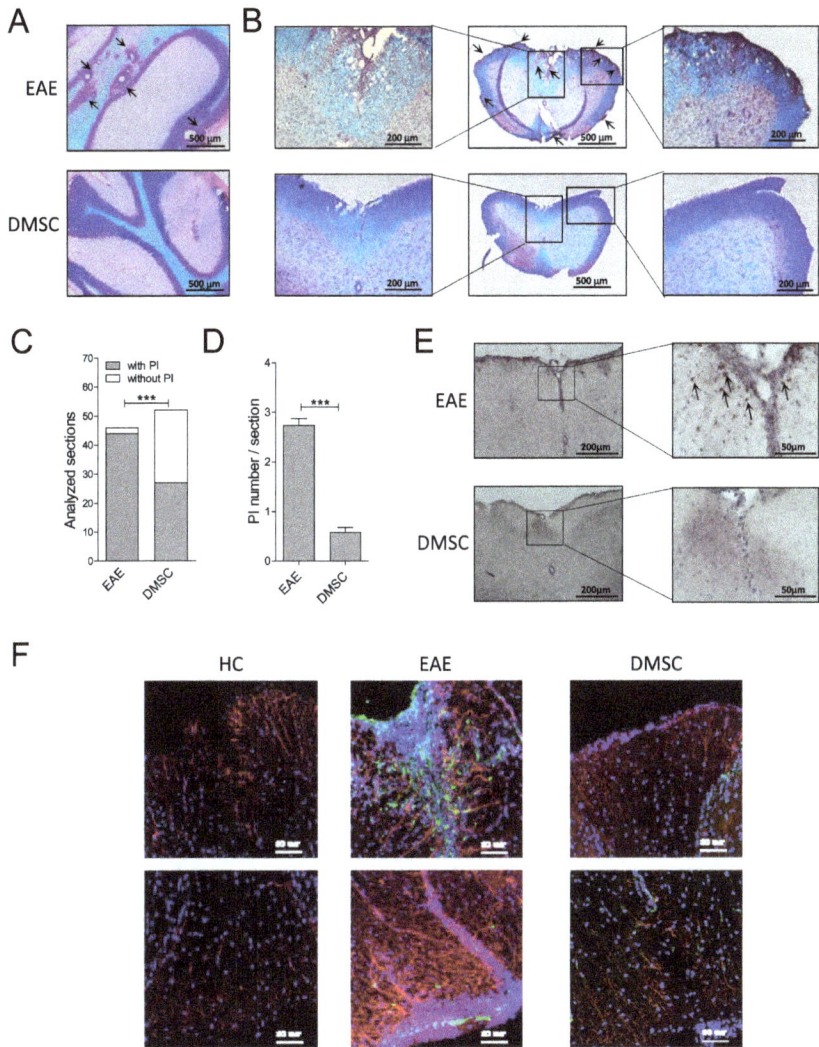

Fig. 2 Decidua-derived mesenchymal stem cell (*DMSC*) treatment of experimental autoimmune encephalomyelitis (*EAE*) decreases inflammation in the CNS. **a** Cerebellum and **b** spinal cord sections from EAE control and DMSC-treated animals were LFB–PAS–hematoxylin stained; *arrows* show perivascular infiltrates (*PI*). Hematoxylin–eosin stained slices were classified as positive (with PI) or negative (without PI) for presence of perivascular infiltrates; the difference between the EAE group (n = 47) and the DMSC group (n = 52) was analyzed by Chi-square test and shown as the histogram representation of the contingency table (**c**). **d** Each section was also classified according to the number of PI that they contained and the averages of PI/section for each group were compared by *t*-test; standard error of the means are shown. Immunohistochemistry with **e** anti-CD4 antibody (*arrows* show CD4[+] cells) and **f** immunofluorescence for astrocytes with anti-GFAP (*red*), anti-PCNA (*green*) and DAPI (*blue*) are illustrative. Scale bars for magnifications are indicated. ***p < 0.001. *HC* Healthy control, *EAE* Untreated EAE group, *DMSC* DMSC-treated group

To explore whether the effect of DMSCs on IL-17 production was dependent on a direct cell interaction between DMSCs and T cells, we used the supernatant of DMSC cultures as conditioned medium for anti-CD3/anti-CD28 T-cell stimulation. Inhibition of IL-17 secretion was consistently found for both nonpolarizing (Fig. 6a) and pro-Th17 (Fig. 6b) culture conditions, even after DMSC supernatant removal (Fig. 6c), indicating that cellular interaction is not required, but that one or more soluble factors produced by DMSCs are involved in IL-17 inhibition.

Influence of DMSC treatment on the content of pro-inflammatory cell types in EAE CNS infiltrates

The above data demonstrate that DMSCs can modulate both CD4[+] T-cell activity and Th subset definition

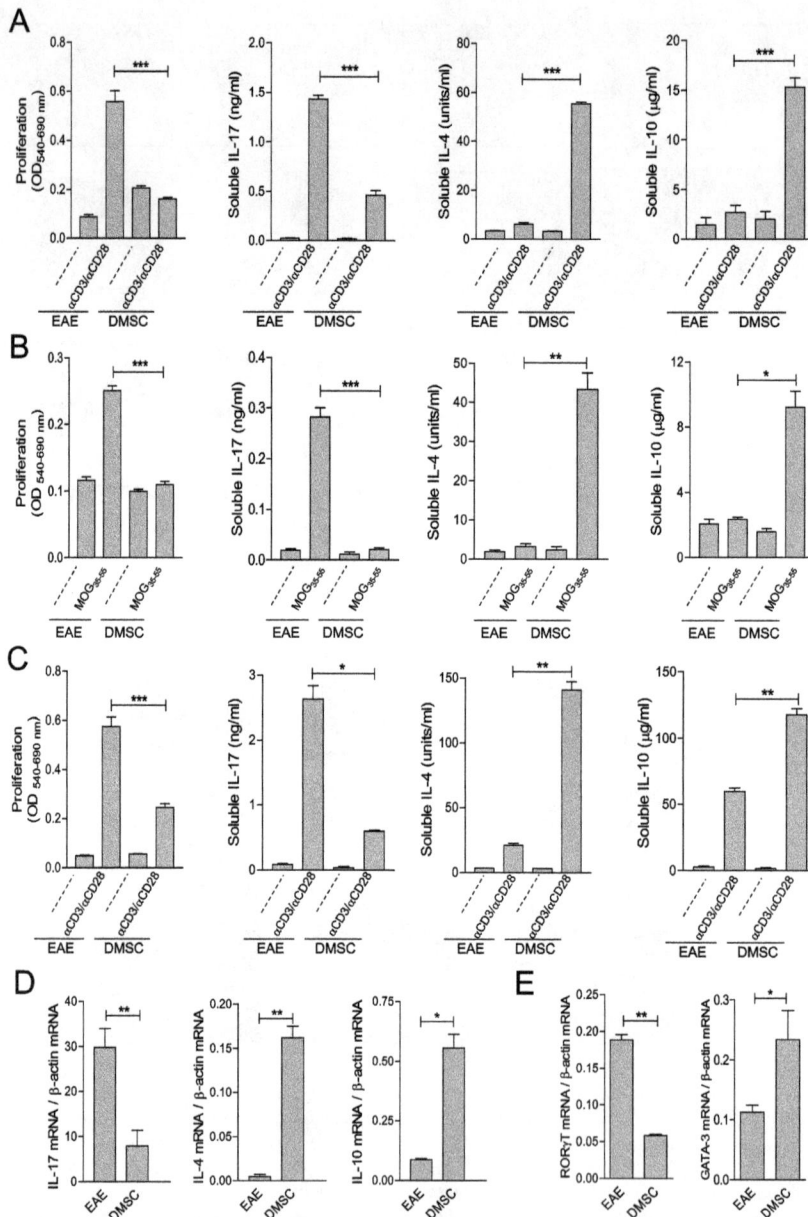

Fig. 3 Decidua-derived mesenchymal stem cell (*DMSC*) treatment of experimental autoimmune encephalomyelitis (*EAE*) promotes anti-inflammatory T-cell profile. Total spleen cell population (**a, b**) or purified CD4$^+$ cells (**c**) from EAE animals were obtained at day 10 p.i. Cells from each group were pooled and stimulated in vitro by anti-CD3 and anti-CD28 antibodies or by MOG$_{35-55}$ peptide as indicated; *dashed lines* designate unstimulated cell cultures. Each sample was assayed in quintuplicate. Proliferation and soluble cytokines released to the medium were measured by the MTT colorimetric method and ELISA, respectively. RNA from anti-CD3/anti-CD28-stimulated CD4$^+$ cells was used for RT-qPCR reactions to evaluate mRNA expression levels of cytokines (**d**) or RORγT and GATA-3 transcription factors (**e**). Results are shown from one representative out of five independent experiments. Significance was analyzed by *t*-test; standard error of the means are shown. *$p < 0.05$, **$p < 0.01$, ***$p < 0.001$. *IL* Interleukin, *MOG* Myelin oligodendrocyte glycoprotein, *OD* Optical density

in lymphoid organs. To address whether such modulation correlates with the severity of inflammatory infiltration in the CNS, the content of different immune cell types with recognized inflammatory contribution was analyzed in CNS infiltrates. CD4$^+$IL-17$^+$ cell analysis showed that this cell type was present in a smaller percentage in DMSC-treated mice (Fig. 7a and d).

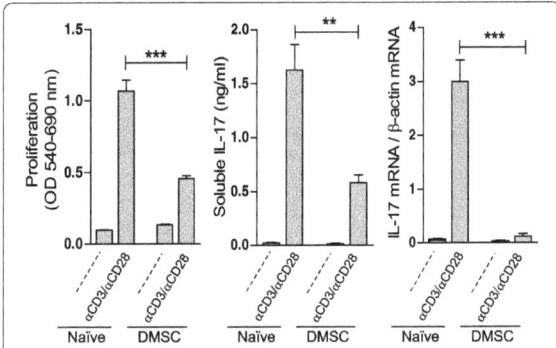

Fig. 4 Decidua-derived mesenchymal stem cell (*DMSC*) administration to healthy mice reduces IL-17 production. Healthy C57BL/6 mice (5 mice/group) were exposed to three successive doses of DMSCs (*DMSC* group) or PBS (*Naïve* group) at 3-day intervals. Seven days after the first DMSC inoculation, total spleen cells from each group were pooled and stimulated in vitro by anti-CD3/anti-CD28. Each sample was assayed in quintuplicate. Results are shown from one representative out of three independent experiments. Proliferation, soluble IL-17 released to the medium, and IL-17 mRNA were measured by the MTT colorimetric method, ELISA and RT-qPCR, respectively. Significance was analyzed by *t*-test; standard error of the means are shown. $**p < 0.01$, $***p < 0.001$. *IL* Interleukin, *OD* Optical density

Two CD11b$^+$ cell subtypes, Ly6G$^+$ and Ly6C$^+$SSC^{low}, which identify neutrophils and inflammatory monocytes, respectively [58, 59], were also quantified. Treatment of EAE mice with DMSCs resulted in a reduction of infiltrating neutrophils (Fig. 7b and d), a significant component of immune infiltration in the CNS during EAE [60–62] whose recruitment is related to IL-17 activity [63]. In our experiments, CD11b$^+$Ly6C$^+$ inflammatory monocytes (other pathogenic players in inflammation [64–66]) could be divided into two subsets (Ly6C^{int} and Ly6C^{high}), as described by Vainchtein et al. [67]. Both Ly6C^{int} and Ly6C^{high} cell percentages were lower in the infiltrates of the DMSC-treated group than in those of the control animals (Fig. 7c and d).

Long-term DMSC treatment provides a more lasting therapeutic effect on EAE

All the above mentioned EAE experiments were performed following the stipulated preventive approach, based on three doses of DMSCs at days –1, 3 and 6 with respect to the day of MOG inoculation. We wished to investigate further the effects of DMSCs on EAE by continuous dosages of DMSCs over a longer period after MOG inoculation (Fig. 8a). In contrast to the result yielded by the brief treatment used beforehand, extended administration of DMSCs showed significant differences in clinical scores between untreated and DMSC-treated groups for at least 52 days after MOG$_{35-55}$ inoculation (Fig. 8b). In addition, disease severity as measured by the AUC was significantly lower

for the DMSC group, even at advanced phases of the clinical course of the disease (Fig. 8c).

We also tested different post-MOG inoculation approaches, once the peripheral immune response had been triggered. Administration of DMSCs once the mice had reached EAE scores higher than 2.5 failed to yield any beneficial effects (data not shown). However, continuous administration of DMSCs to MOG$_{35-55}$-primed mice with lower scores showed that DMSCs could also limit the disease progression when the immune response had already been triggered. For the experiment depicted in Fig. 9, the mice received the first DMSC dosage at day 6 p.i. or at day 10 p.i. (with score values higher than 1), and four additional doses were administered every 3–4 days. For as long as DMSC continued to be dispensed, the clinical signs were significantly milder for both treated groups as compared to the EAE control group (Fig. 9a and b). In addition, the AUC showed lower values for the DMSC-treated animals (Fig. 9c). Moreover, as for the preventive treatment, disease incidence at day 20 decreased by DMSCs, even if treatment had begun when the disease symptoms were already developed (Fig. 9d).

Discussion

The results of this study, showing that EAE is significantly ameliorated by human decidua-derived MSCs (Figs. 1, 8 and 9), suggest that these cells could be seen as a promising cell therapy for MS. We describe here that in vivo treatment of EAE mice with DMSCs inhibits T-cell proliferation driven by antigen presentation and downregulates IL-17 production (Fig. 3). The role of IL-17 in the severity of EAE has been extensively demonstrated [3, 4]. It is likely that the decrease in IL-17 secretion after DMSC treatment is related to impaired differentiation of CD4$^+$ cells into the Th17 phenotype, since expression of the master regulator for Th17 development, RORγT, is downregulated in DMSC-treated animals. This view is reinforced by data from cocultures of T cells with DMSCs, showing diminished percentages of IL-17$^+$ and RORγT$^+$ cells after TCR stimulation (Fig. 5). In spite of the fact that Th17 and Foxp3$^+$ Treg cells are CD4$^+$ subsets mutually exclusive [68], we did not find differences between treated and untreated EAE mice in the levels of Foxp3 mRNA in CD4$^+$ cells nor in the percentages of spleen CD4$^+$Foxp3$^+$ cells (Figure S1 in Additional file 1). However, we cannot rule out that other regulatory cells, as myeloid-derived suppressor cells, could be involved in the lack of proliferation of spleen cells from DMSC-treated EAE mice (Fig. 3).

On the other hand, DMSCs were able to inhibit Th17 establishment even in the Th17-skewing culture conditions generated by IL-6 and TGFβ. Suppression of Th17 cells during EAE by BM-MSCs has also been observed [14, 24, 32], and a recent report has shown that placenta-derived adherent cells led to diminished numbers of IL-

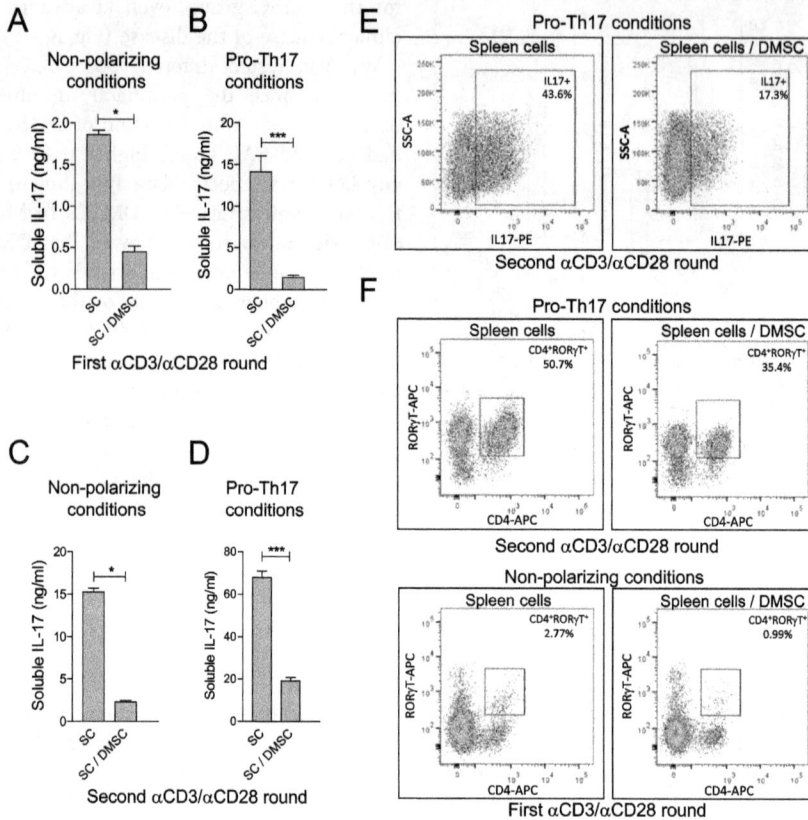

Fig. 5 In vitro treatment of T cells with decidua-derived mesenchymal stem cells (*DMSC*) interferes with Th17 phenotype definition. Spleen cells (*SC*) from C57BL/6 mice were stimulated in vitro by anti-CD3/anti-CD28 antibodies under nonpolarizing condition (**a, c**) or under pro-Th17 pressure in the presence of IL-6 and TGFβ (**b, d, e**). Three-day cultures were analyzed for IL-17 expression (**a, b**) or used for a second round of anti-CD3/anti-CD28 stimulation after DMSC removal during 3 subsequent days before new analysis (**c, d, e**). Soluble IL-17 measurements were quantified by ELISA (**a–d**). Each sample was assayed in quintuplicate and significance was analyzed by *t*-test; standard error of the means are shown. Percentages of IL-17[+] (**e**) or RORγT[+] (**f**) cells were determined by FACS analysis of intracellular staining with anti-IL17 and anti-RORγT antibodies, respectively (**e**). Results are shown from one representative out of three independent experiments. *$p < 0.05$, ***$p < 0.001$. *IL* Interleukin, *Th* T helper

17-producing cells in spinal cord infiltrates [41]. Our data from experiments with DMSC-conditioned medium suggested that the inhibiting activity of the Th17 differentiation process is mediated by one or more soluble factors produced by DMSCs (Fig. 6). In contrast, a cellular crosstalk requirement has been reported for the attenuation of IL-17 expression on T cells by BM-derived MSCs [69]. Such differences in the mechanisms involved in Th17 control might be linked to the MSC source. Most probably this discrepancy could be related to the different maturation state of T cells. Ghannam et al. [69] analyzed the effect of MSCs on fully polarized Th17 cells, whereas we studied the behavior of T cells during the Th phenotype polarization. Indeed, Luz-Crowford et al. [70] found that BM-MSCs displayed different mechanisms, based on soluble factors or on direct cell interaction for regulation of IL-17 production during Th17 development or in fully polarized Th17 cells, respectively.

Concomitant with the reduced levels of IL-17, splenocytes from DMSC-treated animals produced higher levels of IL-4 and IL-10 than did EAE control mice (Fig. 3). Such cytokines have a critical role in tolerance induction, resistance to, and recovery from, EAE [8, 9, 71, 72]. Judging by the results of the analysis of the master transcription factors and the cytokine profiles in isolated CD4[+] spleen cells, DMSCs act directly on this cell type, resulting in a deviation of the Th phenotype in favor of the Th2 versus Th17 subset. However, as IL-4 and IL-10 are not solely produced by CD4[+] cells, DMSCs could also induce other spleen cell types to produce them. In fact, DMSC-induced increase in IL-10 levels was more noticeable in total spleen cell cultures than in isolated CD4[+] cells, which might suggest that, besides CD4[+] cells, other cell populations could also be induced by DMSCs to deliver this anti-inflammatory cytokine. Similarly, other studies in different disease animal models have reported IL-4 or

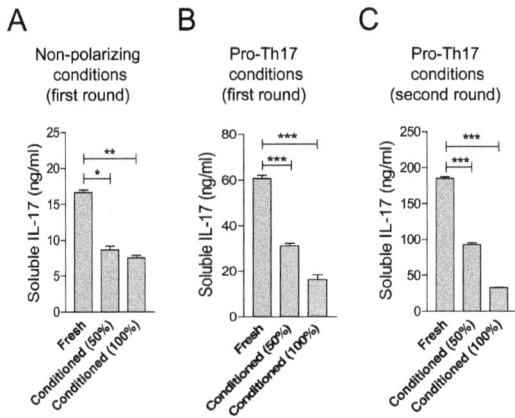

Fig. 6 DMSC effect on Th17 phenotype definition is mediated by soluble factors. DMSC-culture supernatant from one EGF-1-free passage was used as conditioned medium for C57BL/6 spleen cells. Spleen cells were stimulated by anti-CD3 and anti-CD28 antibodies in fresh Click's medium, in Click's medium diluted one-half with conditioned medium (50 %), or in whole conditioned medium (100 %). Soluble IL-17 levels were evaluated by ELISA after 3 days under nonpolarizing condition (**a**), in the presence of IL-6 and TGF-β (**b**), or after a second round of anti-CD3/anti-CD28 stimulation in the absence of DMSC supernatant (**c**). Each sample was assayed in quintuplicate and significance was analyzed by t-test; standard error of the means are shown. Results are shown from one representative out of three independent experiments. Conditioned medium was supplemented with 10 % fresh fetal bovine serum before use. *$p < 0.05$, **$p < 0.01$, ***$p < 0.001$. *IL* Interleukin, *Th* T helper

IL-10 increases by MSCs from different sources [14, 24, 37, 41, 73, 74].

In addition to the Th2 phenotype deviation in peripheral immune organs during DMSC treatment of EAE, we found lower numbers of infiltrating IL17$^+$ cells in the spinal cord of treated animals (Fig. 7). We cannot currently discern whether this difference in cellular infiltration is due to reduced migration to the CNS or to a direct effect of DMSCs on the target organ. Regardless of the cause for the decreased IL17$^+$ cell infiltration in the CNS, it could affect the recruitment of CD11b$^+$Ly6G$^+$ neutrophils, as indeed was found by the cell composition analysis of CNS infiltrates (Fig. 7). A main role for IL-17 is to co-operate in the chemoattraction of polymorphonuclear leukocytes, mainly neutrophils, to inflammatory sites through induction of CXCL-8, CXCL1 and CXCL2 [75–77], which is critical for the disruption of the blood–brain barrier [60]. Alternatively, the diminished number of neutrophils in inflammatory infiltrates could be due to a direct effect of DMSCs unrelated to IL-17 levels. In this regard, some immunomodulatory properties of MSCs have been ascribed to secretion of TSG-6 [78, 79], a molecule recently involved in inhibition of neutrophil migration by interaction with CXCL-8 [80]. We also detected a reduction in CD11b$^+$Ly6C$^+$ cell infiltration in the CNS of DMSC-treated animals. CD11b$^+$Ly6C$^+$ cell subtype, which identifies inflammatory monocytes [58], is another typical component of CNS infiltrates in EAE [64, 66]. This myeloid subset constitutes the precursor of macrophages and dendritic cells able to produce high levels of tissue damage mediators, such as TNFα and IL-1 [58]. These also act as antigen-presenting cells, reactivating T cells to contribute to the inflammatory cascade in the CNS. A reduction in the number of neutrophils and mature macrophages has been described in MSC therapy for other disease models, such as traumatic brain injury [81], ischemia [82], acute kidney injury [83], and allergic inflammation induced by *Aspergillus* [84]. However, to our knowledge, there are no previous reports on MSC treatment of EAE showing diminished presence of neutrophils or monocytes in inflammatory infiltrates. No effect of DMSCs was found in the number of spleen neutrophils or monocytes (data not shown), suggesting that a migration deficiency might underlie the diminished presence of these cell types in the CNS. Any or all of the immunomodulatory effects triggered by the DMSC treatment described here could contribute to a less pro-inflammatory environment in the target organ, which might be involved in the restraining of tissue damage observed in DMSC-treated mice (Fig. 2).

DMSCs offer several advantages over MSCs from other tissues to be used as a therapy, such as easy isolation of cells without any invasive procedures. DMSCs are of maternal origin but express factors involved in pluripotency and organogenesis though not embryonic markers (SSEA-1, -3, -4 and TRA-1-81) [42] expressed by other placental-derived MSCs [46]. These features allow high plasticity and differentiation capacity into derivatives of all germ layers, with reduced ethical problems with respect to embryonic stem cells. In addition, DMSCs also display high genomic stability after proliferation in culture, and low and limited telomerase activity [42, 43, 45]. DMSCs, like BM-derived MSCs, do not express the major histocompatibility complex class II nor the T-cell costimulatory molecules, conferring them an intrinsically hypo-immunogenic and immunomodulating stem cell character [41, 42, 45]. These properties would allow DMSCs to be tolerated, potentially effective and clinically useful in allogeneic receptors. Supporting this discernment, MSCs derived from healthy, full-term human placentas have been administered to relapsing-remitting and secondary progressive MS patients via intravenous infusion. The results showed that placental MSC injections in these patients were safe and well tolerated [23]. In addition, DMSCs show higher proliferation rates and differentiation capacity than do BM-derived MSCs [47, 48]. Furthermore, DMSCs can be easily cryopreserved over the long term without losing their original phenotype, exponential growth, and differentiation characteristics. Indeed, all the effects of DMSCs on EAE described above were observed with thawed cryopreserved cells.

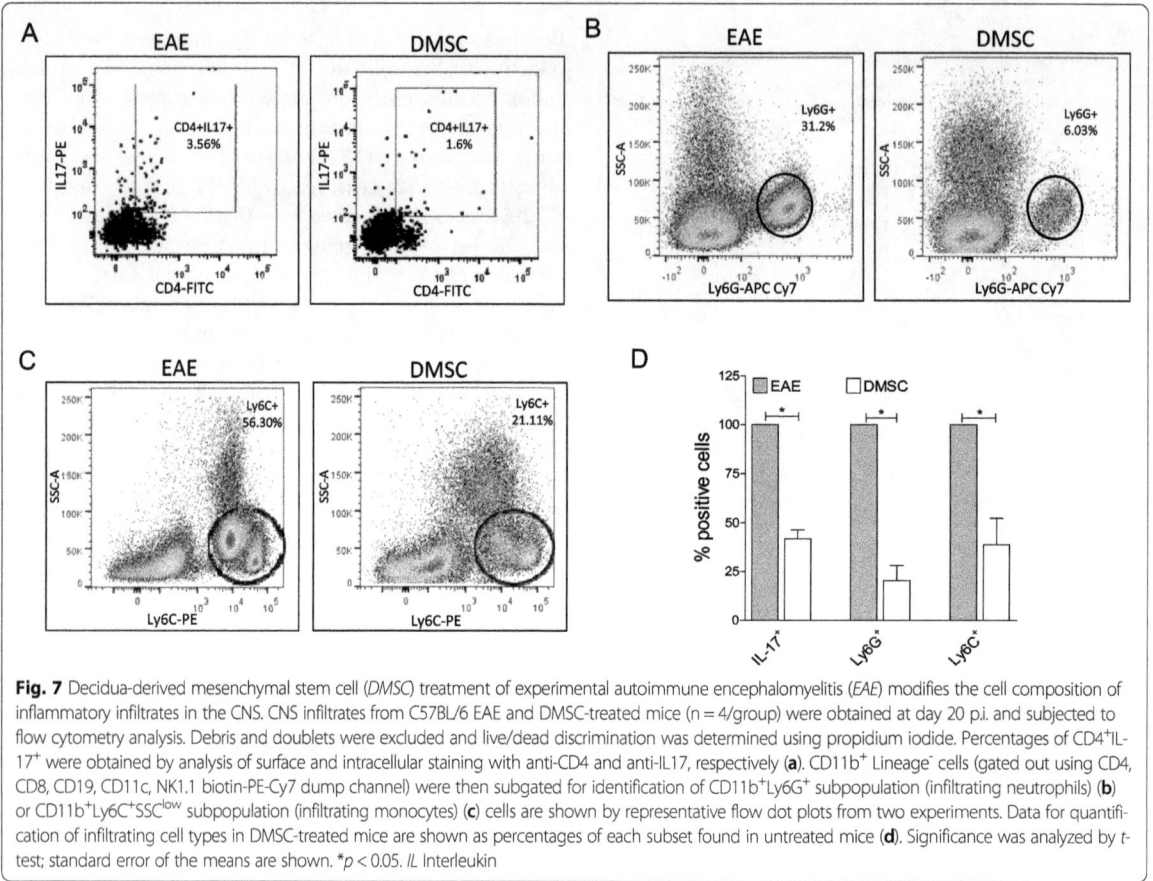

Fig. 7 Decidua-derived mesenchymal stem cell (*DMSC*) treatment of experimental autoimmune encephalomyelitis (*EAE*) modifies the cell composition of inflammatory infiltrates in the CNS. CNS infiltrates from C57BL/6 EAE and DMSC-treated mice (n = 4/group) were obtained at day 20 p.i. and subjected to flow cytometry analysis. Debris and doublets were excluded and live/dead discrimination was determined using propidium iodide. Percentages of CD4$^+$IL-17$^+$ were obtained by analysis of surface and intracellular staining with anti-CD4 and anti-IL17, respectively (**a**). CD11b$^+$ Lineage$^-$ cells (gated out using CD4, CD8, CD19, CD11c, NK1.1 biotin-PE-Cy7 dump channel) were then subgated for identification of CD11b$^+$Ly6G$^+$ subpopulation (infiltrating neutrophils) (**b**) or CD11b$^+$Ly6C$^+$SSC^{low} subpopulation (infiltrating monocytes) (**c**) cells are shown by representative flow dot plots from two experiments. Data for quantification of infiltrating cell types in DMSC-treated mice are shown as percentages of each subset found in untreated mice (**d**). Significance was analyzed by *t*-test; standard error of the means are shown. *$p < 0.05$. *IL* Interleukin

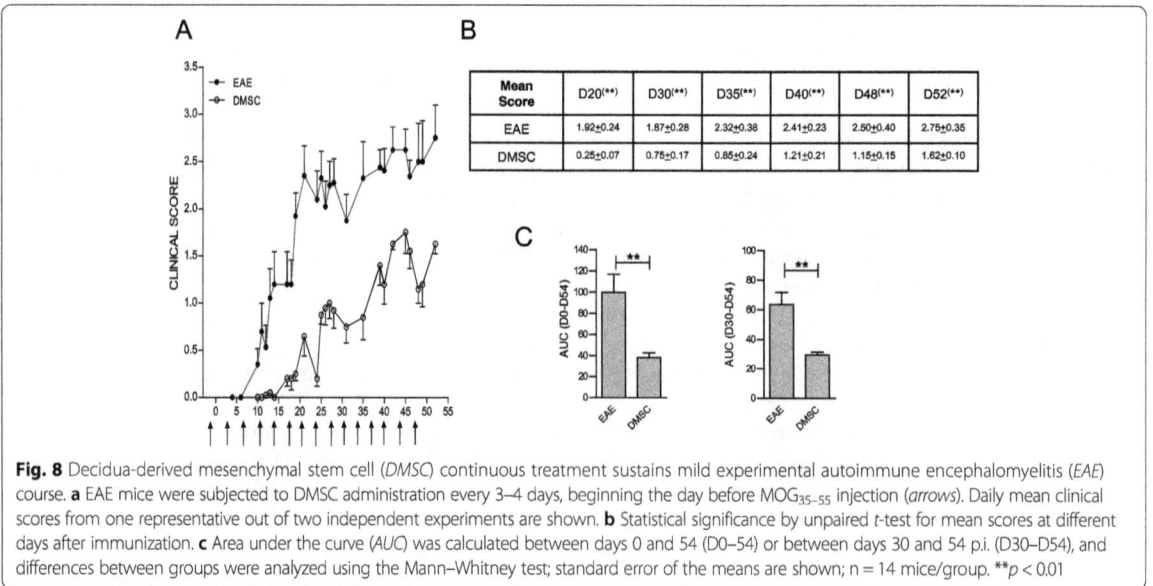

Fig. 8 Decidua-derived mesenchymal stem cell (*DMSC*) continuous treatment sustains mild experimental autoimmune encephalomyelitis (*EAE*) course. **a** EAE mice were subjected to DMSC administration every 3–4 days, beginning the day before MOG$_{35-55}$ injection (*arrows*). Daily mean clinical scores from one representative out of two independent experiments are shown. **b** Statistical significance by unpaired *t*-test for mean scores at different days after immunization. **c** Area under the curve (*AUC*) was calculated between days 0 and 54 (D0–54) or between days 30 and 54 p.i. (D30–D54), and differences between groups were analyzed using the Mann–Whitney test; standard error of the means are shown; n = 14 mice/group. **$p < 0.01$

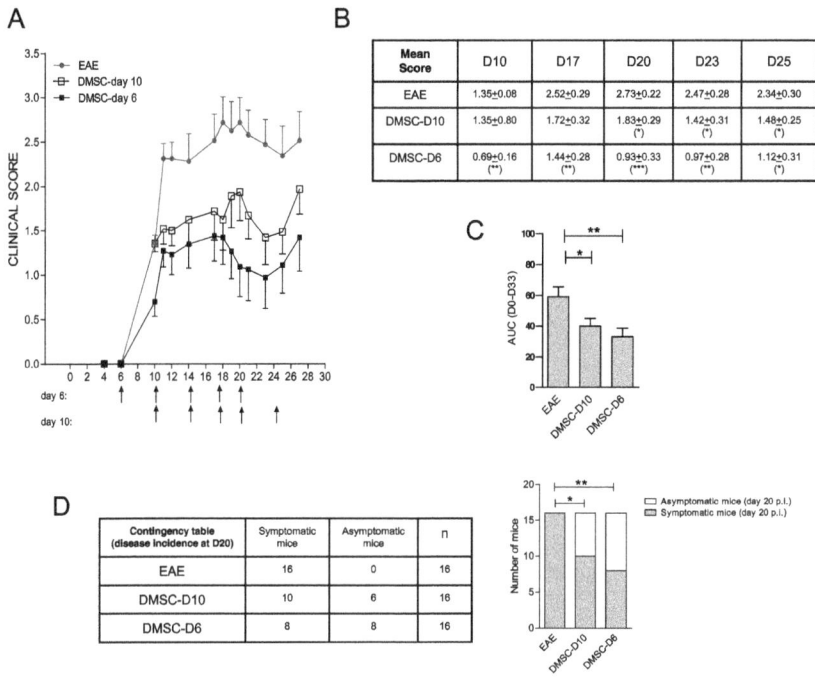

A

B

Mean Score	D10	D17	D20	D23	D25
EAE	1.35±0.08	2.52±0.29	2.73±0.22	2.47±0.28	2.34±0.30
DMSC-D10	1.35±0.80	1.72±0.32	1.83±0.29 (*)	1.42±0.31 (*)	1.48±0.25 (*)
DMSC-D6	0.69±0.16 (**)	1.44±0.28 (**)	0.93±0.33 (***)	0.97±0.28 (**)	1.12±0.31 (*)

C

D

Contingency table (disease incidence at D20)	Symptomatic mice	Asymptomatic mice	n
EAE	16	0	16
DMSC-D10	10	6	16
DMSC-D6	8	8	16

Fig. 9 Decidua-derived mesenchymal stem cell (*DMSC*) treatment after experimental autoimmune encephalomyelitis (*EAE*) triggering. **a** EAE mice were subjected to DMSC administration every 3–4 days, beginning at day 6 (D6) or at day 10 (D10) after MOG$_{35-55}$ injection (*arrows*). Daily mean clinical scores from one representative out of two independent experiments are shown. **b** Statistical significance by unpaired *t*-test for mean scores at different days after immunization. **c** Area under the curve (*AUC*) was calculated between days 0 and 33 post-immunization (*p.i.*) (D0–33), and differences between groups were analyzed using the Mann–Whitney test; standard error of the means are shown; n = 16 mice/group. **d** Disease incidence at day 20 p.i. was analyzed by Fisher's test. The bar graphs for representation of the disease incidence contingency table show the numbers of symptomatic and asymptomatic mice at day 20 p.i. for each group. *$p < 0.05$, **$p < 0.01$, ***$p < 0.001$

Thus, based on the high proliferation rate of placental-derived MSCs, a single donor could be used for multiple patients after cell storage. Finally, these maternal-derived mesenchymal stromal cells could be also an autologous source for cell therapy on several diseases developed by the mother [42]. All these advantages make DMSCs a potentially safe and useful product for future use in humans.

Our results show effective prevention of the disease through short-term DMSC treatment, albeit of limited durability. Interestingly, periodical administration of the cells results in mild clinical course of the disease, even if animals receive the first treatment when they are already presenting moderate symptoms (Figs. 8 and 9). Along with preliminary studies with autologous BM-derived MSC transplantation in MS patients yielding promising results, the clinical use of alternative cell sources, such as placenta-derived DMSCs, warrants further investigation to address whether the long-term safety and potential clinical efficacy of using this new cell therapy could actually provide improvements over other sources of MSCs. Since MS is a chronic disease, several authors have suggested that, to ensure a sustained therapeutic benefit, clinical application of MSCs might require repeated administration of cells instead of just one dosage [18, 29]. MSCs are thought to function through a 'hit-and-run' mechanism, whereby they release an array of cytokines and trophic factors without any significant engraftment. Under conditions of chronic autoimmunity and CNS injury, multiple doses of MSCs are likely to be necessary for the sustained production of immunomodulatory and trophic factors in order to exceed a therapeutic threshold [29]. This scenario is in line with our results in EAE mice, which strongly support a cell therapy strategy based on repeated administration of DMSCs, by adapting the dose schedule to the clinical level, which might substantially modulate the long-term progression of MS, the main treatment goal in this progressive and disabling demyelinating disease.

Conclusions

This study reveals immunomodulatory effects of DMSCs able to modulate EAE. The beneficial properties of such treatment on the disease signs, and on the frequency in infiltration foci in the CNS, correlate with impairment of the Th17 phenotype in favor of the promotion of IL-4 and IL-10 production at the early immune response.

Therefore, human decidua seems to be a valuable source of MSCs, with therapeutic potential for MS and for other immunological diseases in which IL-17 has a pathogenic role.

Additional files

Additional file 1: Figure S1. Foxp3 expression by T cells from untreated or DMSC-treated EAE mice (n = 9/group). Spleen cells from EAE animals were obtained at day 10 p.i. Total RNA samples were obtained from purified CD4+ cells and used to quantify Foxp3 mRNA by RT-qPCR (A). Total spleen populations from individual mice were analyzed by cytometry analysis of surface and intracellular staining with anti-CD4-APC and anti-Foxp3-PE, respectively (B). Percentages of CD4+Foxp3+ cells are shown by representative flow dot plots and by the average of the values obtained for each individual mouse from each group. Standard error of the means are shown. (PDF 150 kb)

Additional file 2: Figure S2. In vitro treatment of CD4+ T cells with DMSCs interferes with Th17 phenotype definition. CD4+ cells purified from C57BL/6 mice spleens were stimulated in vitro by anti-CD3/anti-CD28 antibodies under nonpolarizing condition (A) or under pro-Th17 pressure in the presence of IL-6 and TGFβ (B). Three-day cultures were analyzed for IL-17 expression. Soluble IL-17 measurements were quantified by ELISA. Each sample was assayed in quintuplicate and significance was analyzed by t-test; standard error of the means are shown. Results are shown from one representative out of three independent experiments. (PDF 17 kb)

Abbreviations

AUC: Area under the curve; BM: Bone marrow; CNS: Central nervous system; D: Day; DMEM: Dulbecco's modified Eagle medium; DMSC: Decidua-derived mesenchymal stem cell; EAE: Experimental autoimmune encephalomyelitis; EGF-1: Epidermal growth factor 1; ELISA: Enzyme-linked immunosorbent assay; IFN: Interferon; IL: Interleukin; LFB: Luxol fast blue; MOG: Myelin oligodendrocyte glycoprotein; MSC: Mesenchymal stem cell; MS: Multiple sclerosis; PAS: Periodic acid-Schiff; PBS: Phosphate-buffered saline; p.i.: Post-immunization; RT-qPCR: Retrotranscription and quantitative real-time polymerase chain reaction; TGF: Transforming growth factor; Th: T helper; TNFα: Tumor necrosis factor-alpha; Treg: T regulatory.

Competing interests

The authors declare that they have no competing interests.

Authors' contributions

BB contributed to the design, written description and performance of the in vivo experiments, EAE induction, flow cytometry assays and proliferation, and cytokine measurements. JH was involved in the design, written description and performance of the cocultures of DMSC-murine spleen cells, and determined proliferation and cytokine measurements. MIG, APB, EGZ and RPT were involved in the design of the histopathology studies, contributed to the processing of the tissue samples and to the writing of the manuscript. AIF and PdIT carried out the isolation and culture of DMSCs, and collaborated in the study design, discussion of results, writing of the manuscript, and obtaining funding. JG collaborated in the study design and was in charge of selecting the mothers for obtaining human placentas for DMSC isolation. RB collaborated in the study design, discussion of results, writing of the manuscript, and obtaining funding. AB was involved in the study design, supervising the work, EAE experiments, analysis of the histologic and flow cytometry data, interpretation and discussion of results, and writing of the manuscript. SB was involved in the study design, supervising the work, statistical analysis, interpretation and discussion of results, writing of the manuscript, and obtaining funding. All authors read, critically revised and approved the final manuscript.

Acknowledgments

This work was sponsored by grants from Acción Estratégica en Salud (PI13/00297 and PI11/00581), the Neurosciences and Aging Foundation, the Francisco Soria Melguizo Foundation, Octopharma, and Parkinson Madrid (PI2012/0032). The authors are grateful to Dr. R. Murillas for his critical reading of the manuscript.

Author details

[1]Instituto de Salud Carlos III, Unidad Funcional de Investigación en Enfermedades Crónicas, Laboratory of Gene Regulation, Carretera de Majadahonda-Pozuelo Km 2, 28220 Madrid, Spain. [2]Instituto de Salud Carlos III, Unidad Funcional de Investigación en Enfermedades Crónicas, Laboratory of Mammary Gland Pathology, Carretera de Majadahonda-Pozuelo Km 2, 28220 Madrid, Spain. [3]Grupo de Medicina Regenerativa, Instituto de Investigación Hospital 12 de Octubre, Avda. Córdoba s/n, 28041 Madrid, Spain. [4]Hospital Central de Cruz Roja, Servicio de Hematología y Hemoterapia, Avenida de Reina Victoria 24, 28003 Madrid, Spain. [5]Instituto de Salud Carlos III, Unidad Funcional de Investigación en Enfermedades Crónicas, Histology Core Unit, Carretera de Majadahonda-Pozuelo Km 2, 28220 Madrid, Spain.

References

1. Constantinescu CS, Farooqi N, O'Brien K, Gran B. Experimental autoimmune encephalomyelitis (EAE) as a model for multiple sclerosis (MS). Br J Pharmacol. 2011;164:1079–106.
2. Chen SJ, Wang YL, Fan HC, Lo WT, Wang CC, Sytwu HK. Current status of the immunomodulation and immunomediated therapeutic strategies for multiple sclerosis. Clin Dev Immunol. 2012;2012:970789.
3. Rostami A, Ciric B. Role of Th17 cells in the pathogenesis of CNS inflammatory demyelination. J Neurol Sci. 2013;333:76–87.
4. Pierson E, Simmons SB, Castelli L, Goverman JM. Mechanisms regulating regional localization of inflammation during CNS autoimmunity. Immunol Rev. 2012;248:205–15.
5. O'Connor RA, Anderton SM. Foxp3+ regulatory T cells in the control of experimental CNS autoimmune disease. J Neuroimmunol. 2008;193:1–11.
6. Cua DJ, Hinton DR, Stohlman SA. Self-antigen-induced Th2 responses in experimental allergic encephalomyelitis (EAE)-resistant mice. Th2-mediated suppression of autoimmune disease. J Immunol. 1995;155:4052–9.
7. Liblau RS, Singer SM, McDevitt HO. Th1 and Th2 CD4+ T cells in the pathogenesis of organ-specific autoimmune diseases. Immunol Today. 1995;16:34–8.
8. Falcone M, Rajan AJ, Bloom BR, Brosnan CF. A critical role for IL-4 in regulating disease severity in experimental allergic encephalomyelitis as demonstrated in IL-4-deficient C57BL/6 mice and BALB/c mice. J Immunol. 1998;160:4822–30.
9. Bettelli E, Das MP, Howard ED, Weiner HL, Sobel RA, Kuchroo VK. IL-10 is critical in the regulation of autoimmune encephalomyelitis as demonstrated by studies of IL-10- and IL-4-deficient and transgenic mice. J Immunol. 1998;161:3299–306.
10. Swanborg RH. Experimental autoimmune encephalomyelitis in the rat: lessons in T-cell immunology and autoreactivity. Immunol Rev. 2001;184: 129–35.
11. Batten M, Li J, Yi S, Kljavin NM, Danilenko DM, Lucas S, et al. Interleukin 27 limits autoimmune encephalomyelitis by suppressing the development of interleukin 17-producing T cells. Nat Immunol. 2006;7:929–36.
12. Fitzgerald DC, Ciric B, Touil T, Harle H, Grammatikopolou J, Das Sarma J, et al. Suppressive effect of IL-27 on encephalitogenic Th17 cells and the effector phase of experimental autoimmune encephalomyelitis. J Immunol. 2007;179: 3268–75.
13. Stumhofer JS, Hunter CA. Advances in understanding the anti-inflammatory properties of IL-27. Immunol Lett. 2008;117:123–30.
14. Wang J, Wang G, Sun B, Li H, Mu L, Wang Q, et al. Interleukin-27 suppresses experimental autoimmune encephalomyelitis during bone marrow stromal cell treatment. J Autoimmun. 2008;30:222–9.
15. Wang X, Kimbrel EA, Ijichi K, Paul D, Lazorchak AS, Chu J, et al. Human ESC-derived MSCs outperform bone marrow MSCs in the treatment of an EAE model of multiple sclerosis. Stem Cell Reports. 2014;3:115–30.
16. Uccelli A, Laroni A, Freedman MS. Mesenchymal stem cells as treatment for MS—progress to date. Mult Scler. 2013;19:515–9.
17. Cohen JA. Mesenchymal stem cell transplantation in multiple sclerosis. J Neurol Sci. 2013;333:43–9.
18. Auletta JJ, Bartholomew AM, Maziarz RT, Deans RJ, Miller RH, Lazarus HM, et al. The potential of mesenchymal stromal cells as a novel cellular therapy for multiple sclerosis. Immunotherapy. 2012;4:529–47.

19. Scolding N, Marks D, Rice C. Autologous mesenchymal bone marrow stem cells: practical considerations. J Neurol Sci. 2008;265:111–5.

20. Karussis D, Karageorgiou C, Vaknin-Dembinsky A, Gowda-Kurkalli B, Gomori JM, Kassis I, et al. Safety and immunological effects of mesenchymal stem cell transplantation in patients with multiple sclerosis and amyotrophic lateral sclerosis. Arch Neurol. 2010;67:1187–94.

21. Yamout B, Hourani R, Salti H, Barada W, El-Hajj T, Al-Kutoubi A, et al. Bone marrow mesenchymal stem cell transplantation in patients with multiple sclerosis: a pilot study. J Neuroimmunol. 2010;227:185–9.

22. Connick P, Kolappan M, Crawley C, Webber DJ, Patani R, Michell AW, et al. Autologous mesenchymal stem cells for the treatment of secondary progressive multiple sclerosis: an open-label phase 2a proof-of-concept study. Lancet Neurol. 2012;11:150–6.

23. Lublin FD, Bowen JD, Huddlestone J, Kremenchutzky M, Carpenter A, Corboy JR, et al. Human placenta-derived cells (PDA-001) for the treatment of adults with multiple sclerosis: a randomized, placebo-controlled, multiple-dose study. Mult Scler Relat Disord. 2014;3:696–704.

24. Bai L, Lennon DP, Eaton V, Maier K, Caplan AI, Miller SD, et al. Human bone marrow-derived mesenchymal stem cells induce Th2-polarized immune response and promote endogenous repair in animal models of multiple sclerosis. Glia. 2009;57:1192–203.

25. Jaramillo-Merchan J, Jones J, Ivorra JL, Pastor D, Viso-Leon MC, Armengol JA, et al. Mesenchymal stromal-cell transplants induce oligodendrocyte progenitor migration and remyelination in a chronic demyelination model. Cell Death Dis. 2013;4:e779.

26. Zhang J, Li Y, Lu M, Cui Y, Chen J, Noffsinger L, et al. Bone marrow stromal cells reduce axonal loss in experimental autoimmune encephalomyelitis mice. J Neurosci Res. 2006;84:587–95.

27. Kassis I, Grigoriadis N, Gowda-Kurkalli B, Mizrachi-Kol R, Ben-Hur T, Slavin S, et al. Neuroprotection and immunomodulation with mesenchymal stem cells in chronic experimental autoimmune encephalomyelitis. Arch Neurol. 2008;65:753–61.

28. Crigler L, Robey RC, Asawachaicharn A, Gaupp D, Phinney DG. Human mesenchymal stem cell subpopulations express a variety of neuro-regulatory molecules and promote neuronal cell survival and neuritogenesis. Exp Neurol. 2006;198:54–64.

29. Harris VK, Yan QJ, Vyshkina T, Sahabi S, Liu X, Sadiq SA. Clinical and pathological effects of intrathecal injection of mesenchymal stem cell-derived neural progenitors in an experimental model of multiple sclerosis. J Neurol Sci. 2012;313:167–77.

30. Benkhoucha M, Santiago-Raber ML, Schneiter G, Chofflon M, Funakoshi H, Nakamura T, et al. Hepatocyte growth factor inhibits CNS autoimmunity by inducing tolerogenic dendritic cells and CD25 + Foxp3+ regulatory T cells. Proc Natl Acad Sci U S A. 2010;107:6424–9.

31. Matysiak M, Orlowski W, Fortak-Michalska M, Jurewicz A, Selmaj K. Immunoregulatory function of bone marrow mesenchymal stem cells in EAE depends on their differentiation state and secretion of PGE2. J Neuroimmunol. 2011;233:106–11.

32. Rafei M, Campeau PM, Aguilar-Mahecha A, Buchanan M, Williams P, Birman E, et al. Mesenchymal stromal cells ameliorate experimental autoimmune encephalomyelitis by inhibiting CD4 Th17 T cells in a CC chemokine ligand 2-dependent manner. J Immunol. 2009;182:5994–6002.

33. Gerdoni E, Gallo B, Casazza S, Musio S, Bonanni I, Pedemonte E, et al. Mesenchymal stem cells effectively modulate pathogenic immune response in experimental autoimmune encephalomyelitis. Ann Neurol. 2007;61:219–27.

34. Zappia E, Casazza S, Pedemonte E, Benvenuto F, Bonanni I, Gerdoni E, et al. Mesenchymal stem cells ameliorate experimental autoimmune encephalomyelitis inducing T-cell anergy. Blood. 2005;106:1755–61.

35. Yousefi F, Ebtekar M, Soleimani M, Soudi S, Hashemi SM. Comparison of in vivo immunomodulatory effects of intravenous and intraperitoneal administration of adipose-tissue mesenchymal stem cells in experimental autoimmune encephalomyelitis (EAE). Int Immunopharmacol. 2013;17:608–16.

36. Constantin G, Marconi S, Rossi B, Angiari S, Calderan L, Anghileri E, et al. Adipose-derived mesenchymal stem cells ameliorate chronic experimental autoimmune encephalomyelitis. Stem Cells. 2009;27:2624–35.

37. Peron JP, Jazedje T, Brandao WN, Perin PM, Maluf M, Evangelista LP, et al. Human endometrial-derived mesenchymal stem cells suppress inflammation in the central nervous system of EAE mice. Stem Cell Rev. 2012;8:940–52.

38. Payne NL, Sun G, McDonald C, Layton D, Moussa L, Emerson-Webber A, et al. Distinct immunomodulatory and migratory mechanisms underpin the therapeutic potential of human mesenchymal stem cells in autoimmune demyelination. Cell Transplant. 2013;22:1409–25.

39. Donders R, Vanheusden M, Bogie JF, Ravanidis S, Thewissen K, Stinissen P, et al. Human Wharton's jelly-derived stem cells display immunomodulatory properties and transiently improve rat experimental autoimmune encephalomyelitis. Cell Transplant. 2014;24:2077–98.

40. Fisher-Shoval Y, Barhum Y, Sadan O, Yust-Katz S, Ben-Zur T, Lev N, et al. Transplantation of placenta-derived mesenchymal stem cells in the EAE mouse model of MS. J Mol Neurosci. 2012;48:176–84.

41. Liu W, Morschauser A, Zhang X, Lu X, Gleason J, He S, et al. Human placenta-derived adherent cells induce tolerogenic immune responses. Clin Transl Immunology. 2014;3:e14.

42. Macias MI, Grande J, Moreno A, Dominguez I, Bornstein R, Flores AI. Isolation and characterization of true mesenchymal stem cells derived from human term decidua capable of multilineage differentiation into all 3 embryonic layers. Am J Obstet Gynecol. 2010;203:495. e9–e23.

43. Bornstein R, Macias MI, de la Torre P, Grande J, Flores AI. Human decidua-derived mesenchymal stromal cells differentiate into hepatic-like cells and form functional three-dimensional structures. Cytotherapy. 2012;14:1182–92.

44. Cerrada A, de la Torre P, Grande J, Haller T, Flores AI, Perez-Gil J. Human decidua-derived mesenchymal stem cells differentiate into functional alveolar type II-like cells that synthesize and secrete pulmonary surfactant complexes. PLoS One. 2014;9:e110195.

45. Vegh I, Grau M, Gracia M, Grande J, de la Torre P, Flores AI. Decidua mesenchymal stem cells migrated toward mammary tumors in vitro and in vivo affecting tumor growth and tumor development. Cancer Gene Ther. 2013;20:8–16.

46. Huang YC, Yang ZM, Chen XH, Tan MY, Wang J, Li XQ, et al. Isolation of mesenchymal stem cells from human placental decidua basalis and resistance to hypoxia and serum deprivation. Stem Cell Rev. 2009;5:247–55.

47. Barlow S, Brooke G, Chatterjee K, Price G, Pelekanos R, Rossetti T, et al. Comparison of human placenta- and bone marrow-derived multipotent mesenchymal stem cells. Stem Cells Dev. 2008;17:1095–107.

48. Brooke G, Tong H, Levesque JP, Atkinson K. Molecular trafficking mechanisms of multipotent mesenchymal stem cells derived from human bone marrow and placenta. Stem Cells Dev. 2008;17:929–40.

49. Martin-Saavedra FM, Gonzalez-Garcia C, Bravo B, Ballester S. Beta interferon restricts the inflammatory potential of CD4+ cells through the boost of the Th2 phenotype, the inhibition of Th17 response and the prevalence of naturally occurring T regulatory cells. Mol Immunol. 2008;45:4008–19.

50. Click RE, Benck L, Alter BJ. Enhancement of antibody synthesis in vitro by mercaptoethanol. Cell Immunol. 1972;3:156–60.

51. Portolés P, Rojo J, Golby A, Bonneville M, Gromkowski S, Greenbaum L, et al. Monoclonal antibodies to murine CD3e define distinct epitopes, one of which may interact with CD4 during T cell activation. J Immunol. 1989;142:4169–75.

52. Gonzalez-Garcia C, Bravo B, Ballester A, Gomez-Perez R, Eguiluz C, Redondo M, et al. Comparative assessment of PDE 4 and 7 inhibitors as therapeutic agents in experimental autoimmune encephalomyelitis. Br J Pharmacol. 2013;170:602–13.

53. Mosmann T. Rapid colorimetric assay for cellular growth and survival: application to proliferation and cytotoxicity assays. J Immunol Methods. 1983;65:55–63.

54. Martin-Saavedra FM, Flores N, Dorado B, Eguiluz C, Bravo B, Garcia-Merino A, et al. Beta-interferon unbalances the peripheral T cell proinflammatory response in experimental autoimmune encephalomyelitis. Mol Immunol. 2007;44:3597–607.

55. Pfaffl MW. A new mathematical model for relative quantification in real-time RT-PCR. Nucleic Acids Res. 2001;29:e45.

56. Goto N. Discriminative staining methods for the nervous system: luxol fast blue–periodic acid-Schiff–hematoxylin triple stain and subsidiary staining methods. Stain Technol. 1987;62:305–15.

57. Reiner SL. Development in motion: helper T cells at work. Cell. 2007;129:33–6.

58. Auffray C, Sieweke MH, Geissmann F. Blood monocytes: development, heterogeneity, and relationship with dendritic cells. Annu Rev Immunol. 2009;27:669–92.

59. Lee PY, Wang JX, Parisini E, Dascher CC, Nigrovic PA. Ly6 family proteins in neutrophil biology. J Leukoc Biol. 2013;94:585–94.

60. Aube B, Levesque SA, Pare A, Chamma E, Kebir H, Gorina R, et al. Neutrophils mediate blood-spinal cord barrier disruption in demyelinating neuroinflammatory diseases. J Immunol. 2014;193:2438–54.

61. Christy AL, Walker ME, Hessner MJ, Brown MA. Mast cell activation and neutrophil recruitment promotes early and robust inflammation in the meninges in EAE. J Autoimmun. 2013;42:50–61.

62. Wojkowska DW, Szpakowski P, Ksiazek-Winiarek D, Leszczynski M, Glabinski A. Interactions between neutrophils, Th17 cells, and chemokines during the initiation of experimental model of multiple sclerosis. Mediators Inflamm. 2014;2014:590409.

63. Roussel L, Houle F, Chan C, Yao Y, Berube J, Olivenstein R, et al. IL-17 promotes p38 MAPK-dependent endothelial activation enhancing neutrophil recruitment to sites of inflammation. J Immunol. 2010;184:4531–7.

64. King IL, Dickendesher TL, Segal BM. Circulating Ly-6C+ myeloid precursors migrate to the CNS and play a pathogenic role during autoimmune demyelinating disease. Blood. 2009;113:3190–7.

65. Mishra MK, Wang J, Silva C, Mack M, Yong VW. Kinetics of proinflammatory monocytes in a model of multiple sclerosis and its perturbation by laquinimod. Am J Pathol. 2012;181:642–51.

66. Saederup N, Cardona AE, Croft K, Mizutani M, Cotleur AC, Tsou CL, et al. Selective chemokine receptor usage by central nervous system myeloid cells in CCR2-red fluorescent protein knock-in mice. PLoS One. 2010;5:e13693.

67. Vainchtein ID, Vinet J, Brouwer N, Brendecke S, Biagini G, Biber K, et al. In acute experimental autoimmune encephalomyelitis, infiltrating macrophages are immune activated, whereas microglia remain immune suppressed. Glia. 2014;62:1724–35.

68. Bettelli E, Carrier Y, Gao W, Korn T, Strom TB, Oukka M, et al. Reciprocal developmental pathways for the generation of pathogenic effector TH17 and regulatory T cells. Nature. 2006;441:235–8.

69. Ghannam S, Pene J, Moquet-Torcy G, Jorgensen C, Yssel H. Mesenchymal stem cells inhibit human Th17 cell differentiation and function and induce a T regulatory cell phenotype. J Immunol. 2010;185:302–12.

70. Luz-Crawford P, Noel D, Fernandez X, Khoury M, Figueroa F, Carrion F, et al. Mesenchymal stem cells repress Th17 molecular program through the PD-1 pathway. PLoS One. 2012;7:e45272.

71. Bettelli E, Nicholson LB, Kuchroo VK. IL-10, a key effector regulatory cytokine in experimental autoimmune encephalomyelitis. J Autoimmun. 2003;20:265–7.

72. Chitnis T, Khoury SJ. Cytokine shifts and tolerance in experimental autoimmune encephalomyelitis. Immunol Res. 2003;28:223–39.

73. Nemeth K, Leelahavanichkul A, Yuen PS, Mayer B, Parmelee A, Doi K, et al. Bone marrow stromal cells attenuate sepsis via prostaglandin E(2)-dependent reprogramming of host macrophages to increase their interleukin-10 production. Nat Med. 2009;15:42–9.

74. Kim S, Chang KA, Kim J, Park HG, Ra JC, Kim HS, et al. The preventive and therapeutic effects of intravenous human adipose-derived stem cells in Alzheimer's disease mice. PLoS One. 2012;7:e45757.

75. Cua DJ, Tato CM. Innate IL-17-producing cells: the sentinels of the immune system. Nat Rev Immunol. 2010;10:479–89.

76. Kalyan S, Kabelitz D. When neutrophils meet T cells: beginnings of a tumultuous relationship with underappreciated potential. Eur J Immunol. 2014;44:627–33.

77. Pelletier M, Maggi L, Micheletti A, Lazzeri E, Tamassia N, Costantini C, et al. Evidence for a cross-talk between human neutrophils and Th17 cells. Blood. 2010;115:335–43.

78. Lee RH, Pulin AA, Seo MJ, Kota DJ, Ylostalo J, Larson BL, et al. Intravenous hMSCs improve myocardial infarction in mice because cells embolized in lung are activated to secrete the anti-inflammatory protein TSG-6. Cell Stem Cell. 2009;5:54–63.

79. Oh JY, Lee RH, Yu JM, Ko JH, Lee HJ, Ko AY, et al. Intravenous mesenchymal stem cells prevented rejection of allogeneic corneal transplants by aborting the early inflammatory response. Mol Ther. 2012;20:2143–52.

80. Dyer DP, Thomson JM, Hermant A, Jowitt TA, Handel TM, Proudfoot AE, et al. TSG-6 inhibits neutrophil migration via direct interaction with the chemokine CXCL8. J Immunol. 2014;192:2177–85.

81. Zhang R, Liu Y, Yan K, Chen L, Chen XR, Li P, et al. Anti-inflammatory and immunomodulatory mechanisms of mesenchymal stem cell transplantation in experimental traumatic brain injury. J Neuroinflammation. 2013;10:106.

82. Zhang B, Adesanya TM, Zhang L, Xie N, Chen Z, Fu M, et al. Delivery of placenta-derived mesenchymal stem cells ameliorates ischemia induced limb injury by immunomodulation. Cell Physiol Biochem. 2014;34:1998–2006.

83. Luo CJ, Zhang FJ, Zhang L, Geng YQ, Li QG, Hong Q, et al. Mesenchymal stem cells ameliorate sepsis-associated acute kidney injury in mice. Shock. 2014;41:123–9.

84. Lathrop MJ, Brooks EM, Bonenfant NR, Sokocevic D, Borg ZD, Goodwin M, et al. Mesenchymal stromal cells mediate Aspergillus hyphal extract-induced allergic airway inflammation by inhibition of the Th17 signaling pathway. Stem Cells Transl Med. 2014;3:194–205.

11

Comparative study of equine mesenchymal stem cells from healthy and injured synovial tissues: an in vitro assessment

Joice Fülber[1], Durvanei A. Maria[2], Luis Cláudio Lopes Correia da Silva[3], Cristina O. Massoco[4], Fernanda Agreste[1] and Raquel Y. Arantes Baccarin[1]* (iD)

Abstract

Background: Bone marrow and adipose tissues are known sources of mesenchymal stem cells (MSCs) in horses; however, synovial tissues might be a promising alternative. The aim of this study was to evaluate phenotypic characteristics and differentiation potential of equine MSCs from synovial fluid (SF) and synovial membrane (SM) of healthy joints (SF-H and SM-H), joints with osteoarthritis (SF-OA and SM-OA) and joints with osteochondritis dissecans (SF-OCD and SM-OCD) to determine the most suitable synovial source for an allogeneic therapy cell bank.

Methods: Expression of the markers CD90, CD105, CD44, and CD34 in SF-H, SM-H, SF-OA, SM-OA, SF-OCD and SM-OCD was verified by flow cytometry, and expression of cytokeratin, vimentin, PGP 9.5, PCNA, lysozyme, nanog, and Oct4 was verified by immunocytochemistry. MSCs were cultured and evaluated for their chondrogenic, osteogenic and adipogenic differentiation potential. Final quantification of extracellular matrix and mineralized matrix was determined using AxioVision software. A tumorigenicity test was conducted in Balb-$C^{nu/nu}$ mice to verify the safety of the MSCs from these sources.

Results: Cultured cells from SF and SM exhibited fibroblastoid morphology and the ability to adhere to plastic. The time elapsed between primary culture and the third passage was approximately 73 days for SF-H, 89 days for SF-OCD, 60 days for SF-OA, 68 days for SM-H, 57 days for SM-OCD and 54 days for SM-OA. The doubling time for SF-OCD was higher than that for other cells at the first passage ($P < 0.05$). MSCs from synovial tissues showed positive expression of the markers CD90, CD44, lysozyme, PGP 9.5, PCNA and vimentin and were able to differentiate into chondrogenic (21 days) and osteogenic (21 days) lineages, and, although poorly, into adipogenic lineages (14 days). The areas staining positive for extracellular matrix in the SF-H and SM-H groups were larger than those in the SF-OA and SM-OA groups ($P < 0.05$). The positive mineralized matrix area in the SF-H group was larger than those in all the other groups ($P < 0.05$). The studied cells exhibited no tumorigenic effects.

Conclusions: SF and SM are viable sources of equine MSCs. All sources studied provide suitable MSCs for an allogeneic therapy cell bank; nevertheless, MSCs from healthy joints may be preferable for cell banking purposes because they exhibit better chondrogenic differentiation capacity.

Keywords: Equine, Mesenchymal stem cell, Synovial fluid, Synovial membrane, Allogeneic cell bank

* Correspondence: baccarin@usp.br
[1]Department of Internal Medicine, School of Veterinary Medicine and Animal Science, University of São Paulo (USP), Avenida Prof. Orlando Marques de Paiva, 87, 05508-270 São Paulo, SP, Brazil
Full list of author information is available at the end of the article

Background

Osteoarticular diseases have received a substantial amount of scientific attention in recent years, primarily because of their high prevalence and significant impact on the equine industry. These diseases irreparably damage articular cartilage and negatively influence athletic performance in horses. Proper treatment has therefore been sought to facilitate the regeneration of hyaline cartilage and to maintain the integrity of its structure. For this purpose, the use of cellular therapies, including mesenchymal stem cells (MSCs) from various sources, is a promising tool for the treatment of osteoarticular disease.

MSCs are characterized by their proliferative ability and their capacity to differentiate into several mesenchymal lineages, such as osteoblasts, chondrocytes, adipocytes, tenocytes, and myocytes; therefore, they are classified as multipotent progenitor cells [1–3]. Regenerative medicine provides an opportunity to control the evolution of the disease due to the immunomodulatory, anti-inflammatory, and tissue regenerative properties of MSCs. In this context, the use of appropriate populations appears to be crucial for the successful regeneration of damaged articular structures [4].

Regarding horses, MSCs have been obtained from bone marrow, adipose tissue [5], umbilical cord [6–8], umbilical cord blood [9], amniotic membrane [10], peripheral blood [11], and recently from synovial fluid (SF) and synovial membrane (SM) [12, 13].

Although MSCs from synovial tissues have abilities comparable with those of MSCs from other sources, they have also been shown to possess high chondrogenic potential. Additionally, it was inferred that these cells are already predisposed to differentiate into chondrocytes, suggesting that the ancestral microenvironment directs the "destination" cell upon differentiation [14]. These observations support the hypothesis that these cells may be prime candidates for the regeneration of cartilage [15–18]. SM collection can be performed during arthroscopy [4, 19–22], and SF can easily be collected through arthrocentesis [23].

Although autologous therapy with MSCs does not result in any deleterious effects, its use in horses still has limitations, such as the inability to initiate treatment immediately after arthroscopic diagnosis because the expansion of MSCs in culture takes 15–26 days [3]. The treatment of older horses is also limited because there is an apparent decrease in the abilities of MSCs in this population. Allogeneic treatment eliminates the long timeframe that is required to isolate and expand MSCs.

Typically, SF or SM is harvested in cases of osteoarthritis (OA) or osteochondritis dissecans (OCD) during an arthroscopic procedure that is being conducted for prognostic purposes, and these samples can also be used to create a cell bank for allogeneic therapy. However, it is not currently known whether these cell sources exhibit characteristics similar to those of cells from healthy joint tissues.

Even in the case of allogeneic therapy, the harvest of synovial tissue during arthroscopic treatments of joints with OA or OCD may not the best choice, as SF or SM could instead be harvested from contralateral healthy joints. However, there have been few reports of the biological characterization of equine MSCs from synovial tissues, so concomitant quantitative and qualitative assessment should therefore be encouraged to identify these synovial-derived cells and to provide additional information about them.

Based on this research scenario, we outlined a study to compare the phenotype, morphology, and multilineage differential potential of MSCs from synovial fluid (SF-MSCs) and from synovial membrane (SM-MSCs) of horses, using healthy joints, joints with OA, and joints with OCD. Further, to verify that these SF-MSCs and SM-MSCs would not differentiate into tumoral cells, we used a mouse tumorigenicity test.

Methods

Animal ethics

This study was conducted in accordance with the Ethics Committee on the Use of Animals of the School of Veterinary Medicine and Animal Science of the University of São Paulo; the protocol number was 2871/2013.

Collection of SF and SM

In this study, a total of 97 joints from 68 horses were examined. The horses ranged in age from 2 to 10 years, and no restrictions were placed on breed, sex, or joint. SF and SM of healthy joints (SF = 14; SM = 16), of joints with OCD (SF = 21; SM = 16), and of joints with OA (SF = 16; SM = 11) were collected and used in this experiment.

Horses with joint diseases were examined at the Veterinary Hospital of the School of Veterinary Medicine and Animal Science of the University of São Paulo (FMVZ/USP), and arthroscopic surgery was indicated for use as their treatment. In the surgical center, SF and SM samples were collected from joints with OCD and from joints with OA. Immediately prior to the procedure, SF samples were obtained by arthrocentesis using 40×10 hypodermic needles. SM samples were collected during surgery using conventional arthroscopic forceps.

Samples of SF and SM from healthy joints were collected at the beginning of the arthroscopic procedures in concurrent experiments.

Isolation and culture of SF-derived cells

First, 2 ml of harvested synovial fluid from healthy joints (SF-H), from joints with OCD (SF-OCD), and from joints with OA (SF-OA) was suspended in 3 ml of

Dulbecco's modified Eagle medium (DMEM; LGC Biotechnology, Cotia, São Paulo, Brasil) and 10 % (v/v) fetal bovine serum (FBS; Life Technologies, São Paulo, SP, Brasil) supplemented with 1 % penicillin/streptomycin, 1 % glutamine (200 mM), 1 % pyruvic acid, and 0.25 % amphotericin B (Life Technologies, São Paulo, SP, Brasil) and then plated in a 25 cm^2 flask at a cell density of 3.5×10^2/ml (SF-H), 1.91×10^6/ml (SF-OCD), or 1.93×10^6/ml (SF-OA). Next, the samples were allowed to attach during incubation at 37 °C in a humidified atmosphere containing 5 % CO_2. On the fourth day, the medium was aspirated to remove nonadherent cells and was replaced with fresh medium. The cell cultures were maintained for sufficient time to monitor cell growth via inverted microscopy, and fresh medium was provided every 48 hours until the cells reached 80 % confluence.

Isolation and culture of SM-derived cells

Harvested synovial membrane from healthy joints (SM-H), from joints with OCD (SM-OCD), and from joints with OA (SM-OA) were washed with phosphate-buffered saline (PBS) containing 1 % penicillin under sterile conditions to remove debris and blood. Next, approximately 400 mg of SM were gently debrided with a sterile syringe plunger and distributed in a 25 cm^2 flask with 200 μl of FBS per flask and were incubated for 20 minutes. After incubation, 5 ml of supplemented DMEM as already described was added to each culture, and the culture conditions were maintained as described for the SF samples.

Trypsinization and doubling time

Cells in all cultures were grown in monolayers under standard sterile conditions until reaching >80 % confluence and were then trypsinized. DMEM was removed, and the cells were washed with 2 ml of PBS. Thereafter, 1 ml of 0.25 % trypsin was added to each flask, and the samples were incubated at 37 °C for 5 minutes. Subsequently, 2 ml of culture medium supplemented with FBS was added to inactivate the trypsin. The cell suspension was then aspirated and transferred into a 15 ml conical tube for centrifugation at $287 \times g$ for 5 minutes to remove the trypsin. For each sample, the cell pellet was resuspended in 1 ml of supplemented DMEM, and an aliquot of 10 μl was used for cell counting in a Neubauer chamber. The remaining cells were transferred into a 75 cm^2 flask to which 9 ml of medium was added, and cells were incubated under the conditions already described (considered first passage (P1)). Calculation of the doubling time (DT) of the mesenchymal cells from SF-H, SF-OCD, and SF-OA was performed using an algorithm available online [24], accounting for cell number at P1, second passage (P2), and third passage (P3) during the exponential growth phase. The formula used by the online tool was:

$$DT = t \times \log 2 / (\log Nt / \log N0)$$

where $N0$ is the initial number of cells plated, Nt is the number of cells at the end of the incubation time, and t is the incubation time in hours. For SMs (SM-H, SM-OCD, and SM-OA), only the size of the fragment (in milligrams) was known, rather than the initial numbers of cells, so the initial cell numbers were estimated based on the days required for passages (>80 % confluence).

Immunophenotyping characterization
Flow cytometry

Using a FACSCalibur® cytometer (Becton Dickinson, San Jose, CA, USA) and Cell-Quest software (Becton Dickinson, San Jose, CA, USA), phenotypic assessment of SF-H ($n = 14$), SF-OCD ($n = 21$), SF-OA ($n = 16$), SM-H ($n = 16$), SM-OCD ($n = 16$), and SM-OA ($n = 11$) was performed analyzing 5000 cells per group at P3 for all joint conditions. Mouse anti-rat CD90-phycoerythrin (PE) (clone OX-7; BD, San Jose, CA, USA), mouse anti-horse CD44-fluorescein isothiocyanate (FITC) (clone CVS18; AbD Serotec, Oxford, UK), mouse anti-human CD105-RPE (clone SN6; AbD Serotec, Oxford, UK), and mouse anti-human CD34-FITC (clone 581; BD) antibodies were used. Anti-IgG1-PE and anti-IgG1-FITC were used as control isotypes to calibrate the cytometer. The protocols were performed according to the manufacturer's instructions.

Immunocytochemistry

At P3, samples were plated in six-well plates (TPP; Trasadingen, Switzerland), and 3 ml of supplemented DMEM was added per well. After the cells reached ≥80 % confluence, the DMEM was removed, and the plates were washed twice with 2 ml of PBS. Next, the cells were fixed in 4 % paraformaldehyde at 4 °C for 30 minutes. After fixation, the plates were washed again and then incubated with the following primary antibodies: rabbit anti-human lysozyme (Dako; Carpinteria, California, USA), rabbit anti-human PGP 9.5 (Spring Bioscience; Pleasanton, California, USA), rabbit anti-human Oct4 (Biorbyt; Berkeley, California, USA), goat anti-human nanog (clone N-17; Santa Cruz Biotechnology; Santa Cruz, California, USA), mouse anti-human vimentin (clone V9; Dako; Carpinteria, California, USA), mouse anti-human cytokeratin (clones EA-1 and AE3; Dako; Carpinteria, California, USA), and mouse anti-human proliferating cell nuclear antigen (PCNA) (clone PC10; Dako; Carpinteria, California, USA). The plates were incubated at 4 °C overnight. Super-Picture polymer was used to detect the primary antibodies, and the reactions were revealed using diaminobenzidine solution

(DAB, Sigma, St. Louis, MO, USA) and counterstained with Harris hematoxylin.

Chondrogenic, osteogenic, and adipogenic cell differentiation (SF and SM)

The chondrogenic induction was prepared from a solution containing 2×10^6 cells of pellet cultures during P3, and cells were cultured in a conical tube with 2 ml of DMEM for 48 hours. The inductive phase was initiated after the medium was changed from maintenance medium to a commercial chondrogenic inducer medium (StemPro chondrogenesis kit; GIBCO, Carlsbad, California, USA). The differentiation medium was changed every 48 hours for a course of 21 days. After this period, the pellets were washed with PBS and fixed in 4 % paraformaldehyde for 24 hours. To confirm chondrogenic differentiation, histological slides were prepared, and pellets were stained with toluidine blue, alcian blue, and hematoxylin and eosin (H & E; Sigma-Aldrich Corp., St. Louis, MO, USA).

For osteogenic and adipogenic differentiation, cells at P3 were placed in plastic six-well plates at a concentration of 10^5 cells per well. After the cells had adhered to the plastic, 2 ml of supplemented DMEM was added per well for a period of 48 hours. Next, the DMEM was replaced with either the commercial osteogenic inducer medium (StemPro osteogenesis kit; GIBCO) or the commercial adipogenic inducer medium (StemPro adipogenesis kit; GIBCO), and the medium was changed every 48 hours for 21 or 14 days, respectively. After the differentiation period, the plates were washed twice with PBS. Osteogenic differentiation was confirmed by positive staining of the extracellular calcium matrix using 2 % Alizarin Red at pH 4.2 (Sigma-Aldrich Corp.). Adipogenic differentiation was confirmed by the deposition of lipid droplets in the cytoplasm using Oil Red O (Sigma-Aldrich Corp.) and staining of the cell nuclei using H & E. The analysis of control cells, which received no inducing medium, was conducted following their culture in DMEM under the same timing and staining conditions as already described.

Quantification of positive matrix area was performed using AxioVision LE64 software (Carl Zeiss, Oberkochen, Germany). The program analyzed 10 photographs of each plate (magnification = 10×) and calculated the area of positive matrix in square micrometers (μm^2) for osteogenic differentiation and square centimeters (cm^2)

Table 1 Semiquantitative scoring system used in the evaluation of adipogenic differentiation

Score	% of differentiated cells	Size and arrangement of lipid droplets
0	0–5	No droplets
1	>5–50	Isolated and small
2	>50–80	Medium sized
3	>80–100	Predominantly large

for chondrogenic differentiation. The intensity of adipogenic differentiation was assessed using a scoring system based on Oil Red O staining (Table 1) [25].

Tumorigenicity test

A tumorigenicity test was performed using nine immunosuppressed Balb-C$^{nu/nu}$ female mice, each approximately 6 months old and weighing 19–28 g.

Cells that were cultured from SF and SM taken from healthy and diseased joints (OCD and OA) were grown in culture to P3 and then injected into dorsal subcutaneous tissue in mice at a density of 10^6 cells per animal. After 3 months, the mice were sacrificed by intraperitoneal administration of xylazine, followed by a 10-minute-long exposure to CO_2. Necropsy was performed, and lung, kidney, liver, subcutaneous tissue, and spleen samples were collected, weighed, fixed in 10 % formalin, and sent for histological analysis. The samples were processed by a paraffin inclusion technique and were stained with H & E.

Statistical analysis

The data were analyzed using GraphPad Prism 6 (GraphPad Software Inc., San Diego, CA, USA). Significant differences between groups were determined using one-way analysis of variance (ANOVA) followed by Dunnett's test. All data are expressed as the mean ± standard deviation (SD), and the level of significance was set at $P < 0.05$.

Results

Cell culture and doubling time

MSCs that were cultured from SF exhibited the capacity to adhere to plastic after 4–7 days in culture. Meanwhile, MSCs that were derived from SM adhered to the flasks after 15 days of culture. Both populations had monolayer growth profiles, morphologically resembled fibroblasts (Fig. 1), and maintained this appearance after long-term culture (data not shown).

The doubling times for SF-H, SF-OCD, and SF-OA were, respectively, 334 ± 64, 585 ± 73, and 333 ± 70 hours at P1; 144 ± 24, 162 ± 23, and 134 ± 20 hours at P2; and 108 ± 12, 144 ± 13, and 98 ± 8 hours at P3. At P1, one-way ANOVA revealed a significant difference in doubling time, and the Tukey–Kramer test indicated a significant increase in the doubling time of SF-OCD compared with the SF-H and SF-OA ($P < 0.05$). However, there were no evident differences at P2 or P3 (Fig. 2).

The timing to reach 80 % confluence during primary culture varied among the SM samples: 45 days for SM-H, 38 days for SM-OCD, and 35 days for SM-OA. The doubling time of SF and the days for passage of SM could not be compared because the methods for analysis differed between these conditions. After P1, following the trypsinization protocol, 80 % confluence was

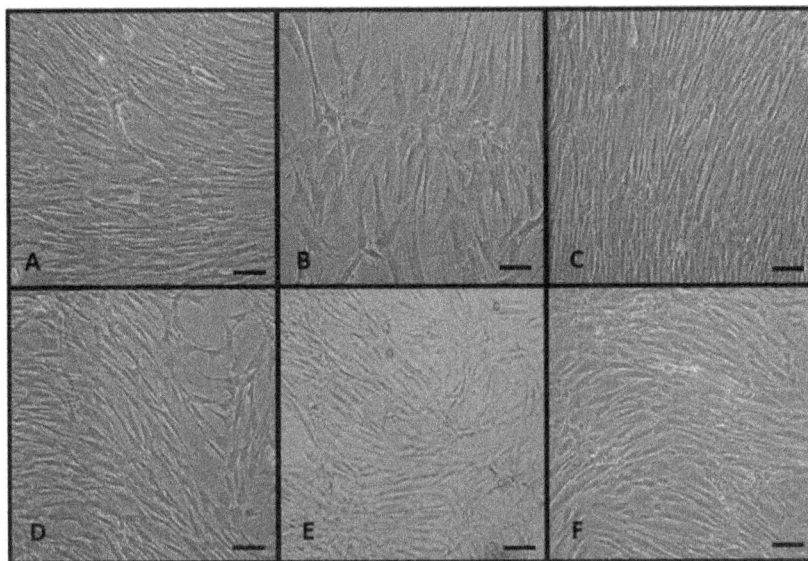

Fig. 1 MSCs from synovial tissues during cell culture (P3) showing ≥80 % confluence. SF-H **a**, SF-OCD **b**, SF-OA **c**, SM-H **d**, SM-OCD **e**, and SM-OA **f**. 100× magnification

achieved at an average of 11 days for both groups (SF and SM).

The time that elapsed between primary culture and P3, when phenotypic characterization and cell differentiation were performed, was approximately 73 days for SF-H, 89 days for SF-OCD, 60 days for SF-OA, 68 days for SM-H, 57 days for SM-OCD, and 54 days for SM-OA.

Phenotypic characterization
Flow cytometry
Flow cytometric analysis at P3 indicated that the cells that were cultured from SF and SM exhibited phenotypic characteristics that were consistent with those of MSCs, and there were no cells of hematopoietic origin because the cells

exhibited positive expression of the markers CD90 and CD44 and no expression of the markers CD105 and CD34.

Table 2 presents the average (SD) percentages for each of the markers from the different sources of MSCs. Significantly higher proportions of double-positive cells

Table 2 Average percentages of MSCs from SF and SM that exhibited positive or negative expression of CD90, CD44, CD105, and CD34 markers by flow cytometry

Source	Expression			
	CD90$^+$CD44$^-$	CD90$^-$CD44$^+$	CD90$^-$CD44$^-$	CD44$^+$CD90$^+$
SF-H	64.9 ± 23.8^a	1.18 ± 1.4^a	27.3 ± 22.3	6.65 ± 8.86^a
SF-OCD	48.3 ± 26.3^b	3.98 ± 6^b	31.2 ± 24.7	16.5 ± 16.8^b
SF-OA	48.1 ± 23^b	14.2 ± 25.7^c	26.5 ± 19.2	11.2 ± 10.8^b
SM-H	66.6 ± 30.1^a	1.49 ± 3.2^a	29.7 ± 27.6	2.2 ± 2.96^a
SM-OCD	40.2 ± 27.2^b	2.17 ± 2.6^a	48.7 ± 30.1	8.9 ± 18.5^{a,b}
SM-OA	40.3 ± 22.1^b	8.56 ± 9.2^b	39.9 ± 15.4	11.2 ± 10^b
	CD105$^+$CD34$^-$	CD105$^-$CD34$^+$	CD105$^-$CD34$^-$	CD105$^+$CD34$^+$
SF-H	0.25 ± 0.49	0.02 ± 0.04	99.7 ± 0.50	0.03 ± 0.04
SF-OCD	0.22 ± 0.48	0.09 ± 0.29	98.7 ± 3.07	0.96 ± 2.85
SF-OA	0.25 ± 0.50	0.11 ± 0.37	99.3 ± 1.36	0.37 ± 0.82
SM-H	0.28 ± 0.76	0.02 ± 0.03	99.7 ± 0.79	0.02 ± 0.03
SM-OCD	0.36 ± 0.83	0.10 ± 0.27	99.4 ± 1.02	0.17 ± 0.28
SM-OA	0.36 ± 0.54	0.04 ± 0.12	99.4 ± 0.56	0.11 ± 0.14

Data presented as mean ± standard deviation. Different superscript letters in the same column denote statistically significant differences (P <0.05)
SF-H healthy synovial fluid, SF-OA osteoarthritis synovial fluid, SF-OCD osteochondritis dissecans synovial fluid, SM-H healthy synovial membrane, SM-OA osteoarthritis synovial membrane, SM-OCD osteochondritis dissecans synovial membrane

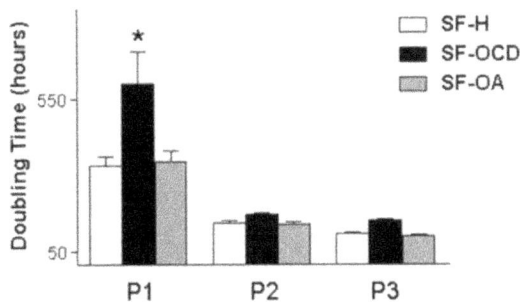

Fig. 2 Graph showing the DT (mean ± SD) from SFs (SF-H, SF-OCD, and SF-OA) during P1, P2, and P3. *P <0.05. P1 first passage, P2 second passage, P3 third passage, SF-H synovial fluid from healthy joints, SF-OA synovial fluid from joints with osteoarthritis, SF-OCD synovial fluid from joints with osteochondritis dissecans

(CD44$^+$CD90$^+$) were observed for SF-OCD (16.5 ± 16.8) and SF-OA (11.2 ± 10.8) than for SF-H (6.65 ± 8.86) (*P* <0.05). For SM groups, the same increase was observed but for SM-OA (11.2 ± 10) compared with SM-OCD (8.9 ± 18.5) and SM-H (2.2 ± 2.96). Figure 3 shows a dot plot of the population chosen (gated cell population) and overlaid histogram analysis representatives from different synovial source.

Immunocytochemistry

Immunocytochemistry analysis confirmed positive immunostaining for lysozyme, PGP 9.5, PCNA, and vimentin in the cells from the SF and SM groups, which ensured the existence of type A and B synoviocytes, intense cell proliferation, and the presence of MSCs, respectively (Fig. 4). The absence of pluripotent cells and fibroblasts was confirmed by negative immunostaining for Oct4, cytokeratin, and nanog.

Differentiation potential

Chondrogenic differentiation was observed after 21 days in MSCs from both of the groups (SF and SM), at which point the formation of spherical pellets of hardened appearance was observed. Chondrogenic potential was confirmed by alcian blue, toluidine blue (Fig. 5), and H & E staining, which enabled the identification of an extracellular matrix rich in proteoglycans and of cells such as chondrocytes by optical microscopy.

Differences in staining were noticeable in both groups (SF and SM). Slides containing healthy SF material exhibited much more evident staining and better morphology than slides containing OCD material, which in

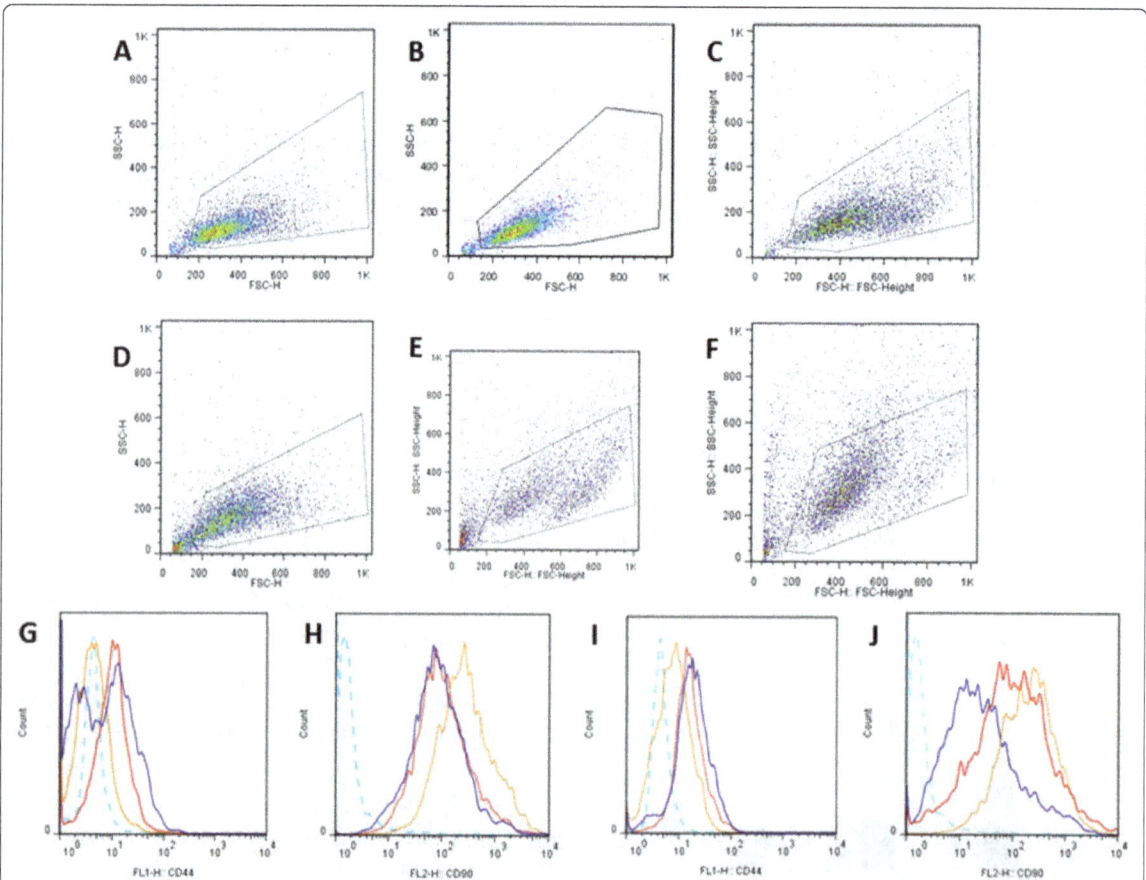

Fig. 3 Flow cytometric analysis of the expression of cell surface markers CD90 and CD44 by MSCs. Representative dot plots and histograms (overlaid) of six different synovial sources: SF-H **a**, SF-OCD **b**, SF-OA **c**, SM-H **d**, SM-OCD **e**, and SM-OA **f** analyzed during the third passage; overlaid CD44$^+$ cells from SF-H, SF-OCD, and SF-OA **g**; overlaid CD90$^+$ cells from SF-H, SF-OCD, and SF-OA **h**; overlaid CD44$^+$ cells from SM-H, SM-OCD, and SM-OA **i**; and overlaid CD90$^+$ cells from SM-H, SM-OCD, and SM-OA **j** (*orange line*, SF-H and SM-H; *red line*, SF-OCD and SM-OCD; *blue line*, SF-OA and SM-OA). Isotype control antibodies were used (*blue dotted line*). All sources showed significant expression of mesenchymal markers (CD90 and CD44). (Color figure online). SSC: side scatter; FSC: forward scatter

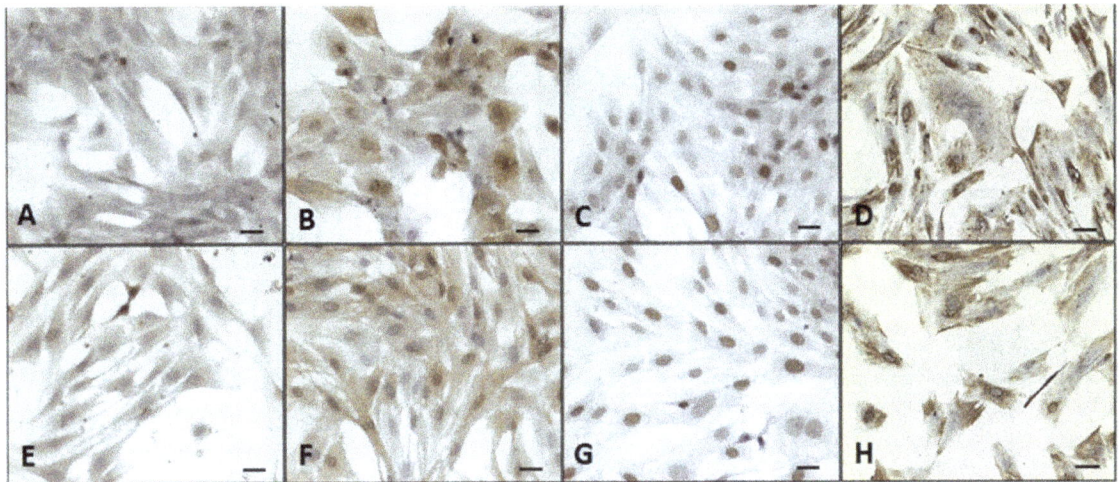

Fig. 4 Representative images of lysozyme, PGP 9.5, PCNA, and vimentin expression in synovial tissues using immunocytochemistry. Immunocytochemistry of cells from equine SF **a–d** and SM **e–h** during their third passage to evaluate the positive expression of the cell surface markers lysozyme **a**, **e**, PGP 9.5 **b**, **f**, PCNA **c**, **g**, and vimentin **d**, **h**. 200× magnification

turn exhibited more evident staining than slides containing OA material.

The SF-H (149 ± 103 cm^2) and SM-H (78 ± 7 cm^2) groups showed larger average areas of positive staining for extracellular matrix than did the SF-OA (32 ± 22 cm^2) and SM-OA (43 ± 20 cm^2) groups ($P <0.05$), respectively, but the SF-OCD (49 ± 41 cm^2) and SM-OCD (32 ± 20 cm^2) groups appeared similar to the SF-H and SM-H groups (Fig. 5).

In both groups (SF and SM), osteogenic differentiation occurred after 21 days of induction. Osteogenic differentiation potential was confirmed by positive staining of the calcium matrix by Alizarin Red (Fig. 5). The cells in control culture did not form a calcium matrix, as certified by negative staining with Alizarin Red.

The average and standard deviation of mineralized matrix areas for SF-H ($1,105,447 \pm 1,415,829$ μm^2) were larger than for SF-OA ($83,765 \pm 48,589$ μm^2), SF-OCD ($295,566 \pm 120,472$ μm^2), SM-H ($166,783 \pm 193,938$ μm^2), SM-OA ($141,648 \pm 123,734$ μm^2), and SM-OCD ($265,098 \pm 174,578$ μm^2) (Fig. 5).

Few cell colonies underwent adipogenic differentiation by 14 days after induction. This ability appeared to be limited for this lineage, and the results were similar among the groups. Adipogenic differentiation was visualized at small isolated points at the edges of the plates, but cell death was observed after the induction of differentiation (i.e., the cells detached from the plates). Adipogenic differentiation potential was confirmed after observation of a morphology change from fusiform to polygon and by the deposition of lipid droplets in the cytoplasm, which were stained by Oil Red O (Fig. 5), and each group reached a score of 1 (showing <20 % positive cells). The control population did not undergo the morphological change and exhibited negative staining.

Tumorigenic potential

None of the mice that received subcutaneous injections of MSCs from synovial tissues exhibited any changes in behavior, appetite, body temperature, or local inoculation temperature throughout the experimental period.

A necropsy evaluation revealed no macroscopic changes; organs and tissue samples were collected (liver, lung, kidney, spleen, and subcutaneous tissue) and were sent for histological analysis. No changes in tissue characteristics or cell morphology were observed (Fig. 6). These results demonstrated that MSCs from synovial tissues are unable to induce tumor formation and indicate the safety of their clinical applicability.

Discussion

In this study, cells from synovial tissues were evaluated as possible sources of MSCs, and their suitability was assessed based on their expression of surface markers and their ability to differentiate into various mesenchymal lineages. Subsequently, cells that were cultured from healthy joints, joints with OA, and joints with OCD were compared to identify the best candidate for future clinical applications.

De Bari et al. [4] were the first to isolate MSCs from synovial tissues. They harvested SM from human patients with degenerative joint disease and used collagenase to extract cells from tissues. Our study was similar with respect to the aseptic harvesting of cells from diseased (OCD and OA) and healthy joints during surgery. We observed cell growth in monolayers in all of the

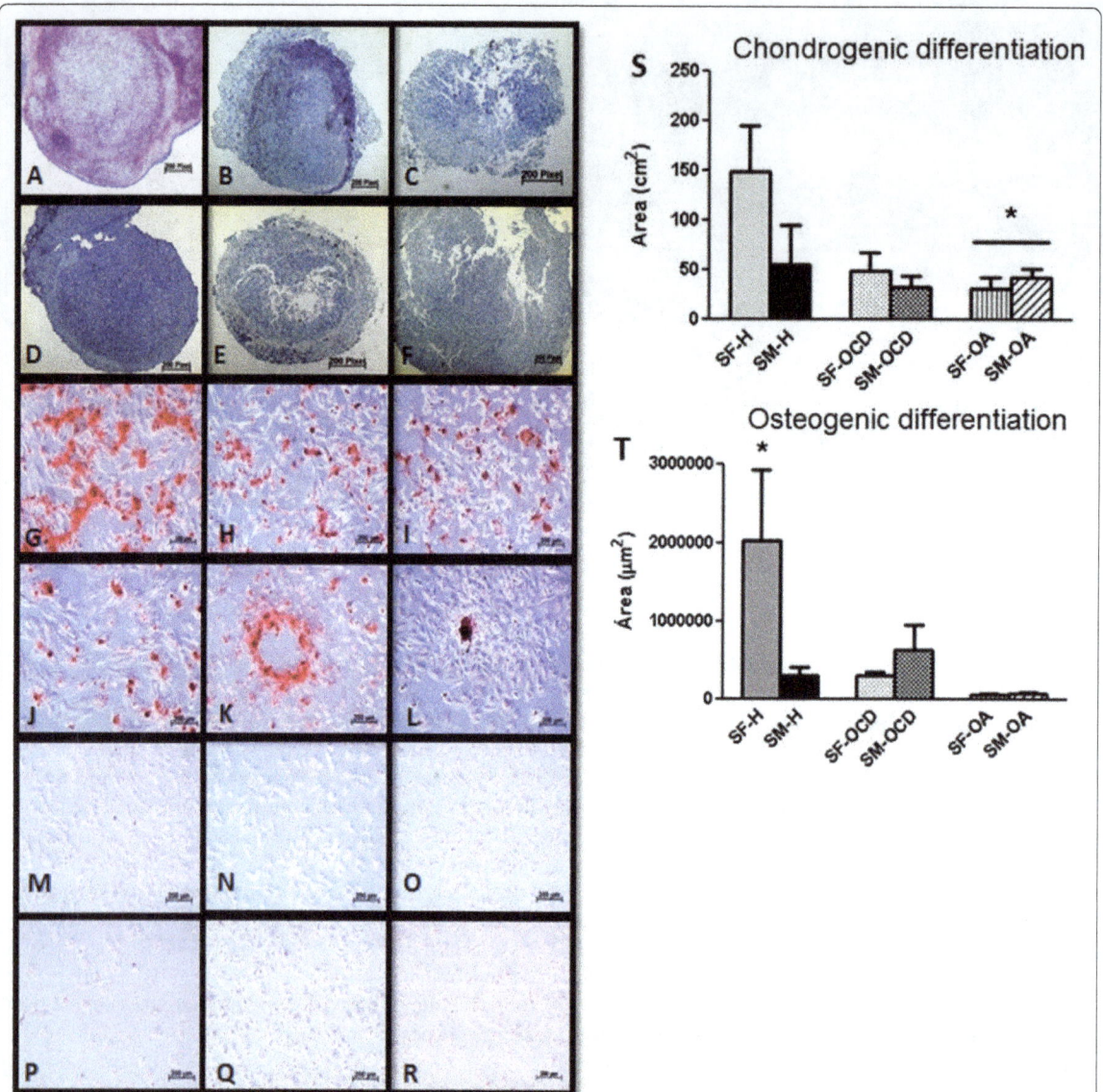

Fig. 5 Differentiation potential of MSCs from equine SF-H, SF-OCD, SF-OA, SM-H, SM-OCD, and SM-OA. Differentiation of MSCs from SF and SM into mesenchymal lineages during P3. MSCs after chondrogenic differentiation stained with toluidine blue showing hyaline matrix (*blue*): SF-H **a**, SF-OCD **b**, SF-OA **c**, SM-H **d**, SM-OCD **e**, and SM-OA **f** (400× magnification). MSCs stained with Alizarin Red showing matrix calcium formation, confirming the osteogenic lineage (*red*): SF-H **g**, SF-OCD **h**, SF-OA **i**, SM-H **j**, SM-OCD **k**, and SM-OA **l**. MSCs showing intracytoplasmic lipid droplets confirming the adipogenic lineage: SF-H **m**, SF-OCD **n**, SF-OA **o**, SM-H **p**, SM-OCD **q**, and SM-OA **r** (1000×). Chondrogenic differentiation area **s**. Osteogenic differentiation area **t**. Quantification of positive matrix area was performed using AxioVision LE64 software (Carl Zeiss). The program analyzed 10 photographs of each plate (magnification = 10×) and calculated the area of positive matrix when blue and red stains for chondrogenic and osteogenic differentiation, respectively, were observed. *P *0.05. *SF-H* synovial fluid from healthy joints, *SF-OA* synovial fluid from joints with osteoarthritis, *SF-OCD* synovial fluid from joints with osteochondritis dissecans, *SM-H* synovial membrane from healthy joints, *SM-OA* synovial membrane from joints with osteoarthritis, *SM-OCD* synovial membrane from joints with osteochondritis dissecans (Color figure online)AU Query: Confirm Fig 5 caption after editing to style OK

samples from SM, although enzymatic digestion with collagenase was not performed.

In our experiment, the first passage of MSCs from SM took approximately 15 days, and another 45 days of incubation time was require for the cells to reach 80 % confluence. This delay could be associated with the fact that collagenase was not used—as collagenase has been used in all related studies, in which cell expansion and cell confluence were observed by the second week [26, 27].

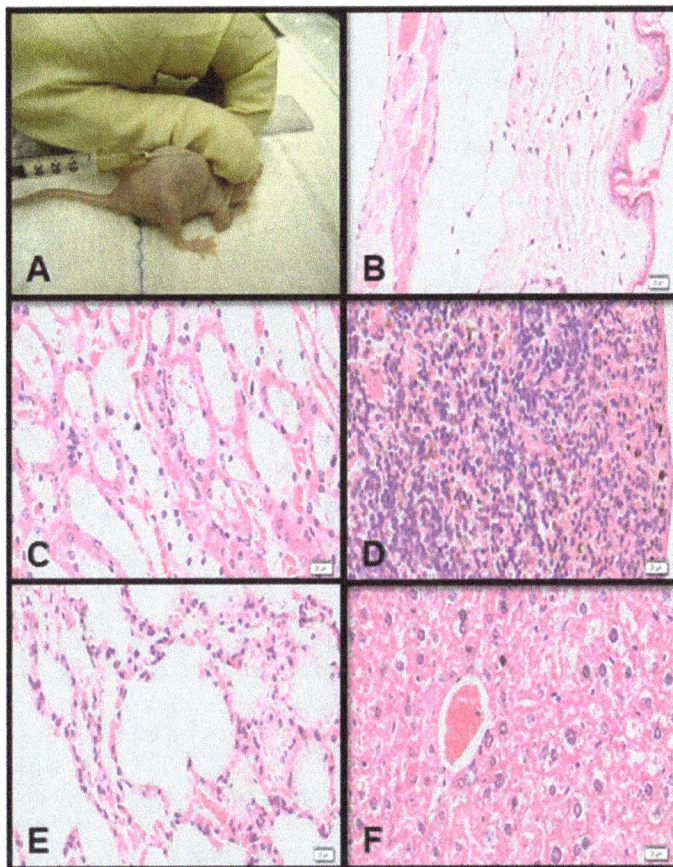

Fig. 6 Histologic evaluation of tumorigenicity test after subcutaneous injection of P3 cells into Balb-C$^{nu/nu}$ mice. MSC inoculation **a**, and subcutaneous tissue **b**, kidney **c**, spleen **d**, lung **e** and liver **f** tissue. The results of tumorigenicity tests in mice showed no compromise of any internal organ, assuring the applicability of the studied cells. 1000× magnification

All of the MSCs evaluated in this study required a long period of time from primary culture until P3. Typically, equine MSCs are derived from other sources, such as bone marrow and adipose tissue, and require 15–25 days to reach P3 [4]. A relevant result observed was an increased time to reach confluence for the SF-OCD group at P1 in relation to other synovial sources, but this increased time was not observed for later passages.

All of the cultures from synovial tissues exhibited plastic adherence capacity, monolayer growth, and fibroblastoid morphology. The processing was easier for SF cells than for SM cells. These results are consistent with findings from other studies [17, 26, 28, 29].

The criteria for characterization of MSCs from horses are based on a marker panel [30] and include several of the criteria that are used to characterize human MSCs, as determined by the International Society for Cellular Therapy [31]. The current study examined cells for the expression of the surface markers CD90, CD44, CD34,

CD105, Oct4, nanog, vimentin, and cytokeratin to establish phenotypic expression patterns of SF and SM cells. Furthermore, the choice of the markers was based on previous studies on equine MSCs from other sources that used several of these markers [4, 7, 32, 33].

MSCs from both SF and SM sources exhibited positive expression for the marker CD90. Higher proportions of cells showing positive expression for CD90 were found in SF-H (64.9 %) and SM-H (66.6 %) (P <0.05), suggesting that cells from healthy SF or SM are better sources of MSCs than are cells from SF-OCD (48.3 %), SF-OA (48.1 %), SM-OCD (40.2 %), or SM-OA (40.3 %).

A surprisingly low expression rate of the marker CD44 was observed. Similar results have been observed by Ranera et al. [7], who demonstrated an absence of CD44 expression on cells from bone marrow and adipose tissue. In the present study, it is important to note that double staining was performed and that competition for epitopes could have occurred, potentially showing greater expression of CD90 and lower expression of CD44. Nevertheless, all of the cell populations exhibited positive expression of CD44, which was more strongly expressed in the OCD and OA groups than in the groups from healthy joints (P <0.05).

Interestingly, cells derived from OA exhibited greater proliferation potential and higher proportions of CD44$^+$CD90$^+$ double-positive cell expression than cells derived from healthy tissue. According to Kobayashi et al. [34], the identification of multipotent CD44$^+$/CD90$^+$ stem cells with high proliferative potential suggested that these cells could provide a basis for continuously self-renewing cartilage.

Many markers have been frequently tested for use in horses because of a lack of specific equine antibodies and the low reactivity levels of markers from other species against equine proteins [30]. Therefore, the lack of CD105 expression observed in the present study may be related to a low specificity to equine CD105. Additionally, De Schauwer et al. [30] observed large variations in CD105 expression (0.1–20 %) among umbilical cord blood samples, and another study on umbilical cord tissue reported negative expression of CD105 [35], possibly because the antibody that was used did not work for the cells that were studied.

In the present study, flow cytometry analysis of the marker CD34 was performed to investigate the presence of hematopoietic cells in SF and SM cultures. Expression of CD34 was not found in any of the analyses (0.02 % SF-H, 0.09 % SF-OCD, 0.11 % SF-OA, 0.02 % SM-H, 0.10 % SM-OCD, and 0.04 % SM-OA). These results are similar to findings from several previously conducted studies on human cells in which MSCs from SF and SM have exhibited either low or no expression of CD34 [36–39]. Only one study has been published that used equine MSCs from SF, and it indicated that there was no expression of CD34 in these cells [12].

Kitamura et al. [40] reported positive immunostaining of the marker PGP 9.5 in type B synoviocytes, clearly documenting their distribution in SM. This specificity in identifying type B synoviocytes was allowed by the comparison of immunoreactions in nerve fibers that are distributed throughout the SM with results obtained from human brain and horse samples. Type A synoviocytes are similar to macrophages and are present in SF [41].

The immunocytochemistry data showed conservation of type A and B synoviocytes in cell cultures from SF and SM from healthy and diseased joints (OCD and OA) by positive immunostaining for lysozyme and PGP 9.5. Type A and B synoviocytes were also observed by Sakaguchi et al. [36] during culture of cells isolated from synovium; their data suggested that nucleated cells after isolation or digestion may lose their original profiles and acquire MSC profiles during the expansion of the MSCs.

Immunocytochemistry analysis mainly showed positive expression of vimentin and PCNA and an absence of expression of cytokeratin. These results suggest mesenchymal origin and proliferative abilities for all of the samples.

Additionally, there was an absence of Oct4 expression in our study. These results were similar to those found in a study by De Vita et al. [42] that evaluated horse amniotic fluid; however, they conflicted with results reported in a study that employed MSCs from goat amniotic fluid [43] and results from a study that employed MSCs from horse bone marrow, umbilical cord matrix, and amniotic fluid [34]. Our results suggest that the studied cells lack pluripotency. MSCs from intervascular and perivascular umbilical cord matrix in horses showed positive expression of Oct4 [35], as did MSCs from human bone marrow and amniotic membrane [10]. Gao et al. [44] demonstrated positive expression of Oct4 and nanog in MSCs from human adipose tissue by immunocytochemistry and reverse transcription PCR.

MSCs are characterized by an extensive proliferative ability, as well as by the ability to differentiate in vitro into various mesenchymal lineages in response to appropriate stimuli [4]. In this study, both types of synovial tissues (SF and SM) were able to differentiate into osteoblasts and chondrocytes.

Slides containing healthy SF material exhibited much more evident staining and better morphology than slides containing OCD and OA materials, but the average areas showing positive staining for extracellular matrix in the SF-H and SM-H groups were only higher than those of the SF-OA and SM-OA groups. These data were consistent with the flow cytometric findings showing higher positive expression of CD90 in SF and SM from healthy joints compared with tissues from OCD or OA joints. These results suggest that healthy equine SF and SM represent an attractive source of MSCs for therapeutic use in osteoarticular diseases.

Currently, several sources have been reported for obtaining MSCs; however, few studies have compared synovial tissue with other type of sources. Yoshimura et al. [45] compared the performance, proliferation capacity, and chondrogenic differentiation potential of MSCs derived from bone marrow, SM, periosteum, fat, and muscle. These authors demonstrated a higher chondrogenic potential in MSCs from SM due to the increased production of cartilage matrix in relation to other sources. Similarly, Mochizuki et al. [14] compared MSCs from SM and adipose tissue (self-renewal and differentiation capacity), which also showed a better chondrogenic capacity of MSCs derived from SM. It was inferred that these cells are already predisposed to differentiate into chondrocytes, suggesting that the ancestral microenvironment directs the "destination" cell upon differentiation.

The cells cultured from SF and SM showed little ability in adipogenic differentiation, as demonstrated by their limited capacity to produce intracytoplasmic lipid droplets, in addition to undergoing cell death during the process of induction. Nevertheless, there were changes in cell morphology which suggested that they have the capacity for differentiation into adipocytes.

Regardless of the source of MSCs, the process of adipogenic differentiation in equine cells appears to be dependent on components found in rabbit serum [25]. Koch et al. [9] and Giovannini et al. [46] found that adipogenic differentiation in equine MSCs which were derived from umbilical cord blood or peripheral blood was limited when they were induced in standard culture medium but that the differentiation occurred when the medium was supplemented with rabbit serum. These observations may explain the low differentiation rate found in the current study, as the adipogenic differentiation was induced using a standard commercial environment that was not supplemented with rabbit serum.

Tumorigenic analysis showed that MSCs from equine SF and SM were not able to induce tumors in nude mice, which was confirmed during necropsy and in histological slides. This is the first study to demonstrate that MSCs from equine SF and SM are safe, as they were unable to cause tumor growth in a laboratory animal model. Previous studies that have shown the safety and immunomodulatory characteristics of MSCs from other sources were based on experiments that were conducted in vitro and that focused on the suppression of B-lymphocyte activity [47–50]. The negative results obtained in the tumorigenicity test conducted in this study further encourage safety of studies in equines.

Regarding the clinical routine, SF is easier to obtain than SM; furthermore, SF exhibited negative tumorigenicity and favorable results during phenotypic characterization and cell differentiation. Based on these results, we consider that the clinical applicability of allogeneic MSCs from healthy SF must be tested in healthy horses. However, other authors have observed that allogeneic MSCs are capable of eliciting antibody responses in vivo (intradermal injection). It has been suggested that such responses could limit the effectiveness of repeated use of allogeneic MSCs in a single horse and could also result in harmful inflammatory responses in recipients [51]. Further studies are necessary to analyze whether allogeneic MSCs injected into equine joint can also be immunogenic.

Conclusions

SF and SM from healthy or diseased joints (OA and OCD) are feasible sources for harvesting equine MSCs based on results confirming the phenotypic and multi-potentiality characteristics of these cells. All sources studied provide suitable MSCs for an allogeneic therapy cell bank; nevertheless, MSCs from healthy joints may be preferable for cell banking purposes because they exhibit better chondrogenic differentiation capacity than MSCs derived from diseased joints.

The tumorigenicity test showed that MSCs from SF and SM can be used in clinical trials because they lack the potential to form teratomas.

Abbreviations

ANOVA: Analysis of variance; DMEM: Dulbecco's modified Eagle medium; DT: Doubling time; FBS: Fetal bovine serum; FITC: Fluorescein isothiocyanate; H % E: Hematoxylin and eosin; MSC: Mesenchymal stem cell; OA: Osteoarthritis; OCD: Osteochondritis dissecans; P1: First passage; P2: Second passage; P3: Third passage; PBS: Phosphate-buffered saline; PCNA: Proliferating cell nuclear antigen; PE: Phycoerythrin; SD: Standard deviation; SF: Synovial fluid; SF-H: Synovial fluid from healthy joints; SF-MSC: Mesenchymal stem cell from synovial fluid; SF-OA: Synovial fluid from joints with osteoarthritis; SF-OCD: Synovial fluid from joints with osteochondritis dissecans; SM: Synovial membrane; SM-H: Synovial membrane from healthy joints; SM-MSC: Mesenchymal stem cell from synovial membrane; SM-OA: Synovial membrane from joints with osteoarthritis; SM-OCD: Synovial membrane from joints with osteochondritis dissecans.

Competing interests

The authors declare that they have no competing interests.

Authors' contributions

JF, DAM, and RYAB designed the study and wrote the manuscript. RYAB and LCLCdS were responsible for obtaining funds. LCLCdS performed all of the arthroscopies and revised the manuscript. JF and RYAB collected tissue samples and conducted the experimental analysis. JF, COM, and FA performed the phenotypic characterization of MSCs and participated in the writing and revision of the manuscript. JF, DAM, and FA performed the histological analysis of mice tissue samples. All authors read and approved the final manuscript.

Acknowledgements

This research was supported by Fundação Coordenação de Aperfeiçoamento de Pessoal de Nível Superior (CAPES), Brasília, SP, Brazil, and by Fundação de Amparo à Pesquisa do Estado de São Paulo (FAPESP), São Paulo, SP, Brazil. These sponsors did not have any influence on the study design, on the collection, analysis and interpretation of data, or on the writing of the manuscript and decision to submit for publication.

Author details

[1]Department of Internal Medicine, School of Veterinary Medicine and Animal Science, University of São Paulo (USP), Avenida Prof. Orlando Marques de Paiva, 87, 05508-270 São Paulo, SP, Brazil. [2]Laboratory of Biochemistry and Biophysics, Butantan Institute, Avenida Vital Brasil 1500, São Paulo 05503-900SP, Brazil. [3]Department of Surgery, School of Veterinary Medicine and Animal Science, University of São Paulo (USP), Avenida Prof. Orlando Marques de Paiva, 87, SP 05508-270SP, Brazil. [4]Department of Pathology, School of Veterinary Medicine and Animal Science, University of São Paulo (USP), Avenida Prof. Orlando Marques de Paiva, 87, São Paulo 05508-270SP, Brazil.

References

1. Pittenger MF, Mackay AM, Beck SC, Jaiswal RK, Douglas R, Mosca JD, et al. Multilineage potential of adult human mesenchymal stem cells. Science. 1999;284:143–7.
2. Rozemuller H, Prins HJ, Naaijkens B, Staal J, Bühring HJ, Martens AC. Prospective isolation of mesenchymal stem cells from multiple mammalian species using cross-reacting anti-human monoclonal antibodies. Stem Cells Dev. 2010;19:1911–21.
3. Barberini DJ, Freitas NPP, Magnoni MS, Maia L, Listoni AJ, Heckler MC, et al. Equine mesenchymal stem cells from bone marrow, adipose tissue and

umbilical cord: immunophenotypic characterization and differentiation potential. Stem Cell Res Ther. 2014;5:25.

4. De Bari C, Dell'Accio F, Tylzanowski P, Luyten FP. Multipotent mesenchymal stem cells from adult human synovial membrane. Arthritis Rheum. 2001;44: 1928–42.

5. De Mattos AC, Alves ALG, Golim MA, Moroz A, Hussni CA, De Oliveira PGG, et al. Isolation and immunophenotypic characterization of mesenchymal stem cells derived from equine species adipose tissue. Vet Immunol Immunopathol. 2009;132:303–6.

6. Toupadakis CA, Wong A, Genetos DC, Cheung WK, Borjesson DL, Ferraro GL, et al. Comparison of the osteogenic potential of equine mesenchymal stem cells from bone marrow, adipose tissue, umbilical cord blood, and umbilical cord tissue. Am J Vet Res. 2010;71:1237–45.

7. Ranera B, Lyahyai J, Romero A, Vázquez FJ, Remacha AR, Bernal ML, et al. Immunophenotype and gene expression profiles of cell surface markers of mesenchymal stem cells derived from equine bone marrow and adipose tissue. Vet Immunol Immunopathol. 2011;144:147–54.

8. Carrade DD, Lame MW, Kent MS, Clark KC, Walker NJ, Borjesson DL. Comparative analysis of the immunomodulatory properties of equine adult-derived mesenchymal stem cells. Cell Med. 2012;4:1–11.

9. Koch TG, Heerkens T, Thomsen PD, Betts DH. Isolation of mesenchymal stem cells from equine umbilical cord blood. BMC Biotechnol. 2007;7:26.

10. Lange-Consiglio A, Corradetti B, Meucci A, Perego R, Bizzaro D, Cremonesi F. Characteristics of equine mesenchymal stem cells derived from amnion and bone marrow: in vitro proliferative and multilineage potential assessment. Equine Vet J. 2013;45:737–44.

11. Koerner J, Nesic D, Romero JD, Brehn W, Mainil-Varlet P, Grogan SP. Equine peripheral blood-derived progenitors in comparison to bone marrow-derived mesenchymal stem cells. Stem Cells. 2006;24:1613–9.

12. Murata D, Miyakoshi D, Hatazoe T, Miura N, Tokunaga S, Fujiki M, et al. Multipotency of equine mesenchymal stem cells derived from synovial fluid. Vet J. 2014;202:53–61.

13. Prado AAF, Favaron PO, Silva LCLC, Baccarin RYA, Miglino MA, Maria DA. Characterization of mesenchymal stem cells derived from the equine synovial fluid and membrane. BMC Vet Res. 2015;11:281.

14. Mochizuki T, Muneta T, Sakaguchi Y, Nimura A, Yokoyama A, Koga H, et al. Higher chondrogenic potential of fibrous synovium- and adipose synovium-derived cells compared with subcutaneous fat-derived cells: Distinguishing properties of mesenchymal stem cells in humans. Arthritis Rheum. 2006;54: 843–53.

15. Arufe MC, De La Fuente A, Fuentes-Boquete I, De Toro FJ, Blanco FJ. Differentiation of synovial CD-105+ human mesenchymal stem cells into chondrocyte-like cells through spheroid formation. J Cell Biochem. 2009; 108:145–55.

16. Jones E. Synovial mesenchymal stem cells in vivo: potential key players for joint regeneration. World J Rheumatol. 2011;1:4.

17. Krawetz RJ, Wu YE, Martin L, Rattner JB, Matyas JR, Hart DA. Synovial fluid progenitors expressing CD90+ from normal but not osteoarthritic joints undergo chondrogenic differentiation without micro-mass culture. PLoS One. 2012;7:1–10.

18. McGonagle D, Jones E. A potential role for synovial fluid mesenchymal stem cells in ligament regeneration. Rheumatology. 2008;47:1114–6.

19. Hermida-Gómez T, Fuentes-Boquete I, Gimeno-Longas MJ, Muiños-López E, Díaz-Prado S, De Toro FJ, et al. Quantification of cells expressing mesenchymal stem cell markers in healthy and osteoarthritic synovial membranes. J Rheumatol. 2011;38:339–49.

20. Lee DH, Sonn CH, Han SB, Oh Y, Lee KM, Lee SH. Synovial fluid CD34–CD44 + CD90+ mesenchymal stem cell levels are associated with the severity of primary knee osteoarthritis. Osteoarthritis Cartilage. 2012;20:106–9.

21. Lee JC, Lee SY, Min HJ, Han SA, Jang J, Lee S, et al. Synovium-derived mesenchymal stem cells encapsulated in a novel injectable gel can repair osteochondral defects in a rabbit model. Tissue Eng Part A. 2012;18: 2173–86.

22. Santhagunam A, Santos FD, Madeira C, Salgueiro JB, Cabral JMS. Isolation and ex vivo expansion of synovial mesenchymal stromal cells for cartilage repair. Cytotherapy. 2014;16:440–53.

23. Morito T, Muneta T, Hara K, Ju YJ, Mochizuki T, Makino H, et al. Synovial fluid-derived mesenchymal stem cells increase after intra-articular ligament injury in humans. Rheumatology. 2008;47:1137–43.

24. Roth V. 2006. http://www.doubling-time.com/compute.php. Accessed 3 Nov 2015.

25. Burk J, Ribitsch I, Gittel C, Juelke H, Kasper C, Staszyk C, Brehm W. Growth and differentiation characteristics of equine mesenchymal stroma cells derived from different sources. Vet J. 2013;195:98–106.

26. Koga H, Muneta T, Ju YJ, Nagase T, Nimura A, Mochizuki T, et al. Synovial stem cells are regionally specified according to local microenvironments after implantation for cartilage regeneration. Stem Cells. 2007;25:689–96.

27. Kurth T, Hedbom E, Shintani N, Sugimoto M, Chen FH, Haspl M, et al. Chondrogenic potential of human synovial mesenchymal stem cells in alginate. Osteoarthritis Cartilage. 2007;15:1178–89.

28. Jones EA, English A, Henshaw K, Kinsey SE, Markham AF, Emery P, et al. Enumeration and phenotypic characterization of synovial fluid multipotential mesenchymal progenitor cells in inflammatory and degenerative arthritis. Arthritis Rheum. 2004;50:817–27.

29. Matsukura Y, Muneta T, Tsuji K, Koga H, Sekiya I. Mesenchymal stem cells in synovial fluid increase after meniscus injury. Clin Orthop Relat Res. 2014;472: 1357–64.

30. De Schauwer C, Meyer E, Van De Walle GR, Van Soom A. Markers of stemness in equine mesenchymal stem cells: a plea for uniformity. Theriogenology. 2011;75:1431–43.

31. Dominici M, Le Blanc K, Mueller I, Slaper-Cortenbach I, Marini F, Krause D, et al. Minimal criteria for defining multipotent mesenchymal stromal cells. The International Society for Cellular Therapy position statement. Cytotherapy. 2006;8:315–17.

32. Vidal MA, Robinson SO, Lopez MJ, Paulsen DB, Borkhsenious O, Johnson JR, et al. Comparison of chondrogenic potential in equine mesenchymal stromal cells derived from adipose tissue and bone marrow. Vet Surg. 2008; 37:713–24.

33. Lovati AB, Corradetti B, Lange-Consiglio A, Recordati C, Bonacina E, Bizzaro D, et al. Comparison of equine bone marrow-, umbilical cord matrix and amniotic fluid-derived progenitor cells. Vet Res Commun. 2011;35:103–21.

34. Kobayashi S, Takebe T, Inui M, Iwai S, Kan H, Zheng YW, et al. Reconstruction of human elastic cartilage by a CD44+ CD90+ stem cell in the ear perichondrium. Proc Natl Acad Sci USA. 2011;108:14479–84.

35. Corradeti B, Lange-Consiglio A, Barucca M, Cremonesi F, Bizzaro D. Size-sieved subpopulations of mesenchymal stem cells from intervascular and perivascular equine umbilical cord matrix. Cell Prolif. 2011;44:330–42.

36. Sakaguchi Y, Sekiya I, Yagishita K, Muneta T. Comparison of human stem cells derived from various mesenchymal tissues: superiority of synovium as a cell source. Arthritis Rheum. 2005;52:2521–9.

37. Zhang S, Muneta T, Morito T, Mochizuki T, Sekiya I. Autologous synovial fluid enhances migration of mesenchymal stem cells from synovium of osteoarthritis patients in tissue culture system. J Orthop Res. 2008;26: 1413–8.

38. Sekiya I, Ojima M, Suzuki S, Yamaga M, Horie M, Koga H, et al. Human mesenchymal stem cells in synovial fluid increase in the knee with degenerated cartilage and osteoarthritis. J Orthop Res. 2012;30:943–9.

39. Sun Y, Zheng Y, Liu W, Zheng Y, Zhang Z. Synovium fragment-derived cells exhibit characteristics similar to those of dissociated multipotent cells in synovial fluid of the temporomandibular joint. PLoS One. 2014;9:e101896.

40. Kitamura HP, Yanase H, Kitamura H, Iwanaga T. Unique localization of protein gene product 9.5 in type B synoviocytes in the joints of the horse. J Histochem Cytochem. 1999;47:343–52.

41. Mapp PI, Revell PA. Ultrastructural localisation of muramidase in the human synovial membrane. Ann Rheum Dis. 1987;46:30–7.

42. De Vita B, Campo LL, Listoni AJ, Maia L, Sudano MJ, Curcio BR, et al. Isolamento, caracterização e diferenciação de células-tronco mesenquimais do líquido amniótico equino obtido em diferentes idades gestacionais. Pesq Vet Bras. 2013;33:535–42.

43. Pratheesh MD, Gade NE, Katiyar AN, Dubey PK, Sharma B, Saikumar G, et al. Isolation, culture and characterization of caprine mesenchymal stem cells derived from amniotic fluid. Res Vet Sci. 2013;94:313–9.

44. Gao S, Zhao P, Lin C, Sun Y, Wang Y, Zhou Z. Differentiation of human adipose-derived stem cells into neuron-like cells which are compatible. Tissue Eng Part A. 2014;20:1271–84.

45. Yoshimura H, Muneta T, Nimura A, Yokoyama A, Koga H, Sekiya I. Comparison of rat mesenchymal stem cells derived from bone marrow, synovium, periosteum, adipose tissue, and muscle. Cell Tissue Res. 2007;327:449–62.

46. Giovannini S, Brehm W, Mainil-Varlet P, Nesic D. Multilineage differentiation potential of equine blood-derived fibroblast-like cells. Differentiation. 2008; 76:18–129.

47. Aggarwal S, Pittenger MF. Human mesenchymal stem cells modulate allogeneic immune cell responses. Blood. 2005;105:1815–22.
48. Nauta AJ, Fibbe WE. Review in translational hematology: immunomodulatory properties of mesenchymal stromal cells. Library. 2008;110:3499–506.
49. Peroni JF, Borjesson DL. Anti-inflammatory and immunomodulatory activities of stem cells. Vet Clin North Am Equine Pract. 2011;27:351–62.
50. van Lent PL, van den Berg WB. Mesenchymal stem cell therapy in osteoarthritis: advanced tissue repair or intervention with smouldering synovial activation? Arthritis Res Ther. 2013;15:112.
51. Pezzanite LM, Fortier LA, Antczak DF, Cassano JM, Brosnahan MM, Miller D, et al. Equine allogeneic bone marrow-derived mesenchymal stromal cells elicit antibody responses in vivo. Stem Cell Res Ther. 2015;6:1–11.

12

Mesenchymal stem cells in cardiac regeneration: a detailed progress report of the last 6 years (2010–2015)

Aastha Singh[1], Abhishek Singh[1] and Dwaipayan Sen[1,2*]

Abstract

Mesenchymal stem cells have been used for cardiovascular regenerative therapy for decades. These cells have been established as one of the potential therapeutic agents, following several tests in animal models and clinical trials. In the process, various sources of mesenchymal stem cells have been identified which help in cardiac regeneration by either revitalizing the cardiac stem cells or revascularizing the arteries and veins of the heart. Although mesenchymal cell therapy has achieved considerable admiration, some challenges still remain that need to be overcome in order to establish it as a successful technique. This in-depth review is an attempt to summarize the major sources of mesenchymal stem cells involved in myocardial regeneration, the significant mechanisms involved in the process with a focus on studies (human and animal) conducted in the last 6 years and the challenges that remain to be addressed.

Keywords: Mesenchymal stem cells, Cardiac regeneration, Niche hypothesis, Cell therapy, Cell transplantation

Background

Stem cells are capable of differentiating into cells of the same type, which in turn give rise to other kinds of cells [1]. Stem cells can be classified on the basis of their origin and potential to differentiate. Based on origin, these cells are of two types: embryonic stem cells (ESCs) and non-ESCs. The non-ESCs are present in two forms: haematopoietic stem cells (HSCs) that differentiate into different blood cells and are CD34$^+$; and the less differentiated mesenchymal stem cells (MSCs). Under the second classification system, stem cells can be categorized as totipotent, pluripotent and multipotent, based on their potential to differentiate into different cell types. All stem cells have three common features, namely boundless self-renewal capacity, potential for asymmetric divisions and an irreversible differentiation process [2].

Cardiovascular diseases account for the highest mortality in the western countries of the world [3]. Unlike lower vertebrates like zebrafish [4], adult mammals do not possess the capacity for natural heart regeneration throughout their lifetime [5] and hence several therapeutic measures have been investigated for myocardial regeneration and repair. Out of these numerous approaches, the first clinical trials about a decade ago bolstered stem cell therapy as one of the potential strategies utilized in the cure of these disorders. The current research in the field of cardiac regenerative medicine thus attempts to stimulate the endogenous regenerative mechanisms via cell therapy for conditions such as myocardial infarction (MI). This is achieved by intermingling of two components: a cardiomyocyte source as the target for regeneration; and a non-myocardial tissue acting as a source for regeneration in an effective cardiac environment [5].

This review focuses on summarizing all studies concerning MSCs in terms of in-vivo and clinical observations in the last 6 years (2010–2015), following a critical evaluation of its cardiomyogenic potential as well as the clinical trials.

* Correspondence: dwaipayan.sen@vit.ac.in
[1]School of Bio Sciences and Technology, VIT University, Vellore, India
[2]Cellular and Molecular Therapeutics Laboratory, Centre for Biomaterials, Cellular and Molecular Theranostics (CBCMT), VIT University, Vellore 632014, Tamil Nadu, India

Main text

Importance of the MSC niche for cardiac regeneration

The Niche hypothesis [6] proposes the existence of an optimal microenvironment for stem cells. This concept has been pledged to explain the hierarchy of stem cells, with different degrees of differentiation capacity [2].

In 2011, Vunjak-Novakovic and Scadden [7] categorized the cellular and acellular components into key factors such as regulatory molecules (cytokines, O_2, nutrients), extracellular matrix (ECM) (structure, stiffness, immobilized and released factors), other cells (cell–cell contact, paracrine and autocrine signals) and physical factors (stretch, electrical signals). Many studies have concentrated on the hypoxic environment of the MSC niche [8]. Since oxygen tension (i.e. O_2 levels below 8–9 %) [9] can lead to cellular damage and apoptosis, hypoxia preconditioning of MSCs and pro-survival gene overexpression (e.g. *Akt* gene) can lead to reduction in hypoxia-induced cell death [10]. Hypoxia stimulation can be attained by transducing hypoxia-inducible factor *(HIF)-1α* [11] lentivirus vector into the MSCs, which increases proliferation and differentiation rates of the mesenchymal lineages. Cellular repressor of E1A-stimulated genes (*CREG*) also plays a role in activating *HIF-1α*, but not *HIF-1β*, by degrading a key protein that degrades *HIF-1α* [12]. This in turn modulates the paracrine signalling, resulting in upregulation of angiogenic factors such as vascular endothelial growth factor (*VEGF*) [13, 14], stromal cell-derived factor-1α (*SDF-1α*) [14], hepatocyte growth factor (*HGF*) [15] and *IL-6* [10]. *CREG* also leads to reduction in fibrotic tissue and cardiomyocyte proliferation [11]. MSCs have also been studied to release extracellular vesicles under hypoxic conditions, resulting in neoangiogenesis and enhanced cardiac functioning [16]. Human tissue kallikrein (*TK*) gene [17], trimetazidine (*TMZ*) [18] and midkine [19], when transduced or overexpressed in MSCs and transplanted into rat hearts, were found to provide more resistance to hypoxia-induced apoptosis, inflammatory damage and cardiac injury. Overall the MSCS promoted enhanced neovascularization and cardiac functional recovery. TK-MSCs have also been shown to exhibit enhanced *VEGF* expression and reduced *caspase-3* activity [17], while *TMZ* preconditioning of MSCs led to increased levels of the anti-apoptotic protein *Bcl-2* [20]. However, *TMZ* has been observed to induce adverse drug reactions associated with Parkinson's syndrome [21] and thus requires careful evaluation before being established as a promising therapeutic agent. *Let7b*-transfected MSCs also target the *caspase-3* expression for upregulating the pro-survival genes such as *p-ERK*, *Bcl-2* and *p-MEK* and result in improved left ventricular ejection fraction (LVEF) in the rat MI model [22].

Adult stem cells in regenerative medicine
Adult stem cells
Adult stem cells were thought to have a multipotent lineage, but recent research has highlighted their pluripotent nature, transdifferentiating into various progenies

[23]. The progenies in turn form cells of multipotent lineages, such as HSCs and MSCs [24]. HSCs are pluripotent cells that further differentiate into blood cells of lymphoid (B, T and NK cells) and myeloid (monocyte, granulocyte, megakaryocyte and erythrocyte) lineages [25]. They are therefore mainly involved in haematopoiesis and treatment of related diseases. MSCs have shown promising regenerative abilities in stimulating cardiomyocyte formation, in association with a Notch ligand, Jagged 1 [26]. MSCs along with other pluripotent stem cells have been said to be an effective tool for angiogenesis, cardiac regeneration and hence cardiac tissue revitalization [27], and they have also been established to be more effective than HSCs for treatment of MI in nude rat model [28].

Cardiac stem cells (CSCs) are multipotent in nature, and are capable of differentiating into vascular cells and cardiomyocytes [29]. These can be differentiated from hMSCs on the basis of their inability to differentiate into osteocytes and adipocytes [30]. The presence of *c-kit* marker is used as an interpretation for cardiac progenitor cells (CPCs) [31]. The cardiac regenerative capacity of CSCs was studied against that of MSCs and enhanced levels of histone acetylation at the promoter regions of the cardiac specific genes were found to be higher in CSCs than in MSCs [32]. This observation indicates that CSCs have a higher potential to differentiate into cardiomyocytes than MSCs and has further been supported by animal studies showing higher modulatory characteristics of CSCs, such as reduced scar size and vascular overload [33, 34]. Fetal cardiac MSCs (fC-MSCs) are said to be primitive stem cell types with the ability to differentiate into osteocytes, adipocytes, neuronal cells and hepatocytic cells [35]. These cells demonstrate a high degree of plasticity and have a wide spectrum of therapeutic applications. Cardiac colony-forming unit fibroblasts (CFU-Fs) are another population of cells which are pro-epicardium derived and resemble MSCs. According to a study by Williams et al. [36], combination of hCSCs and hMSCs enhance the therapeutic response by producing greater infarct size reduction post MI. Yet another study highlighted the prospect of cardiac CFU-Fs holding higher therapeutic potential than bone marrow-derived MSCs (BM-MSCs) for cardiac repair [37]. The formation of CFU-Fs has been said to be enhanced by treatment of BM-MSCs with 1,25-dihydroxy vitamin D_3 [38]. Adult stem cells tend to undergo cardiomyogenesis due to stimulation by oxytocin [39] (Fig. 1c) and paracrine factors released by human cardiac explants which leads to expression of cardiac-specific markers and differentiation of the MSCs into cardiomyocyte-like cells [40]. In a study conducted to estimate the efficacies of different stem cells, the results suggested that unrestricted somatic stem cells are more effective in providing cardiac functionality to the damaged tissue post MI than the BM-MSCs, even though their

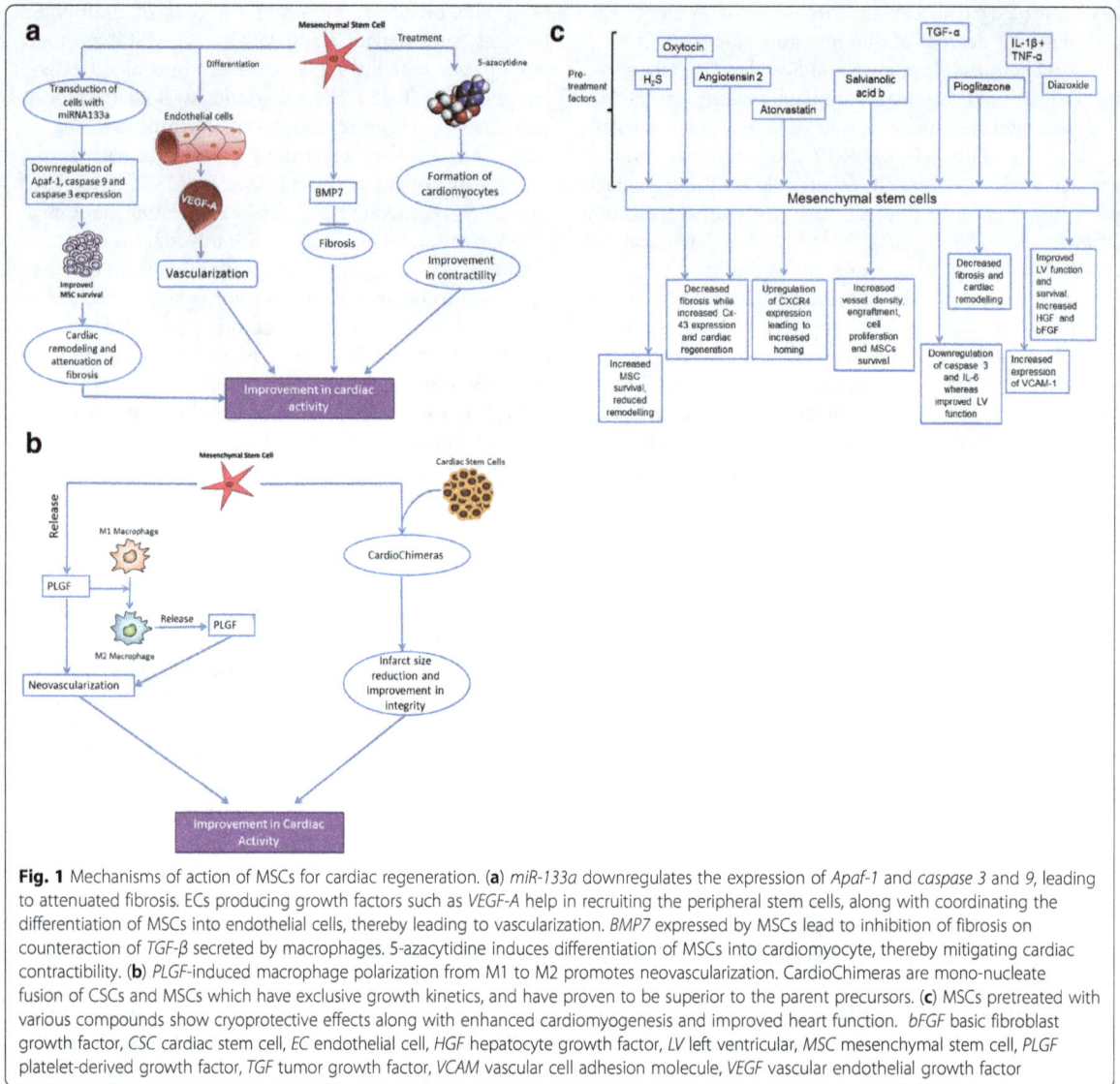

Fig. 1 Mechanisms of action of MSCs for cardiac regeneration. (**a**) *miR-133a* downregulates the expression of *Apaf-1* and *caspase 3* and *9*, leading to attenuated fibrosis. ECs producing growth factors such as *VEGF-A* help in recruiting the peripheral stem cells, along with coordinating the differentiation of MSCs into endothelial cells, thereby leading to vascularization. *BMP7* expressed by MSCs lead to inhibition of fibrosis on counteraction of *TGF-β* secreted by macrophages. 5-azacytidine induces differentiation of MSCs into cardiomyocyte, thereby mitigating cardiac contractibility. (**b**) *PLGF*-induced macrophage polarization from M1 to M2 promotes neovascularization. CardioChimeras are mono-nucleate fusion of CSCs and MSCs which have exclusive growth kinetics, and have proven to be superior to the parent precursors. (**c**) MSCs pretreated with various compounds show cryoprotective effects along with enhanced cardiomyogenesis and improved heart function. *bFGF* basic fibroblast growth factor, *CSC* cardiac stem cell, *EC* endothelial cell, *HGF* hepatocyte growth factor, *LV* left ventricular, *MSC* mesenchymal stem cell, *PLGF* platelet-derived growth factor, *TGF* tumor growth factor, *VCAM* vascular cell adhesion molecule, *VEGF* vascular endothelial growth factor

capacity to repair the damage is moderate [41]. Another interesting subfamily of the CSCs is cardiac resident stem cells (CRSCs) which can be obtained from adult human atrial appendages. These stem cells when administered with W8B2 antigen exhibit cardiogenic differentiation capacity, along with secretion of a variety of angiogenic, inflammatory, chemotaxic and cell growth and survival cytokines [42].

MSCs: a promising source of cell-based therapy
General characteristics
Mesenchymal cells, being multipotent stem cells, can differentiate into several cell types such as mesodermal lineage cells (adipocyte, osteoblast, chondrocyte) [43] and myogenic lineage [44]. This feature of the MSC makes it

an alluring therapeutic agent. According to the Tissue Stem Cell Committee of the International Society of Cellular Therapy [45], the basic criteria to categorize stem cell as MSC include following three key features:

(a) The cells must be plastic adherent under basic culture conditions.
(b) The cells should express CD73, CD90 and CD105, lacking the expression of CD11b, CD19 or CD79α, CD14, CD34, CD45 and HLA-DR surface molecules.
(c) The cells must be able to differentiate to adipocytes, chondrocytes and osteoblasts in vitro.

MSCs are said to exhibit immunomodulatory effects by virtue of their inhibitory effect towards both B-cell and T-cell proliferation [46], along with dendritic and NK cells, to promote allograft survival. In contrast, some studies have suggested the immunogenicity of MSCs, leading to proliferation of T cells towards infused MSCs and rejection of skin allografts by engendering functional memory T cells [47]. A very recent study has established the characteristics of human MSCs to be phenotypically and physiologically similar to human cardiac myofibroblasts. This study was concluded based on the positive staining of hMSCs for α-SMA, NMMIIB, ED-A fibronectin, vimentin and sp1D8 (collagen type I), which was similar to that of cardiac myofibroblasts [48].

Cardiomyogenic potential of MSCs obtained from different sources

MSCs are present in almost all tissues of the body and are mainly located in the perivascular alcove [49]. These can be derived from disparate adult (e.g. peripheral blood, adipose tissue, bone marrow) and neonatal (umbilical cord, amnion, cord blood and placenta) tissues [49], based on their therapeutic application. Although bone marrow represents the major source of MSCs in the body, it does not qualify as a viable isolation source of the cells due to high-grade viral infection and a substantial reduction in the proliferative capacity of the cells with age [50]. Also, MSC extraction from bone marrow is an invasive procedure, which causes immense pain to the patients and can also cause an infection [49]. Thus, MSCs derived from peripheral blood [51], heart [29], lung [52] and adipose tissue [53] have been explored for their biological properties, differentiation capacities and surface marker expression. Also, the cells obtained from neonatal tissues have been found to have superior biological properties as compared with BM-MSCs due to their ready availability, use of non-invasive techniques and avoidance of ethical problems [49]. A study categorized BM-MSCs based on their surface differentiation antigens and found that SCA-1$^+$/CD31$^+$/CD45$^+$ subgroups displayed substantial cardiac improvement capacity, as compared to other BM-MSC subgroups such as SCA-1$^+$/CD45$^-$/CD31$^-$, SCA-1$^+$/CD45$^+$/CD31$^-$ or SCA-1$^+$/CD45$^-$/CD31$^+$ [54]. Several animal cell lines have also been established to be used as biological tools for ex-vivo expansion and MSC differentiation into a definite lineage. One such MSC cell line was obtained using a porcine model, which when treated with 5-aza differentiated into cells containing positive cardiac phenotypic markers such as connexin-43 (Cx-43) and α-actin [55]. A study using porcine model demonstrated the use of histological staining as a feasible method to study the effect of these MSCs in myocardial regeneration [56]. MSCs obtained from patients with either coronary artery disease (CAD) or diabetes mellitus (DM), or both, help ameliorate cardiac function on transplantation but diabetes

in a patient reduces the myocardial protection and proliferative capacity in hMSCs, as compared with CAD [57]. Bcl-2 is family of proteins having a critical role in regulating anti-apoptotic pathways and cell death inhibition [58]. This feature of higher protective and proliferative capacity has been attributed to the lower expression of Bcl-2 in CAD + DM patients compared with the CAD-only group [57].

According to one study [59], the traditional therapy techniques have been effective in treatment of acute diseases and improving a patient's lifespan, but they do not serve to provide a permanent cure, thereby leaving the patients with protracted disease. On the contrary, cardiovascular regenerative medicine prevents further disease advancement by replacing the damaged cells with cardiac myocytes obtained from stem cells [60]. This is possible because stem cells are responsible for the generation and maintenance of terminally differentiated cell populations in tissues that undergo continuous turnover [2]. For instance, a study conducted by Brunt et al. investigated the myogenic differentiation based on age, where bone marrow MSCs were obtained from cardiovascular patients and a protein evaluation was conducted to estimate the β-catenin nuclear translocation in these patients. The study concluded with a first-time discovery of increased β-catenin bioavailability leading to myogenic differentiation and the WNT/β-catenin network as a potential target for reinvigoration of MSCs [61]. Regenerative medicine has explored several options in order to establish the use of MSCs as an expedient and more pragmatic technique towards cardiac regeneration from the various possible sources of regenerative tissue.

Bone marrow

Differentiation of scar tissue into cardiomyocytes can be instigated by transplanting bone marrow cells into the tissue and thereby restoring the myocardial function [62]. BM-MSCs have shown promising potential in cardiac repair due to their powerful proliferative capacity [63, 64], their ability to reduce the infarct size [65] and their ability to change the milieu of the damaged cardiac tissue to upregulate VEGF [66]. These have also been studied specifically for differentiation of CSCs [67]. For the first time, Cai et al. [68] demonstrated the use of these MSCs for the treatment of isopreterenol-induced myocardial hypertrophy. Another interesting observation in this study included the significance of inhibition of VEGF, and not fibroblast growth factor (FGF) or insulin-like growth factor, which restricted the protective effects of BM-MSCs on the hypertrophic condition [68]. Having mentioned this, the combined effect of BM-MSCs along with basic fibroblast growth factor (bFGF)-binding ECM has been observed to improve the left ventricular (LV) function and enhance myocardial regeneration [69]. One of the most effective delivery methods for the treatment has been observed to be via the retrograde infusion of the two [70]. Mixed treatment of BM-MSCs with endothelial progenitor cells (EPCs) pre-treated with salvianolic acid B results in reduced infarct area and enhanced stem cell proliferation [71]. BM-MSCs can be tracked by labelling them with superparamagnetic iron oxide (SPIO) nanoparticles in any MI rat [72] or swine model [73] and locating the shortened T2 value on the MRI scan. Similarly, quantum dots have been recently identified as another medium to label and track the cells, both in vitro and in vivo [74]. Emmert et al. [75] aligned a series of methods of cell tracking and imaging, including micron-sized iron-oxide labelling (MPIO), MRI, micro-CT flow cytometry and PCR followed by immunohistochemstry, in intra-uterine and intramyocardial (i.m.) BM-MSC transplantation pre-immune sheep models. The multipotency of these cells has been confirmed by a study based on human MSCs, which led to their differentiation into adipocytes, chondrocytes and osteoblasts [48]. Despite the similarity of these cells with cardiac myofibroblasts, they remain different due to their proliferative and differentiation properties, which are characteristic of MSCs. The repair coordinated by BM-MSCs is mainly mediated by causing relief from heart failure symptoms, and improving blood flow to the myocytes [76]. Also, bone marrow was the first source identified for MSCs, but several alternatives are being explored due to the invasive and painful extraction process.

BM-MSCs have been studied to transdifferentiate into cardiomyocytes, which involves a negative regulation by histone deacetylase 1 (HDAC1) [77]. HDAC1 when knocked down leads to directed differentiation of the MSCs into cardiac cells. Multipotent BM-MSCs when reprogrammed into pluripotent cells result in MSC-derived induced pluripotent stem cells (MiPS), which express cardiac-specific transcription factors and form spontaneously beating cardiac progenitors [78]. These MiPS-derived progenitors engender infarcted heart and lead to improvement in global heart function. Bone marrow MSC/silk fibroin/hyaluronic acid (BMSC/SH) was implanted into myocardial infarcted rat hearts, where the condition was obtained by cryo-injury technique [79]. In comparison with the control and the other experimental models, BMSC-SH proved to improve the thickness of the LV wall, reduce apoptosis, promote neovascularization and stimulate several paracrine factors (e.g. VEGF), thereby compiling the advantages of the bioactive SH patches and stem cell therapy. In another study, the BM-MSCs were transplanted with induced (iBM-MSC) and uninduced (uBM-MSC) BM-MSCs in MI-induced rat hearts. As per the results obtained, the iBM-MSC-treated hearts showed improved fractional shortening as compared with any of the other models. Thus, iBM-MSC implantation has been considered as another potential therapeutic strategy for post-infarcted heart failure [80].

A combined therapy of BM-MSCs with Tanshinone IIA (Tan IIA) increased the migratory rate of the cells to the ischaemic region by promoting SDF-1α expression in the area, which was suppressed by AMD3100 (a CXC chemokine receptor 4 blocker (CXCR4)) [81]. This finding indicated the role of SDF1/CXCR4 in BM-MSC migration. SDF-1 recruits the MSCs from bone marrow through a CXCR4-dependent mechanism [82] and when transfected into MSCs results in improved viability of the cells in infarcted hearts, thereby preserving the contractile function along with improving the paracrine action of the cells [83]. Similarly, TG-0054, a CXCR4 antagonist, was studied in debilitating MI and cardiac dysfunction after 12 weeks of the treatment. This functional improvement is attributed to the ability of TG-0054 to mobilize the CD271-MSCs and reduce both plasma and myocardial cytokine levels [84]. BM-MSCs overexpressing myocardin-related transcription factor-A (MRTF-A) prevent primary cardiomyocyte apoptosis caused by H_2O_2, and thus help in reversing the cardiac damage after MI [85]. Similarly, overexpression of CREG in intramyocardially implanted BM-MSCs resulted in increased angiogenesis and reduced apoptosis and fibrosis [12]. Also, BM-MSCs treated with 5-aza along with exposure to 2G-hypergravity, when transplanted into a rat MI model, showed positive cardiac markers such as Nkx2.5, Mef-2 and GATA-4 indicating cardiac differentiation and functional recovery [86]. When GATA-4 and Nkx2.5 are transfected into BM-MSCs which are then co-cultured in the myocardial environment, the differentiation capacity of the cells increases along with the reparative capacity [87]. Another study compared rat BM-MSCs transfused with 5-aza to those exposed to electrical stimulation [88]. The results obtained showed higher levels of Cx-43 and Mef-2c in the second group as compared with the first. This instigated the idea of electrically-stimulated MSC differentiation into cardiomyocytes. Similar results were obtained when a recombinant cocktail consisting of IL-6, FGF-2, α-thrombin, BMP-4, TGF-β1 [89], retinoic acid, activin-A and insulin-like growth factor was transduced into hMSCs in order to guide cardiopoiesis [90]. Bone marrow mononuclear cells (BM-MNC) are an attractive source of MSCs [91] due to the ease of extraction of the cells. Comparing both of these bone-marrow-derived populations, MSCs result in higher vascularization, smaller infarct size [92] and improved LVEF [93] with respect to mononuclear cells [94]. BM-MSCs have been shown to degrade functionally and quantitatively with increase in age of patients undergoing successful reperfusion treatment, and hence this aspect of the MSCs needs to be explored further [95]. Prostaglandin E_1 protects BM-MSCs against serum-deprived induced apoptosis by decreasing *Bax* and *caspase-3* expression levels and increasing *Bcl-2* expression [96]. In one study, bone marrow cells derived from heart failure patients were shown to express higher levels of remodelling enzymes and pathways regulating tissue remodelling, scar formation and maturation. This was attributed the increase in $CD146^+$/*SMA-α* myofibroblast frequency [97]. Beyond this, BM-MSCs have shown to promote c-kit^+ CSC differentiation via the tumour growth factor beta (*TGF-β*) signalling pathway, through paracrine activity [98]. Inflow of endogenous c-kit^+ cells is also possible by thymosin β4 (*Tβ4*) administration, which in turn can lead to significant increase in survival of the transplanted cells and the vascular growth [99].

Umbilical cord

The MSCs derived from different compartments of the umbilical cord such as vein, arteries, Wharton's jelly, umbilical cord lining and so forth have been observed to accumulate in damaged tissues and bolster the repair of the tissues [100]. The umbilical cord-MSCs (UC-MSCs) are said to have faster self-renewal capacity than the BM-MSCs and a lower potential of forming teratomas [101]. A very first study was performed on an animal model where the cord lining-derived MSCs combined with a vascularized omental flap ameliorated cardiac dysfunction by myocardial revascularization and attenuated remodelling [102]. Polycaprolactone nanofibres immobilized with UC-MSC-seeded fibronectin demonstrated enhanced LVEF and improved cardiac function [103]. Wharton's jelly-derived MSCs (WJ-MSCs), obtained from embryonic epiblasts, have been identified to have properties of hESCs and adult stem cells, thereby serving as an alternative source for stem cells with significant barriers of immunorejection, tumorigenesis, teratoma formation and so forth [104, 105]. WJ-MSCs are highly specific for cardiac tissue due to their natural chemoattractive nature [105] and production of pro-angiogenic factors such as *IIGF, VEGF, angiopoietin* and *TGF-β1* [106], inducing recruitment of CSCs [107]. Overexpression of N-cadherin, a cell surface gene present in UC-MSCs, leads to upregulation of *VEGF*, via the *ERK* signalling pathway [108]. Intracoronary infusion of WJ-MSCs has also been considered an alternative to BM-MSCs on the basis of their increased LVEF and decreased incidence of adverse events [109]. H_2O_2-preconditioned WJ-MSCs have an enhanced therapeutic effect possibly due to *IL-6* production, which leads to migration and proliferation of endothelial cells (ECs) and increased neovascularization [110]. Konstantinou et al. [111] have for the first time demonstrated the formation of cardiac polymicrotissue by differentiating hUC-MSCs

using a combination of growth factors suramin and sphignosine-1-phosphate. This generated the possibility of using the polymicrotissue as a therapeutic patch over the infarct cardiac area. Similarly, umbilical-cord-derived exosome resulted in improved cardiac function by angiogenesis and their protective nature towards the myocardial tissue [112]. Also, 5-aza-induced hUC-MSCs have been observed to express *GATA-4* and *Nkx2.5* genes, and to differentiate into myocardial cells [113, 114], better than myocardial-induced fluids [115].

Cord blood

The haematopoietic stem progenitor cells obtained from umbilical cord blood have been studied to be very useful for clinical therapy [116–118]. However the presence of MSCs in umbilical cord blood is disputable because of the inability to obtain these cells from the gestation term cord blood [116]. On the contrary, studies suggest the presence of MSCs in fetal organs [119], with circulation in pre-term fetus blood, along with the haematopoietic precursors [120, 121]. This conflicting result has been attributed to the use of a different percentage of umbilical cord blood harvests in the two studies [116]. In the results obtained by Lee et al. [122], it is possible to extract MSCs from the cord blood that would further differentiate into mesodermal lineages. Cardiac muscles, being of mesodermal origin, can therefore also be obtained from cord blood-derived MSCs. Oxytocin exerts a promigratory effect on umbilical cord blood-derived MSCs (UCB-MSCs) [123], and the supplementation of UCB-MSCs with oxytocin results in lowered cardiac fibrosis, macrophage infiltration and restoration of *Cx-43* expression, along with a sustained ejection fraction [39]. A study established that co-transplantation of hUCB-$CD34^+$ and hUC-MSCs leads to reduction in collagen deposition and improved cardiac function in MI rabbits [124].

Adipose tissue

The colony frequency of cells obtained from adipose tissue is higher than those of bone marrow [125] and cord blood, and these adipose tissue-derived MSCs (ASCs) can differentiate into adipocytes, chondrocytes and osteoblasts [125]. Although these cells can differentiate into vascular ECs leading to angiogenesis, along with demonstrating a paracrine effect in animal models with MI [126], cardiomyocyte differentiation is not quite feasible [127]. Under hypoxic conditions, ASCs secrete large amounts of *VEGF, SDF-1* and *HGF*, increasing the migration and proliferation of cardiomyocytes and reducing the apoptosis and infarct size [128]. ASCs can be isolated from the subcutaneous adipose tissue region or omental region [129]. Liver X receptor (LXR) is helpful in improving the retention and survival of the injected

ASCs post MI, and when combined with ASCs leads to improvement of the cardiac function [130]. This has been studied to be possible though the toll-like receptor (*TLR*)-4/*NF-kB* and *Keap-1/Nrf-2* pathways [131]. Also, ASCs secrete various cytokines with different immuno-modulatory effects which contribute a great deal in tissue regeneration [132, 133]. ASCs with overexpressed granulocyte chemotactic protein (*GCP*)-2 have resulted in enhanced angiogenic potential and survival properties [134]. Similar results were obtained for dimethyl sulfoxide-induced ASCs which differentiated into cardiomyocyte-like cells, eventually resulting in cardiac function recovery [135]. These cells have thus attracted great attention in terms of therapeutic approach towards skeletal tissue repair [132]. ASCs transplanted with hydrogel and β-galactose-caged nitric oxide donor showed improved cardiac function and enhanced cell survival [136]. ASCs embedded in scaffold containing platelet-rich fibrin are functionally superior to direct ASC transplantation, in terms of expression of *IL-10*, *Bcl-2* and *TGF-β* [137, 138]. Quite recently, another very interesting discovery made was in relation to the human adult epicardial fat surrounding the heart which served as a reservoir for mesenchymal-like pro-genitor cells (cardiac ATDPCs) [139]. These cells show cardiac-like phenotype despite their residence in an adipocytic environment. Also, increasing the num-ber of cardiac ATDPCs has been shown to exert great immunosuppression [139] because of increased T-cell proliferation.

Skeletal muscle

Muscle-derived stem cells (MDSCs) are not restricted to myogenic or mesenchymal tissues, and can regenerate bone and muscle along with cartilage healing [140]. Satel-lite cells have been considered to be skeletal muscle stem cells, but they have been identified as myogenic precursors with a committed differentiation lineage that act as a reservoir of regenerative cells in case of injury [141]. Stud-ies provide evidence for the formation of myotubes by transplantation of the satellite cell-containing myoblast into a MI model [76]. Thus, the muscle precursor cells derived from satellite cells can be considered as a viable option for regeneration of myopathic skeletal muscle [141]. MSCs obtained from skeletal muscle showed significant improvement in the LVEF of acute MI rat models, comparable with that of ASCs, but they did not transdifferentiate into cardiomyocytes or any vas-cular cells [142]. MDSCs have been a recent focus of study and these cells can be harvested either from orthopaedic reconstruct wastes [143] or from healthy muscle tissue biopsies [144]. The general delivery ap-proach used for MDSCs is a tissue engineering strat-egy such as the use of a scaffold.

Placenta

The study by Vellasamy et al. substantiated the presence of MSCs in the placenta (p-SC) and suggested them as feasible regenerative medicine. Stem cells can be derived from two different parts of the placenta, namely chorionic villi and chorionic plate [145, 146]. These cells demon-strate the ability to differentiate into osteocytes and adipo-cytes, and show typical features of MSCs [146]. Along with their non-tumorigenic property, these cells have characteristics of both ESCs and MSCs, thereby exhibiting the capacity to differentiate into the three germ layers [147]. The major advantage of using this as a source of MSCs is that they are available in abundance as medical waste after delivery. The limitation of using p-SCs is the occurrence of high chances of impurity, since the placenta is the common medium of exchange between a mother and the baby.

Amnion

Amniotic mesenchymal cells (AMCs) are derived from fetal mesoderm and can be peeled off the chorionic membrane mechanically by blunt dissection [148]. These are considered a fitting cell source for cellular cardio-myoplasty by both integrating and differentiating into cardiac tissue [149]. An in-vivo study assessing the effect of AMC transplantation in a damaged myocardial tissue, in comparison with UCB-MSCs and ASCs, showed com-parable results with respect to decreased infarct size, cardiomyocyte-like cell differentiation and improved car-diac function [150]. Also, these cells serve as potential curative agents due to their chemotactic characteristic [151], ample availability, lack of ethical concerns and low immune response [150]. The cardiomyogenic differ-entiation capacity of AMC has been shown to improve by administration of *IL-10* or progesterone [148].

Fibroblast

Fibroblasts are mesenchymal precursor cells that express CD34 and CD45 surface markers [152]. They migrate to the tissues via blood circulation [153], differentiating into myofibroblasts (contractile cells involved in secretion of ECM for tissue remodelling and wound healing) [152]. MSCs have been studied to promote myofibroblast con-gregation in the infarcted area through *TGF-β(1)-Smad2* signalling pathway [154]. An important factor discovered for myofibroblast differentiation is transient receptor potential cation channel (*TRPC6*) activity [155]. This study was conducted in vitro as well as in vivo in an experimental mice model (*TRPC6* knockout mice). The knockout mice had debilitated myofibroblast differenti-ation, resulting in increased ventricular dilation and re-duced cardiac function [156].

Table 1 summarizes some additional information about the sources of MSCs based on frequency of

Table 1 Comparison between different stem cells

	ESCs	iPSCs	HSCs	MSCs
Potency	Totipotent: zygote – morula Pluripotent: inner cell mass of blastocyst	Pluripotent	Pluripotent	Multipotent
Major sources	Inner cell mass of blastocyst	Reprogramming of adult cells	Bone marrow, peripheral blood, umbilical cord blood	Bone marrow, adipose tissues, umbilical cord matrix
Cell surface markers	hESC lines: *SSEA-4, Tra 1-60, Tra 1-81* [273] mESC lines: *NANOG, OCT4, SOX2, SSEA-1* [274]	Cell surface antigenic markers expressed on ESCs, e.g. *SSEA-3* in human, *SSEA-1* in mouse [274]	CD34 [275], CD133$^+$ [276]	CD70$^+$, CD90$^+$, CD105$^+$ [277] CD34$^-$ [278]
Potential clinical application in cardiac regeneration	• Yield a variety of cardiomyocyte-atrial, ventricular and sinus-nodal like cells [279] • Isolation of pure ventricular cardiomyocyte population using adenovirus vectors [280]	Generation of cardiomyocyte sheet along with endothelial cells using angiogenic. factors (VEGF) [281]	No transdifferentiation into cardiac cells in ischaemic tissues [282]	• Improves heart function • Increase in augmented angiogenesis • Reduction in fibrosis [283]
Advantages	Differentiates into three germ layers: ectoderm, mesoderm, endoderm	Produced using adult cells, hence avoids ethical issues	Proliferation and migration to site of injury	• Allogenic grafting possible without immunosuppressive agents • Limited inclination towards mutation
Limitations	• Availability of cell lines for federally funded research • Risk of producing teratomas from transplanting undifferentiating stem cells	• Generation and safe delivery of iPSC-derived cardiomyocytes is strenuous [284] • Tumour formation possible [285]	• Insufficiency in the DNA repair system caused by ageing, thereby limiting the function of HSCs [286] • Insufficient information on signalling pathway [21] • Possibility of gonadal dysfunction and infertility [287]	• Insufficient information on which MSC source to be used for the therapeutic strategy concerning a disease [19] • Route of administration is uncertain for different diseases [19]
Ethical concerns	• Involves human blastocyst • Consent for blastocyst/egg donation is required	None specifically	• Need for clinical parity • Consideration required for cure of children with ess severe sickle cell disease [287]	None specifically

ESC embryonic stem cell, *HSC* hematopoietic stem cell, *iPSC* induced pluripotent stem cell, *MSC* mesenchymal stem cell, *VEGF* vascular endothelial growth factor

production and proliferation potential in comparison with BM-MSCs, along with the techniques of administration to the intended location. The frequency of MSCs in tissues is estimated by assay of the CFU-Fs which serve as the hallmark of these cells [157]. Apart from the comparison presented in Table 1, a very interesting study by Ramkisoensing et al. investigated the differentiation potential of hMSCs derived from ESCs, fetal umbilical cord, amniotic membrane, bone marrow, adult adipose tissue and bone marrow. The results proved hESC-MSCs and fetal hMSCs to be superior to all the other MSCs co-cultured with neonatal rat cardiomyocytes, in terms of expression of most cardiac-specific genes, positive staining for α-actinin, higher basal levels of *Cx-43* and formation of capillary-like structures. Additionally, hESCs and fetal MSCs, when co-cultured with neonatal

rat cardiac fibroblasts, showed no expression of α-actin and decreased *Cx-43* expression. Also unlike adult MSCs, the MSCs derived from hESCs and fetal tissue were found to differentiate into three cardiac lineages, which highlights the developmental stage of the donor tissue as a significant factor in differentiation study [158]. The MSCs derived from rat fetal heart also resulted in upregulation of anti-apoptotic, anti-fibrotic and cardiogenic growth factors when intravenously injected in a MI rat model [119]. The human fetal liver-derived MSCs have also been shown to differentiate into cardiomyocyte-like cells with a combined treatment of retinoic acid, dimethyl sulfoxide and 5-aza in high dose [159]. These cells expressed *Nkx2.5*, cardiac troponin I (*cTnT*), *Oct4* and desmin after harvesting them in the mixture.

Delivery methods of MSCs into host myocardium

Delivery of MSCs into a damaged myocardium is affected by three key factors: nature of the injury, timing of the treatment and ability of the cells to implant into the host myocardium [160]. MSCs can be delivered via several routes such as intravenous (i.v.) and i.m. injections. A study concluded improved LV function [161], improved cardiac function and higher efficiency of cell engraftment post MI in the case of i.m. injection of MSCs [162]. Also, the MSCs transplanted intramyocardially have been thought to improve myocardial lymphatic system due to their property of integrating into the lymphatic endothelium [163]. BM-MSCs when administered via intracoronary injection have been very effective in angiogenesis and improvement of cardiac function [164]. An early study for MSC delivery investigated a tissue engineering approach where two strategies are mainly applied: engineering of a stem cell-containing tissue construct or a beating cardiomyocyte-containing tissue construct [160]. For instance, to give rise to a stem cell-populated tissue construct poly(lactic–co–glycolic acid) (PLGA) [165] can be used as a scaffold and BM-MSC-derived cardiomyocyte-like cells can be used for seeded cells [166], which mimicked the structural and functional aspects of a myocardium [167, 168]. This construct was found to substantially stimulate MSC differentiation into cardiac tissue. PLGA loaded with *SDF-1α* and fabricated with coaxial electrospraying limits the contact between the protein and organic phase. When bovine serum albumin is incorporated as a carrier protein, the chemotactic effect of *SDF-1α* is enhanced and the synergistic effect leads to higher growth and proliferation of the cells [169]. Various biomaterials have been used for development of scaffold in order for it to be an ECM analogue of the host tissue. In 2014, high-density cardiac fibroblast was proposed for the development of ECM scaffolds from cardiac fibroblasts [170]. When seeded with hESC-derived MSCs, these scaffolds can be used as a delivery medium for the stem cells. In the same year Vashi et al. [171] assessed a commercial pericardial material, CarioCel, which served as a scaffold to cling onto the seeded stem cells and act as a template for formation of the new issue. A study on collagen-1 scaffold seeded with autologous MSCs demonstrated reverse modelling in rat models of chronic MI [172]. There has been limited study on the number of cells that remain localized at the site of transplantation. One such study using a hyaluronan-based scaffold for MSCs showed that although most of the cells had moved to the border leaving the scaffold, the treatment did manage to alleviate fibrosis in the area along with enhanced vascularization [173]. Hydrogel is a 3D polymeric network that swells up on exposure to water and can be of various types like collagen, fibrin, gelatin, alginate and so

forth [174]. BM-MSCs with hydrogel composite have been studied to improve the cardiac functioning by preventing LV remodelling [175]. Gelatin-coated ECM dishes have also been determined as a suitable method for MSC differentiation into beating cardiomyocytes [176]. Along with preserving the structure of the matrix, this method also yields greater amounts of collagen and protein [177]. Decellularized ECMs are also used as biological scaffolds because of their advantage of being able to mimic the host ECM properties [178]. Several other ECM proteins have been identified which further lead to cardiomyocyte differentiation, protection, proliferation and angiogenesis [177]. Genipin, a natural cross-linking agent, has been utilized in various studies to fabricate biocompatible and stable hydrogels with increased stiffness and prolonged degradation. This technique does not harm the possibility of minimally invasive catheter delivery of the hydrogel [179]. Thermosensitive hydrogel has proved to be a novel method for delivering MSCs and is based on *N*-acryloxysuccinimide, *N*-isopropylacrylamide, poly(trimethylene carbonate)-hydroxyethyl methacrylate and acrylic acid [180, 181]. This hydrogel-based delivery results in higher differentiation efficiency of MSCs than co-culturing of cardiomyocytes and MSCs or chemical induction. Similarly, polytetrafluoroethylene (PTFE) and porcine small intestinal submucosa (pSIS) have been found to account for varying cell proliferation capacity of CPCs as compared with MSCs [182]. Another study determined a self-assembling polypeptide *RAD16-II*, which when mixed with cardiac marker-positive MSCs yielded a stable nanofibre scaffold, promoting cardiac regeneration at the site of tissue damage [183]. Some polymeric scaffolds lack structural integrity and thus prove to be inefficient in their delivery capacity. Thus, the use of hMSCs encapsulated in arginine–glycine–asparagine (RGD)-modified alginate microspheres helps to restore the LV function and increase the cell survival after an MI, along with enhanced angiogenesis [184]. A non-invasive cell delivery system was explored by Xu et al. where they used ultrasound-mediated bubble destruction for the delivery of drugs, genes and stem cells by upregulating *SDF-1/CXCR4* [185], and this could be used as an efficient delivery system [186]. Lee et al. developed spheroid 3D bullets from hUCB-MSCs to deliver these stem cells without the use of any cytokines [187]. The factors that seemed essential during the formation of the bullets were Ca^{2+}-dependent cell–cell interaction and presence of E-cadherin as an adhesion molecule. E-cadherin activation was found to switch on the *ERK/Akt* signalling pathway required for the proliferative and paracrine activity of MSCs [187].

Mechanisms of action of MSCs

In normal conditions of a non-injured heart, the MSCs are found to exist in low numbers, and on induction of

MI these cells start proliferating rapidly for participation in wound healing, by generation of fibroblasts and myofibroblasts.

Homing of MSCs

The transplantation of MSCs after MI has shown that the cells infiltrate the injured tissue by trafficking through the ECM [188] and considerably repairing the cardiac function [189]. To understand the general mechanism of MSC infiltration into the damaged cardiac tissue, some studies have demonstrated the production of *HGF* by apoptotic cardiomyocytes, and not by necrotic cardiomyocytes [190]. The recruitment of MSCs has been credited to the presence of *HGF* receptor MET, which activates a wide range of signalling pathways, one of which leads to attraction of MSCs to the apoptotic cell death site [191]. This study also concluded the involvement of platelets in the migration of MSCs to the apoptotic cardiac cells through the interaction of high mobility group box-1 (*HMGB1*), which is a nuclear protein with *TLR-4* expressed on MSCs. On activation of platelet, *HMGB1/TLR-4* downregulate MET on MSCs, thereby impairing the recruitment of the cells. As a result, gene-knockout or blocking of *TLR-4* on MSCs can lead to improved infiltration of MSCs to the damaged tissue, thereby increasing the efficacy of MSC-based therapy [191].

In case of any damaged myocardium, *SDF-1α* mediates the homing of the endogenous MSCs [169]. Although the chemokine receptor *CXCR4* has not been found to be expressed in large amounts on the MSC surface, about 80–90 % of hMSCs have an intracellular storage of the receptor [192]. Following overexpression by mRNA nucleofection, the receptor stimulates Ca^{2+} signalling through its ligand *SDF-1α* [193]. SDF-1 functions as a $CD34^+$ progenitor cell-recruiting agent at the site of damage in an organ [194]. However in conditions such as dilated cardiomyopathy (DCM), monocyte-chemotactic protein-1 (*MCP-1*) has been established as a homing factor of MSCs because of the presence of chemokine receptor type 2 (CCR2), a *MCP-1* receptor, on the cell surface [195]. Having said this, the further alignment of these migrated MSCs has been established and therefore additional study is required to determine whether the MSCs cause transdifferentiation, have a paracrine effect or themselves differentiate into cardiomyocytes [195]. There have been several in-vitro and in-vivo studies to understand the mechanism of MSC recruitment to the site of the damaged tissue for the reparative process to occur, along with its protective characteristic. MSCs either differentiate into beating cardiomyocytes [196], transdifferentiate or induce a paracrine effect for the regenerative process to occur.

Structural organization for cardiomyogenesis

Cardiac actin is the main component of thin filaments of cardiac myofibrils and sarcomere. The contraction of cardiac muscle is mediated by sarcomere [197] and troponin is an essential protein required for the cardiac muscle contractility [198] as demonstrated by a study on familial hypertrophic cardiomyopathy [199]. Beta myosin is predominantly expressed in the normal human ventricle [200]. In 2011, Wei et al. [201] conducted a study to investigate the biological characteristics of the subpopulation of MSCs that served as the therapeutic agent in heart injury and established these cells to be CPCs, due to expression of cardiac-specific markers α-actin and *cTnT* on them. The studies which used 5-aza to convert MSCs to cardiomyocytes [202], whether BM-MSCs [203] or UCB-MSCs [204], have shown the expression of all of the genes in the differentiated cardiomyocytes, such as desmin, β-myosin heavy chain, Nkx2.5 and *cTnT* A [204]. Such studies support the hypothesis that 5-aza can be useful in the reparative process of heart ventricle as well as in the amelioration of heart muscle contractility [205] (Fig. 1a). From earlier studies, *cTnT* [206] and tropomyosin [207] have been shown to play a role in Ca^{2+} regulation during contraction. Results obtained by Asumda and Chase [208] also anticipate the presence of actin in BM-MSCs, in addition to the other cardiac isoforms of troponin such as troponin I (*cTnI*), *cTnT*, troponin C (*cTnC*) and that of tropomyosin (*cTm*) which appear in the early stages of cardiomyogenic differentiation.

Paracrine effect

MSCs insulate the cardiac tissue from any kind of damage by reprogramming the molecular wiring of the cardiac myocytes, thereby protecting them from any hazardous compound. For instance, Rogers et al. [209] studied the therapeutic aspect of hMSCs by co-culturing them with injured myocytes from a neonatal mouse. The mouse myocytes were subjected to stress by incubating them with either toxin cytokine, *IL-1β*, or with endotoxin, lipopolysaccharide (LPS). These two compounds act as pro-inflammatory cytokines [210]. The hMSCs blocked the activation of cardiac transcription factor *NF-kB*, which is dependent on LPS, *IL-1β* [209] and *IL-6* [211, 212], thereby inhibiting the adverse effect and rendering protection to the neonatal mouse myocytes. Co-treatment of MSCs with various inflammatory factors such as *TNF-α* and *IL-1β* leads to the upregulation of vascular cell adhesion molecule-1 (*VCAM-1*) [213]. With the increase in cell adhesion ability, cardiac function was also enhanced. Several attempts have been made to protect the myocardium against ischaemia through preconditioning, which has further led to an increase in the levels of *TNF-α, VEGF* and *IL-8*, along with migration and recruitment of MSCs to the injured tissue [214].

In normal conditions, the cardiac fibroblasts regulate the ECM by two mechanisms: synthesis and degradation of the matrix molecules [215]. The matrix-degrading enzymes are matrix metalloproteinases (MMPs) which help the infiltrated myofibroblasts in sequential degrading of the matrix, followed by ECM synthesis. According to Wang et al. [216], MSCs affect MMP expression via the *ERK 1/2* signalling pathway, where erythropoietin may act as a paracrine factor. When MSCs of an old human, transfected with tissue inhibitor of MMP-3 (*TIMP3*) and *VEGF*, was transplanted into a rat model of MI, they showed a similar degree of angiogenic capacity to that demonstrated by young MSCs [217]. However, when young MSCs were injected into aged rat recipients, the results showed a significant decrease in scar deposition. This study thus opened up the possibility of allotransplantation of MSCs from young donors to older patients suffering from MI [218]. Neuropeptide Y (*NPY*) is a neurotransmitter present in the human central and peripheral nervous system which helps to regulate the endocrine and autonomic functions. It has been shown to promote angiogenesis with similar efficacy as fetal basic fibroblast growth factor (*fbFGF*) and *VEGF* [219]. *NPY*-induced differentiation of BM-MSCs into cardiomyocytes leads to improved angiogenesis and cardiac function along with reduced fibrosis via upregulation of *FGF-2*, cycline A2 and eukaryotic initiation factor (*EIF*)-4E genes [220]. Glycogen synthase kinase (*GSK*)-3β, when overexpressed in MSCs and injected into a coronary ligated heart, resulted in improved mortality, reduced infarct size, LV remodelling and a higher cardiomyocyte differentiation rate [221]. *GSK-3β*-MSCs also upregulated the paracrine factor *VEGF-A*, which led to increased capillary density and survival of MSCs in the tissue [221]. Similarly, genetically engineered MSCs with enhanced prostaglandin I synthase (*PGIS*) gene expression have been shown to improve cardiac function by reducing apoptosis and limiting the cardiac remodelling and increasing the VEGF-A levels, as found in a *GSK-3β* study [222]. Injection of MSCs results in activation of the *JAK/* signal transducer and activator of transcription 3 (*STAT3*) signalling pathway which has a role in the upregulation of growth factors in both diseased hearts and skeletal muscles [223]. This became evident from a study where BM-MSCs improved ventricular function in cardiomyopathic hamsters [224, 225]. The *STAT3* pathway increases the *caspase-4* level in the transplanted MSCs, and improves the post-ischaemic function by reducing pro-inflammatory and pro-apoptotic signalling in the tissue [226].

Macrophages have been another target of study to initiate the neovascularization along with MSCs [227]. Earlier studies have established that increased levels of VEGF,

produced by *STAT3*, are the driving force behind angiogenesis in order to alleviate conditions like DCM [221, 228] and ischaemic reperfusion injury [229]. Additionally, myocardial mRNA expressions of *AT1*, *TGF-β1* and *CYP11B2* have been found to be lower in a doxorubicin-induced DCM-MSC group as compared with placebo or blank groups, where doxorubicin is administered by intraperitoneal injection in the rat model [230]. Additionally, the doxorubicin-induced injury is also possible to mitigate through BM-MSC or ASC injection [231]. The *VEGF* expression is also induced by a combined therapy of granulocyte growth factor (G-CSF) and BM-MSCs, carrying *HGF* for angiogenesis in MI rat models [232]. However, recent studies have emphasized secretion of platelet-derived growth factor (*PLGF*) factor by MSCs to promote neovascularization [233]. Hence, *PLGF* was used to check the proliferation or apoptosis of macrophages. Although no change was observed, however, a dose-dependent polarization of M1 macrophage to M2 macrophage was found to take place which released *PLGF* 50 times more than M1. This study suggested that *PLGF*, not *VEGF* secreted by MSCs, stimulates the polarization of macrophages which further secrete *PLGF* to promote neovascularization and enhance cardiac muscle repair [234] (Fig. 1b). Also, *PLGF* has been shown to directly stimulate neovascularization and hence help in cardiac repair [233].

Previous studies have underlined a significant interaction between *TGF-β1* and bone morphogenetic protein *BMP7* in the epithelial-to-mesenchymal transition for fibrosis [235, 236] (Fig. 1a). Macrophages express high *TGF-β1* [237] and MSCs express a high level of *BMP7* [238] which have a contradictory fibrogenic effect of the *TGF-β* secreted by macrophages. Another study showed improved functional recovery of the ischaemic cardiac tissue when the MSCs were co-treated with *TGF-β1* and *IL-1β*, due to an increased *VEGF* level [239].

Surgical treatment methods are mainly employed only after a patient suffers MI. This was studied in mammals for the first time based on a study of neonatal mice undergoing a 10-min surgery to induce MI [240]. This procedure leads to vascular injury [241] following which ECs synthesize cytokines, chemokines and growth factors such as *VEGF-A* [242], all of which play a protective role and stimulate the ECs along with recruitment of peripheral stem cells [242, 243]. *VEGF-A* also coordinates the differentiation of MSCs into ECs in vitro [243, 244] (Fig. 1a) and factors such as *IL-6* and *TNF-α* inhibit *VEGF-A*-induced differentiation of MSCs into ECs and subsequent capillary tube formation [245]. However, this fact has been negated in a study by Mohri et al. [246], where the authors claimed activation of the *JAK/STAT* pathway in CSCs by *IL-6* cytokines, which in turn leads to vasculogenesis of vascular endothelial precursor cells. Combined treatment of angiotensin II (*AngII*) and

VEGF-A effectively increases the marker expression of ECs despite the presence of IL-6 and TNF-α.

Pre-treatment and conditioning of MSCs
MSCs induced with ischaemic cardiac conditioned media showed positive reaction for GATA-4, Nkx 2.5 and MLC-2a, suggesting cardiomyogenic differentiation of MSCs, as compared with the negligible effect by a non-ischaemic environment exerted on the MSCs [247]. Cardiomyogenic media-primed MSCs enhanced expressions of sarcomeric α-actinin and Cx-43, establishing themselves as better therapeutic agents than direct MSC transplantation [248].

Diazoxide is an ATP-sensitive potassium channel regulator present in the mitochondria and plays a role in suppressing apoptosis and promoting cell survival. Selected MSCs preconditioned with diazoxide resulted in an improved cell survival rate by upregulating the expression of bFGF and HGF and protecting the cells from oxidative stress injury [249].

One of the earlier studies established MSCs pre-treated with Ang receptor blockers (ARB) as an agent involved in improvement of cardiac function and also as a potential CSC source for cardiomyogenesis [250]. In a more recent study, it was demonstrated that Ang II, through an angiotensin II type-2 receptor (AT2R)-dependent mechanism, promoted the differentiation of MSCs into functional ECs [251] and upregulated the expression of Cx-43 for gap junction formation [252] (Fig. 1c). Hence AT2R agonists and inflammatory compounds are considered key candidates for angiogenesis or vessel repair. G9a is a mammalian histone methyltransferase which acts as a transcriptional repressor [253]. Thus, use of BIX01294, which is a G9a HMT inhibitor, induced the expression of cardiac transcription factors such as GATA-4, Nkx2.5 and myocardin on BM-MSCs when the cells were exposed to cardiogenic stimulating factor WNT11 [254]. Islet-1 is considered another cardiac cell marker [255], and thus progenitors with Islet-1 can differentiate into various cardiac lineages. C3H10T1/2 MSCs were used for the study of cells that differentiated into cardiomyocyte-like cells via histone acetylation [256]. These cardiomyocyte-like cells when present in the proximity of myofibres expressing collagen V show escalated integration and recovery of the infracted myocardium [257].

Effects of modification in MSCs
Heme oxygenase-1 (HO-1) when transduced into MSCs using an adenoviral vector has been shown to induce angiogenic effects [258], with enhanced anti-oxidative and anti-apoptotic capabilities [259], leading to improvement in cardiac function post MI. Human receptor activity-modifying protein 1 (hRAMP1) gene when overexpressed in MSCs using the same vector [260] and tagged with

enhance green fluorescent protein (EGFP) resulted in smaller infarct size and enhanced cardiac function [261] by decreasing the TNF-α level, inhibiting NF-kB expression and enhancing the IL-10 level [260]. hRAMP1-expressing MSCs are otherwise also noted to inhibit the vascular smooth muscle cell proliferation [262]. CXCR4-overexpressed hypoxic MSCs were also shown to enhance neovascularization, enhance EC differentiation, reduce infarct size and restore cardiac function [263]. MSCs transduced with lentiviral CXCR4 lead to downregulation of the caspase 3 pathways and upregulation of pAkt and IGF-1α levels [264].

An animal study used integrin-linked kinase (ILK)-transfected MSCs to investigate the effect on collagen synthesis and cardiac fibroblast proliferation. The study demonstrated inhibition of cardiac fibroblast proliferation and a few other factors, thereby leading to a decrease in infarct size and a reduction in fibrosis in these animals [265] along with increased cardiomyocyte proliferation [266]. Also, MSC transplantation in infarcted area has been shown to enhance the synthesis of collagen and this could be the mechanism behind attenuated ventricular remodelling post transplantation [267].

MSCs are valued for their paracrine effects in reducing inflammation [188] and promoting growth of the surrounding cells [268]. MSC injection promotes the recruitment of CPCs and helps in the improvement of myocardium [67]. Studies were performed to check the efficiency of dual cell transplantation on cardiac repair. These cells were fused to form CardioChimeras (CCs) which proved to be more efficient than single cell delivery. CPC phenotype expression dominates CCs and mediates the cardiomyogenic factors [269]. These cells also demonstrated the same phenotypic properties of commitment and high paracrine effect as those of MSCs along with increased basal expression of cardiomyogenic factors [269]. To check the effects of CCs and their parental cells, neonatal rat cardiac myocytes were incubated with them. Addition of CCs increased the expression of stromal-derived factor, a cardioprotective agent, and also acted as a ligand to CXCR4+ stem cells [270]. The study also showed an increase in capillary density in the area incubated with the CCs (Fig. 1b). Furthermore, the ejection fraction (fraction of blood being pumped out of the heart per heartbeat) and the anterior wall thickness of the heart also showed an improvement [269].

MicroRNA regulation in modified MSCs
Researchers have been investigating several other techniques to accelerate cardiac regeneration, keeping in mind the feasibility of the process. microRNAs (miRs) are approximately 22-nucleotide RNAs [271], found endogenously and involved in post-transcriptional regulation of gene expression. Some of these miRNAs are

said to be cell specific or tissue specific, helping to fathom the underlying pathophysiological condition [272]. *miR-133a* is muscle specific and is proposed as a novel therapeutic target in cardiovascular disease [273]. Patients suffering from MI have been shown to have lower levels of *miR-133a* [274, 275]. *miR-133a* is known to play an important role in terminating embryonic car- diomyocyte proliferation [276], attenuating fibrosis [277] and promoting cardiac remodelling [278]. To assess its role in survival of MSCs, *miR-133a* was made to express in these cells. Researchers found that improvement in MSC survival was due to the attenuation of expression of *Apaf-1* and *caspase 9* and *3* (Fig. 1a). In contrast, depleting or blocking of *miR-133a* by its antagonist resulted in upregulation of these proteins [279]. Another significant study illustrated the overexpression of *miR-16* in cardiac-niche-induced hMSCs, when co-cultured with rat ventricular myocytes [280]. *miR-16* was found to in- hibit cell proliferation, modulate the cell cycle, promote cell apoptosis and abolish tumorigenicity both in vitro and in vivo [281]. The induced cardiac niche led to dysregula- tion of the miRNA and increased G1 phase arrest in hMSCs, leading to their differentiation into myogenic phenotypes in the cardiac niche [280]. Similarly, *miR- 499* is an embedded miRNA present within a ventricular-specific myosin heavy chain gene [282]. When overexpressed in rat BM-MSCs, *miR-499* acti- vates the *WNT/β-catenin* signalling pathway, inducing cardiac differentiation [283]. Another mechanism of cardiac protection used by *miR-499* is calcineurin- mediated dynamin-related protein-1 (*Drp1*) activation, which prevents cardiomyocyte apoptosis [284]. *miR-34* acts as a crucial cell death regulator and its deletion or silencing reduces the age-associated cardiac cell death [285]. This occurs due to inactivation or knockdown of the stem cell factor (*SCF*), which serves as the main target of *miR-34* and thus lead to inhibition of angio- genesis [286]. *miR-23a* is also studied to regulate the *caspase 7*-induced apoptosis, involving the *TNF-α* path- way, along with a reduction in infarct size and improve- ment of the LV function [287].

Effects of treatment of MSCs on cardiac regeneration
Several compounds such as pioglitazone [288], rosuvastatin [289], *TMZ* [18, 21], gingko biloba extract 761 [290] and hydrogen sulfide [291] have been demonstrated to enhance the repair of cardiac tissue in MI models. Pioglitazone is generally used to increase the insulin sensitivity in diabetic type 2 patients. Oral intake of this drug after BM-MSC transplantation has been studied to improve cardiac func- tion. When used in pre-treatment of MSCs, pioglitazone yielded significantly upgraded cardiac function and was even put forward as a promising CSC source for cardio- myogenesis [288] (Fig. 1c). Combined treatment of MSCs

with pioglitazone showed higher levels of peroxisome proliferator-activated receptor gamma (*PPAR-γ*), which in turn led to increased *Cx-43* levels [292]. Similarly rosuvas- tatin, when administered with ASCs, reduced fibrosis and safeguarded the cardiac function by decreasing pro-apoptotic proteins (*Bim* and *Bam*) and increasing anti-apoptotic proteins (*Bcl-2* and *Bcl-xL*), thereby inhi- biting cardiomyocyte apoptosis [289]. Hydrogen sulfide led to increased levels of phosphorylated *Erk1/2*, *Akt* and *GSK-3β*, and resulted in an increased survival rate of the transplanted MSCs, enhanced LV function and reduced infarct size [291] (Fig. 1c). Similarly, atorva- statin treatment increased the expression of *CXCR4* in MSCs, leading to enhanced migration of *SDF-1* and low levels of *IL-6* and *TNF-α* [293] (Fig. 1c). The drug also facilitated MSC survival along with improvement of LV function and decrease in the infarct size, inflammation, fibrosis and apoptosis [294]. Salvianolic acid B pre- treatment of MSCs has been found to be very effective in a rat model of MI following transplantation [295]. Improved survival of the transplanted MSCs was ob- served along with increase in angiogenic factors such as *VEGF*, *bFGF* and *SCF* with concomitant reduction in fi- brosis and infarcted area [295] (Fig. 1c). Another study concluded that a combination of angiogenic factor genes, chemokine and stem cells could increase the angiogenesis rate and improve cardiac function [296].

MSCs have also been studied to modulate electro- physiological properties including the excitability and conduction of cardiomyocytes by two mechanisms. First, by intercellular coupling through the gap junction for reduction in instinctive activity of cardiomyocytes; and second, by increase in the conduction velocity of cardio- myocyte by paracrine signalling, via upregulation of *Cx- 43* [158] and nerve growth factor [297], without any amendments in the beating frequency [298]. This thera- peutic action of MSCs has been studied in a swine model, which resulted in decreased heart rate turbulence, amelio- rated repolarization time and higher slope of action poten- tial durations indicating improved cardiac functioning and reduced risk of ventricular arrhythmias [299]. In addition, MSCs have also shown to improve the contractile func- tion and compensate for a 50 % loss of cardiomyocytes after any cardiac damage by supplementing the engineered cardiac tissues (ECTs), which serve as a 3D in-vitro model system to appraise stem cell therapies [300].

Clinical trials using MSCs in cardiac disease performed between 2010 and 2015 and their shortcomings
There have been about 41 clinical trials (Additional file 1: Table S1) performed between 2010 and 2015 for the study of MSCs in relation to cardiac injury and repair. These tri- als were performed in distinct locations and were mostly found to have completed phase II, where some of them

even managed to reach phase III of the study. The trials could be characterized in several ways based on their focus of study. Most of the trials focused on the injection and infusion of MSCs from different sources into the injured cardiac tissue, via different sites of injection. The second type of study compared the different kinds of MSCs (i.e. autologous and allogeneic MSCs) in context of both ischaemic as well as non-ischaemic cardiomyopathy in patients. A third type of study was performed in order to focus the safety and efficacy of these MSCs when implanted into patients, as done in a study by Da Silva and Hare [301] with the focus on the role of BM-MSCs in the treatment of chronically injured heart.

In these studies, a number of candidate cells such as neonatal and fetal cardiomyocytes, ESC-derived myocytes, skeletal myoblasts, cell types from adult BM and cardiac precursor cells have been considered. Autologous BM progenitor cells (mononuclear or MSCs) when administered myocardially resulted in improved regional contractility of the myocardial scar within 3 months of treatment [302] (ClinicalTrials.gov NCT01392625). The trial comparing the two BM preparations conducted TAC-HFT [303] and POSEIDON-DCM [304] studies to estimate the optimal cell type, delivery method, dose, mechanism of action of cell delivery and so forth. MSCs have been the main focus of these studies due to their paracrine effect, high regeneration capacity, ability to perpetuate potency and ability to avoid adverse reactions to autologous versus allogeneic transplant. A study conducted in Korea proved MSC therapy to be safe and quite efficient in terms of LVEF improvement for the treatment of acute MI [305] (ClinicalTrials.gov NCT01392105). A similar study comparing the two types of bone marrow transplants for patients with LV dysfunction due to ischaemic cardiomyopathy showed low alloimmune reactions in allogeneic MSCs and improved functional as well as structural measures when both were administered together [306] (ClinicalTrials.gov NCT01087996). Another study conducted on nine acute MI patients, following a 5-year follow-up plan to check the feasibility and safety of i.m. infection, gave a positive outcomes on MSC expansion and safety of the method and justified the possibility of placebo-controlled trials for i.m. MSC injections [307]. A similar study was conducted for i.v. allogeneic BM-MSCs in MI patients which proved to be equally efficient in improving the ejection fraction and the LV volumes [308] (Clinicaltrials.gov NCT00114452). These clinical trials have concluded the safety and feasibility of BM-MSCs, but after MI the functional recovery of the cardiac cells remains ambiguous [309]. A similar study conducted to investigate the safety and efficacy of WJ-MSCs administered via an intracoronary route demonstrated no trigger in troponin concentration as observed with BM-

MSCs, indicating no coronary artery occlusion after the treatment [109] (ClinicalTrials.gov NCT01291329). Another very interesting study was conducted to observe the combined effects of stem cell implantation and mechanical circulatory support which resulted in synergistic symptomatic improvement in LV functioning [310].

Although only a few of the 41 trials have been completed and the status of some remains unknown, these trials have established various results which help bolster the upcoming clinical trials and research. On evaluating these trials, a definite trend of limitations is evident which probably played a crucial role in undermining some of the studies. Firstly, the number of patients in some of the trials was extremely lower, which could have compromised the efficacy of the study. Thus, further large-scale randomized trials are required to establish successful results. Secondly, the number of patients excluded in some studies was extremely high because of several reasons such as poor image quality after randomization. This leads to generation of unreliable and non-interpretable data. Thirdly, there is also a possibility of ignoring inter-observer and intra-observer variability in evaluating the obtained data at different intervals either due to manual error or due to unavailability of required tools such as contrast-enhanced magnetic resonance imaging (CE-MRI). Hence, Lee et al. [305] suggested the use of SPECT to minimize the inter-observer error during data analysis (ClinicalTrials.gov NCT01392105). Some trials lacked use of diverse assessment tools such as exercise tolerance, 6-min walking distance test, pulmonary function test and so forth. This could have restricted the possible varied outcomes of the trials. Another technical limitation of the trials using autologous MSCs is the inability to use them immediately since they take at least 3 weeks [311] to harvest and culture to reach an effective confluence, thereby limiting the efficiency of stem cell therapy in an acute setting. The randomized clinical trials concerning these cells need to define the period of treatment. Some experimental limitations include lack of placebo comparison groups for the study [302], which then makes the study prone to observational bias (ClinicalTrials.gov NCT01392625). Some trials evaluated only the functional aspect of the stem cell therapy, whereas others focused only on the safety and efficacy of the study, keeping the other factors constant. We anticipate more reliable outcomes if important criteria such as donor source, cell type, delivery method, dosage, cohort size and optimal time of treatment are taken into consideration.

Challenges faced in stem cell therapy

The main challenges faced in the use of stem cells, including CSCs, for cardiovascular repair revolve around isolation of adequate stem cells, ex-vivo expansion frequency, appropriate delivery strategy and adequate differentiation and functional improvement in vivo [27]. In order to overcome the afore-mentioned challenges, MSCs have proved to be extremely efficient. Isolation of MSCs is comparatively easier; for example, the bone marrow cells can be extracted from the peripheral blood or the bone marrow itself [27]. In order to meet the increasing demand of MSCs, a microcarrier-based stirred culture system technique has been evolved for the efficient ex-vivo expansion of the stem cells, for different sources of MSCs using the various kinds of microcarriers [312]. Mesenchymal cells have thus attracted immense attention due to their therapeutic characteristics and lack of both ethical concerns and teratogenic properties [313].

Cell therapy precautions
Stem cell therapy has been used for the treatment of cancer, repair of damaged tissue and various degenerative diseases. The potential of such therapies was recognized long ago, leading to further developments in the field of stem-cell-based therapeutics. The success of these therapies depends on several factors such as the type of stem cell being used, its proliferative capacity and differentiation status, the route and site of administration, survival capability of the engrafted cells and so forth. On compiling these factors, a risk profile is generated that then evaluates the potential risks of the technique which can include tumour formation along with some other unwanted immune responses. As far as pluripotent cells like ESCs and induced pluripotent stem cells are concerned, they have not demonstrated any clinical risks in any of the trials. Theoretically, the high proliferation rate and unlimited self-renewal capacity of these cells constitute the risk of tumour formation. On the contrary, multipotent MSCs have not reported any major health concerns, implying the safety of MSC therapy. However, some trials have recounted serious adverse events [314], such as malignant tumour formation on transplantation of unmodified BM-MSCs in the peri-infarct area of a mouse model [315]. This calls for further investigation of the mechanisms involving MSCs. For instance, in a study conducted to observe an infarcted heart region, several calcified or ossified encapsulated structures were identified after the injection of MSCs [316]. A study on arrhythmic mechanisms established the pro-arrhythmic effects of hMSCs in neonatal rat cardiomyocytes and the pattern of the MSCs was said to be determinant of the arrhythmic severity of the myocardial tissue [317]. Another study concluded the possibility of primary cardiac sarcoma formation from MSCs, which can further develop into tumours with multi-lineage differentiation [318]. According to a study conducted by Huang et al. [319], allogeneic MSC transplantation in the myocardium exhibited a biphasic immune response of these cells,

resulting in a shift from an immune-privileged state to an immunogenic phenotype after differentiation leading to characteristics such as fractional shortening and progressive ventricular dysfunction. Also, the recent investigation on electrically stimulated cardiomyocyte-like cell differentiation needs to be explored in depth [88]. Thus, evaluation of these processes tops the list of upcoming research on MSCs. Another important consideration in cell therapy is the number of passages studied in any experiment. For instance, a study based on commercially available murine MSCs showed altering expression patterns over a period of time, and this was further established by comparing the early and late passages of the model [320].

Conclusion

To evaluate the safety of MSCs in regenerative medicine, 41 clinical trials and more than 120 animal model studies have been performed since 2010 and these studies have shown MSCs to have the potential to differentiate into various mesodermal (e.g. osteoblast, adipocyte and chondrocyte) [43] and myeloid lineages [44]. The immunomodulatory characteristic of MSCs makes them a worthy competitor in the field of regenerative therapeutics. However, many pathways and underlying processes concerning MSCs still exist that remain unexplored in the field of reparative medicine (Fig. 2). Despite the therapeutic effects of MSCs, Dayan et al. [321] observed no improvements in cardiac function in a chronic ischaemic heart failure model, with no difference in the scar area, fractional shortening and so forth. A study illustrated induced and spontaneous transformation of MSCs into sarcomas in mouse, whereas in humans only induced transformation of MSCs has been observed [322]. The spontaneous transformation of hMSCs in vitro was found to be caused by the contamination of the cells by tumour cell lines [323], and studies have negated the idea of MSC transformation into tumours, even after long-term culturing of cells [324]. In contrast, in-vivo spontaneous transformation has been shown to lead to osteosarcoma genesis in patients with infused BM-MSCs for some other disease [325].

This brings us to the prospective studies in relation to the therapeutic competency of MSCs. These cells after transplantation have been shown to demonstrate paracrine effects which can prove to be of great advantage in future medical therapies (Fig. 2). Liang et al. [326] evaluated, for the first time, pigment epithelium-derived factor (*PEDF*), a paracrine factor, as a target for modifying and improving the impaired aged MSCs and thereby enhancing the cellular profile. The same is possible by overexpressing silent mating type information regulation 2 homolog 1 (*SIRT1*) in aged MSCs to restore pro-angiogenic factors, *bFGF* and so forth [327]. The regeneration process can be severely compromised by the lack of suitable MSC delivery

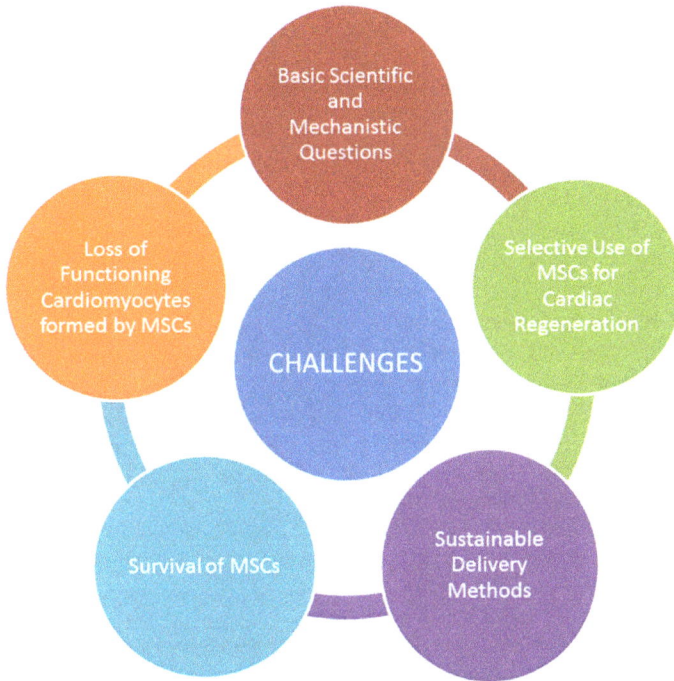

Fig. 2 Challenges in use of MSCs for cardiac regeneration. Tumour formation in MSCs has been considered inconceivable, but there have been instances of osteosarcoma in patients infused with BM-MSCs for some other disease. Hence, in the context of MSCs in cardiac regeneration, some pathways and processes might exist that still remain unexplored. Additionally, these pathways comprise MSCs obtained from different sources, out of which only a few such as BM-MSCs have been used extensively for clinical applications, in spite of evidences of more proliferative capacity in MSCs obtained from umbilical cord, peripheral blood, etc. This limitation arises due to the lack of an efficient delivery method of MSCs to the target site. Another challenge that has seemed to come in the way of researchers is the prolonged survival of MSCs post engraftment into the host myocardium. This challenge has been overcome to a large extent by using miRNAs and CCs, but more sustainable methods need to be studied further. Studies have gained several advancements in the field of safety and efficacy of the MSC therapy, but success rates in terms of the functional regeneration of cardiac tissue for the loss of functioning cardiomyocytes after any damage remain mediocre. *MSC* mesenchymal stem cell

methods to the intended site of regeneration and reduced survival of transplanted MSCs. The delivery methods for specific MSCs to the specific site of injury have yet not been established, although several delivery systems such as engineered tissue constructs and biomaterials have been explored for the same in order to gain maximum efficiency. For improving the survival of MSCs, researchers have been scrutinizing various methods which have proved beneficial under different conditions. For instance, the use of alginate-encapsulated MSCs secreting paracrine factors [328], miRNA [279] and CCs [269] has increased the survival rate of these cells. Hence, in future, the major areas of focus should involve figuring out more sustainable/evolved solutions to the afore-mentioned challenge than

Table 2 Frequency of MSC production, proliferation potential and delivery methods for therapeutic targets in different body organs, as compared with BM-MSCs

Different sources of MSCs	Frequency of production[a]	Potential of proliferation[a]	Delivery methods for regeneration	References
Bone marrow	1 in 3.4×10^4 cells	–	Intravenously	[332]
Umbilical cord matrix	Low	High	Not specified	[128, 333]
Amnion	High	Low	Not specified	[334]
Placenta	High	High	Not specified	[128]
Adipose tissue	High	High	Not specified	[55, 105, 128]
Peripheral blood	High	High	Intravenously	[128, 335]
Cord blood	Low	High	Intramyocardial, intravenous, intracoronary	[103, 336, 337]

[a]In comparison with the BM-MSCs

BM-MSC bone marrow-derived mesenchymal stem cell, *MSC* mesenchymal stem cell

those under current implementation and more investigation is required in order to corroborate the efficacy of the therapies (Fig. 2).

MSCs can also be obtained from different sources in the body, but the studies in cardiac regeneration are mainly done using only a few of them. Referring to information presented in Table 2, the bone marrow has been established as one of the most promising sources of MSCs, but there have been studies indicating a higher MSC production and proliferation capacity in other parts of the body such as the umbilical cord, placenta and peripheral blood. Similarly hUC-MSCs have been found to improve motor function, reduce abnormal levels of the concerned enzymes such as lactate dehydrogenase (*LDH*), creatine kinase (*CK*), and so forth, and increase the muscle strength (ClinicalTrials.gov NCT01610440). Thus, hUC-MSCs become an important source of treatment for genetic conditions like Duchenne muscular dystrophy (DMD). A very important aspect that plays a crucial role in the treatment of cardiac disorders is the ability of any treatment strategy to compensate for the loss of the functioning cardiomyocytes [329]. Thus, one of the future challenges of cardiovascular therapies is to strategize the functional regeneration of myocardial contractility using tissue engineering, cell-based therapy or reprogramming of scar fibroblasts [330, 331].

Throughout this review we came across compounds such as pioglitazone [288], rosuvastatin [289] and so forth that were studied in the initial years of the developmental era of MSCs but have not received much attention in recent years, despite the promising results obtained in cardiac therapy. There thus needs to be more research carried out on such compounds in order to not lose out on some extremely propitious therapeutic agents. Cell therapy has been adopted as a novel therapeutic strategy for treatment of cardiac disorders such as severe heart failure and CAD. Unfortunately, although these approaches have led to advancements in the field of safety and efficacy of these cell therapies, the mediocre success rates in terms of functional improvement serve as a disappointment in the field [3]. Thus we need to further investigate the sources of MSCs that can help benefit the treatment of any disorder accordingly with 'true' reparative potential, in order to help focus on the field of regenerative medicine.

Additional file

Additional file 1: Table S1 Clinical trials executed during 2010–2015 which used MSCs to treat heart diseases. (DOCX 36 kb)

Abbreviations
AMC, amniotic mesenchymal cell; AngII, angiotensin II; ASC, adipose tissue-derived mesenchymal stem cell; Aza, azacytidine; BM-MSC, bone marrow-derived mesenchymal stem cell; CAD, coronary artery disease; CC, CardioChimera; CFU-F, colony-forming unit fibroblast; CPC, cardiac progenitor cell; CSC, cardiac stem cell; cTnT, Cardiac troponin T; Cx, Connexin; DCM, dilated cardiomyopathy; DM, diabetes mellitus; Drp, dynamin-related protein; EC, endothelial cell; ECM, extracellular matrix; ESC, embryonic stem cell; FGF, fibroblast growth factor; GCP, granulocyte chemotactic protein; HGF, hepatocyte growth factor; HIF, hypoxia-inducible factor; HSC, haematopoietic stem cell; i.m., intramyocardial; i.v., intravenous; LPS, lipopolysaccharide; LV, left ventricular; LVEF, left ventricular ejection fraction; MDSC, muscle-derived stem cell; MI, myocardial infarction; MMP, matrix metalloproteinase; MSC, mesenchymal stem cell; PEDF, pigment epithelium-derived factor; PLGA, poly(lactic–co-glycolic acid); PLGF, platelet-derived growth factor; PPAR-γ, peroxisome proliferator-activated receptor gamma; p-SC, placenta-derived stem cell; SDF, stromal cell-derived factor; SIRT1, silent mating type information regulation 2 homolog 1; TGF-β, tumour growth factor beta; TK, tissue kallikrein; TLR, Toll-like receptor; TMZ, trimetazidine; UCB-MSC, umbilical cord blood-derived mesenchymal stem cell; UC-MSC, umbilical cord-derived mesenchymal stem cell; VEGF, vascular endothelial growth factor; WJ-MSC, Wharton's Jelly-derived mesenchymal stem cell.

Acknowledgements
The authors would like to thank Ms Erfath Thanjeem Begum, Dr Venkat Katari and Mr Vinod Reddy for their enduring help throughout the period of the completion of this review article.

Funding
DS is supported by a 'Fast Track Young Scientist' grant (YSS/2014/000027) from the Department of Science and Technology (DST), Government of India and an investigator initiated grant (H15-27983) from Baxalta, USA. The funding sources did not play any role in the research and/or preparation of the article, study design, data collection, analysis/interpretation of data, writing of the report and decision to submit the article for publication.

Authors' contributions
AaS and AbS prepared the outline and wrote the manuscript. DS conceptualized, wrote and edited the manuscript. All authors read and approved the final manuscript.

Competing interests
The authors declare that they have no competing interests.

References
1. Tuch BE. Stem cells—a clinical update. Aust Fam Physician. 2006;35:719–21.
2. Hall PA, Watt FM. Stem cells: the generation and maintenance of cellular diversity. Development. 1989;106:619–33.
3. Doppler SA, Deutsch MA, Lange R, Krane M. Cardiac regeneration: current therapies–future concepts. J Thorac Dis. 2013;5:683–97.
4. Jopling C, Sleep E, Raya M, Marti M, Raya A, Izpisua Belmonte JC. Zebrafish heart regeneration occurs by cardiomyocyte dedifferentiation and proliferation. Nature. 2010;464:606–9.
5. Choi WY, Poss KD. Cardiac regeneration. Curr Top Dev Biol. 2012;100:319–44.
6. Schofield R. The relationship between the spleen colony-forming cell and the haemopoietic stem cell. Blood Cells. 1978;4:7–25.
7. Vunjak-Novakovic G, Scadden DT. Biomimetic platforms for human stem cell research. Cell Stem Cell. 2011;8:252–61.
8. Mohyeldin A, Garzon-Muvdi T, Quinones-Hinojosa A. Oxygen in stem cell biology: a critical component of the stem cell niche. Cell Stem Cell. 2010;7:150–61.
9. Sharp FR, Ran R, Lu A, Tang Y, Strauss KI, Glass T, et al. Hypoxic preconditioning protects against ischemic brain injury. NeuroRx. 2004;1:26–35.
10. Das R, Jahr H, van Osch GJ, Farrell E. The role of hypoxia in bone marrow-derived mesenchymal stem cells: considerations for regenerative medicine approaches. Tissue Eng Part B Rev. 2010;16:159–68.
11. Cerrada I, Ruiz-Sauri A, Carrero R, Trigueros C, Dorronsoro A, Sanchez-Puelles JM, et al. Hypoxia-inducible factor 1 alpha contributes to cardiac healing in mesenchymal stem cells-mediated cardiac repair. Stem Cells Dev. 2013;22:501–11.
12. Peng C, Pei H, Wei F, Tian X, Deng J, Yan C, et al. Cellular repressor of E1A-stimulated gene overexpression in bone mesenchymal stem cells protects against rat myocardial infarction. Int J Cardiol. 2015;183:232–41.
13. Kim SH, Moon HH, Kim HA, Hwang KC, Lee M, Choi D. Hypoxia-inducible vascular endothelial growth factor-engineered mesenchymal stem cells prevent myocardial ischemic injury. Mol Ther. 2011;19:741–50.

14. Huang B, Qian J, Ma J, Huang Z, Shen Y, Chen X, et al. Myocardial transfection of hypoxia-inducible factor-1alpha and co-transplantation of mesenchymal stem cells enhance cardiac repair in rats with experimental myocardial infarction. Stem Cell Res Ther. 2014;5:22.

15. Sullivan KE, Quinn KP, Tang KM, Georgakoudi I, Black 3rd LD. Extracellular matrix remodeling following myocardial infarction influences the therapeutic potential of mesenchymal stem cells. Stem Cell Res Ther. 2014;5:14.

16. Bian S, Zhang L, Duan L, Wang X, Min Y, Yu H. Extracellular vesicles derived from human bone marrow mesenchymal stem cells promote angiogenesis in a rat myocardial infarction model. J Mol Med (Berl). 2014;92:387–97.

17. Gao L, Bledsoe G, Yin H, Shen B, Chao L, Chao J. Tissue kallikrein-modified mesenchymal stem cells provide enhanced protection against ischemic cardiac injury after myocardial infarction. Circ J. 2013;77:2134–44.

18. Xu H, Zhu G, Tian Y. Protective effects of trimetazidine on bone marrow mesenchymal stem cells viability in an ex vivo model of hypoxia and in vivo model of locally myocardial ischemia. J Huazhong Univ Sci Technolog Med Sci. 2012;32:36–41.

19. Zhao SL, Zhang YJ, Li MH, Zhang XL, Chen SL. Mesenchymal stem cells with overexpression of midkine enhance cell survival and attenuate cardiac dysfunction in a rat model of myocardial infarction. Stem Cell Res Ther. 2014;5:37.

20. Wisel S, Khan M, Kuppusamy ML, Mohan IK, Chacko SM, Rivera BK, et al. Pharmacological preconditioning of mesenchymal stem cells with trimetazidine (1-[2,3,4-trimethoxybenzyl]piperazine) protects hypoxic cells against oxidative stress and enhances recovery of myocardial function in infarcted heart through Bcl-2 expression. J Pharmacol Exp Ther. 2009;329:543–50.

21. Marti Masso JF, Marti I, Carrera N, Poza JJ, Lopez de Munain A. Trimetazidine induces parkinsonism, gait disorders and tremor. Therapie. 2005;60:419–22.

22. Ham O, Lee SY, Lee CY, Park JH, Lee J, Seo HH, et al. let-7b suppresses apoptosis and autophagy of human mesenchymal stem cells transplanted into ischemia/reperfusion injured heart 7by targeting caspase-3. Stem Cell Res Ther. 2015;6:147.

23. Wagers AJ, Weissman IL. Plasticity of adult stem cells. Cell. 2004;116:639–48.

24. Pittenger MF, Mackay AM, Beck SC, Jaiswal RK, Douglas R, Mosca JD, et al. Multilineage potential of adult human mesenchymal stem cells. Science. 1999;284:143–7.

25. Gunsilius E, Gastl G, Petzer AL. Hematopoietic stem cells. Biomed Pharmacother. 2001;55:186–94.

26. Kolf CM, Cho E, Tuan RS. Mesenchymal stromal cells. Biology of adult mesenchymal stem cells: regulation of niche, self-renewal and differentiation. Arthritis Res Ther. 2007;9:204.

27. Sun Q, Zhang Z, Sun Z. The potential and challenges of using stem cells for cardiovascular repair and regeneration. Genes Dis. 2014;1:113–9.

28. Arminan A, Gandia C, Garcia-Verdugo JM, Lledo E, Trigueros C, Ruiz-Sauri A, et al. Mesenchymal stem cells provide better results than hematopoietic precursors for the treatment of myocardial infarction. J Am Coll Cardiol. 2010; 55:2244–53.

29. Beltrami AP, Barlucchi L, Torella D, Baker M, Limana F, Chimenti S, et al. Adult cardiac stem cells are multipotent and support myocardial regeneration. Cell. 2003;114:763–76.

30. Koninckx R, Daniels A, Windmolders S, Carlotti F, Mees U, Steels P, et al. Mesenchymal stem cells or cardiac progenitors for cardiac repair? A comparative study. Cell Mol Life Sci. 2011;68:2141–56.

31. Bolli R, Chugh AR, D'Amario D, Loughran JH, Stoddard MF, Ikram S, et al. Cardiac stem cells in patients with ischaemic cardiomyopathy (SCIPIO): initial results of a randomised phase 1 trial. Lancet. 2013;378:1847–57.

32. Wang M, Yu Q, Wang L, Gu H. Distinct patterns of histone modifications at cardiac-specific gene promoters between cardiac stem cells and mesenchymal stem cells. Am J Physiol Cell Physiol. 2013;304:C1080–90.

33. Zheng SX, Weng YL, Zhou CQ, Wen ZZ, Huang H, Wu W, et al. Comparison of cardiac stem cells and mesenchymal stem cells transplantation on the cardiac electrophysiology in rats with myocardial infarction. Stem Cell Rev. 2013;9:339–49.

34. Oskouei BN, Lamirault G, Joseph C, Treuer AV, Landa S, Da Silva J, et al. Increased potency of cardiac stem cells compared with bone marrow mesenchymal stem cells in cardiac repair. Stem Cells Transl Med. 2012;1:116–24.

35. Srikanth GV, Tripathy NK, Nityanand S. Fetal cardiac mesenchymal stem cells express embryonal markers and exhibit differentiation into cells of all three germ layers. World J Stem Cells. 2013;5:26–33.

36. Williams AR, Hatzistergos KE, Addicott B, McCall F, Carvalho D, Suncion V, et al. Enhanced effect of combining human cardiac stem cells and bone marrow mesenchymal stem cells to reduce infarct size and to restore cardiac function after myocardial infarction. Circulation. 2013;127:213–23.

37. Singh MK, Epstein JA. Epicardium-derived cardiac mesenchymal stem cells: expanding the outer limit of heart repair. Circ Res. 2012;110:904–6.

38. Jin J, Zhao Y, Tan X, Guo C, Yang Z, Miao D. An improved transplantation strategy for mouse mesenchymal stem cells in an acute myocardial infarction model. PLoS One. 2011;6:e21005.

39. Kim YS, Ahn Y, Kwon JS, Cho YK, Jeong MH, Cho JG, et al. Priming of mesenchymal stem cells with oxytocin enhances the cardiac repair in ischemia/reperfusion injury. Cells Tissues Organs. 2012;195:428–42.

40. Schittini AV, Celedon PF, Stimamiglio MA, Krieger M, Hansen P, da Costa FD, et al. Human cardiac explant-conditioned medium: soluble factors and cardiomyogenic effect on mesenchymal stem cells. Exp Biol Med (Maywood). 2010;235:1015–24.

41. Flynn A, Chen X, O'Connell E, O'Brien T. A comparison of the efficacy of transplantation of bone marrow-derived mesenchymal stem cells and unrestricted somatic stem cells on outcome after acute myocardial infarction. Stem Cell Res Ther. 2012;3:36.

42. Zhang Y, Sivakumaran P, Newcomb AE, Hernandez D, Harris N, Khanabdali R, et al. Cardiac repair with a novel population of mesenchymal stem cells resident in the human heart. Stem Cells. 2015;33:3100–13.

43. Minguell JJ, Erices A, Conget P. Mesenchymal stem cells. Exp Biol Med (Maywood). 2001;226:507–20.

44. Short B, Brouard N, Occhiodoro-Scott T, Ramakrishnan A, Simmons PJ. Mesenchymal stem cells. Arch Med Res. 2003;34:565–71.

45. Dominici M, Le Blanc K, Mueller I, Slaper-Cortenbach I, Marini F, Krause D, et al. Minimal criteria for defining multipotent mesenchymal stromal cells. The International Society for Cellular Therapy position statement. Cytotherapy. 2006;8:315–7.

46. Herrmann R, Sturm M, Shaw K, Purtill D, Cooney J, Wright M, et al. Mesenchymal stromal cell therapy for steroid-refractory acute and chronic graft versus host disease: a phase 1 study. Int J Hematol. 2012;95:182–8.

47. Hoogduijn MJ, Roemeling-van Rhijn M, Engela AU, Korevaar SS, Mensah FK, Franquesa M, et al. Mesenchymal stem cells induce an inflammatory response after intravenous infusion. Stem Cells Dev. 2013;22:2825–35.

48. Ngo MA, Muller A, Li Y, Neumann S, Tian G, Dixon IM, et al. Human mesenchymal stem cells express a myofibroblastic phenotype in vitro: comparison to human cardiac myofibroblasts. Mol Cell Biochem. 2014; 392:187–204.

49. Hass R, Kasper C, Bohm S, Jacobs R. Different populations and sources of human mesenchymal stem cells (MSC): a comparison of adult and neonatal tissue-derived MSC. Cell Commun Signal. 2011;9:12.

50. Romanov YA, Svintsitskaya VA, Smirnov VN. Searching for alternative sources of postnatal human mesenchymal stem cells: candidate MSC-like cells from umbilical cord. Stem Cells. 2003;21:105–10.

51. Cao C, Dong Y. Study on culture and in vitro osteogenesis of blood-derived human mesenchymal stem cells. Zhongguo Xiu Fu Chong Jian Wai Ke Za Zhi. 2005;19:642–7.

52. Griffiths MJ, Bonnet D, Janes SM. Stem cells of the alveolar epithelium. Lancet. 2005;366:249–60.

53. Fraser JK, Wulur I, Alfonso Z, Hedrick MH. Fat tissue: an underappreciated source of stem cells for biotechnology. Trends Biotechnol. 2006;24:150–4.

54. He JG, Shen ZY, Teng XM, Yu YS, Huang HY, Ye WX, et al. Efficacy of subgroup mouse bone mesenchymal stem cells on mobilizing autologous cardiac stem cells and repairing ischemic myocardial tissue. Zhonghua Xin Xue Guan Bing Za Zhi. 2013;41:210–4.

55. Moscoso I, Rodriguez-Barbosa JI, Barallobre-Barreiro J, Anon P, Domenech N. Immortalization of bone marrow-derived porcine mesenchymal stem cells and their differentiation into cells expressing cardiac phenotypic markers. J Tissue Eng Regen Med. 2012;6:655–65.

56. Alestalo K, Lehtonen S, Yannopoulos F, Makela T, Makela J, Ylitalo K, et al. Activity of mesenchymal stem cells in a nonperfused cardiac explant model. Tissue Eng Part A. 2013;19:1122–31.

57. Liu Y, Li Z, Liu T, Xue X, Jiang H, Huang J, et al. Impaired cardioprotective function of transplantation of mesenchymal stem cells from patients with diabetes mellitus to rats with experimentally induced myocardial infarction. Cardiovasc Diabetol. 2013;12:40.

58. Willis S, Day CL, Hinds MG, Huang DC. The Bcl-2-regulated apoptotic pathway. J Cell Sci. 2003;116:4053–6.

59. Nadal-Ginard B, Torella D, Ellison G. Cardiovascular regenerative medicine at the crossroads. Clinical trials of cellular therapy must now be based on

reliable experimental data from animals with characteristics similar to human's. Rev Esp Cardiol. 2006;59:1175–89.

60. Kovacic JC, Harvey RP, Dimmeler S. Cardiovascular regenerative medicine: digging in for the long haul. Cell Stem Cell. 2007;1:628–33.

61. Brunt KR, Zhang Y, Mihic A, Li M, Li SH, Xue P, et al. Role of WNT/beta-catenin signaling in rejuvenating myogenic differentiation of aged mesenchymal stem cells from cardiac patients. Am J Pathol. 2012;181:2067–78.

62. Tomita S, Li RK, Weisel RD, Mickle DA, Kim EJ, Sakai T, et al. Autologous transplantation of bone marrow cells improves damaged heart function. Circulation. 1999;100:II247–56.

63. Wen Z, Zheng S, Zhou C, Wang J, Wang T. Repair mechanisms of bone marrow mesenchymal stem cells in myocardial infarction. J Cell Mol Med. 2011;15:1032–43.

64. Lin X, Peng P, Cheng L, Chen S, Li K, Li ZY, et al. A natural compound induced cardiogenic differentiation of endogenous MSCs for repair of infarcted heart. Differentiation. 2012;83:1–9.

65. Zhao JJ, Liu XC, Kong F, Qi TG, Cheng GH, Wang J, et al. Bone marrow mesenchymal stem cells improve myocardial function in a swine model of acute myocardial infarction. Mol Med Rep. 2014;10:1448–54.

66. Rahbarghazi R, Nassiri SM, Ahmadi SH, Mohammadi E, Rabbani S, Araghi A, et al. Dynamic induction of pro-angiogenic milieu after transplantation of marrow-derived mesenchymal stem cells in experimental myocardial infarction. Int J Cardiol. 2014;173:453–66.

67. Hatzistergos KE, Quevedo H, Oskouei BN, Hu Q, Feigenbaum GS, Margitich IS, et al. Bone marrow mesenchymal stem cells stimulate cardiac stem cell proliferation and differentiation. Circ Res. 2010;107:913–22.

68. Cai B, Tan X, Zhang Y, Li X, Wang X, Zhu J, et al. Mesenchymal stem cells and cardiomyocytes interplay to prevent myocardial hypertrophy. Stem Cells Transl Med. 2015;4:1425–35.

69. Zhang GW, Gu TX, Guan XY, Sun XJ, Qi X, Li XY, et al. bFGF binding cardiac extracellular matrix promotes the repair potential of bone marrow mesenchymal stem cells in a rabbit model for acute myocardial infarction. Biomed Mater. 2015;10:065018.

70. Wang X, Zhen L, Miao H, Sun Q, Yang Y, Que B, et al. Concomitant retrograde coronary venous infusion of basic fibroblast growth factor enhances engraftment and differentiation of bone marrow mesenchymal stem cells for cardiac repair after myocardial infarction. Theranostics. 2015;5:995–1006.

71. Zhao GF, Fan YC, Jiang XJ. Effects of the proliferation state of the endothelial progenitor cells preconditioned with salvianolic acid B and bone marrow mesenchymal stem cells transplanted in acute myocardial infarction rats. Zhongguo Zhong Xi Yi Jie He Za Zhi. 2012;32:671–5.

72. Hua P, Wang YY, Liu LB, Liu JL, Liu JY, Yang YQ, et al. In vivo magnetic resonance imaging tracking of transplanted superparamagnetic iron oxide-labeled bone marrow mesenchymal stem cells in rats with myocardial infarction. Mol Med Rep. 2015;11:113–20.

73. Yang K, Xiang P, Zhang C, Zou L, Wu X, Gao Y, et al. Magnetic resonance evaluation of transplanted mesenchymal stem cells after myocardial infarction in swine. Can J Cardiol. 2011;27:818–25.

74. Collins MC, Gunst PR, Muller-Borer BJ. Functional integration of quantum dot labeled mesenchymal stem cells in a cardiac microenvironment. Methods Mol Biol. 2014;1199:141–54.

75. Emmert MY, Weber B, Wolint P, Frauenfelder T, Zeisberger SM, Behr L, et al. Intramyocardial transplantation and tracking of human mesenchymal stem cells in a novel intra-uterine pre-immune fetal sheep myocardial infarction model: a proof of concept study. PLoS One. 2013;8:e57759.

76. Sawa Y. Current status of myocardial regeneration therapy. Gen Thorac Cardiovasc Surg. 2013;61:17–23.

77. Lu DF, Yao Y, Su ZZ, Zeng ZH, Xing XW, He ZY, et al. Downregulation of HDAC1 is involved in the cardiomyocyte differentiation from mesenchymal stem cells in a myocardial microenvironment. PLoS One. 2014;9:e93222.

78. Buccini S, Haider KH, Ahmed RP, Jiang S, Ashraf M. Cardiac progenitors derived from reprogrammed mesenchymal stem cells contribute to angiomyogenic repair of the infarcted heart. Basic Res Cardiol. 2012;107:301.

79. Chi NH, Yang MC, Chung TW, Chen JY, Chou NK, Wang SS. Cardiac repair achieved by bone marrow mesenchymal stem cells/silk fibroin/hyaluronic acid patches in a rat of myocardial infarction model. Biomaterials. 2012;33:5541–51.

80. Li XH, Fu YH, Lin QX, Liu ZY, Shan ZX, Deng CY, et al. Induced bone marrow mesenchymal stem cells improve cardiac performance of infarcted rat hearts. Mol Biol Rep. 2012;39:1333–42.

81. Tong Y, Xu W, Han H, Chen Y, Yang J, Qiao H, et al. Tanshinone IIA increases recruitment of bone marrow mesenchymal stem cells to infarct region via up-

regulating stromal cell-derived factor-1/CXC chemokine receptor 4 axis in a myocardial ischemia model. Phytomedicine. 2011;18:443–50.

82. Penn MS, Pastore J, Miller T, Aras R. SDF-1 in myocardial repair. Gene Ther. 2012;19:583–7.

83. Tang J, Wang J, Guo L, Kong X, Yang J, Zheng F, et al. Mesenchymal stem cells modified with stromal cell-derived factor 1 alpha improve cardiac remodeling via paracrine activation of hepatocyte growth factor in a rat model of myocardial infarction. Mol Cells. 2010;29:9–19.

84. Hsu WT, Jui HY, Huang YH, Su MY, Wu YW, Tseng WY, et al. CXCR4 antagonist TG-0054 mobilizes mesenchymal stem cells, attenuates inflammation, and preserves cardiac systolic function in a porcine model of myocardial infarction. Cell Transplant. 2015;24:1313–28.

85. Zhong Z, Hu JQ, Wu XD, Sun Y, Jiang J. Myocardin-related transcription factor-A-overexpressing bone marrow stem cells protect cardiomyocytes and alleviate cardiac damage in a rat model of acute myocardial infarction. Int J Mol Med. 2015;36:753–9.

86. Ling SK, Wang R, Dai ZQ, Nie JL, Wang HH, Tan YJ, et al. Pretreatment of rat bone marrow mesenchymal stem cells with a combination of hypergravity and 5-azacytidine enhances therapeutic efficacy for myocardial infarction. Biotechnol Prog. 2011;27:473–82.

87. Li P, Zhang L. Exogenous Nkx2.5- or GATA-4-transfected rabbit bone marrow mesenchymal stem cells and myocardial cell co-culture on the treatment of myocardial infarction in rabbits. Mol Med Rep. 2015;12:2607–21.

88. Tang M, Yang G, Jiang J, He X, Li H, Zhang M, et al. Expression of myocardial specificity markers MEF-2C and Cx43 in rat bone marrow-derived mesenchymal stem cells induced by electrical stimulation in vitro. Sheng Wu Yi Xue Gong Cheng Xue Za Zhi. 2015;32:629–34.

89. Herrmann JL, Abarbanell AM, Weil BR, Wang Y, Poynter JA, Manukyan MC, et al. Postinfarct intramyocardial injection of mesenchymal stem cells pretreated with TGF-alpha improves acute myocardial function. Am J Physiol Regul Integr Comp Physiol. 2010;299:R371–8.

90. Behfar A, Yamada S, Crespo-Diaz R, Nesbitt JJ, Rowe LA, Perez-Terzic C, et al. Guided cardiopoiesis enhances therapeutic benefit of bone marrow human mesenchymal stem cells in chronic myocardial infarction. J Am Coll Cardiol. 2010;56:721–34.

91. Alvarez-Viejo M, Menendez-Menendez Y, Blanco-Gelaz MA, Ferrero-Gutierrez A, Fernandez-Rodriguez MA, Gala J, et al. Quantifying mesenchymal stem cells in the mononuclear cell fraction of bone marrow samples obtained for cell therapy. Transplant Proc. 2013;45:434–9.

92. Mazo M, Gavira JJ, Abizanda G, Moreno C, Ecay M, Soriano M, et al. Transplantation of mesenchymal stem cells exerts a greater long-term effect than bone marrow mononuclear cells in a chronic myocardial infarction model in rat. Cell Transplant. 2010;19:313–28.

93. van der Spoel TI, Gathier WA, Koudstaal S, van Slochteren F, Of Lorkeers SJ, Sluijter JP, et al. Autologous mesenchymal stem cells show more benefit on systolic function compared to bone marrow mononuclear cells in a porcine model of chronic myocardial infarction. J Cardiovasc Transl Res. 2015;8:393–403.

94. Tao B, Cui M, Wang C, Ma S, Wu F, Yi F, et al. Percutaneous intramyocardial delivery of mesenchymal stem cells induces superior improvement in regional left ventricular function compared with bone marrow mononuclear cells in porcine myocardial infarcted heart. Theranostics. 2015;5:196–205.

95. Gao LR, Pei XT, Ding QA, Chen Y, Zhang NK, Chen HY, et al. A critical challenge: dosage-related efficacy and acute complication intracoronary injection of autologous bone marrow mesenchymal stem cells in acute myocardial infarction. Int J Cardiol. 2013;168:3191–9.

96. Zeng K, Deng BP, Jiang HQ, Wang M, Hua P, Zhang HW, et al. Prostaglandin E(1) protects bone marrow-derived mesenchymal stem cells against serum deprivation-induced apoptosis. Mol Med Rep. 2014;12:5723–9.

97. Dmitrieva RI, Revittser AV, Klukina MA, Sviryaev YV, Korostovtseva LS, Kostareva AA, et al. Functional properties of bone marrow derived multipotent mesenchymal stromal cells are altered in heart failure patients, and could be corrected by adjustment of expansion strategies. Aging (Albany NY). 2015;7:14–25.

98. Cao Q, Wang F, Lin J, Xu Q, Chen S. Mesenchymal stem cells enhance the differentiation of c-kit+ cardiac stem cells. Front Biosci (Landmark Ed). 2012; 17:1323–8.

99. Ye L, Zhang P, Duval S, Su L, Xiong Q, Zhang J. Thymosin beta4 increases the potency of transplanted mesenchymal stem cells for myocardial repair. Circulation. 2013;128:S32–41.

100. Nagamura-Inoue T, He H. Umbilical cord-derived mesenchymal stem cells: their advantages and potential clinical utility. World J Stem Cells. 2014;6:195–202.

101. Fong CY, Chak LL, Biswas A, Tan JH, Gauthaman K, Chan WK, et al. Human Wharton's jelly stem cells have unique transcriptome profiles compared to human embryonic stem cells and other mesenchymal stem cells. Stem Cell Rev. 2011;7:1–16.

102. Lilyanna S, Martinez EC, Vu TD, Ling LH, Gan SU, Tan AL, et al. Cord lining-mesenchymal stem cells graft supplemented with an omental flap induces myocardial revascularization and ameliorates cardiac dysfunction in a rat model of chronic ischemic heart failure. Tissue Eng Part A. 2013;19:1303–15.

103. Kang BJ, Kim H, Lee SK, Kim J, Shen Y, Jung S, et al. Umbilical-cord-blood-derived mesenchymal stem cells seeded onto fibronectin-immobilized polycaprolactone nanofiber improve cardiac function. Acta Biomater. 2014;10:3007–17.

104. Gao LR, Zhang NK, Ding QA, Chen HY, Hu X, Jiang S, et al. Common expression of stemness molecular markers and early cardiac transcription factors in human Wharton's jelly-derived mesenchymal stem cells and embryonic stem cells. Cell Transplant. 2013;22:1883–900.

105. Musialek P, Mazurek A, Jarocha D, Tekieli L, Szot W, Kostkiewicz M, et al. Myocardial regeneration strategy using Wharton's jelly mesenchymal stem cells as an off-the-shelf 'unlimited' therapeutic agent: results from the Acute Myocardial Infarction First-in-Man Study. Postepy Kardiol Interwencyjnej. 2015;11:100–7.

106. Bilal M, Haseeb A, Sher Khan MA. Intracoronary infusion of Wharton's jelly-derived mesenchymal stem cells: a novel treatment in patients of acute myocardial infarction. J Pak Med Assoc. 2015;65:1369.

107. Zhang W, Liu XC, Yang L, Zhu DL, Zhang YD, Chen Y, et al. Wharton's jelly-derived mesenchymal stem cells promote myocardial regeneration and cardiac repair after miniswine acute myocardial infarction. Coron Artery Dis. 2013;24:549–58.

108. Lee EJ, Choi EK, Kang SK, Kim GH, Park JY, Kang HJ, et al. N-cadherin determines individual variations in the therapeutic efficacy of human umbilical cord blood-derived mesenchymal stem cells in a rat model of myocardial infarction. Mol Ther. 2012;20:155–67.

109. Gao LR, Chen Y, Zhang NK, Yang XL, Liu HL, Wang ZG, et al. Intracoronary infusion of Wharton's jelly-derived mesenchymal stem cells in acute myocardial infarction: double-blind, randomized controlled trial. BMC Med. 2015;13:162.

110. Zhang J, Chen GH, Wang YW, Zhao J, Duan HF, Liao LM, et al. Hydrogen peroxide preconditioning enhances the therapeutic efficacy of Wharton's Jelly mesenchymal stem cells after myocardial infarction. Chin Med J (Engl). 2012;125:3472–8.

111. Konstantinou D, Lei M, Xia Z, Kanamarlapudi V. Growth factors mediated differentiation of mesenchymal stem cells to cardiac polymicrotissue using hanging drop and bioreactor. Cell Biol Int. 2015;39:502–7.

112. Zhao Y, Sun X, Cao W, Ma J, Sun L, Qian H, et al. Exosomes derived from human umbilical cord mesenchymal stem cells relieve acute myocardial ischemic injury. Stem Cells Int. 2015;2015:761643.

113. Ruan ZB, Zhu L, Yin YG, Chen GC. The mechanism underlying the differentiation of human umbilical cord-derived mesenchymal stem cells into myocardial cells induced by 5-azacytidine. Indian J Med Sci. 2010;64:402–7.

114. Latifpour M, Nematollahi-Mahani SN, Deilamy M, Azimzadeh BS, Eftekhar-Vaghefi SH, Nabipour F, et al. Improvement in cardiac function following transplantation of human umbilical cord matrix-derived mesenchymal cells. Cardiology. 2011;120:9–18.

115. Ma N, Ding F, Zhang J, Bao C, Zhong H, Mei J. Myocardial structural protein expression in umbilical cord blood mesenchymal stem cells after myogenic induction. Cell Biol Int. 2013;37:899–904.

116. Mareschi K, Biasin E, Piacibello W, Aglietta M, Madon E, Fagioli F. Isolation of human mesenchymal stem cells: bone marrow versus umbilical cord blood. Haematologica. 2001;86:1099–100.

117. Huss R. Isolation of primary and immortalized CD34-hematopoietic and mesenchymal stem cells from various sources. Stem Cells. 2000;18:1–9.

118. Hows JM. Status of umbilical cord blood transplantation in the year 2001. J Clin Pathol. 2001;54:428–34.

119. Garikipati VN, Jadhav S, Pal L, Prakash P, Dikshit M, Nityanand S. Mesenchymal stem cells from fetal heart attenuate myocardial injury after infarction: an in vivo serial pinhole gated SPECT-CT study in rats. PLoS One. 2014;9:e100982.

120. Campagnoli C, Roberts IA, Kumar S, Bennett PR, Bellantuono I, Fisk NM. Identification of mesenchymal stem/progenitor cells in human first-trimester fetal blood, liver, and bone marrow. Blood. 2001;98:2396–402.

121. Erices A, Conget P, Minguell JJ. Mesenchymal progenitor cells in human umbilical cord blood. Br J Haematol. 2000;109:235–42.

122. Lee OK, Kuo TK, Chen WM, Lee KD, Hsieh SL, Chen TH. Isolation of multipotent mesenchymal stem cells from umbilical cord blood. Blood. 2004;103:1669–75.

123. Kim YS, Kwon JS, Hong MH, Kim J, Song CH, Jeong MH, et al. Promigratory activity of oxytocin on umbilical cord blood-derived mesenchymal stem cells. Artif Organs. 2010;34:453–61.

124. Li T, Ma Q, Ning M, Zhao Y, Hou Y. Cotransplantation of human umbilical cord-derived mesenchymal stem cells and umbilical cord blood-derived CD34(+) cells in a rabbit model of myocardial infarction. Mol Cell Biochem. 2014;387:91–100.

125. Kern S, Eichler H, Stoeve J, Kluter H, Bieback K. Comparative analysis of mesenchymal stem cells from bone marrow, umbilical cord blood, or adipose tissue. Stem Cells. 2006;24:1294–301.

126. Lee HW, Lee HC, Park JH, Kim BW, Ahn J, Kim JH, et al. Effects of intracoronary administration of autologous adipose tissue-derived stem cells on acute myocardial infarction in a porcine model. Yonsei Med J. 2015;56:1522–9.

127. Yang D, Wang W, Li L, Peng Y, Chen P, Huang H, et al. The relative contribution of paracrine effect versus direct differentiation on adipose-derived stem cell transplantation mediated cardiac repair. PLoS One. 2013;8:e59020.

128. He J, Cai Y, Luo LM, Liu HB. Hypoxic adipose mesenchymal stem cells derived conditioned medium protects myocardial infarct in rat. Eur Rev Med Pharmacol Sci. 2015;19:4397–406.

129. Van Harmelen V, Rohrig K, Hauner H. Comparison of proliferation and differentiation capacity of human adipocyte precursor cells from the omental and subcutaneous adipose tissue depot of obese subjects. Metabolism. 2004;53:632–7.

130. Li CH, Duan HL, Fan WW, Wang YB, Zhang Z, Zhang RQ, et al. Beneficial effects of liver X receptor agonist on adipose-derived mesenchymal stem cells transplantation in mice with myocardial infarction. Zhonghua Xin Xue Guan Bing Za Zhi. 2012;40:723–8.

131. Wang Y, Li C, Cheng K, Zhang R, Narsinh K, Li S, et al. Activation of liver X receptor improves viability of adipose-derived mesenchymal stem cells to attenuate myocardial ischemia injury through TLR4/NF-kappaB and Keap-1/Nrf-2 signaling pathways. Antioxid Redox Signal. 2014;21:2543–57.

132. Arnhold S, Wenisch S. Adipose tissue derived mesenchymal stem cells for musculoskeletal repair in veterinary medicine. Am J Stem Cells. 2015;4:1–12.

133. Timmers L, Lim SK, Hoefer IE, Arslan F, Lai RC, van Oorschot AA, et al. Human mesenchymal stem cell-conditioned medium improves cardiac function following myocardial infarction. Stem Cell Res. 2011;6:206–14.

134. Kim SW, Lee DW, Yu LH, Zhang HZ, Kim CE, Kim JM, et al. Mesenchymal stem cells overexpressing GCP-2 improve heart function through enhanced angiogenic properties in a myocardial infarction model. Cardiovasc Res. 2012;95:495–506.

135. Okura H, Matsuyama A, Lee CM, Saga A, Kakuta-Yamamoto A, Nagao A, et al. Cardiomyoblast-like cells differentiated from human adipose tissue-derived mesenchymal stem cells improve left ventricular dysfunction and survival in a rat myocardial infarction model. Tissue Eng Part C Methods. 2010;16:417–25.

136. Yao X, Liu Y, Gao J, Yang L, Mao D, Stefanitsch C, et al. Nitric oxide releasing hydrogel enhances the therapeutic efficacy of mesenchymal stem cells for myocardial infarction. Biomaterials. 2015;60:130–40.

137. Sun CK, Zhen YY, Leu S, Tsai TH, Chang LT, Sheu JJ, et al. Direct implantation versus platelet-rich fibrin-embedded adipose-derived mesenchymal stem cells in treating rat acute myocardial infarction. Int J Cardiol. 2014;173:410–23.

138. Chen YL, Sun CK, Tsai TH, Chang LT, Leu S, Zhen YY, et al. Adipose-derived mesenchymal stem cells embedded in platelet-rich fibrin scaffolds promote angiogenesis, preserve heart function, and reduce left ventricular remodeling in rat acute myocardial infarction. Am J Transl Res. 2015;7:781–803.

139. Perea-Gil I, Monguio-Tortajada M, Galvez-Monton C, Bayes-Genis A, Borras FE, Roura S. Preclinical evaluation of the immunomodulatory properties of cardiac adipose tissue progenitor cells using umbilical cord blood mesenchymal stem cells: a direct comparative study. Biomed Res Int. 2015;2015:439808.

140. Jankowski RJ, Deasy BM, Huard J. Muscle-derived stem cells. Gene Ther. 2002;9:642–7.

141. Morgan JE, Partridge TA. Muscle satellite cells. Int J Biochem Cell Biol. 2003;35:1151–6.

142. Otto Beitnes J, Oie E, Shahdadfar A, Karlsen T, Muller RM, Aakhus S, et al. Intramyocardial injections of human mesenchymal stem cells following acute myocardial infarction modulate scar formation and improve left ventricular function. Cell Transplant. 2012;21:1697–709.

143. Nesti LJ, Jackson WM, Shanti RM, Koehler SM, Aragon AB, Bailey JR, et al. Differentiation potential of multipotent progenitor cells derived from war-traumatized muscle tissue. J Bone Joint Surg Am. 2008;90:2390–8.

144. Zheng B, Cao B, Crisan M, Sun B, Li G, Logar A, et al. Prospective identification of myogenic endothelial cells in human skeletal muscle. Nat Biotechnol. 2007; 25:1025–34.

145. Passipieri JA, Kasai-Brunswick TH, Suhett G, Martins AB, Brasil GV, Campos DB, et al. Improvement of cardiac function by placenta-derived mesenchymal stem cells does not require permanent engraftment and is independent of the insulin signaling pathway. Stem Cell Res Ther. 2014;5:102.

146. Vellasamy S, Sandrasaigaran P, Vidyadaran S, George E, Ramasamy R. Isolation and characterisation of mesenchymal stem cells derived from human placenta tissue. World J Stem Cells. 2012;4:53–61.

147. Oliveira MS, Barreto-Filho JB. Placental-derived stem cells: culture, differentiation and challenges. World J Stem Cells. 2015;7:769–75.

148. Lindenmair A, Hatlapatka T, Kollwig G, Hennerbichler S, Gabriel C, Wolbank S, et al. Mesenchymal stem or stromal cells from amnion and umbilical cord tissue and their potential for clinical applications. Cells. 2012;1:1061–88.

149. Zhao P, Ise H, Hongo M, Ota M, Konishi I, Nikaido T. Human amniotic mesenchymal cells have some characteristics of cardiomyocytes. Transplantation. 2005;79:528–35.

150. Fang CH, Jin J, Joe JH, Song YS, So BI, Lim SM, et al. In vivo differentiation of human amniotic epithelial cells into cardiomyocyte-like cells and cell transplantation effect on myocardial infarction in rats: comparison with cord blood and adipose tissue-derived mesenchymal stem cells. Cell Transplant. 2012;21:1687–96.

151. Kim SW, Zhang HZ, Kim CE, Kim JM, Kim MH. Amniotic mesenchymal stem cells with robust chemotactic properties are effective in the treatment of a myocardial infarction model. Int J Cardiol. 2013;168:1062–9.

152. Chamberlain G, Fox J, Ashton B, Middleton J. Concise review: mesenchymal stem cells: their phenotype, differentiation capacity, immunological features, and potential for homing. Stem Cells. 2007;25: 2739–49.

153. Frid MG, Brunetti JA, Burke DL, Carpenter TC, Davie NJ, Reeves JT, et al. Hypoxia-induced pulmonary vascular remodeling requires recruitment of circulating mesenchymal precursors of a monocyte/macrophage lineage. Am J Pathol. 2006;168:659–69.

154. Du YY, Yao R, Pu S, Zhao XY, Liu GH, Zhao LS, et al. Mesenchymal stem cells implantation increases the myofibroblasts congregating in infarct region in a rat model of myocardial infarction. Zhonghua Xin Xue Guan Bing Za Zhi. 2012;40:1045–50.

155. Zhao J, Hang P, Li Y. TRPC6, a potential novel target for enhancing cardiac repair of bone marrow mesenchymal stem cells. Int J Cardiol. 2012;155:497–8.

156. Turner NA, Porter KE. Function and fate of myofibroblasts after myocardial infarction. Fibrogenesis Tissue Repair. 2013;6:5.

157. Bourin P, Gadelorge M, Peyrafitte JA, Fleury-Cappellesso S, Gomez M, Rage C, et al. Mesenchymal progenitor cells: tissue origin, isolation and culture. Transfus Med Hemother. 2008;35:160–7.

158. Ramkisoensing AA, Pijnappels DA, Askar SF, Passier R, Swildens J, Goumans MJ, et al. Human embryonic and fetal mesenchymal stem cells differentiate toward three different cardiac lineages in contrast to their adult counterparts. PLoS One. 2011;6:e24164.

159. Deng F, Lei H, Hu Y, He L, Fu H, Feng R, et al. Combination of retinoic acid, dimethyl sulfoxide and 5-azacytidine promotes cardiac differentiation of human fetal liver-derived mesenchymal stem cells. Cell Tissue Bank. 2015;17: 147-59 .

160. Dixit P, Katare R. Challenges in identifying the best source of stem cells for cardiac regeneration therapy. Stem Cell Res Ther. 2015;6:26.

161. Li Q, Turdi S, Thomas DP, Zhou T, Ren J. Intra-myocardial delivery of mesenchymal stem cells ameliorates left ventricular and cardiomyocyte contractile dysfunction following myocardial infarction. Toxicol Lett. 2010; 195:119–26.

162. Li Y, Yao Y, Sheng Z, Yang Y, Ma G. Dual-modal tracking of transplanted mesenchymal stem cells after myocardial infarction. Int J Nanomed. 2011;6: 815–23.

163. Perin EC, Tian M, Marini 3rd FC, Silva GV, Zheng Y, Baimbridge F, et al. Imaging long-term fate of intramyocardially implanted mesenchymal stem cells in a porcine myocardial infarction model. PLoS One. 2011;6: e22949.

164. Hao L, Hao J, Fang W, Han C, Zhang K, Wang X. Dual isotope simultaneous imaging to evaluate the effects of intracoronary bone marrow-derived mesenchymal stem cells on perfusion and metabolism in canines with acute myocardial infarction. Biomed Rep. 2015;3:447–52.

165. Tay CY, Yu H, Pal M, Leong WS, Tan NS, Ng KW, et al. Micropatterned matrix directs differentiation of human mesenchymal stem cells towards myocardial lineage. Exp Cell Res. 2010;316:1159–68.

166. Xing Y, Lv A, Wang L, Yan X, Zhao W, Cao F. Engineered myocardial tissues constructed in vivo using cardiomyocyte-like cells derived from bone marrow mesenchymal stem cells in rats. J Biomed Sci. 2012;19:6.

167. Zhang J, Zhi W, Tan M, Chen X, Li X, Deng L. An experimental study on rabbit bone marrow mesenchymal stem cells double-labeled with PKH26 and 5-bromo-2'-deoxyuridine in vitro and application in cardiac patch. Zhongguo Xiu Fu Chong Jian Wai Ke Za Zhi. 2010;24:828–33.

168. Guan J, Wang F, Li Z, Chen J, Guo X, Liao J, et al. The stimulation of the cardiac differentiation of mesenchymal stem cells in tissue constructs that mimic myocardium structure and biomechanics. Biomaterials. 2011;32:5568–80.

169. Zamani M, Prabhakaran MP, Thian ES, Ramakrishna S. Controlled delivery of stromal derived factor-1alpha from poly lactic-co-glycolic acid core-shell particles to recruit mesenchymal stem cells for cardiac regeneration. J Colloid Interface Sci. 2015;451:144–52.

170. Schmuck EG, Mulligan JD, Ertel RL, Kouris NA, Ogle BM, Raval AN, et al. Cardiac fibroblast-derived 3D extracellular matrix seeded with mesenchymal stem cells as a novel device to transfer cells to the ischemic myocardium. Cardiovasc Eng Technol. 2014;5:119–31.

171. Vashi AV, White JF, McLean KM, Neethling WM, Rhodes DI, Ramshaw JA, Werkmeister JA. Evaluation of an established pericardium patch for delivery of mesenchymal stem cells to cardiac tissue. J Biomed Mater Res A. 2014; 103:1999–2005.

172. Maureira P, Marie PY, Yu F, Poussier S, Liu Y, Groubatch F, et al. Repairing chronic myocardial infarction with autologous mesenchymal stem cells engineered tissue in rat promotes angiogenesis and limits ventricular remodeling. J Biomed Sci. 2012;19:93.

173. Fiumana E, Pasquinelli G, Foroni L, Carboni M, Bonafe F, Orrico C, et al. Localization of mesenchymal stem cells grafted with a hyaluronan-based scaffold in the infarcted heart. J Surg Res. 2013;179:e21–9.

174. Radhakrishnan J, Krishnan UM, Sethuraman S. Hydrogel based injectable scaffolds for cardiac tissue regeneration. Biotechnol Adv. 2014;32:449–61.

175. Chen J, Guo R, Zhou Q, Wang T. Injection of composite with bone marrow-derived mesenchymal stem cells and a novel synthetic hydrogel after myocardial infarction: a protective role in left ventricle function. Kaohsiung J Med Sci. 2014;30:173–80.

176. Miskon A, Mahara A, Uyama H, Yamaoka T. A suspension induction for myocardial differentiation of rat mesenchymal stem cells on various extracellular matrix proteins. Tissue Eng Part C Methods. 2010;16:979–87.

177. Santhakumar R, Vidyasekar P, Verma RS. Cardiogel: a nano-matrix scaffold with potential application in cardiac regeneration using mesenchymal stem cells. PLoS One. 2014;9:e114697.

178. Singelyn JM, Christman KL. Injectable materials for the treatment of myocardial infarction and heart failure: the promise of decellularized matrices. J Cardiovasc Transl Res. 2010;3:478–86.

179. Jeffords ME, Wu J, Shah M, Hong Y, Zhang G. Tailoring material properties of cardiac matrix hydrogels to induce endothelial differentiation of human mesenchymal stem cells. ACS Appl Mater Interfaces. 2015;7:11053–61.

180. Li Z, Guo X, Palmer AF, Das H, Guan J. High-efficiency matrix modulus-induced cardiac differentiation of human mesenchymal stem cells inside a thermosensitive hydrogel. Acta Biomater. 2012;8:3586–95.

181. Xu Y, Li Z, Li X, Fan Z, Liu Z, Xie X, et al. Regulating myogenic differentiation of mesenchymal stem cells using thermosensitive hydrogels. Acta Biomater. 2015;26:23–33.

182. Hodonsky C, Mundada L, Wang S, Witt R, Raff G, Kaushal S, et al. Effects of scaffold material used in cardiovascular surgery on mesenchymal stem cells and cardiac progenitor cells. Ann Thorac Surg. 2014;99:605–11.

183. Cui XJ, Xie H, Wang HJ, Guo HD, Zhang JK, Wang C, et al. Transplantation of mesenchymal stem cells with self-assembling polypeptide scaffolds is conducive to treating myocardial infarction in rats. Tohoku J Exp Med. 2010; 222:281–9.

184. Yu J, Du KT, Fang Q, Gu Y, Mihardja SS, Sievers RE, et al. The use of human mesenchymal stem cells encapsulated in RGD modified alginate microspheres in the repair of myocardial infarction in the rat. Biomaterials. 2010;31:7012–20.

185. Li L, Wu S, Liu Z, Zhuo Z, Tan K, Xia H, et al. Ultrasound-targeted microbubble destruction improves the migration and homing of mesenchymal stem cells after myocardial infarction by upregulating SDF-1/ CXCR4: a pilot study. Stem Cells Int. 2015;2015:691310.

186. Xu YL, Gao YH, Liu Z, Tan KB, Hua X, Fang ZQ, et al. Myocardium-targeted transplantation of mesenchymal stem cells by diagnostic ultrasound-mediated microbubble destruction improves cardiac function in myocardial infarction of New Zealand rabbits. Int J Cardiol. 2010;138:182–95.

187. Lee EJ, Park SJ, Kang SK, Kim GH, Kang HJ, Lee SW, et al. Spherical bullet formation via E-cadherin promotes therapeutic potency of mesenchymal stem cells derived from human umbilical cord blood for myocardial infarction. Mol Ther. 2012;20:1424–33.

188. Van Linthout S, Stamm C, Schultheiss HP, Tschope C. Mesenchymal stem cells and inflammatory cardiomyopathy: cardiac homing and beyond. Cardiol Res Pract. 2011;2011:757154.

189. Shi B, Liu ZJ, Zhao RZ, Long XP, Wang DM, Wang ZL. Effect of mesenchymal stem cells on cardiac function and restenosis of injured artery after myocardial infarction. Zhonghua Yi Xue Za Zhi. 2011;91: 2269–73.

190. Vogel S, Trapp T, Borger V, Peters C, Lakbir D, Dilloo D, et al. Hepatocyte growth factor-mediated attraction of mesenchymal stem cells for apoptotic neuronal and cardiomyocytic cells. Cell Mol Life Sci. 2010;67:295–303.

191. Vogel S, Chatterjee M, Metzger K, Borst O, Geisler T, Seizer P, et al. Activated platelets interfere with recruitment of mesenchymal stem cells to apoptotic cardiac cells via high mobility group box 1/Toll-like receptor 4-mediated down-regulation of hepatocyte growth factor receptor MET. J Biol Chem. 2014;289:11068–82.

192. Wynn RF, Hart CA, Corradi-Perini C, O'Neill L, Evans CA, Wraith JE, et al. A small proportion of mesenchymal stem cells strongly expresses functionally active CXCR4 receptor capable of promoting migration to bone marrow. Blood. 2004;104:2643–5.

193. Wiehe JM, Kaya Z, Homann JM, Wohrle J, Vogt K, Nguyen T, et al. GMP-adapted overexpression of CXCR4 in human mesenchymal stem cells for cardiac repair. Int J Cardiol. 2012;167:2073–81.

194. Lau TT, Wang DA. Stromal cell-derived factor-1 (SDF-1): homing factor for engineered regenerative medicine. Expert Opin Biol Ther. 2011;11:189–97.

195. Guo J, Zhang H, Xiao J, Wu J, Ye Y, Li Z, et al. Monocyte chemotactic protein-1 promotes the myocardial homing of mesenchymal stem cells in dilated cardiomyopathy. Int J Mol Sci. 2013;14:8164–19.

196. Wang T, Sun S, Wan Z, Weil MH, Tang W. Effects of bone marrow mesenchymal stem cells in a rat model of myocardial infarction. Resuscitation. 2012;83:1391–6.

197. Ebashi S. Ca2+ and the contractile proteins. J Mol Cell Cardiol. 1984;16:129–36.

198. Zot AS, Potter JD. Structural aspects of troponin-tropomyosin regulation of skeletal muscle contraction. Annu Rev Biophys Biophys Chem. 1987;16:535–59.

199. Szczesna D, Zhang R, Zhao J, Jones M, Guzman G, Potter JD. Altered regulation of cardiac muscle contraction by troponin T mutations that cause familial hypertrophic cardiomyopathy. J Biol Chem. 2000;275:624–30.

200. Matsakas A. Molecular advances shed light on cardiac myosin heavy chain expression in health and disease. Exp Physiol. 2009;94:1161–2.

201. Wei F, Wang T, Liu J, Du Y, Ma A. The subpopulation of mesenchymal stem cells that differentiate toward cardiomyocytes is cardiac progenitor cells. Exp Cell Res. 2011;317:2661–70.

202. Barry FP, Murphy JM. Mesenchymal stem cells: clinical applications and biological characterization. Int J Biochem Cell Biol. 2004;36:568–84.

203. Xu W, Zhang X, Qian H, Zhu W, Sun X, Hu J, et al. Mesenchymal stem cells from adult human bone marrow differentiate into a cardiomyocyte phenotype in vitro. Exp Biol Med (Maywood). 2004;229:623–31.

204. Qian Q, Qian H, Zhang X, Zhu W, Yan Y, Ye S, et al. 5-Azacytidine induces cardiac differentiation of human umbilical cord-derived mesenchymal stem cells by activating extracellular regulated kinase. Stem Cells Dev. 2012;21:67–75.

205. Antonitsis P, Ioannidou-Papagiannaki E, Kaidoglou A, Papakonstantinou C. In vitro cardiomyogenic differentiation of adult human bone marrow mesenchymal stem cells. The role of 5-azacytidine. Interact Cardiovasc Thorac Surg. 2007;6:593–7.

206. Sharma S, Jackson PG, Makan J. Cardiac troponins. J Clin Pathol. 2004;57:1025–6.

207. Michele DE, Metzger JM. Physiological consequences of tropomyosin mutations associated with cardiac and skeletal myopathies. J Mol Med (Berl). 2000;78:543–53.

208. Asumda FZ, Chase PB. Nuclear cardiac troponin and tropomyosin are expressed early in cardiac differentiation of rat mesenchymal stem cells. Differentiation. 2012;83:106–15.

209. Rogers TB, Pati S, Gaa S, Riley D, Khakoo AY, Patel S, et al. Mesenchymal stem cells stimulate protective genetic reprogramming of injured cardiac ventricular myocytes. J Mol Cell Cardiol. 2011;50:346–56.

210. Lawrence T. The nuclear factor NF-kappaB pathway in inflammation. Cold Spring Harb Perspect Biol. 2009;1:a001651.

211. Weil BR, Herrmann JL, Abarbanell AM, Manukyan MC, Poynter JA, Meldrum DR. Intravenous infusion of mesenchymal stem cells is associated with improved myocardial function during endotoxemia. Shock. 2011;36:235–41.

212. Weil BR, Manukyan MC, Herrmann JL, Wang Y, Abarbanell AM, Poynter JA, et al. Mesenchymal stem cells attenuate myocardial functional depression and reduce systemic and myocardial inflammation during endotoxemia. Surgery. 2010;148:444–52.

213. Wang CM, Guo Z, Xie YJ, Hao YY, Sun JM, Gu J, et al. Co-treating mesenchymal stem cells with IL1beta and TNF-alpha increases VCAM-1 expression and improves post-ischemic myocardial function. Mol Med Rep. 2014;10:792–8.

214. Gyongyosi M, Posa A, Pavo N, Hemetsberger R, Kvakan H, Steiner-Boker S, et al. Differential effect of ischaemic preconditioning on mobilisation and recruitment of haematopoietic and mesenchymal stem cells in porcine myocardial ischaemia-reperfusion. Thromb Haemost. 2010;104:376–84.

215. Mias C, Lairez O, Trouche E, Roncalli J, Calise D, Seguelas MH, et al. Mesenchymal stem cells promote matrix metalloproteinase secretion by cardiac fibroblasts and reduce cardiac ventricular fibrosis after myocardial infarction. Stem Cells. 2009;27:2734–43.

216. Wang Y, Hu X, Xie X, He A, Liu X, Wang JA. Effects of mesenchymal stem cells on matrix metalloproteinase synthesis in cardiac fibroblasts. Exp Biol Med (Maywood). 2011;236:1197–204.

217. Yao J, Jiang SL, Liu W, Liu C, Chen W, Sun L, et al. Tissue inhibitor of matrix metalloproteinase-3 or vascular endothelial growth factor transfection of aged human mesenchymal stem cells enhances cell therapy after myocardial infarction. Rejuvenation Res. 2012;15:495–506.

218. Nayan M, Paul A, Chen G, Chiu RC, Prakash S, Shum-Tim D. Superior therapeutic potential of young bone marrow mesenchymal stem cells by direct intramyocardial delivery in aged recipients with acute myocardial infarction: in vitro and in vivo investigation. J Tissue Eng. 2011;2011:741213.

219. Movafagh S, Hobson JP, Spiegel S, Kleinman HK, Zukowska Z. Neuropeptide Y induces migration, proliferation, and tube formation of endothelial cells bimodally via Y1, Y2, and Y5 receptors. FASEB J. 2006;20:1924–6.

220. Wang Y, Zhang D, Ashraf M, Zhao T, Huang W, Ashraf A, et al. Combining neuropeptide Y and mesenchymal stem cells reverses remodeling after myocardial infarction. Am J Physiol Heart Circ Physiol. 2010;298:H275–86.

221. Cho J, Zhai P, Maejima Y, Sadoshima J. Myocardial injection with GSK-3beta-overexpressing bone marrow-derived mesenchymal stem cells attenuates cardiac dysfunction after myocardial infarction. Circ Res. 2011; 108:478–89.

222. Lian WS, Cheng WT, Cheng CC, Hsiao FS, Chen JJ, Cheng CF, et al. In vivo therapy of myocardial infarction with mesenchymal stem cells modified with prostaglandin I synthase gene improves cardiac performance in mice. Life Sci. 2011;88:455–64.

223. Mohri T, Iwakura T, Nakayama H, Fujio Y. JAK-STAT signaling in cardiomyogenesis of cardiac stem cells. JAKSTAT. 2012;1:125–30.

224. Shabbir A, Zisa D, Lin H, Mastri M, Roloff G, Suzuki G, et al. Activation of host tissue trophic factors through JAK-STAT3 signaling: a mechanism of mesenchymal stem cell-mediated cardiac repair. Am J Physiol Heart Circ Physiol. 2010;299:H1428–38.

225. Shabbir A, Zisa D, Suzuki G, Lee T. Heart failure therapy mediated by the trophic activities of bone marrow mesenchymal stem cells: a noninvasive therapeutic regimen. Am J Physiol Heart Circ Physiol. 2009;296:H1888–97.

226. Poynter JA, Herrmann JL, Manukyan MC, Wang Y, Abarbanell AM, Weil BR, et al. Intracoronary mesenchymal stem cells promote postischemic myocardial functional recovery, decrease inflammation, and reduce apoptosis via a signal transducer and activator of transcription 3 mechanism. J Am Coll Surg. 2011;213:253–60.

227. Autiero M, Waltenberger J, Communi D, Kranz A, Moons L, Lambrechts D, et al. Role of PlGF in the intra- and intermolecular cross talk between the VEGF receptors Flt1 and Flk1. Nat Med. 2003;9:936–43.

228. Mu Y, Cao G, Zeng Q, Li Y. Transplantation of induced bone marrow mesenchymal stem cells improves the cardiac function of rabbits with dilated cardiomyopathy via upregulation of vascular endothelial growth factor and its receptors. Exp Biol Med (Maywood). 2011;236:1100–7.

229. Gong S, Seng Z, Wang W, Lv J, Dong Q, Yan B, et al. Bosentan protects the spinal cord from ischemia reperfusion injury in rats through vascular endothelial growth factor receptors. Spinal Cord. 2015;53:19–23.

230. Yu Q, Li Q, Na R, Li X, Liu B, Meng L, et al. Impact of repeated intravenous bone marrow mesenchymal stem cells infusion on myocardial collagen

network remodeling in a rat model of doxorubicin-induced dilated cardiomyopathy. Mol Cell Biochem. 2014;387:279–85.

231. Ammar HI, Sequiera GL, Nashed MB, Ammar RI, Gabr HM, Elsayed HE, et al. Comparison of adipose tissue- and bone marrow-derived mesenchymal stem cells for alleviating doxorubicin-induced cardiac dysfunction in diabetic rats. Stem Cell Res Ther. 2015;6:148.

232. Guo Y, Liu C, He J. Effect of combined therapy of granulocyte colony stimulating factor and bone marrow mesenchymal stem cells carrying hepatocyte growth factor gene on angiogenesis of myocardial infarction in rats. Zhongguo Xiu Fu Chong Jian Wai Ke Za Zhi. 2011;25:736–40.

233. Zhang J, Chen A, Wu Y, Zhao Q. Placental growth factor promotes cardiac muscle repair via enhanced neovascularization. Cell Physiol Biochem. 2015; 36:947–55.

234. Zhang J, Wu Y, Chen A, Zhao Q. Mesenchymal stem cells promote cardiac muscle repair via enhanced neovascularization. Cell Physiol Biochem. 2015; 35:1219–29.

235. Buijs JT, Henriquez NV, van Overveld PG, van der Horst G, ten Dijke P, van der Pluijm G. TGF-beta and BMP7 interactions in tumour progression and bone metastasis. Clin Exp Metastasis. 2007;24:609–17.

236. Shen B, Liu X, Fan Y, Qiu J. Macrophages regulate renal fibrosis through modulating TGFbeta superfamily signaling. Inflammation. 2014;37:2076–84.

237. Standiford TJ, Kuick R, Bhan U, Chen J, Newstead M, Keshamouni VG. TGF-beta-induced IRAK-M expression in tumor-associated macrophages regulates lung tumor growth. Oncogene. 2011;30:2475–84.

238. Wang M, Zhang G, Wang Y, Liu T, Zhang Y, An Y, et al. Crosstalk of mesenchymal stem cells and macrophages promotes cardiac muscle repair. Int J Biochem Cell Biol. 2014;58:53–61.

239. Luo Y, Wang Y, Poynter JA, Manukyan MC, Herrmann JL, Abarbanell AM, et al. Pretreating mesenchymal stem cells with interleukin-1beta and transforming growth factor-beta synergistically increases vascular endothelial growth factor production and improves mesenchymal stem cell-mediated myocardial protection after acute ischemia. Surgery. 2012;151:353–63.

240. Mahmoud AI, Porrello ER, Kimura W, Olson EN, Sadek HA. Surgical models for cardiac regeneration in neonatal mice. Nat Protoc. 2014;9:305–11.

241. Doostzadeh J, Clark LN, Bezenek S, Pierson W, Sood PR, Sudhir K. Recent progress in percutaneous coronary intervention: evolution of the drug-eluting stents, focus on the XIENCE V drug-eluting stent. Coron Artery Dis. 2010;21:46–56.

242. Coultas L, Chawengsaksophak K, Rossant J. Endothelial cells and VEGF in vascular development. Nature. 2005;438:937–45.

243. Neufeld G, Cohen T, Gengrinovitch S, Poltorak Z. Vascular endothelial growth factor (VEGF) and its receptors. FASEB J. 1999;13:9–22.

244. Pankajakshan D, Kansal V, Agrawal DK. In vitro differentiation of bone marrow derived porcine mesenchymal stem cells to endothelial cells. J Tissue Eng Regen Med. 2013;7:911–20.

245. Ikhapoh IA, Pelham CJ, Agrawal DK. Atherogenic cytokines regulate VEGF-A-induced differentiation of bone marrow-derived mesenchymal stem cells into endothelial cells. Stem Cells Int. 2015;2015:498328.

246. Mohri T, Fujio Y, Obana M, Iwakura T, Matsuda K, Maeda M, et al. Signals through glycoprotein 130 regulate the endothelial differentiation of cardiac stem cells. Arterioscler Thromb Vasc Biol. 2009;29:754–60.

247. Ramesh B, Bishi DK, Rallapalli S, Arumugam S, Cherian KM, Guhathakurta S. Ischemic cardiac tissue conditioned media induced differentiation of human mesenchymal stem cells into early stage cardiomyocytes. Cytotechnology. 2012;64:563–75.

248. Carvalho JL, Braga VB, Melo MB, Campos AC, Oliveira MS, Gomes DA, et al. Priming mesenchymal stem cells boosts stem cell therapy to treat myocardial infarction. J Cell Mol Med. 2013;17:617–25.

249. Cui X, Wang H, Guo H, Wang C, Ao H, Liu X, et al. Transplantation of mesenchymal stem cells preconditioned with diazoxide, a mitochondrial ATP-sensitive potassium channel opener, promotes repair of myocardial infarction in rats. Tohoku J Exp Med. 2010;220:139–47.

250. Numasawa Y, Kimura T, Miyoshi S, Nishiyama N, Hida N, Tsuji H, et al. Treatment of human mesenchymal stem cells with angiotensin receptor blocker improved efficiency of cardiomyogenic transdifferentiation and improved cardiac function via angiogenesis. Stem Cells. 2011;29:1405–14.

251. Ikhapoh IA, Pelham CJ, Agrawal DK. Synergistic effect of angiotensin II on vascular endothelial growth factor-A-mediated differentiation of bone marrow-derived mesenchymal stem cells into endothelial cells. Stem Cell Res Ther. 2015;6:4.

252. Liu C, Fan Y, Zhou L, Zhu HY, Song YC, Hu L, et al. Pretreatment of mesenchymal stem cells with angiotensin II enhances paracrine effects,

angiogenesis, gap junction formation and therapeutic efficacy for myocardial infarction. Int J Cardiol. 2015;188:22–32.

253. Tachibana M, Sugimoto K, Nozaki M, Ueda J, Ohta T, Ohki M, et al. G9a histone methyltransferase plays a dominant role in euchromatic histone H3 lysine 9 methylation and is essential for early embryogenesis. Genes Dev. 2002;16:1779–91.

254. Yang J, Kaur K, Ong LL, Eisenberg CA, Eisenberg LM. Inhibition of G9a histone methyltransferase converts bone marrow mesenchymal stem cells to cardiac competent progenitors. Stem Cells Int. 2015;2015:270428.

255. Carvalho PH, Daibert AP, Monteiro BS, Okano BS, Carvalho JL, Cunha DN, et al. Differentiation of adipose tissue-derived mesenchymal stem cells into cardiomyocytes. Arq Bras Cardiol. 2013;100:82–9.

256. Yin N, Lu R, Lin J, Zhi S, Tian J, Zhu J. Islet-1 promotes the cardiac-specific differentiation of mesenchymal stem cells through the regulation of histone acetylation. Int J Mol Med. 2014;33:1075–82.

257. Tan G, Shim W, Gu Y, Qian L, Chung YY, Lim SY, et al. Differential effect of myocardial matrix and integrins on cardiac differentiation of human mesenchymal stem cells. Differentiation. 2010;79:260–71.

258. Zeng B, Lin G, Ren X, Zhang Y, Chen H. Over-expression of HO-1 on mesenchymal stem cells promotes angiogenesis and improves myocardial function in infarcted myocardium. J Biomed Sci. 2010;17:80.

259. Tsubokawa T, Yagi K, Nakanishi C, Zuka M, Nohara A, Ino H, et al. Impact of anti-apoptotic and anti-oxidative effects of bone marrow mesenchymal stem cells with transient overexpression of heme oxygenase-1 on myocardial ischemia. Am J Physiol Heart Circ Physiol. 2010;298:H1320–9.

260. Zhao RZ, Long XP, Liu ZJ, Wang DM, Shi B. Effect of gene modified mesenchymal stem cells overexpression human receptor activity modified protein 1 on inflammation and cardiac repair in a rabbit model of myocardial infarction. Zhonghua Xin Xue Guan Bing Za Zhi. 2012;40:736–41.

261. Shi B, Long X, Zhao R, Liu Z, Wang D, Xu G. Transplantation of mesenchymal stem cells carrying the human receptor activity-modifying protein 1 gene improves cardiac function and inhibits neointimal proliferation in the carotid angioplasty and myocardial infarction rabbit model. Exp Biol Med (Maywood). 2014;239:356–65.

262. Long XP, Zhao RZ, Shi B, Xu GX, Chen CY. Effects of hRAMP1 modified mesenchymal stem cells on restenosis and heart function in rabbit model of carotid angioplasty and myocardial infarction. Zhonghua Yi Xue Za Zhi. 2012;92:2134–9.

263. Liang J, Huang W, Yu X, Ashraf A, Wary KK, Xu M, et al. Suicide gene reveals the myocardial neovascularization role of mesenchymal stem cells overexpressing CXCR4 (MSC(CXCR4)). PLoS One. 2012;7:e46158.

264. Kang K, Ma R, Cai W, Huang W, Paul C, Liang J, et al. Exosomes secreted from CXCR4 overexpressing mesenchymal stem cells promote cardioprotection via Akt signaling pathway following myocardial infarction. Stem Cells Int. 2015;2015:659890.

265. Mao Q, Lin CX, Liang XL, Gao JS, Xu B. Mesenchymal stem cells overexpressing integrin-linked kinase attenuate cardiac fibroblast proliferation and collagen synthesis through paracrine actions. Mol Med Rep. 2013;7:1617–23.

266. Mao Q, Lin C, Gao J, Liang X, Gao W, Shen L, et al. Mesenchymal stem cells overexpressing integrin-linked kinase attenuate left ventricular remodeling and improve cardiac function after myocardial infarction. Mol Cell Biochem. 2014;397:203–14.

267. Du YY, Yao R, Hu XQ, Chen QH, Zhou T, Liu QM, et al. Dural modulation effects of mesenchymal stem cells implantation on myocardial collagen remodeling in a rat model of myocardial infarction. Zhonghua Xin Xue Guan Bing Za Zhi. 2011;39:840–6.

268. Wang Y, Chen X, Cao W, Shi Y. Plasticity of mesenchymal stem cells in immunomodulation: pathological and therapeutic implications. Nat Immunol. 2014;15:1009–16.

269. Quijada P, Salunga HT, Hariharan N, Cubillo JD, El-Sayed FG, Moshref M, et al. Cardiac stem cell hybrids enhance myocardial repair. Circ Res. 2015;117:695–706.

270. Zaruba MM, Franz WM. Role of the SDF-1-CXCR4 axis in stem cell-based therapies for ischemic cardiomyopathy. Expert Opin Biol Ther. 2010;10:321–35.

271. Bartel DP. MicroRNAs: genomics, biogenesis, mechanism, and function. Cell. 2004;116:281–97.

272. Weckbach LT, Grabmaier U, Clauss S, Wakili R. MicroRNAs as a diagnostic tool for heart failure and atrial fibrillation. Curr Opin Pharmacol. 2016;27:24–30.

273. Meder B, Katus HA, Rottbauer W. Right into the heart of microRNA-133a. Genes Dev. 2008;22:3227–31.

274. Bostjancic E, Zidar N, Stajner D, Glavac D. MicroRNA miR-1 is up-regulated in remote myocardium in patients with myocardial infarction. Folia Biol (Praha). 2010;56:27–31.

275. Ye Y, Perez-Polo JR, Qian J, Birnbaum Y. The role of microRNA in modulating myocardial ischemia-reperfusion injury. Physiol Genomics. 2011;43:534–42.

276. Liu N, Bezprozvannaya S, Williams AH, Qi X, Richardson JA, Bassel-Duby R, et al. microRNA-133a regulates cardiomyocyte proliferation and suppresses smooth muscle gene expression in the heart. Genes Dev. 2008;22:3242–54.

277. Matkovich SJ, Wang W, Tu Y, Eschenbacher WH, Dorn LE, Condorelli G, et al. MicroRNA-133a protects against myocardial fibrosis and modulates electrical repolarization without affecting hypertrophy in pressure-overloaded adult hearts. Circ Res. 2010;106:166–75.

278. Duisters RF, Tijsen AJ, Schroen B, Leenders JJ, Lentink V, van der Made I, et al. miR-133 and miR-30 regulate connective tissue growth factor: implications for a role of microRNAs in myocardial matrix remodeling. Circ Res. 2009;104:170–8. 176p following 178.

279. Dakhlallah D, Zhang J, Yu L, Marsh CB, Angelos MG, Khan M. MicroRNA-133a engineered mesenchymal stem cells augment cardiac function and cell survival in the infarct heart. J Cardiovasc Pharmacol. 2015;65:241–51.

280. Liu JL, Jiang L, Lin QX, Deng CY, Mai LP, Zhu JN, et al. MicroRNA 16 enhances differentiation of human bone marrow mesenchymal stem cells in a cardiac niche toward myogenic phenotypes in vitro. Life Sci. 2012;90:1020–6.

281. Yan X, Liang H, Deng T, Zhu K, Zhang S, Wang N, et al. The identification of novel targets of miR-16 and characterization of their biological functions in cancer cells. Mol Cancer. 2013;12:92.

282. Shieh JT, Huang Y, Gilmore J, Srivastava D. Elevated miR-499 levels blunt the cardiac stress response. PLoS One. 2011;6:e19481.

283. Zhang LL, Liu JJ, Liu F, Liu WH, Wang YS, Zhu B, et al. MiR-499 induces cardiac differentiation of rat mesenchymal stem cells through wnt/beta-catenin signaling pathway. Biochem Biophys Res Commun. 2012;420:875–81.

284. Wang JX, Jiao JQ, Li Q, Long B, Wang K, Liu JP, et al. miR-499 regulates mitochondrial dynamics by targeting calcineurin and dynamin-related protein-1. Nat Med. 2011;17:71–8.

285. Boon RA, Iekushi K, Lechner S, Seeger T, Fischer A, Heydt S, et al. MicroRNA-34a regulates cardiac ageing and function. Nature. 2013;495:107–10.

286. Kang HJ, Kang WS, Hong MH, Choe N, Kook H, Jeong HC, et al. Involvement of miR-34c in high glucose-insulted mesenchymal stem cells leads to inefficient therapeutic effect on myocardial infarction. Cell Signal. 2015;27:2241–51.

287. Mao J, Lv Z, Zhuang Y. MicroRNA-23a is involved in tumor necrosis factor-alpha induced apoptosis in mesenchymal stem cells and myocardial infarction. Exp Mol Pathol. 2014;97:23–30.

288. Shinmura D, Togashi I, Miyoshi S, Nishiyama N, Hida N, Tsuji H, et al. Pretreatment of human mesenchymal stem cells with pioglitazone improved efficiency of cardiomyogenic transdifferentiation and cardiac function. Stem Cells. 2011;29:357–66.

289. Zhang Z, Li S, Cui M, Gao X, Sun D, Qin X, et al. Rosuvastatin enhances the therapeutic efficacy of adipose-derived mesenchymal stem cells for myocardial infarction via PI3K/Akt and MEK/ERK pathways. Basic Res Cardiol. 2013;108:333.

290. Liu YL, Zhou Y, Sun L, Wen JT, Teng SJ, Yang L, et al. Protective effects of Gingko biloba extract 761 on myocardial infarction via improving the viability of implanted mesenchymal stem cells in the rat heart. Mol Med Rep. 2014;9:1112–20.

291. Xie X, Sun A, Zhu W, Huang Z, Hu X, Jia J, et al. Transplantation of mesenchymal stem cells preconditioned with hydrogen sulfide enhances repair of myocardial infarction in rats. Tohoku J Exp Med. 2012;226:29–36.

292. Hou J, Wang L, Guo T, Xing Y, Zheng S, Zhou C, et al. Peroxisome proliferator-activated receptor gamma promotes mesenchymal stem cells to express Connexin43 via the inhibition of TGF-beta1/Smads signaling in a rat model of myocardial infarction. Stem Cell Rev. 2015;11:885–99.

293. Li N, Yang YJ, Qian HY, Li Q, Zhang Q, Li XD, et al. Intravenous administration of atorvastatin-pretreated mesenchymal stem cells improves cardiac performance after acute myocardial infarction: role of CXCR4. Am J Transl Res. 2015;7:1058–70.

294. Song L, Yang YJ, Dong QT, Qian HY, Gao RL, Qiao SB, et al. Atorvastatin enhance efficacy of mesenchymal stem cells treatment for swine myocardial infarction via activation of nitric oxide synthase. PLoS One. 2013;8:e65702.

295. Guo HD, Cui GH, Tian JX, Lu PP, Zhu QC, Lv R, et al. Transplantation of salvianolic acid B pretreated mesenchymal stem cells improves cardiac function in rats with myocardial infarction through angiogenesis and paracrine mechanisms. Int J Cardiol. 2014;177:538–42.

296. Tang J, Wang J, Zheng F, Kong X, Guo L, Yang J, et al. Combination of chemokine and angiogenic factor genes and mesenchymal stem cells could enhance angiogenesis and improve cardiac function after acute myocardial infarction in rats. Mol Cell Biochem. 2010;339:107–18.

297. Chen J, Zheng S, Huang H, Huang S, Zhou C, Hou J, et al. Mesenchymal stem cells enhanced cardiac nerve sprouting via nerve growth factor in a rat model of myocardial infarction. Curr Pharm Des. 2014;20:2023–9.

298. Mureli S, Gans CP, Bare DJ, Geenen DL, Kumar NM, Banach K. Mesenchymal stem cells improve cardiac conduction by upregulation of connexin 43 through paracrine signaling. Am J Physiol Heart Circ Physiol. 2013;304:H600–9.

299. Wang D, Jin Y, Ding C, Zhang F, Chen M, Yang B, et al. Intracoronary delivery of mesenchymal stem cells reduces proarrhythmogenic risks in swine with myocardial infarction. Ir J Med Sci. 2011;180:379–85.

300. Serrao GW, Turnbull IC, Ancukiewicz D, Kim do E, Kao E, Cashman TJ, et al. Myocyte-depleted engineered cardiac tissues support therapeutic potential of mesenchymal stem cells. Tissue Eng Part A. 2012;18:1322–33.

301. Da Silva JS, Hare JM. Cell-based therapies for myocardial repair: emerging role for bone marrow-derived mesenchymal stem cells (MSCs) in the treatment of the chronically injured heart. Methods Mol Biol. 2013;1037:145–63.

302. Williams AR, Trachtenberg B, Velazquez DL, McNiece I, Altman P, Rouy D, et al. Intramyocardial stem cell injection in patients with ischemic cardiomyopathy: functional recovery and reverse remodeling. Circ Res. 2011;108:792–6.

303. Trachtenberg B, Velazquez DL, Williams AR, McNiece I, Fishman J, Nguyen K, et al. Rationale and design of the Transendocardial Injection of Autologous Human Cells (bone marrow or mesenchymal) in Chronic Ischemic Left Ventricular Dysfunction and Heart Failure Secondary to Myocardial Infarction (TAC-HFT) trial: a randomized, double-blind, placebo-controlled study of safety and efficacy. Am Heart J. 2011;161:487–93.

304. Mushtaq M, DiFede DL, Golpanian S, Khan A, Gomes SA, Mendizabal A, et al. Rationale and design of the Percutaneous Stem Cell Injection Delivery Effects on Neomyogenesis in Dilated Cardiomyopathy (the POSEIDON-DCM study): a phase I/II, randomized pilot study of the comparative safety and efficacy of transendocardial injection of autologous mesenchymal stem cell vs. allogeneic mesenchymal stem cells in patients with non-ischemic dilated cardiomyopathy. J Cardiovasc Transl Res. 2014;7:769–80.

305. Lee JW, Lee SH, Youn YJ, Ahn MS, Kim JY, Yoo BS, et al. A randomized, open-label, multicenter trial for the safety and efficacy of adult mesenchymal stem cells after acute myocardial infarction. J Korean Med Sci. 2013;29:23–31.

306. Hare JM, Fishman JE, Gerstenblith G, DiFede Velazquez DL, Zambrano JP, Suncion VY, et al. Comparison of allogeneic vs autologous bone marrow-derived mesenchymal stem cells delivered by transendocardial injection in patients with ischemic cardiomyopathy: the POSEIDON randomized trial. JAMA. 2012;308: 2369–79.

307. Rodrigo SF, van Ramshorst J, Hoogslag GE, Boden H, Velders MA, Cannegieter SC, et al. Intramyocardial injection of autologous bone marrow-derived ex vivo expanded mesenchymal stem cells in acute myocardial infarction patients is feasible and safe up to 5 years of follow-up. J Cardiovasc Transl Res. 2013;6:816–25.

308. Hare JM, Traverse JH, Henry TD, Dib N, Strumpf RK, Schulman SP, et al. A randomized, double-blind, placebo-controlled, dose-escalation study of intravenous adult human mesenchymal stem cells (prochymal) after acute myocardial infarction. J Am Coll Cardiol. 2009;54:2277–86.

309. Jeevanantham V, Butler M, Saad A, Abdel-Latif A, Zuba-Surma EK, Dawn B. Adult bone marrow cell therapy improves survival and induces long-term improvement in cardiac parameters: a systematic review and meta-analysis. Circulation. 2012;126:551–68.

310. Anastasiadis K, Antonitsis P, Doumas A, Koliakos G, Argiriadou H, Vaitsopoulou C, et al. Stem cells transplantation combined with long-term mechanical circulatory support enhances myocardial viability in end-stage ischemic cardiomyopathy. Int J Cardiol. 2012;155:e51–3.

311. Chen Y, Teng X, Chen W, Yang J, Yang Z, Yu Y, et al. Timing of transplantation of autologous bone marrow derived mesenchymal stem cells for treating myocardial infarction. Sci China Life Sci. 2014;57: 195–200.

312. dos Santos F, Andrade PZ, Eibes G, da Silva CL, Cabral JM. Ex vivo expansion of human mesenchymal stem cells on microcarriers. Methods Mol Biol. 2011;698:189–98.

313. Wei X, Yang X, Han ZP, Qu FF, Shao L, Shi YF. Mesenchymal stem cells: a new trend for cell therapy. Acta Pharmacol Sin. 2013;34:747–54.

314. Herberts CA, Kwa MS, Hermsen HP. Risk factors in the development of stem cell therapy. J Transl Med. 2011;9:29.

315. Jeong JO, Han JW, Kim JM, Cho HJ, Park C, Lee N, et al. Malignant tumor formation after transplantation of short-term cultured bone marrow mesenchymal stem cells in experimental myocardial infarction and diabetic neuropathy. Circ Res. 2011;108:1340–7.

316. Ko IK, Kim BS. Mesenchymal stem cells for treatment of myocardial infarction. Int J Stem Cells. 2008;1:49–54.

317. Askar SF, Ramkisoensing AA, Atsma DE, Schalij MJ, de Vries AA, Pijnappels DA. Engraftment patterns of human adult mesenchymal stem cells expose electrotonic and paracrine proarrhythmic mechanisms in myocardial cell cultures. Circ Arrhythm Electrophysiol. 2013;6:380–91.

318. Hegyi L, Thway K, Fisher C, Sheppard MN. Primary cardiac sarcomas may develop from resident or bone marrow-derived mesenchymal stem cells: use of immunohistochemistry including CD44 and octamer binding protein 3/4. Histopathology. 2012;61:966–73.

319. Huang XP, Sun Z, Miyagi Y, McDonald Kinkaid H, Zhang L, Weisel RD, et al. Differentiation of allogeneic mesenchymal stem cells induces immunogenicity and limits their long-term benefits for myocardial repair. Circulation. 2010;122:2419–29.

320. Hodgkiss-Geere HM, Argyle DJ, Corcoran BM, Whitelaw B, Milne E, David B, et al. Cardiac specific gene expression changes in long term culture of murine mesenchymal stem cells. Int J Stem Cells. 2011;4:143–8.

321. Dayan V, Yannarelli G, Filomeno P, Keating A. Human mesenchymal stromal cells improve scar thickness without enhancing cardiac function in a chronic ischaemic heart failure model. Interact Cardiovasc Thorac Surg. 2012;14:516–20.

322. Xiao W, Mohseny AB, Hogendoorn PC, Cleton-Jansen AM. Mesenchymal stem cell transformation and sarcoma genesis. Clin Sarcoma Res. 2013;3:10.

323. Torsvik A, Rosland GV, Svendsen A, Molven A, Immervoll H, McCormack E, et al. Spontaneous malignant transformation of human mesenchymal stem cells reflects cross-contamination: putting the research field on track—letter. Cancer Res. 2010;70:6393–6.

324. Bernardo ME, Zaffaroni N, Novara F, Cometa AM, Avanzini MA, Moretta A, et al. Human bone marrow derived mesenchymal stem cells do not undergo transformation after long-term in vitro culture and do not exhibit telomere maintenance mechanisms. Cancer Res. 2007;67:9142–9.

325. Amariglio N, Hirshberg A, Scheithauer BW, Cohen Y, Loewenthal R, Trakhtenbrot L, et al. Donor-derived brain tumor following neural stem cell transplantation in an ataxia telangiectasia patient. PLoS Med. 2009;6:e1000029.

326. Liang H, Hou H, Yi W, Yang G, Gu C, Lau WB, et al. Increased expression of pigment epithelium-derived factor in aged mesenchymal stem cells impairs their therapeutic efficacy for attenuating myocardial infarction injury. Eur Heart J. 2013;34:1681–90.

327. Liu X, Chen H, Zhu W, Hu X, Jiang Z, Xu Y, et al. Transplantation of SIRT1-engineered aged mesenchymal stem cells improves cardiac function in a rat myocardial infarction model. J Heart Lung Transplant. 2014;33:1083–92.

328. de Jong R, van Hout GP, Houtgraaf JH, Kazemi K, Wallrapp C, Lewis A, et al. Intracoronary infusion of encapsulated glucagon-like peptide-1-eluting mesenchymal stem cells preserves left ventricular function in a porcine model of acute myocardial infarction. Circ Cardiovasc Interv. 2014;7:673–83.

329. Steinhauser ML, Lee RT. Regeneration of the heart. EMBO Mol Med. 2011;3: 701–12.

330. Assmus B, Zeiher AM. Early cardiac retention of administered stem cells determines clinical efficacy of cell therapy in patients with dilated cardiomyopathy. Circ Res. 2013;112:6–8.

331. Laflamme MA, Murry CE. Heart regeneration. Nature. 2011;473:326–35.

332. Wexler SA, Donaldson C, Denning-Kendall P, Rice C, Bradley B, Hows JM. Adult bone marrow is a rich source of human mesenchymal 'stem' cells but umbilical cord and mobilized adult blood are not. Br J Haematol. 2003;121: 368–374.

333. Baksh D, Yao R, Tuan RS. Comparison of proliferative and multilineage differentiation potential of human mesenchymal stem cells derived from umbilical cord and bone marrow. Stem Cells. 2007;25:1384–92.

334. Tsuji H, Miyoshi S, Ikegami Y, Hida N, Asada H, Togashi I, Suzuki J, Satake M, Nakamizo H, Tanaka M, et al. Xenografted human amniotic membrane-derived mesenchymal stem cells are immunologically tolerated and transdifferentiated into cardiomyocytes. Circ Res. 2010;106:1613–23.

335. Wan C, He Q, Li G. Allogenic peripheral blood derived mesenchymal stem cells (MSCs) enhance bone regeneration in rabbit ulna critical-sized bone defect model. J Orthop Res. 2006;24:610–18.

336. Acosta SA, Franzese N, Staples M, Weinbren NL, Babilonia M, Patel J, Merchant N, Simancas AJ, Slakter A, Caputo M, et al. Human Umbilical Cord Blood for Transplantation Therapy in Myocardial Infarction. J Stem Cell Res. Ther 2013;(Suppl 4).

337. Bieback K, Kern S, Kluter H, Eichler H. Critical parameters for the isolation of mesenchymal stem cells from umbilical cord blood. Stem Cells. 2004;22: 625–634.

13

Minicircle DNA-mediated endothelial nitric oxide synthase gene transfer enhances angiogenic responses of bone marrow-derived mesenchymal stem cells

Nadeeka Bandara[1,3], Saliya Gurusinghe[1], Haiying Chen[2], Shuangfeng Chen[2], Le-xin Wang[1,2], Shiang Y. Lim[3,4] and Padraig Strappe[1*]

Abstract

Background: Non-viral-based gene modification of adult stem cells with endothelial nitric oxide synthase (eNOS) may enhance production of nitric oxide and promote angiogenesis. Nitric oxide (NO) derived from endothelial cells is a pleiotropic diffusible gas with positive effects on maintaining vascular tone and promoting wound healing and angiogenesis. Adult stem cells may enhance angiogenesis through expression of bioactive molecules, and their genetic modification to express eNOS may promote NO production and subsequent cellular responses.

Methods: Rat bone marrow-derived mesenchymal stem cells (rBMSCs) were transfected with a minicircle DNA vector expressing either green fluorescent protein (GFP) or eNOS. Transfected cells were analysed for eNOS expression and NO production and for their ability to form in vitro capillary tubules and cell migration. Transcriptional activity of angiogenesis-associated genes, CD31, VEGF-A, PDGFRα, FGF2, and FGFR2, were analysed by quantitative polymerase chain reaction.

Results: Minicircle vectors expressing GFP (MC-GFP) were used to transfect HEK293T cells and rBMSCs, and were compared to a larger parental vector (P-GFP). MC-GFP showed significantly higher transfection in HEK293T cells (55.51 ± 3.3 %) and in rBMSC (18.65 ± 1.05 %) compared to P-GFP in HEK293T cells (43.4 ± 4.9 %) and rBMSC (15.21 ± 0.22 %). MC-eNOS vectors showed higher transfection efficiency (21 ± 3 %) compared to P-eNOS (9 ± 1 %) and also generated higher NO levels. In vitro capillary tubule formation assays showed both MC-eNOS and P-eNOS gene-modified rBMSCs formed longer (14.66 ± 0.55 mm and 13.58 ± 0.68 mm, respectively) and a greater number of tubules (56.33 ± 3.51 and 51 ± 4, respectively) compared to controls, which was reduced with the NOS inhibitor L-NAME. In an in vitro wound healing assay, MC-eNOS transfected cells showed greater migration which was also reversed by L-NAME treatment. Finally, gene expression analysis in MC-eNOS transfected cells showed significant upregulation of the endothelial-specific marker CD31 and enhanced expression of VEGFA and FGF-2 and their corresponding receptors PDGFRα and FGFR2, respectively.

Conclusions: A novel eNOS-expressing minicircle vector can efficiently transfect rBMSCs and produce sufficient NO to enhance in vitro models of capillary formation and cell migration with an accompanying upregulation of CD31, angiogenic growth factor, and receptor gene expression.

Keywords: Minicircle, DNA vector, Transfection, Endothelial nitric oxide synthase, Mesenchymal stem cells, Nitric oxide, Angiogenesis

---

* Correspondence: pstrappe@csu.edu.au
[1]School of Biomedical Sciences, Charles Sturt University, Wagga Wagga, NSW 2650, Australia
Full list of author information is available at the end of the article

Background

Development of safe and efficient systems for gene transfer is required for translation of gene-modified stem cells into therapeutic applications. Conventional plasmid DNA (pDNA)-based non-viral vectors contain bacterial sequences and transcriptional units that may contribute to an immune response against bacterial proteins expressed from cryptic upstream eukaryotic expression signals. Furthermore, changes in eukaryotic gene expression may be altered due to the antibiotic resistance marker and immune responses to bacterial CpG sequences [1]. These prokaryotic DNA sequences present in pDNA vectors may lower their biocompatibility and safety. In clinical studies, un-methylated CpG motifs induced inflammatory responses [2] and necrosis- or apoptosis-mediated cell death in target cells, resulting in short-lived transgene expression [3, 4]. Furthermore, during the intracellular trafficking of pDNA, the bacterial sequences of pDNA vectors are rapidly associated with histone proteins, packing the sequences into a dense heterochromatin structure. If these are spread into the adjacent transgene in the vector, the sequences can become inaccessible by transcription factors, leading to reduced transgene expression through silencing of the eukaryotic promoter [5]. The removal of CpG islands by cloning out, or elimination of non-essential sequences, can reduce these undesirable responses but is time-consuming and tedious.

Minicircle (MC) pDNA technology consists of super-coiled DNA molecules for non-viral gene transfer, which has neither a bacterial origin of replication nor an antibiotic resistance gene [6]. MCs can be generated in *E. coli* ZYCY10P3S2T by attachment sites ((*att*P and *att*B), with specific recombination mediated by the phage ΦC31 integrase [1]. As a result of this recombination event between *att*P and *att*B sites, MCs contain only a eukaryotic expression cassette and the *att*R fragments are formed but are devoid of bacterial backbone sequences. Absence of the bacterial backbone sequences leads to a size reduction in the MC relative to the parental pDNA which can enhance in vitro transfection efficiency [7] and in vivo gene delivery [8, 9]. Gene expression from non-viral episomal vectors may also enhance persistence of transgene expression without interrupting to the cellular genome [10].

Endothelial nitric oxide synthase (eNOS), also known as NOS3, is expressed in endothelial cells [11], and is responsible for generating nitric oxide (NO) which plays an important role in vasculogenesis [12, 13]. NO produced from endothelial cells is important for maintaining vascular integrity and may enhance vasculogenesis through fibroblast growth factor (FGF) signalling [14]. Vascular endothelial growth factor (VEGF) is also induced by the NO synthesis pathway [15] contributing to angiogenesis. eNOS knockout mice (eNOS$^{-/-}$) display impaired vasculogenesis [16] and have also demonstrated diminished wound healing due to reduced VEGF-mediated migration of endothelial cells [17] and bone marrow progenitor cells [18] to the sites of injury. eNOS-based gene therapy approaches have shown restoration of impaired angiogenesis in rats [19, 20] and promotion of re-endothelialisation [21] in injured rabbits upon adenovirus-mediated eNOS gene transfer.

Similar to endothelial progenitor cells, mesenchymal stem cells (MSCs) also participate in post-natal angiogenesis [22], and vascular pericytes, which are crucial for maintaining vascular integrity, share similar phenotypic features with MSCs [23]. Exogenously administered, MSCs form new capillaries and medium-sized arteries [24] which are important properties of tissue regeneration by MSCs [25]. MSCs can differentiate into endothelial cells in vitro [26] and contribute to neovascularisation, particularly during tissue ischaemia and tumour vascularisation [27]. In MSCs, VEGF-A binds with platelet-derived growth factor receptor (PDGFR) to initiate VEGF-A/PDGFR signalling and drive vasculogenesis, as opposed to the VEGFR2 in endothelial cells, which is absent on MSCs [28]. NO has been shown to upregulate PDGFRα receptor expression in rat mesangial cells [29], and the induction of tumour angiogenesis has been linked to the NO-induced Notch signalling pathway in PDGFR-activated mouse glioma cells [30]. FGF2 signalling also enhances vasculogenesis through promotion of NO production [31, 32]. eNOS is the only NOS isoform absent in MSCs [13], and hence eNOS-based genetic modification of MSCs may enhance their therapeutic application. In this study, we describe a novel non-viral MC vector to deliver the eNOS transgene to MSCs with higher transfection efficiency than regular plasmids. NO signalling in the gene-modified MSC promotes capillary tube-like network formation and cell motility. Quantitative real time polymerase chain reaction (PCR) data revealed that MC-mediated eNOS gene transfer significantly upregulates endothelial-specific CD31 gene expression. Furthermore, NO upregulates the angiogenic responsive genes VEGF-A and FGF2 and expression of their corresponding receptors, PDGFRα and FGFR2.

Methods

Rat bone marrow-derived mesenchymal stem cell isolation

All experiments involving animals were approved by the Charles Sturt University animal ethics committee. MSCs were isolated from the bone marrow of 8–12 week old male Sprague–Dawley rats as previously described [33].

Tri-lineage differentiation of rat bone marrow-derived mesenchymal stem cells

The ability of the isolated rat bone marrow-derived MSCs (rBMSCs) (Passage 6) to differentiate to adipogenic, osteogenic and chondrogenic lineages was investigated. To induce osteogenic differentiation, rBMSCs at 80–90 % confluency were incubated in osteogenic-defined medium (Dulbecco's modified Eagle medium (DMEM) supplemented with 10 % fetal bovine serum (FBS), 10 mM beta-glycerol phosphate, 10 nM dexamethasone and 0.2 mM L-ascorbic acid 2-phosphate) for 11 days with medium changed twice a week, as described previously [34]. Cells were then fixed with 4 % paraformaldehyde and stained with Alizarin Red S (pH 4.1) as described previously [35].

To induce adipogenic differentiation, rBMSCs at 80–90 % confluency were incubated in adipogenic-defined medium (DMEM supplemented with 10 % FBS, 10 µM indomethacin, 1 µM dexamethasone, 0.8 µM insulin, 0.5 mM rosiglitazone) [36] for 1 week with media changed twice. Adipogenic differentiation was assessed by 0.18 % Oil Red O staining after fixing the cells in 10 % neutral-buffered formalin (NBF) [35].

To induce chondrogenic differentiation, three-dimensional pellet cultures of rBMSCs (2.5×10^5 cells) were formed by centrifugation at $500 \times g$ in 10 ml conical-bottomed sterile tubes. The chondrogenic induction medium consisted of DMEM supplemented with $1 \times ITS + 3$ (Sigma), $1 \times$ non-essential amino acids (Sigma), 10 ng/ml transforming growth factor β (TGF-β3; Peprotech), 100 nM dexamethasone, and 2 µM ascorbic acid (Sigma) [37]. Pellet cultures were incubated in induction medium for 14 days with the medium changed every second day with the lids of the tube loosened to facilitate gas exchange. At day 14 the pellets were fixed in 10 % NBF for 24 h, and the three-dimensional tissues were processed and embedded in paraffin wax for microtome processing. To assess chondrogenic differentiation, embedded pellets were sectioned (5 µm slices) and stained with 1 % Alcian blue to visualise glycosaminoglycan accumulation.

The images for differentiated cells into all three lineages were captured by a colour camera (Nikon Digital Sight Ds-Fi2) attached to a Nikon Eclipse-Ti-U microscope (Nikon).

Production of minicircle plasmid DNA-expressing eNOS

To construct an eNOS expressing minicircle vector, a codon optimized human eNOS cDNA sequence (3633 bp) was cloned into the minicircle parental plasmid consisting of expression cassette CMV–MCS–EF1α–GFP–SV40–PolyA (P-GFP) (System Biosciences, Mountain View, CA, USA). This cloning strategy allowed removal of the EF1α–GFP portion from the final construct (P-eNOS).

The minicircle DNA plasmids expressing eNOS and GFP were produced according to the manufacturer's instructions (System Biosciences). Briefly, *E. coli* ZYCY10P3S2T cells were transformed with P-GFP and

P-eNOS. Following this, single colonies were grown in 2 ml LB (luria broth) media containing 50 µg/ml kanamycin for 1 h at 30 °C with vigorous shaking at 200 rpm. Next, 50 µl of the starter culture was then used to inoculate 200 ml fresh terrific broth (TB; Sigma) in a 1 litre flask with 50 µg/ml kanamycin followed by incubation at 30 °C for 17 h with constant shaking at 200 rpm. Minicircle induction medium consisting of 200 ml LB (luria broth), 8 ml 1 N NaOH and 200 µl 20 % L-arabinose was combined with the TB bacterial culture and incubated for a further 4 h at 30 °C with constant shaking at 200 rpm. Minicircle plasmid DNA (MC-eNOS and MC-GFP) was isolated using a Genomed Jetstar 2.0 midi kit according to the manufacturer's instructions (Genomed, Germany) and treated with plasmid-safe ATP-dependent DNase (Epicentre, USA) to remove bacterial genomic DNA contamination. eNOS- and GFP-containing minicircles were designated as MC-eNOS and MC-GFP, respectively.

Cell culture and transfection

Human embryonic kidney (HEK293T) cells and rBMSCs were maintained in DMEM (Sigma) supplemented with 10 % (v/v) FBS (Sigma), 1 % (v/v) L-glutamate (Sigma) and 1 % (v/v) penicillin/streptomycin antibiotics mix (Sigma). Cells were transfected with the plasmids (P-GFP, MC-GFP, P-eNOS and MC-eNOS) using Lipofectamine 2000 reagent (Life technologies, USA) following the manufacturer's instructions. GFP expression was assessed by fluorescence microscopy at 24 and 48 h after transfection, and flow cytometry analysis (Gallios Instrument, Beckmann).

Immunocytochemistry

Immunocytochemical detection of eNOS expression in P-eNOS and MC-eNOS transfected HEK293T and rBMSCs was performed as follows. Briefly, cells were fixed in 4 % paraformaldehyde for 20 min at room temperature, treated with 0.1% Triton-X100 in phosphate-buffered saline (PBS) for 10 min, and blocked in a 10 % FBS in PBS solution for 30 min at room temperature. This was followed by a 2-h incubation with a primary mouse monoclonal anti-eNOS antibody (BD Bioscience), and subsequently with an anti-mouse IgG secondary antibody conjugated with Alexa 488 (Cell Signalling Technology) for 1 h followed by DAPI (nuclear stain) and phalloidin-TRITC (cytoskeleton stain) (Sigma). eNOS-positive cells were counted by fluorescence microscopy in five randomly selected fields per well in three independent experiments and 500–1000 cells were counted in total; the percentage of eNOS positivity was calculated from the total nuclear stained cells.

Nitric oxide detection

Nitric oxide released from P-eNOS and MC-eNOS transfected cells in cell supernatants was measured using the griess reagent (Promega) following the manufacturer's instructions. NO was also directly detected in transfected cells using a specific fluorescent NO indicator, 4,5-diaminofluorescein diacetate (DAF-2DA; Cayman chemicals, USA), as described previously [13, 38]. Cells were grown to confluence on a 12-well plate and incubated for 30 min with 1 μM DAF-2DA. Subsequently, cells were washed with fresh PBS and viewed by a fluorescence microscope.

In vitro angiogenesis

In vitro capillary formation was performed as described previously [39]. Briefly, Geltrex™ (Life technologies) was thawed on ice overnight and applied evenly over each well (50 μl) of a 96-well plate and incubated for 30 min at 37 °C allowing polymerisation. Transfected rBMSCs or control cells were seeded at 20,000 cells per well and grown in 100 μl angiogenic induction medium (DMEM (Sigma), 1.5 % FBS, 1 % (v/v) L-glutamate (Sigma) and 1 % (v/v) penicillin/streptomycin (Sigma)) and incubated at 37 °C for 5 h. The capillary network was fixed with 4 % paraformaldehyde and visualized by staining with DAPI and Phalloidin (Sigma). The efficiency of in vitro tubule formation was evaluated by measuring the number of nodes and length of the tubules as described previously [13].

In vitro scratch wound healing assay

The effect of nitric oxide on cell migration was assessed using an in vitro scratch wound healing assay as described previously [37]. Briefly, HEK293T cells and rBMSCs were transfected with P-eNOS, MC-eNOS, P-GFP and MC-GFP in 6-well tissue culture plates. Next, 48 h following the transfection when the cells reached 100 % confluence, scratch wounds were made using a sterile 200 μl pipette tip and the boundaries were marked. The cells were then cultured with 2 ml fresh DMEM supplemented with 10 % (v/v) FBS (Sigma), 1 % (v/v), L-glutamate (Sigma), and 1 % (v/v) penicillin/streptomycin (Sigma). Phase-contrast microscopy images

were acquired at 0 and 1 h after scratches were created for rBMSCs and after17 h for HEK293T cells. Cell migration was measured at the indicated times by measuring the distance from the initial boundary edge to the boundary of the migrating cells, followed by calculation of the percentage of wound closure as follows: percentage of wound closure = (distance from the boundary edge at 0 h − distance from the boundary edge at 1 h or 17 h)/(distance from the boundary edge at 0 h) × 100.

Gene expression by quantitative real time PCR

Total RNA from transfected and control cells was isolated using the PureZol reagent (BioRad) according to the manufacturer's instructions and the concentration of isolated RNA was determined using a Nanodrop spectrophotometer (Thermo Scientific) following treatment with RQ1 RNase free DNase (Promega) to remove contaminating DNA. Then, cDNA was synthesized with 1 μg RNA using a High Capacity Reverse Transcription Kit (Life technologies). The quantitative real time PCR assays were performed on a BioRad CFX96 Real-Time system (BioRad) using the SsoFast EvaGreen Supermix (BioRad). Primers used for target amplification are described in Table 1. Assays were performed in triplicate, and target mRNA expression was normalized to rat GAPDH mRNA levels using the ΔCt method.

Western blot analysis

Transfected and control cells were washed with ice-cold PBS (Sigma) twice, and lysates were prepared by homogenization of cells in RIPA buffer (Sigma), following mixing with 4 × NuPAGE LDS sample buffer (Life technologies) and lysed by heating for 10 min at 70 °C. Total proteins were separated by 4–12 % Bis-Tris NuPAGE (Novex, Life technologies) and transferred to PVDF membrane (Millipore). After blocking with odyssey blocking buffer (LI-COR) for 30 min at room temperature, the membrane was incubated with primary antibodies specific to eNOS (1:1000 dilution) and β-actin (LI-COR; 1:1000 dilution) overnight at 4 °C. The membrane was washed with 0.1 % tween in PBS three times for 10 min each, incubated with donkey

Table 1 Primers used in this study

Target	Forward	Reverse	Expected size	Accession number
VEGF-A	GGTGGACATCTTCCAGGAGT	TGATCTGCATGGTGATGTTG	146	NM_001317043
FGF2	GCTGCTGGCTTCTAAGTGTG	TACTGCCCAGTTCGTTTCAG	129	NM_019305
PDGFR α	TTGAGCCCATTACTGTTGGA	CCCATAGCTCCTGAGACCTT	148	NM_011058
FGFR2	GACGACACAGATAGCTCCGA	CAGCGGAACTTCACAGTGTT	134	EF143338
CD31	CATTGGTTACCTCGGGAGTC	GTCTTCACCCAGCCTTTCTC	104	NM_001107202
GAPDH	ACAGCAACAGGGTGGTGGAC	TTTGAGGGTGCAGCAACTT	252	NM_017008.4

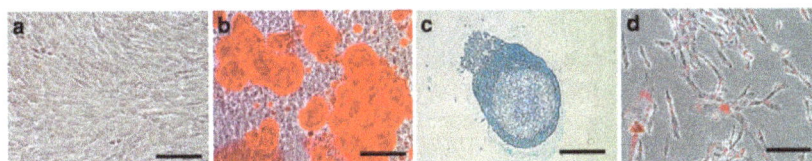

Fig. 1 Characterization of rBMSCs. Tri-lineage differentiation of rBMSC was performed in vitro. **a** Undifferentiated rBMSC. **b** Alizarin red S staining of cells cultured for 14 days in osteogenic induction medium. **c** Alcian blue staining and toluidine blue staining of cells cultured for 14 days in chondrogenic induction medium. **d** Oil red O staining of cells cultured for 7 days in adipogenic induction medium. *Scale bar* = 100 μm

anti-rabbit IgG (H&L) (Alexa Fluor® 680) secondary antibody (Life technologies; 1:20,000) at room temperature for 1 h, and antibody-bound proteins were visualized by fluorescence detection with a LI-COR odyssey system.

Statistical analysis

All experiments were performed in triplicate and at least three times and data analysed by an independent two-tailed Student's t test. A p value <0.05 was regarded as statistically significant.

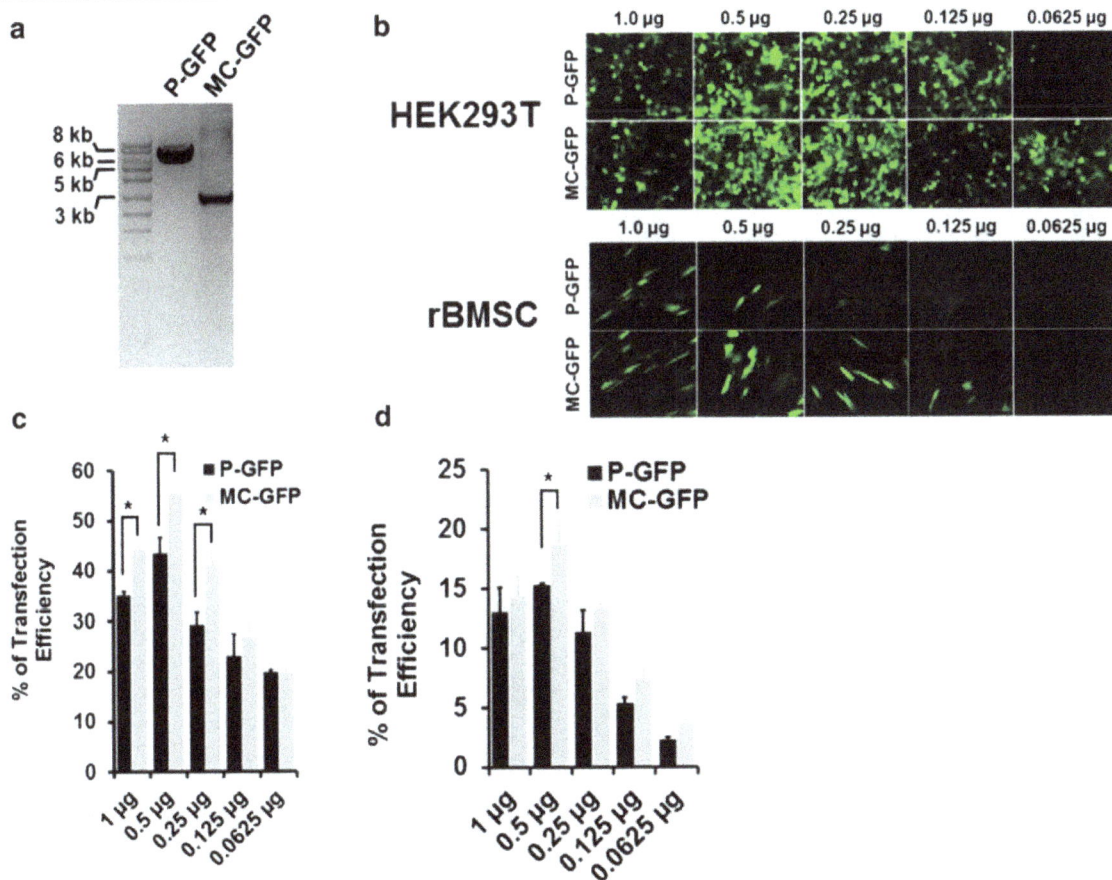

Fig. 2 Gene delivery efficiency of P-GFP and MC-GFP DNA vectors. **a** Gel electrophoresis of P-GFP and MC-GFP plasmids following plasmid purification and enzyme digestion. **b** Fluorescence microscopy of transfected HEK293T cells and rBMSCs with P-GFP and MC-GFP with a range of plasmid concentrations and quantitation by flow cytometry of transfection efficiencies for (**c**) HEK293T cells and (**d**) rBMSCs. *MC-GFP* minicircle vector expressing green fluorescent protein, *P-GFP* plasmid vector expressing green fluorescent protein, *rBMSC* rat bone marrow-derived mesenchymal stem cell

Results

Characterisation of rBMSCs

rBMSCs were isolated from adult Sprague–Dawley rats as previously described [33], and plastic adherent rBMSCs displayed typical fibroblastoid morphology (Fig. 1a) [40]. Tri-lineage differentiation of the rBMSCs was performed in the appropriate media to osteoblasts as demonstrated by Alzarin Red S staining of mineralised extracellular matrix (Fig. 1b), to chondrocytes by Alcian Blue staining of proteoglycans in three-dimensional pellet cultures (Fig. 1c) and to adipocytes as shown by Oil Red O staining of lipid vesicles (Fig. 1d).

Transfection of P-GFP and MC-GFP vectors

The GFP expressing minicircle vector (MC-GFP) was produced from the parental plasmid (P-GFP) as described in the manufacturer's instructions (Systems Bioscience). We observed an approximate 4-kb reduction in plasmid size following minicircle induction using L-arabinose (Fig. 2a). HEK293T cells and rBMSCs were transfected with a range of plasmid

DNA concentrations (1 µg, 0.5 µg, 0.25 µg, 0.125 µg, 0.0625 µg,). After 48 h post-transfection, the cells were visualised by fluorescence microscopy and analysed by flow cytometry to estimate the percentage of GFP-expressing (GFP+) cells (Fig. 2b). The optimum plasmid DNA concentration for transection was 0.5 µg which showed highest transfection efficiency for both P-GFP and MC-GFP in both HEK293T (Fig. 2c) and rBMSC (Fig. 2d) cell types.

Transfection of HEK293T cells with MC-GFP plasmid resulted in a significantly higher number of GFP+ cells ((55.51 ± 3.3 %) compared to P-GFP (43.4 ± 4.9 %). A similar trend was seen in rBMSCs, with MC-GFP resulting in a higher transfection efficiency (18.65 ± 1.05 %) compared to P-GFP (15.21 ± 0.22 %).

Generation of eNOS minicircle vector

To generate an eNOS minicircle expression plasmid vector, a codon optimized cDNA of human eNOS (3633 bp) was synthesised (Geneart) and sub-cloned into the parental plasmid P-GFP (CMV-MCS-EF1-

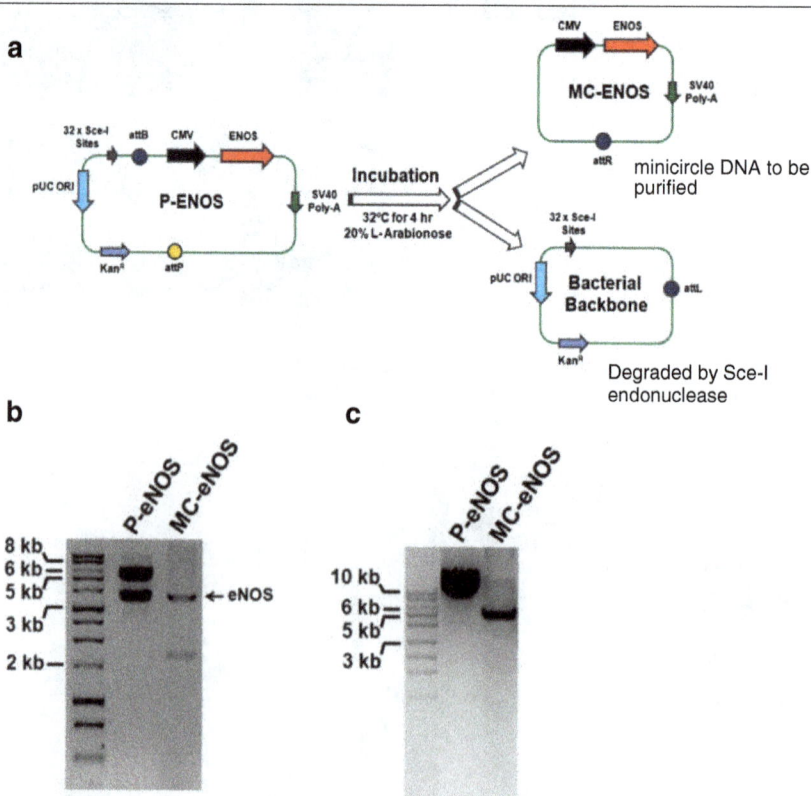

Fig. 3 Construction of eNOS expressing minicircle DNA vector. **a** Schematic representation of in vitro production of MC-eNOS vector. **b** Confirmation of cloning of eNOS gene into P-eNOS and MC-eNOS vectors by restriction enzyme digestion analysis. **c** P-eNOS and MC-eNOS gel electrophoresis following minicircle plasmid purification. *MC-eNOS* minicircle vector expressing endothelial nitric oxide synthase, *P-eNOS* plasmid vector expressing endothelial nitric oxide synthase

GFP-SV40PolyA) (System Biosciences, Mountain View, CA, USA) at the *Bam*HI and *Sal*I restriction sites in the multiple cloning sites downstream to the CMV promoter resulting in removal of the EF1α promoter and eGFP coding sequence (Fig. 3a). The eNOS minicircle vector was constructed as described above for the MC-GFP vector. The cloning was confirmed by double digestion of the parental plasmid encoding eNOS (P-eNOS) with *Bam*HI and *Sal*I yielding a fragment of ~3.7 kb (Fig. 3b). A reduction of the P-eNOS vector size was also observed after the production of MC-eNOS, to approximately 5 kb (Fig. 3c).

Transfection of P-eNOS and MC-eNOS vectors

Transfection of HEK293T cells with P-eNOS and MC-eNOS was assessed by immunofluorescence staining (Fig. 4a) and western blot analysis (Fig. 4b), using an eNOS-specific monoclonal antibody (BD bioscience). Nitric oxide production from transfected cells was measured by the production of nitrite at 24 h and 48 h post-transfection and in un-transfected HEK293T cells (Fig. 4c). Both P-eNOS and MC-eNOS transfected

HEK293T cells showed significantly higher nitrite accumulation in cell culture media (at both 24 h and 48 h) compared to P-GFP, MC-GFP transfected cells and un-transfected HEK293T controls. At 24 h post-transfection, HEK293T cells transfected with P-eNOS and MC-eNOS resulted in 3.8 ± 0.2 μM and 4.46 ± 0.12 μM nitrite concentrations, respectively (Fig. 4c). The NO production increased significantly at 48 h post-transfection, resulting in 4.18 ± 0.12 μM and 5.06 ± 0.13 μM for P-eNOS and MC-eNOS, respectively. Furthermore, detection of nitric oxide produced from transfected cells was also confirmed by DAF-2DA staining in live cells. Both P-eNOS and MC-eNOS transfected HEK293T cells emitted a strong green fluorescence signal compared to no fluorescence in un-transfected cells (Fig. 4d).

eNOS gene transfer to rBMSCs

Transfection of P-eNOS and MC-eNOS vectors into rBMSCs was confirmed by immunostaining (Fig. 5a), and western blot analysis (Fig. 5c) with an eNOS-specific monoclonal antibody (BD bioscience). Both

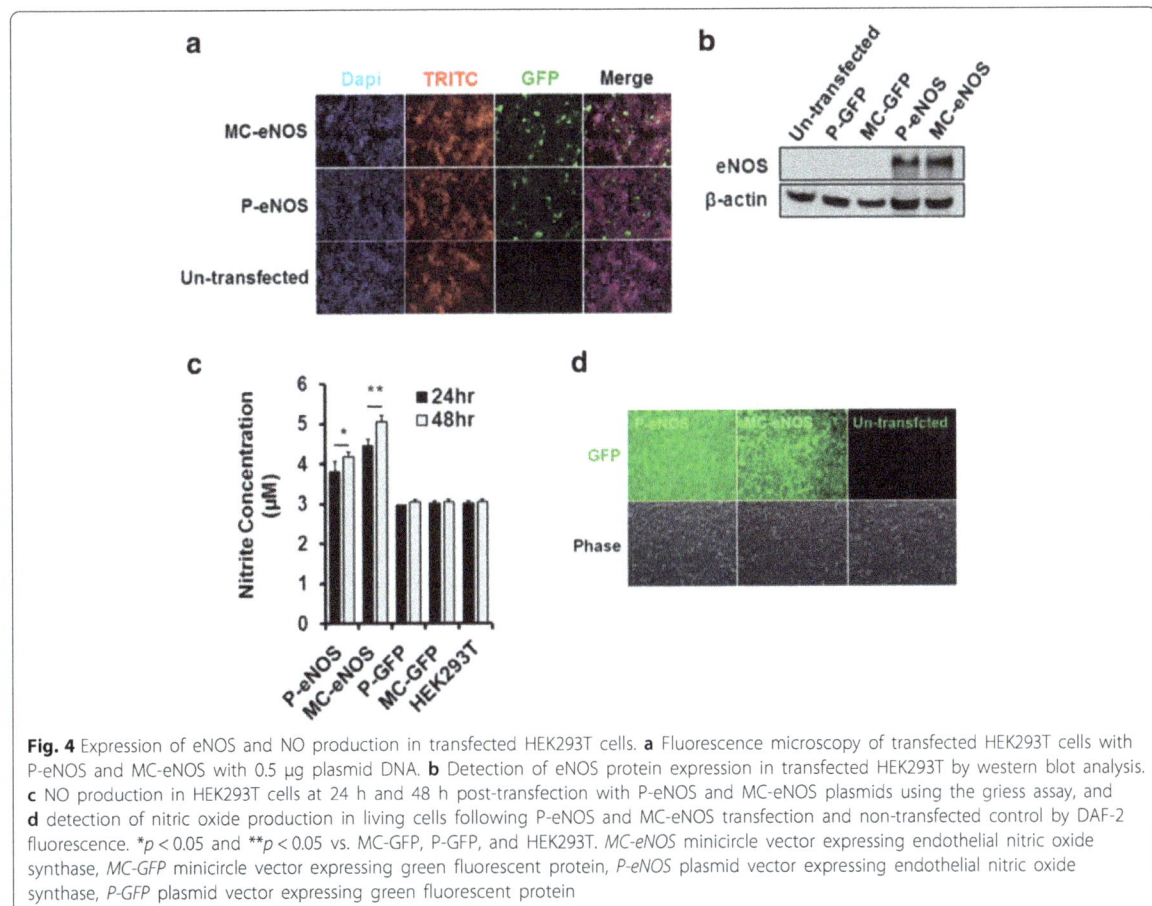

Fig. 4 Expression of eNOS and NO production in transfected HEK293T cells. **a** Fluorescence microscopy of transfected HEK293T cells with P-eNOS and MC-eNOS with 0.5 μg plasmid DNA. **b** Detection of eNOS protein expression in transfected HEK293T by western blot analysis. **c** NO production in HEK293T cells at 24 h and 48 h post-transfection with P-eNOS and MC-eNOS plasmids using the griess assay, and **d** detection of nitric oxide production in living cells following P-eNOS and MC-eNOS transfection and non-transfected control by DAF-2 fluorescence. *$p < 0.05$ and **$p < 0.05$ vs. MC-GFP, P-GFP, and HEK293T. *MC-eNOS* minicircle vector expressing endothelial nitric oxide synthase, *MC-GFP* minicircle vector expressing green fluorescent protein, *P-eNOS* plasmid vector expressing endothelial nitric oxide synthase, *P-GFP* plasmid vector expressing green fluorescent protein

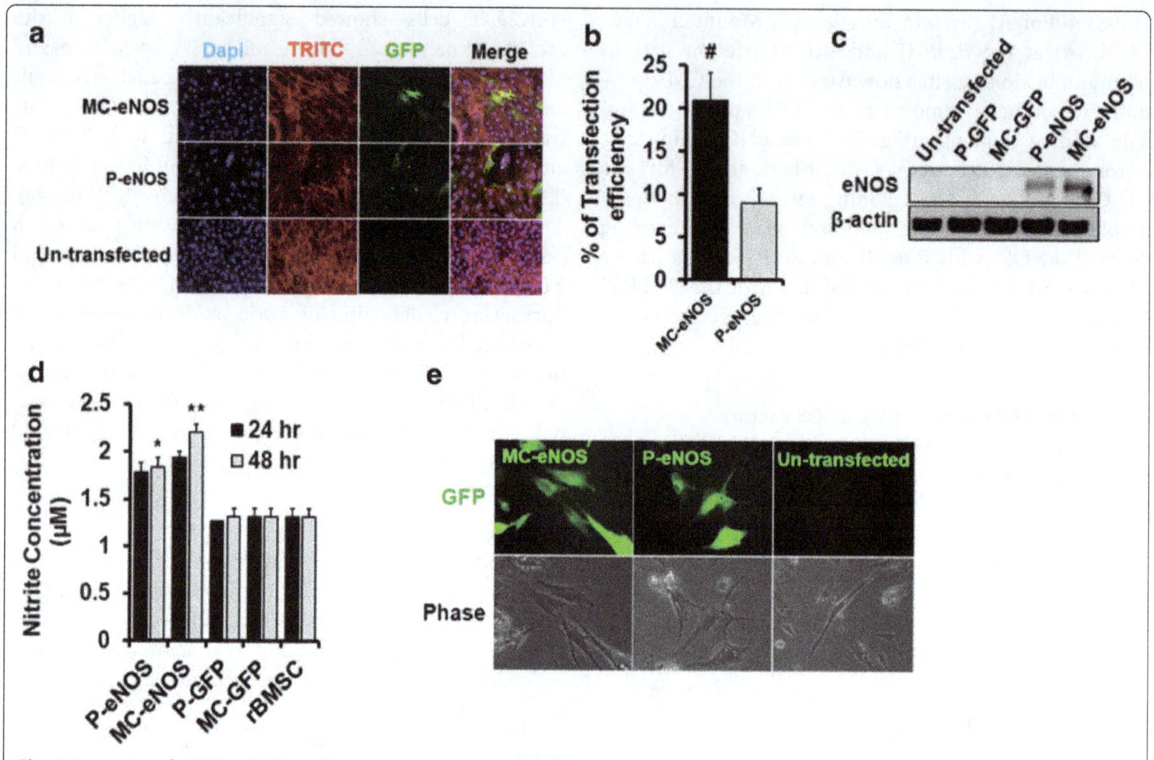

Fig. 5 Expression of eNOS and NO production in transfected rBMSCs. **a** Fluorescence microscopy of transfected of rBMSCs with P-eNOS and MC-eNOS with 0.5 µg plasmid DNA. **b** Transfection efficiency of MC-eNOS and P-eNOS. **c** Detection of eNOS protein expression in transfected rBMSC by western blot analysis. **d** NO production in rBMSC cells at 24 hand 48 h post-transfection with P-eNOS and MC-eNOS plasmids using the griess assay. **e** Detection of nitric oxide production in living cells following P-eNOS and MC-eNOS transfection by DAF-2 fluorescence. $^{#}p < 0.05$ vs. P-eNOS; $^{*}p < 0.05$ and $^{**}p < 0.05$ vs. MC-GFP, P-GFP, and rBMSC. *MC-eNOS* minicircle vector expressing endothelial nitric oxide synthase, *MC-GFP* minicircle vector expressing green fluorescent protein, *P-eNOS* plasmid vector expressing endothelial nitric oxide synthase, *P-GFP* plasmid vector expressing green fluorescent protein, *rBMSC* rat bone marrow-derived mesenchymal stem cell

the assays confirmed that no endogenous eNOS expression was seen in un-transfected rBMSCs (Fig. 5a and c). Significantly higher transfection efficiency for MC-eNOS (21 ± 3 %) compared to P-eNOS (9 ± 3 %) (Fig. 5b) was observed which resulted in higher NO production for MC-eNOS transfected rBMSCs (1.93 ± 0.06 µM) than P-eNOS (1.78 ± 0.1 µM) (Fig. 5d) compared to controls after 24 h of transfection. NO production increased further in MC-eNOS transfected rBMSCs (2.20 ± 0.08 µM) compared to P-eNOS at 48 h post-transfection (1.84 ± 0.1 µM) (Fig. 5d). NO synthesis in transfected rBMSCs was also demonstrated DAF-2DA staining in both P-eNOS and MC-eNOS transfected rBMSCs (Fig. 5e).

eNOS gene delivery enhances in vitro capillary tubule formation

Rat BMSCs were transfected with 0.5 µg P-eNOS, MC-eNOS, P-GFP, and MC-GFP. Un-transfected rBMSCs were used as a control. Transfected cells were then plated on a 96-well cell culture plate coated with an extracellular matrix (Geltrex). Both MC-eNOS and P-eNOS transfected rBMSCs formed significantly longer (14.66 ± 0.55 mm and 13.58 ± 0.68 mm, respectively) tubules and a greater number of tubules (56.33 ± 3.51 and 51 ± 4, respectively) compared to rBMSCs transfected with P-GFP, MC-GFP and non-transfected cells (Fig. 6).

To confirm that capillary-like tubule formation was NO-mediated, eNOS transfected rBMSCs were treated with 2 mM of the nitric oxide synthase inhibitor, L-N^G-nitroarginine methyl ester (L-NAME). L-NAME treatment resulted in a significant impairment of the tubule network, in terms of length (7.33 ± 1.03 mm and 7.06 ± 0.88 mm for MC-eNOS and P-eNOS, respectively) and tubule number (24 ± 4 and 24 ± 2 for MC-eNOS and P-eNOS, respectively) compared to untreated cells (Fig. 6).

Nitric oxide promotes in vitro cell migration

Using the scratch wound healing assay [37], migration of eNOS transfected rBMSCs was assessed. Transfection of

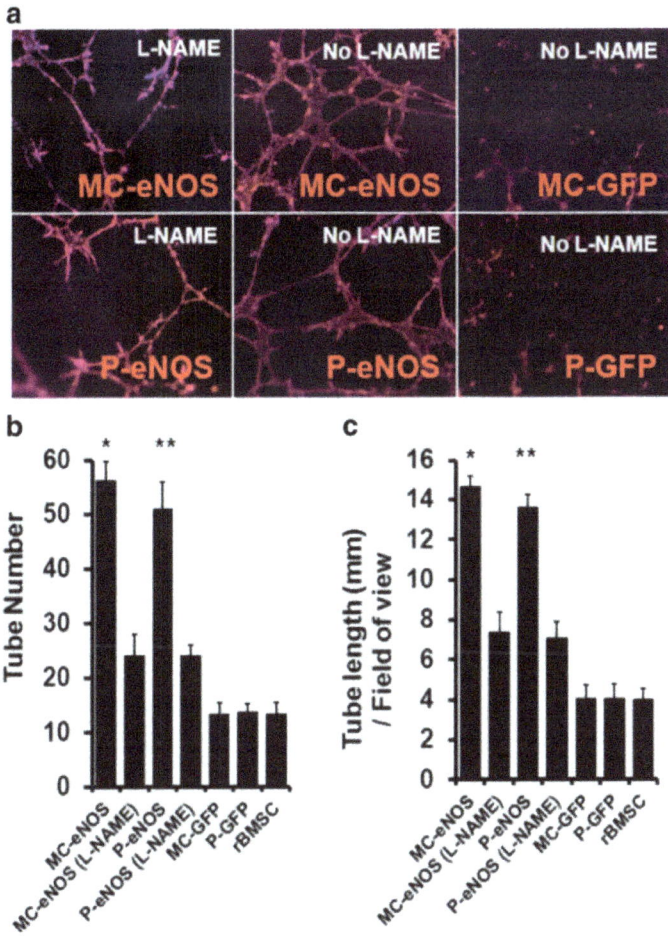

Fig. 6 In vitro tubule formation in eNOS transfected rBMSCs. **a** Capillary tubule formation in rBMSCs transfected with P-eNOS and MC-eNOS and cytoskeletal staining by Phalloidin TRITC; treatment with the NO inhibitor L-NAME reduces capillary formation. **b** Quantitation of tubule number. **c** Measurement of tubule length. $*p < 0.05$ and $**p < 0.05$ vs. MC-eNOS (L-NAME), P-eNOS (L-NAME), MC-GFP, P-GFP, and rBMSC. *L-NAME*, L-N^G-nitroarginine methyl ester, *MC-eNOS* minicircle vector expressing endothelial nitric oxide synthase, *MC-GFP* minicircle vector expressing green fluorescent protein, *P-eNOS* plasmid vector expressing endothelial nitric oxide synthase, *P-GFP* plasmid vector expressing green fluorescent protein, *rBMSC* rat bone marrow-derived mesenchymal stem cell

P-eNOS and MC-eNOS enhanced cell migration compared to P-GFP, MC-GFP and un-transfected rBMSCs (MC-eNOS, 44.05 ± 0.81 %; P-eNOS, 43.13 ± 3.45 %; MC-GFP, 10.43 ± 2.63 %; P-GFP, 11.39 ± 3.03 %; and rMSC, 9.46 ± 4.13 %) (Fig. 7a and c). However, cell migration rates between P-eNOS and MC-eNOS were not significantly different. Inhibition of NO production by treatment with 2 mM L-NAME significantly diminished the cell migration rates of both MC-eNOS and P-eNOS transfected cells (12.18 ± 1.67 % and 15.59 ± 4.69 %, respectively) (Fig. 7a and c). Cell migration rates were not significantly different among MC-GFP, P-GFP and un-transfected cells (Fig. 7a and c). A similar phenomenon was observed with HEK293T cells (Fig. 7b and d).

MC-eNOS gene transfer to rBMSCs induces endothelial CD31 gene expression

We found a significant increase in CD31 mRNA expression by 0.42-fold in P-eNOS transfected cells compared to P-GFP, MC-GFP and un-transfected control, suggesting that eNOS gene transfer may promote endothelial differentiation of rBMSCs (Fig. 8). Interestingly, minicircle-mediated eNOS (MC-eNOS) gene transfer showed a highly significant increase in CD31 mRNA expression by 1.8-fold compared to P-eNOS, P-GFP, MC-GFP, and un-transfected control (Fig. 8). Treatment with 2 mM L-NAME abolished the CD31 expression (Fig. 8), suggesting that expression of endothelial CD31 in rBMSCs through eNOS gene transfer is NO-mediated.

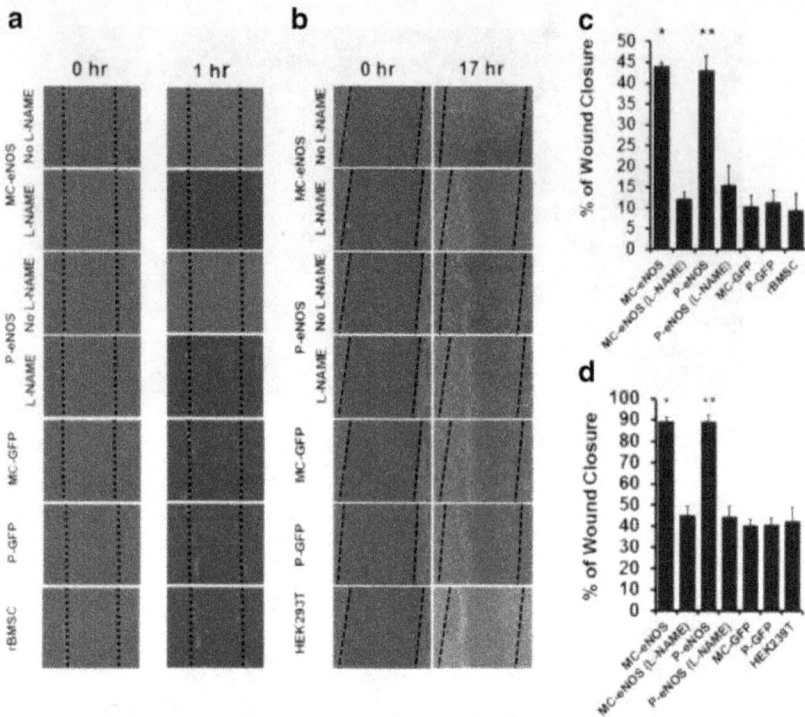

Fig. 7 In vitro cell scratch assay of eNOS transfected rBMSCs and HEK293T cells. Phase-contrast microscopy images of transfected and control cell migration at (**a**) 0 and 1 h for rBMSCs and (**b**) 0 and 17 h for HEK293T post-cell scratch and effect of the NO inhibitor L-NAME. Percentage of cell migration (**c**) at 1 h for rBMSCs and (**d**) 17 h for HEK293T post-cell scratch and effect of the NO inhibitor L-NAME. *$p < 0.05$ and **$p < 0.05$ vs. MC-eNOS (L-NAME), P-eNOS (L-NAME), MC-GFP, P-GFP, rBMSC and HEK293T. *L-NAME*, L-N^G-nitroarginine methyl ester, *MC-eNOS* minicircle vector expressing endothelial nitric oxide synthase, *MC-GFP* minicircle vector expressing green fluorescent protein, *P-eNOS* plasmid vector expressing endothelial nitric oxide synthase, *P-GFP* plasmid vector expressing green fluorescent protein, *rBMSC* rat bone marrow-derived mesenchymal stem cell

NO modulates VEGF-A/PDGFR and FGF2/FGFR2 signalling pathways in eNOS transfected rBMSCs

Expression of two key genes, VEGF-A and FGF2, which are involved in angiogenesis and cell migration were examined by quantitative real time PCR. Upregulation of both VEGF-A by 1.19-fold and 1.0-fold in MC-eNOS and P-eNOS modified rBMSCs, respectively (Fig. 9a), and FGF2 by 1.08-fold in MC-eNOS and 0.74-fold in P-eNOS delivered rBMSCs (Fig. 9b), compared to P-GFP, MC-GFP delivered rBMSCs and un-transfected rBMSCs. Treatment with 2 mM L-NAME reduced both VEGF-A (Fig. 9a) and FGF2 (Fig. 9b) expression in P-eNOS and MC-eNOS transfected cells. Furthermore, delivery of P-GFP and MC-GFP did not affect the VEGF-A and FGF2 expression compared to control rBMSCs (Fig. 9a and 9b). Next, we examined the effect of NO on the expression of PDGFRα and FGFR2 receptors as they are corresponding receptors of VEGF-A and FGF2. Expression of PDGFRα was increased by 1.82-fold and 1.56-fold in MC-eNOS and P-eNOS transfected rBMSCs, respectively (Fig. 9c), and FGFR2 receptor expression was

increased by 1.46-fold in MC-eNOS and 1.14-fold in P-eNOS delivered rBMSCs (Fig. 9d), compared to P-GFP, MC-GFP delivered rBMSCs and un-transfected rBMSCs. Treatment with 2 mM L-NAME abolished both the PDGFRα (Fig. 9c) and FGFR2 (Fig. 9d) expression in P-eNOS and MC-eNOS transfected cells. Furthermore, neither PDGFRα nor FGFR2 receptor expression were affected by the delivery of P-GFP and MC-GFP compared to control rBMSCs (Fig. 9c and d).

Discussion

Minicircle vectors are supercoiled DNA molecules that are devoid of bacterial backbone sequences such as a bacterial origin of replication, antibiotic resistance gene and CpG motifs [41], and primarily consist of a eukaryotic expression cassette [6]. Compared to conventional plasmid DNA, minicircle vectors benefit from higher transfection efficiencies and longer transgene expression, possibly attributed to a lower activation of gene silencing mechanisms [42].

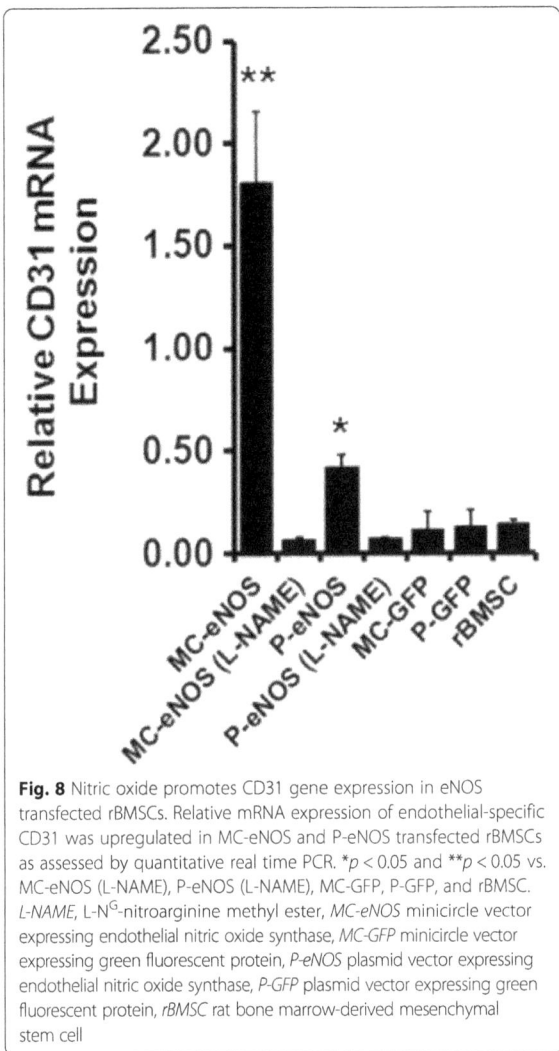

Fig. 8 Nitric oxide promotes CD31 gene expression in eNOS transfected rBMSCs. Relative mRNA expression of endothelial-specific CD31 was upregulated in MC-eNOS and P-eNOS transfected rBMSCs as assessed by quantitative real time PCR. *$p < 0.05$ and **$p < 0.05$ vs. MC-eNOS (L-NAME), P-eNOS (L-NAME), MC-GFP, P-GFP, and rBMSC. *L-NAME*, L-N^G-nitroarginine methyl ester, *MC-eNOS* minicircle vector expressing endothelial nitric oxide synthase, *MC-GFP* minicircle vector expressing green fluorescent protein, *P-eNOS* plasmid vector expressing endothelial nitric oxide synthase, *P-GFP* plasmid vector expressing green fluorescent protein, *rBMSC* rat bone marrow-derived mesenchymal stem cell

In this study, minicircles expressing GFP exhibited higher in vitro gene transfer efficiency than the parental plasmid to both HEK293T cells and rBMSCs (Fig. 3). As expected transfection efficiency was higher in the transformed cell line (HEK293T) compared to primary rBMSCs. eNOS expressing minicircles also showed higher gene transfer efficiency than P-eNOS (Fig. 5b). This higher gene transfer efficiency may also account for the significantly increased level of NO synthesis by MC-eNOS compared to P-eNOS (Figs. 4 and 5). We reasoned that this high level of NO synthesis from MC-eNOS transfected rBMSCs may be attributed to the removal of other plasmid sequences, which can affect gene expression [42], and the smaller size of the minicircle may also provide a more efficient route to the nucleus for transcription. This process involves several steps, including cellular entry of DNA through the cell membrane, DNA diffusion into the cytoplasm, and DNA entry to the nucleus [43]. Importantly, the DNA diffusion step depends on the physicochemical properties of DNA such as its diffusion coefficient, which is inversely proportional to its molecular weight [44, 45]. Endocytosis is a major route for entry of DNA–cationic lipid complexes through the cell membrane in vitro [46] which takes place following specific interactions between DNA and caveolae [47]. This mechanism is also limited by particle size, where larger DNA–cationic lipid complexes are not efficiently taken up by the endocytosis [47]. DNA uptake and transfer to the nucleus via the nuclear membrane results in successful gene transfer [48]. Minicircle plasmid vector may overcome these cellular obstacles more efficiently and, combined with a lack of bacterial backbone sequences, reduced promoter methylation [49] may also contribute to the higher levels of gene transfer compared to larger parental plasmids.

Angiogenesis is a complex process involving endothelial cell proliferation and migration, remodelling of extracellular matrix, and tubular structure formation. These processes are tightly regulated by the actions of angiogenic cytokines such as VEGF-A and FGF [31]. Angiogenesis also requires endothelial cell-to-cell, and cell-to-matrix interactions, which are mediated by various cell adhesion molecules [50]. eNOS plays a key role in angiogenesis mediated by substance P, a potent endothelium-dependent vasodilator (NO releaser) [51]. It has also been demonstrated that eNOS-KO (knockout) mice show impaired angiogenesis [52].

NO has been shown to play an important role in angiogenesis both in vitro and in vivo, and furthermore NO also contributes to endothelial cell migration in vitro [52]. We found that eNOS gene transfer by MC vector remarkably promoted endothelial-specific CD31 gene expression (Fig. 8), contributing to the capillary-like tubule network formation by rBMSCs (Fig. 6) and enhanced cell motility as evident by in vitro wound healing assay (Fig. 7). Noteworthy, in these assays, CD31 mRNA expression, tubule formation and cell migration in transfected cells were significantly abrogated by L-NAME treatment, suggesting NO plays a major role in enhancing endothelial characteristics in rBMSCs. Collectively, our data may suggest that MC-mediated eNOS gene transfer may contribute to the reprogramming of adults stem cells into endothelial cells, which may be used in cell therapy applications involving vascular repair. Interestingly, Gomes and co-workers demonstrated that MSCs from S-nitrosoglutathione reductase (GSNOR)-deficient mice, where NO is produced mainly from iNOS (NOS2) rather than eNOS, exhibited attenuated vasculogenesis both in vitro and in vivo [13]. Furthermore, they revealed that pharmacological inhibition of NO in GSNOR$^{-/-}$ MSCs, or genetic reduction of NO

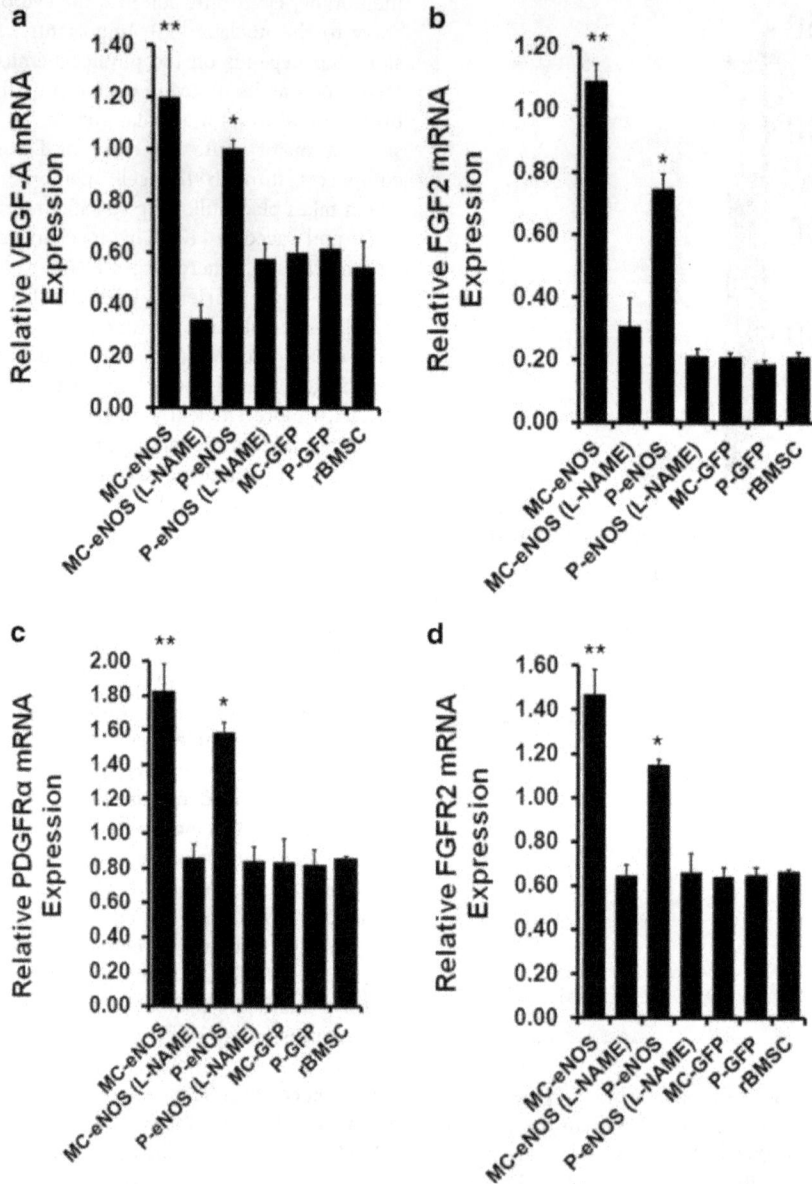

Fig. 9 Nitric oxide modulates VEGF-A/PDGFRα and FGF2/FGFR2 gene expression. Relative mRNA expression of the angiogenesis-related genes (**a**) VEGF-A and (**b**) FGF2 and their corresponding receptors, **c** PDGFRα and **d** FGFR2, were upregulated in MC-eNOS and P-eNOS transfected rBMSCs as assessed by quantitative real time PCR. *$p < 0.05$ and **$p < 0.05$ vs. MC-eNOS (L-NAME), P-eNOS (L-NAME), MC-GFP, P-GFP, and rBMSC. *FGF(R)* fibroblast growth factor (receptor), *L-NAME*, L-N^G-nitroarginine methyl ester, *MC-eNOS* minicircle vector expressing endothelial nitric oxide synthase, *MC-GFP* minicircle vector expressing green fluorescent protein, *PDGFR* platelet-derived growth factor receptor, *P-eNOS* plasmid vector expressing endothelial nitric oxide synthase, *P-GFP* plasmid vector expressing green fluorescent protein, *rBMSC* rat bone marrow-derived mesenchymal stem cell, *VEGF* vascular endothelial growth factor

production in the NOS2$^{-/-}$, enhanced vasculogenesis by MSCs than that for HUVECs, where NO synthesis is driven by eNOS enhanced vascular tube formation. MSCs have not been shown to express endogenous eNOS, unlike endothelial cells [53], and have been shown to participate in pro-angiogenic signalling [54].

Additionally, eNOS plays an important role in endothelial cell-mediated postnatal angiogenesis and vascular tone [55, 56].

NO may contribute to angiogenesis through VEGF and FGF signalling through an angiogenic switch which is preceded by a local increase in VEGF-A and FGF [31].

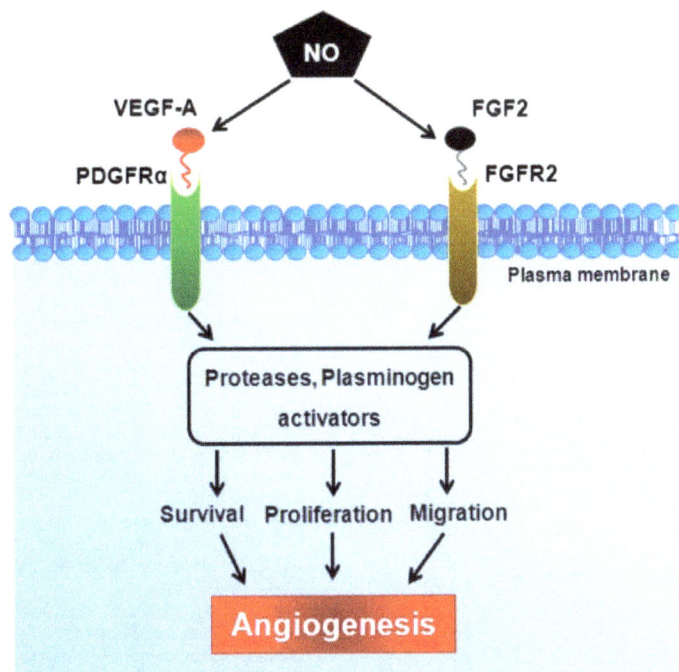

Fig. 10 Proposed molecular mechanism underlying the NO mediated angiogenic responses by MSCs. *FGF(R)* fibroblast growth factor (receptor), *NO* nitric oxide, *PDGFR* platelet-derived growth factor receptor, *VEGF* vascular endothelial growth factor

Nitric oxide can mediate the production of VEGF-A in human adipose-derived stem cells [57] and NO and FGF2 have also been shown to enhance angiogenesis in mouse embryonic stem cells [58]. Furthermore, FGF2 has been shown to induce eNOS expression [32]. Our data proposes that NO signalling through VEGF-A/PDGFRα and FGF2/FGFR2 pathways may directly promote rBMSC vasculogenesis (Fig. 9). We showed that eNOS transfected rBMSCs express increased levels of VEGF-A and FGF2 (Fig. 9) and their corresponding receptors PDGFRα, and FGFR2, respectively (Fig. 9). It is noteworthy that MC-eNOS vector transfection was associated with a significantly higher FGF2 expression compared to the P-eNOS vector. Interestingly, treatment with L-NAME diminished the VEGF-A, PDGFRα, FGF2 and FGFR2 expression levels (Fig. 9) which were observed as being linked to impaired capillary tube-like network formation (Fig. 6). It has been shown that VEGF-A contributes to differentiation of MSCs to endothelial-like cells when co-cultured with endothelial cells expressing eNOS and this process is inhibited by VEGF-A antisera [59].

Angiogenesis is also associated with endothelial cell migration and proliferation [60]. Our results show that eNOS gene transfer into HEK293T and rBMSCs (Fig. 7) can increase cell motility compared to controls, and the effect is diminished by L-NAME treatment,

suggesting that NO plays a role in regulating rBMSC cell migration (Fig. 7) which has been previously demonstrated for endothelial cell migration [61]. Together, these findings show that genetic manipulation of MSCs to enhance bioavailable NO may upregulate VEGF-A/PDGFRα and FGF2/FGFR2 signalling pathways to promote angiogenesis (Fig. 10).

Conclusions
In summary, this study demonstrates that NO derived from a minicircle DNA vector expressing eNOS exerts a positive effect on rBMSCs by promoting in vitro capillary tubule formation and cell migration and significant increases in angiogenesis-related gene expression. Use of MC-eNOS-based vectors may represent an efficient approach to gene therapy applications where enhancing NO bioavailability is beneficial.

Abbreviations
DAF-2DA: 4,5-diaminofluorescein diacetate; DMEM: Dulbecco's modified Eagle medium; eNOS: endothelial nitric oxide synthase; FBS: fetal bovine serum; FGF(R): fibroblast growth factor (receptor); GFP: green fluorescent protein; LB: Luria Broth; L-NAME: L-N^G-nitroarginine methyl ester; MC: minicircle; MSC: mesenchymal stem cell; NBF: neutral-buffered formalin; NO: nitric oxide; PBS: phosphate-buffered saline; PCR: polymerase chain reaction; PDGF(R): platelet-derived growth factor (receptor); pDNA: plasmid DNA; rBMSC: rat bone marrow-derived mesenchymal stem cell; TB: terrific broth; VEGF(R): vascular endothelial growth factor (receptor).

Competing interests
The authors declare that they have no competing interests.

Authors' contributions
NB conceived of the study, collected data, performed data analysis, and prepared the manuscript. SG collected data and performed data analysis and prepared the manuscript. HC and SC collected data and prepared the manuscript. LW performed data analysis and prepared the manuscript. SYL prepared and revised the manuscript. PS conceived of the study, and prepared and revised the manuscript. All authors read and approved the final manuscript.

Acknowledgements
This work was supported by an Australian Government International Post-Graduate Research Scholarship (IPRS) and Australian Post-Graduate Award (APA) to NB. LW, SC, and HC acknowledge funding from the National Natural Science Foundation of China (N. 81270104).

Author details
[1]School of Biomedical Sciences, Charles Sturt University, Wagga Wagga, NSW 2650, Australia. [2]Central laboratory and key Laboratory of Oral and Maxillofacial-Head and Neck Medical Biology, Liaocheng People's Hospital, Liaocheng 252000, PR China. [3]O'Brien Institute Department, St. Vincent's Institute of Medical Research, Fitzroy, VIC 3065, Australia. [4]Department of Surgery, St. Vincent's Hospital, University of Melbourne, Melbourne, VIC 3002, Australia.

References
1. Kay MA, He CY, Chen ZYA. Robust system for production of minicircle DNA vectors. Nat Biotechnol. 2010;28:1287–9.
2. Zhang Q, Raoof M, Chen Y, Sumi Y, Sursal T, Junger W, et al. Circulating mitochondrial DAMPs cause inflammatory responses to injury. Nature. 2010;464:104–7.
3. Mitsui M, Nishikawa M, Zang L, Ando M, Hattori K, Takahashi Y, et al. Effect of the content of unmethylated CpG dinucleotides in plasmid DNA on the sustainability of transgene expression. J Gene Med. 2009;11:435–43.
4. Takahashi Y, Nishikawa M, Takakura Y. Development of safe and effective nonviral gene therapy by eliminating CpG motifs from plasmid DNA vector. Front Biosci. 2012;4:133–41.
5. Riu E, Chen ZY, Xu H, He CY, Kay MA. Histone modifications are associated with the persistence or silencing of vector-mediated transgene expression in vivo. Mol Ther. 2007;15:1348–55.
6. Maniar LEG, Maniar JM, Chen ZY, Lu J, Fire AZ, Kay MA. Minicircle DNA vectors achieve sustained expression reflected by active chromatin and transcriptional level. Mol Ther. 2013;21:131–8.
7. Dad AB, Ramakrishna S, Song M, Kim H. Enhanced gene disruption by programmable nucleases delivered by a minicircle vector. Gene Ther. 2014;21:921–30.
8. Dietz WM, Skinner NE, Hamilton SE, Jund MD, Heitfeld SM, Litterman AJ, et al. Minicircle DNA is superior to plasmid DNA in eliciting antigen-specific CD8+ T-cell responses. Mol Ther. 2013;21:1526–35.
9. Osborn MJ, McElmurry RT, Lees CJ, DeFeo AP, Chen ZY, Kay MA, et al. Minicircle DNA-based gene therapy coupled with immune modulation permits long-term expression of α-l-iduronidase in mice with mucopolysaccharidosis type I. Mol Ther. 2011;19:450–60.
10. Broll S, Oumard A, Hahn K, Schambach A, Bode J. Minicircle performance depending on S/MAR–nuclear matrix interactions. J Mol Biol. 2010;395:950–65.
11. Parathath SR, Gravanis I, Tsirka SE. Nitric oxide synthase isoforms undertake unique roles during excitotoxicity. Stroke. 2007;38:1938–45.
12. Yu J, de Muinck ED, Zhuang Z, Drinane M, Kauser K, Rubanyi GM, et al. Endothelial nitric oxide synthase is critical for ischemic remodeling, mural cell recruitment, and blood flow reserve. PNAS. 2005;102:10999–1004.
13. Gomes SA, Rangel EB, Premer C, Dulce RA, Cao Y, Florea V, et al. S-nitrosoglutathione reductase (GSNOR) enhances vasculogenesis by mesenchymal stem cells. PNAS. 2013;10:2834–9.
14. Oladipupo SS, Smith C, Santeford A, Park C, Sene A, Wiley LA, et al. Endothelial cell FGF signaling is required for injury response but not for vascular homeostasis. PNAS. 2014;111:13379–84.
15. Sunshine SB, Dallabrida SM, Durand E, Ismail NS, Bazinet L, Birsner AE, et al. Endostatin lowers blood pressure via nitric oxide and prevents hypertension associated with VEGF inhibition. PNAS. 2012;109:11306–11.
16. Dai X, Faber JE. Endothelial nitric oxide synthase deficiency causes collateral vessel rarefaction and impairs activation of a cell cycle gene network during arteriogenesis. Circ Res. 2010;106:1870–81.
17. Aicher A, Heeschen C, Mildner-Rihm C, Urbich C, Ihling C, Technau-Ihling K, et al. Essential role of endothelial nitric oxide synthase for mobilization of stem and progenitor cells. Nat Med. 2003;9:1370–6.
18. Lu Y, Xiong Y, Huo Y, Han J, Yang X, Zhang R, et al. Grb-2–associated binder 1 (Gab1) regulates postnatal ischemic and VEGF-induced angiogenesis through the protein kinase A–endothelial NOS pathway. PNAS. 2011;108:2957–62.
19. Smith Jr RS, Lin KF, Agata J, Chao L, Chao J. Human endothelial nitric oxide synthase gene delivery promotes angiogenesis in a rat model of hindlimb ischemia. Arterioscler Thromb Vasc Biol. 2002;22:1279–85.
20. Tai MH, Hsiao M, Chan JY, Lo WC, Wang FS, Liu GS, et al. Gene delivery of endothelial nitric oxide synthase into nucleus tractus solitarii induces biphasic response in cardiovascular functions of hypertensive rats. Am J Hypertens. 2004;17:63–70.
21. Sharif F, Sean OH, Cooney R, Howard L, McMahon J, Daly K, et al. Gene-eluting stents: adenovirus-mediated delivery of eNOS to the blood vessel wall accelerates re-endothelialization and inhibits restenosis. Mol Ther. 2008;16(10):1674–80.
22. Bautch VL. Stem cells and the vasculature. Nat Med. 2011;17:1437–43.
23. Caplan AI. All MSCs are pericytes? Cell Stem Cell. 2008;3:229–30.
24. Quevedo HC, Hatzistergos KE, Oskouei BN, Feigenbaum GS, Rodriguez JE, Valdes D, et al. Allogeneic mesenchymal stem cells restore cardiac function in chronic ischemic cardiomyopathy via trilineage differentiating capacity. PNAS. 2009;106:14022–7.
25. Williams AR, Trachtenberg B, Velazquez DL, McNiece I, Altman P, Rouy D, et al. Intramyocardial stem cell injection in patients with ischemic cardiomyopathy: functional recovery and reverse remodeling. Circ Res. 2011;108:792–6.
26. Oswald J, Boxberger S, Jørgensen B, Feldmann S, Ehninger G, Bornhäuser M, et al. Mesenchymal stem cells can be differentiated into endothelial cells in vitro. Stem Cells. 2004;22:377–84.
27. Sun B, Zhang S, Ni C, Zhang D, Liu Y, Zhang W, et al. Correlation between melanoma angiogenesis and the mesenchymal stem cells and endothelial progenitor cells derived from bone marrow. Stem Cells Dev. 2005;14:292–8.
28. Ball SG, Shuttleworth CA, Kielty CM. Vascular endothelial growth factor can signal through platelet-derived growth factor receptors. J Cell Biol. 2007;177:489–500.
29. Beck KF, Güder G, Schaefer L, Pleskova M, Babelova A, Behrens MH, et al. Nitric oxide upregulates induction of PDGF receptor-expression in rat renal mesangial cells and in anti-Thy-1 glomerulonephritis. J Am Soc Nephrol. 2005;16:1948–57.
30. Charles N, Ozawa T, Squatrito M, Bleau AM, Brennan CW, Hambardzumyan D, et al. Perivascular nitric oxide activates notch signaling and promotes stem-like character in PDGF-induced glioma cells. Cell Stem Cell. 2010;6:141–52.
31. Straume O, Shimamura T, Lampa MJ, Carretero J, Øyan AM, Jia D, et al. Suppression of heat shock protein 27 induces long-term dormancy in human breast cancer. PNAS. 2012;109:8699–704.
32. Murphy PR, Limoges M, Dodd F, Boudreau RT, Too CK. Fibroblast growth factor-2 stimulates endothelial nitric oxide synthase expression and inhibits apoptosis by anitric oxide-dependent pathway in Nb2 lymphoma cells. Endocrinology. 2001;142:81–8.
33. McGinley L, McMahon J, Strappe P, Barry F, Murphy M, O'Toole D, et al. Lentiviral vector mediated modification of mesenchymal stem cells and enhanced survival in an in vitro model of ischaemia. Stem Cell Res Ther. 2011;2:12.
34. Shih YR, Hwang Y, Phadke A, Kang H, Hwang NS, Caro EJ, et al. Calcium phosphate-bearing matrices induce osteogenic differentiation of stem cells through adenosine signalling. PNAS. 2014;11:990–5.
35. Fan G, Wen L, Li M, Li C, Luo B, Wang F, et al. Isolation of mouse mesenchymal stem cells with normal ploidy from bone marrows by reducing oxidative stress in combination with extracellular matrix. BMC Cell Biol. 2011;12:30.

36. Liu C, Feng T, Zhu N, Liu P, Han X, Chen M, et al. Identification of a novel selective agonist of PPARc with no promotion of adipogenesis and less inhibition of osteoblastogenesis. Sci Rep. 2015;5:9530.

37. Ye R, Hao J, Song J, Zhao Z, Fang S, Wang Y, et al. Microenvironment is involved in cellular response to hydrostatic pressures during chondrogenesis of mesenchymal stem cells. J Cell Biochem. 2014;115:1089–96.

38. Iwakiri Y, Satoh A, Chatterjee S, Toomre DK, Chalouni CM, Fulton D, et al. Nitric oxide synthase generates nitric oxide locally to regulate compartmentalized protein S-nitrosylation and protein trafficking. PNAS. 2006;103:19777–82.

39. Faulkner A, Purcell R, Hibbert A, Latham S, Thomson S, Hall WL, et al. A thin layer angiogenesis assay: a modified basement matrix assay for assessment of endothelial cell differentiation. BMC Cell Biol. 2014;15:41.

40. Xu J, Liu X, Jiang Y, Chu L, Hao H, Liua Z, et al. MAPK/ERK signaling mediate VEGF-induced bone marrow stem cell differentiation into endothelial cell. J Cell Mol Med. 2012;12:2395–406.

41. Jia F, Wilson KD, Sun N, Gupta DM, Huang M, Li Z, et al. A nonviral minicircle vector for deriving human iPS cells. Nat Met. 2010;7:197–9.

42. Chen ZY, He CY, Ehrhardt A, Kay MA. Minicircle DNA vectors devoid of bacterial DNA result in persistent and high-level transgene expression in vivo. Mol Ther. 2003;8:495–500.

43. Darquet AM, Rangara R, Kreiss P, Schwartz B, Naimi S, Delaere P, et al. Minicircle: an improved DNA molecule for in vitro and in vivo gene transfer. Gene Ther. 1999;6:209–18.

44. Shen H, Hu Y, Saltzman WM. DNA diffusion in mucus: effect of size, topology of DNAs, and transfection reagents. Biophys J. 2006;91:639–44.

45. Speit G, Vesely A, Schütz P, Linsenmeyer R, Bausinger J. The low molecular weight DNA diffusion assay as an indicator of cytotoxicity for the in vitro comet assay. Mutagenesis. 2014;29:267–77.

46. Le Bihan O, Chèvre R, Mornet S, Garnier B, Pitard B, Lambert O. Probing the in vitro mechanism of action of cationic lipid/DNA lipoplexes at a nanometric scale. Nucleic Acids Res. 2011;39:1595–609.

47. Rejman J, Bragonzi A, Conese M. Role of clathrin- and caveolae-mediated endocytosis in gene transfer mediated by lipo- and polyplexes. Mol Ther. 2005;12:468–74.

48. Zabner J, Fasbender AJ, Moninger T, Poellinger KA, Welsh MJ. Cellular and molecular barriers to gene transfer by a cationic lipid. J Biol Chem. 1995;270:18997–9007.

49. Hong K, Sherley J, Lauffenburger DA. Methylation of episomal plasmids as a barrier to transient gene expression via a synthetic delivery vector. Biomol Eng. 2001;18:185–92.

50. Dejana E. Endothelial adherens junctions: implications in the control of vascular permeability and angiogenesis. J Clin Invest. 1996;98:1949–53.

51. Ziche M, Morbidelli L, Masini E, Amerini S, Granger HJ, Maggi CA, et al. Nitric oxide mediates angiogenesis in vivo and endothelial cell growth and migration in vitro promoted by substance P. J Clin Invest. 1994;94:2036–44.

52. Murohara T, Asahara T, Silver M, Bauters C, Masuda H, Kalka C, et al. Nitric oxide synthase modulates angiogenesis in response to tissue ischemia. J Clin Invest. 1998;101:2567–78.

53. Kuhlencordt PJ, Rosel E, Gerszten RE, Morales-Ruiz M, Dombkowski D, Atkinson WJ, et al. Role of endothelial nitric oxide synthase in endothelial activation: insights from eNOS knockout endothelial cells. Am J Physiol Cell Physiol. 2004;286:C1195–202.

54. Huang PL. Endothelial nitric oxide synthase and endothelial dysfunction. Curr Hypertens Rep. 2003;5:473–80.

55. Beigi F, Gonzalez DR, Minhas KM, Sun QA, Foster MW, Khan SA, et al. Dynamic denitrosylation via S-nitrosoglutathione reductase regulates cardiovascular function. PNAS. 2012;109:4314–9.

56. Lima B, Lam GK, Xie L, Diesen DL, Villamizar N, Nienaber J, et al. Endogenous S-nitrosothiols protect against myocardial injury. PNAS. 2009;106:6297–302.

57. Bassaneze V, Barauna VG, Lavini-Ramos C, Kalil J, Schettert IT, Miyakawa AA, et al. Shear stress induces nitric oxide-mediated vascular endothelial growth factor production in human adipose tissue mesenchymal stem cells. Stem Cells Dev. 2010;19:371–8.

58. Sauer H, Ravindran F, Beldoch M, Sharifpanah F, Jedelská J, Strehlow B, et al. α2-Macroglobulin enhances vasculogenesis/angiogenesis of mouse embryonic stem cells by stimulation of nitric oxide generation and induction of fibroblast growth factor-2 expression. Stem Cells Dev. 2013;22:1443–54.

59. Wu X, Huang L, Zhou Q, Song Y, Li A, Jin J, et al. Mesenchymal stem cells participating in ex vivo endothelium repair and its effect on vascular smooth muscle cells growth. Int J Cardiol. 2005;105:274–82.

60. Carmeliet P. Angiogenesis in health and disease. Nat Med. 2003;9:653–60.

61. Eller-Borges R, Batista WL, da Costa PE, Tokikawa R, Curcio MF, Strumillo ST, et al. Ras, Rac1, and phosphatidylinositol-3-kinase (PI3K) signaling in nitric oxide induced endothelial cell migration. Nitric Oxide. 2015;1:40–51.

14

VCAM-1$^+$ placenta chorionic villi-derived mesenchymal stem cells display potent pro-angiogenic activity

Wenjing Du[1], Xue Li[1], Ying Chi[1], Fengxia Ma[1], Zongjin Li[2], Shaoguang Yang[1], Baoquan Song[1], Junjie Cui[1], Tao Ma[3], Juanjuan Li[1], Jianjian Tian[1], Zhouxin Yang[1], Xiaoming Feng[1], Fang Chen[1], Shihong Lu[1], Lu Liang[2], Zhi-Bo Han[1*] and Zhong-Chao Han[1,2,3*]

Abstract

Introduction: Mesenchymal stem cells (MSCs) represent a heterogeneous cell population that is promising for regenerative medicine. The present study was designed to assess whether VCAM-1 can be used as a marker of MSC subpopulation with superior angiogenic potential.

Methods: MSCs were isolated from placenta chorionic villi (CV). The VCAM-1$^{+/-}$ CV-MSCs population were separated by Flow Cytometry and subjected to a comparative analysis for their angiogenic properties including angiogenic genes expression, vasculo-angiogenic abilities on Matrigel *in vitro and in vivo*, angiogenic paracrine activities, cytokine array, and therapeutic angiogenesis in vascular ischemic diseases.

Results: Angiogenic genes, including HGF, ANG, IL8, IL6, VEGF-A, TGFβ, MMP$_2$ and bFGF, were up-regulated in VCAM-1$^+$CV-MSCs. Consistently, angiogenic cytokines especially HGF, IL8, angiogenin, angiopoitin-2, μPAR, CXCL1, IL-1β, IL-1α, CSF2, CSF3, MCP-3, CTACK, and OPG were found to be significantly increased in VCAM-1$^+$ CV-MSCs. Moreover, VCAM-1$^+$CV-MSCs showed remarkable vasculo-angiogenic abilities by angiogenesis analysis with Matrigel *in vitro and in vivo* and the conditioned medium of VCAM-1$^+$ CV-MSCs exerted markedly pro-proliferative and pro-migratory effects on endothelial cells compared to VCAM-1$^-$CV-MSCs. Finally, transplantation of VCAM-1$^+$CV-MSCs into the ischemic hind limb of BALB/c nude mice resulted in a significantly functional improvement in comparison with VCAM-1$^-$CV-MSCs transplantation.

Conclusions: VCAM-1$^+$CV-MSCs possessed a favorable angiogenic paracrine activity and displayed therapeutic efficacy on hindlimb ischemia. Our results suggested that VCAM-1$^+$CV-MSCs may represent an important subpopulation of MSC for efficient therapeutic angiogenesis.

Keywords: Mesenchymal stem cells, Placenta, Angiogenesis, Paracrine, Vascular cell adhesion molecule-1 (CD106)

Introduction

Peripheral arterial disease (PAD), characterized by the critical limb ischemia (CLI) with high morbidity and mortality risks, is gradually becoming an urgent life-threatening disease in our aging society. To date, the main treatments for PAD are bypass grafting and end-arterectomy. However, surgery has not always been

* Correspondence: zhibohan@163.com; hanzhongchao@hotmail.com
[1]The State Key Laboratory of Experimental Hematology, Institute of Hematology and Hospital of Blood Diseases, Chinese Academy of Medical Sciences & Peking Union Medical College, No.288, Nanjing Road, Heping District, Tianjin 300020, China
Full list of author information is available at the end of the article

allowed [1]. Numerous studies have demonstrated that mesenchymal stem cells (MSCs) derived from different tissue sources exert therapeutic efficacy on ischemia [2–5]. Varieties of reports have highlighted the therapeutic angiogenesis of MSCs by focusing on differentiation and paracrine mechanisms [6]. Several angiogenic cytokines and enzymes secreted by MSCs, including vascular endothelial cell growth factor (VEGF)-A [7], hepatocyte growth factor (HGF) [8], interleukin (IL)-8 [9], transforming growth factor beta (TGFβ) [10], matrix metalloproteases (MMPs) [11], and so forth, have been widely reported to initiate angiogenesis. Based on their angiogenic properties, MSCs are attractive in various clinical trials [12]. However, MSCs have been known to be heterogeneous [13, 14] and it remains to be determined whether some MSC subpopulations exert superior angiogenic activities and are more suitable for therapeutic angiogenesis.

Vascular cell adhesion molecule 1 (VCAM-1), also known as CD106, is extensively expressed on endothelial cells [15], and is also constitutively expressed on some stromal cells, existing in a particular vascular niche [16]. VCAM-1 plays a critical role in early embryonic development since VCAM-1-deficient mice often die early or show multiple severe defects in placental development [17]. In addition, soluble VCAM-1 (sVCAM-1) has shown evidence of mediating angiogenesis in rat cornea [18] and the sVCAM-1/α4 integrin pathway plays an important role in inflammatory stimuli-induced angiogenesis [19]. Recent studies demonstrated that VCAM-1 overexpression was associated with tumor angiogenesis, such as gastric carcinoma [20], breast cancer [21], and renal cancer [22]. These studies suggest VCAM-1 may be involved in angiogenesis.

We have previously isolated a VCAM-1$^+$ MSC subpopulation of placenta chorionic villi (CV) that displayed unique immunomodulation capacity. VCAM-1$^+$CV-MSCs secreted not only inflammatory factors but also angiogenic cytokines [23]. The aim of this work was to assess the angiogenic potential of the VCAM-1$^+$CV-MSC subpopulation, and to explore its therapeutic application in an animal model of vascular ischemic disease.

Methods
Cell isolation and culture
This study was approved by the Ethical Committee and the Institutional Review Board of the Chinese Academy of Medical Science and Peking Union Medical College, Tianjin, China. All volunteers provided informed consent. CV-MSCs were harvested and cultured as described previously [23]. The regular culture medium for CV-MSCs was DF12 medium (Gibco, Grand Island, NY, USA), 10 % fetal bovine serum (FBS), 10 ng/ml epidermal growth factor (EGF; Peprotech, Rocky Hill, NJ, USA), 2 mM glutamine (Sigma, St.Louis, MO, USA), 1 % nonessential amino acids (Gibco), and 100 U/ml penicillin–streptomycin (Invitrogen, Carlsbad, CA, USA). Human umbilical vein endothelial cells (HUVECs) were harvested by digesting umbilical vein with 0.25 % trypsin (Gibco) for 15 minutes at 37 °C. The HUVECs were then cultured in EGM2-MV (Lonza, Walkersville, MD, USA).

Flow cytometry analysis
The phenotype of CV-MSCs was analyzed using the following antibodies: phycoerythrin (PE)-conjugated CD105, CD73, CD166, CD29, CD54, VCAM-1, CD14, CD144, and CD133; and fluorescein isothiocyanate (FITC)-conjugated

CD90, CD45, HLA-ABC, HLA-DR, and CD31. PE or FITC isotype-matched antibodies served as controls. Cells were examined by LSRII flow cytometer (BD Bioscience, San Jose, CA, USA). For cell sorting, CV-MSCs were stained with PE-anti VCAM-1 antibodies for 30 minutes on ice before cell sorting using the BD FACS Aria III cell sorter (BD Biosciences, San Jose, CA, USA). All of the antibodies were purchased from BD Pharmingen (San Diego, CA, USA), and the flow cytometry data were analyzed by FlowJo 7.6 software (San Carlos, CA, USA).

RNA extraction, reverse transcription, and real-time PCR
Total RNA was extracted using the E.Z.N.A. Total RNA Kit I (OMEGA, Norcross, GA, USA), and cDNA synthesis was performed using the MLV RT kit (Invitrogen). All of the procedures followed the manufacturer's instructions. Real-time PCR was performed on an Applied Bio system 7900 Real-Time PCR System (Foster City, CA, USA), using a SYBR Green-based real-time detection method. Primers used are shown in Additional file 1: Table S1. Each sample was performed in triplicate.

Tubular network formation assay in vitro
Pairs of VCAM-1$^+$CV-MSCs and VCAM-1$^-$CV-MSCs were seeded at 2×10^4 cells/well gently on a Matrigel-coated (BD Biosciences, Bedford, MA, USA) 96-well plate. Photographs were taken by Microscope (Olympus, Melville, NY, USA) 12 hours later. Tube numbers in each well were counted. Three pairs of VCAM-1$^+$CV-MSCs and VCAM-1$^-$CV-MSCs were used, and each sample was performed in triplicate.

Matrigel plug angiogenesis assay in vivo
Six-week-old nude male mice were purchased from the Institute of Experimental Animal (Beijing, China). All of the animal experiments followed the Peking Union Medical College Animal Care and Use Committee guidelines. VCAM-1$^{+/-}$CV-MSCs or nonseparated (NS) CV-MSCs (10^6 cells) were suspended in 400 μl Matrigel and injected subcutaneously into the dorsal area of nude mice. Matrigel supplement with phosphate-buffered saline (PBS) served as the negative control. Each group contained three to six mice. Three weeks later, Matrigel implants were harvested, photographed, fixed, sliced, and stained with hematoxylin and eosin (H & E; Sigma). Vessel numbers were counted under the microscope. Frozen slices stained with alpha-smooth muscle actin (α-SMA; Invitrogen) and von Willebrand factor (vWF; Abcam, Cambridge, MA, USA) were employed to detect the neovascular structures in the Matrigel plug. Photographs were taken at ×20 and ×60 objectives by confocal microscopy (UltraView; Perkin-Elmer, Waltham, Massachusetts, USA).

Conditioned medium preparation and proliferation assay

Pairs of 10^6 VCAM-1$^{+/-}$CV-MSCs were incubated in EBM2 medium (Lonza) for 48 hours. Then their conditioned mediums (CMs) were collected, centrifuged at 1800 rpm for 10 minutes to remove cell debris, filtered through 0.2 μm filters (Pall Corporation, Ann Arbor, MI, USA), and frozen at –80 °C. To determine the pro-proliferative effect, VCAM-1$^{+/-}$CV-MSC^{CM} supplemented with 2 % FBS were used to culture HUVECs for 72 hours. EBM2 supplemented with 2 % FBS, and EGM2-MV (endothelial cells commercial culture medium; Lonza) served as the negative and positive control, respectively. The Cell Counting Kit 8 (Dojindo, Rockville, MD, USA) method was used to measure HUVEC proliferation at 24, 48, and 72 hours. ΔOD450 indicated the final data after subtracting the background. Each sample was performed in quadruplicate.

Scratch wound healing assay

When endothelial cells reached confluence, a scratch wound was generated across each well using a pipette tip. After washing with PBS, pairs of CM supplemented with 2 % FBS, EGM2-MV, or EBM2 + 2 % FBS were used to culture endothelial cells for 18 hours. The cleared area of each well was photographed under × 40 magnification at 0 and 18 hours, and measured by ImageJ software (NIH, USA). The percentage of area repopulation was calculated by the following formula:

$$\% \text{ of area repopulation} = (1 - \text{clear area of 18 hours}) / (\text{clear area of 0 hours})$$

Three pairs of CM were used and each sample was performed in triplicate.

Enzyme-linked immunosorbent assay

The VEGF concentration in CM of VCAM-1$^+$CV-MSCs and VCAM-1$^-$CV-MSCs was measured using an enzyme-linked immunosorbent assay (ELISA; Neobioscience Biotech, Shenzhen, China). Each sample was measured in triplicate.

Human cytokine antibody array

The human cytokine antibody array (AAH-CYT-G1000) was performed following the manufacturer's instructions (RayBiotech, Norcross, GA, USA) to detect 120 cytokine expressions in supernatants (SN) of VCAM-1$^{+/-}$CV-MSCs. Cytokine signals above 200 were further studied, and the cytokine signal ratio in VCAM-1$^+$CV-MSCs and VCAM-1$^-$CV-MSCs was calculated. This was statistically significant if the cytokine signal ratio was >1.3 or <0.75. Two pairs of VCAM-1$^+$CV-MSCs and VCAM-1$^-$CV-MSCs were used. Each sample was performed in duplicate. The targeted names of all cytokines involved are presented in Additional file 1: Table S2.

Transplantation of VCAM-1$^{+/-}$CV-MSCs in the hind limb ischemia model

Nude mice (male, 7–8 weeks old, 18–22 g) were intraperitoneally anesthetized with 100 mg/kg sodium pentobarbital (Sigma). Unilateral femoral artery ligation and excision were performed as described previously [24]. Nude mice were randomly divided into three groups (PBS, VCAM-1$^+$CV-MSCs, and VCAM-1$^-$CV-MSCs groups) after arterial ligations, and then 100 μl of a 10^6 cell suspension or PBS was intramuscularly injected into ischemic hind limbs within 6 hours post surgery. Blood perfusion in ischemia and nonischemia limbs was measured by the PeriCam PSI System (PERIMED AB Company, Järfälla, stockholm, Sweden) on day 0, day 7, and day 20. Ischemia damage and functional assessment of ischemic hind limbs in each treatment group were assessed on day 20 according to the semiquantitative scores that had been described previously [24].

Angiography

On day 20, after blood perfusion detection, mice were sacrificed for angiography to evaluate the vessel density in ischemic limbs. Angiographic images of hind limbs in three treatments were acquired by the Kodak In-Vivo FX ProImaging System (Kodak, New Haven, Connecticut, USA), and the angiography score was employed [24] to quantitatively analyze the collateral vessel formation at the ischemia site.

Histological analysis

On day 20, after angiography, the ischemia adductor muscle of nude mice in each group was collected, fixed in 10 % formaldehyde (Sigma) overnight, and embedded in paraffin. To detect capillary densities in ischemic sites, H & E staining was performed and images were taken under × 200 magnification. Vessels containing barium sulfate or erythrocytes were counted, and the vessel density in each group was calculated and compared.

Statistical analysis

Statistical analysis was performed using Graph Pad Prism 6.0 (Graph Pad Software, Inc., San Diego, CA, USA). All data are presented as mean ± standard error of the mean. The Mann–Whitney test and one-way analysis of variance (ANOVA) were performed to determine the significance. Fisher's exact test (Freeman–Halton) was employed to assess the outcome of transplantation via a 3 × 3 contingency table. The difference was considered to be significant if p <0.05.

Results

Characteristics of CV-MSCs

CV-MSCs expressed high levels of CD105 (98.21 % ± 1.28 %), CD73 (99.22 % ± 0.05 %), CD166 (71.72 % ± 13.23 %), CD29 (99.69 % ± 0.14 %), CD90 (97.94 % ± 1.91 %), HLA-ABC (94.32 % ± 2.09 %), CD54 (80.87 % ± 8.25 %), and VCAM-1 (62.9 % ± 5.36 %), but hardly expressed endothelial cells markers (CD144, CD133, and CD31), the hematopoietic cell markers (CD14 and CD45), and immunogenic marker HLA-DR. FACS analysis of a representative sample is shown in Fig. 1a. Phenotypes of CV-MSCs derived from three distinct donors are presented in Additional file 1: Table S3. Cell sorting was carried out to separate the VCAM-1$^+$CV-MSCs and VCAM-1$^-$CV-MSCs (Fig. 1b), and the purity of cell sorting was greater than 90 %. VCAM-1$^+$CV-MSCs and VCAM-1$^-$CV-MSCs cultured in a flask showed typical spindle fibroblast-like shapes; no morphological difference was observed. Photographs of VCAM-1$^+$CV-MSCs and VCAM-1$^-$CV-MSCs are presented in Fig. 1c (scale bar = 200 μm).

Angiogenic genes were highly expressed in VCAM-1$^+$ CV-MSCs

Our previous gene profile result indicated that VCAM-1$^+$CV-MSCs expressed higher levels of angiogenic cytokines than VCAM-1$^-$CV-MSCs, such as IL-6 (2.44-fold) and IL-8 (11.10-fold) [23]. Apart from that, the CXC chemokine family (chemokine (C-X-C motif) ligand (CXCL)1–CXCL3, CXCL5, and CXCL6 and chemokine (C-C motif) ligand (CCL7)), MMPs (including MMP1 and MMP2), several growth factors (VEGFA, HGF, basic fibroblast growth factor (bFGF), TGFβ1, and TGFβ3), hypoxia-induced factor (HIF1A), and angiopoietin-like protein 2 (ANGPTL2) were also highly expressed in VCAM-1$^+$CV-MSCs. Meanwhile, the expressions of lymph-angiogenesis related VEGF-C and intercellular cell adhesion molecule-1 (ICAM-1) were lower in VCAM-1$^+$CV-MSCs (Fig. 2a). Several critical angiogenic genes were further confirmed by real-time PCR. Results showed that HGF, angiogenin (ANG), MMP2, VEGFA, TGFβ, and bFGF expressed in VCAM-1$^+$CV-MSCs were upregulated to varying degrees, with a 3.34-fold, 2.64-fold, 2.34-fold, 1.93-fold, 1.74-fold, and 1.14-fold increase compared with VCAM-1$^-$CV-MSCs, respectively (n = 3–5; Fig. 2b).

VCAM-1$^+$CV-MSCs displayed angiogenic potential on Matrigel assay in vitro and in vivo

To determine the angiogenic potential of VCAM-1$^+$CV-MSCs and VCAM-1$^-$CV-MSCs, a tubular network assay was performed in vitro. To our surprise, without exogenous VEGF, VCAM-1$^+$CV-MSCs spontaneously formed about 4.14-fold intact tubular structures on Matrigel compared with VCAM-1$^-$CV-MSCs (n = 3, p <0.01; Fig. 3a). Matrigel plug angiogenesis assays in vivo [25] were then performed to explore the angiogenic differences. Interestingly, plenty of macroscopic blood vessels were observed in the Matrigel plugs of the VCAM-1$^+$CV-MSCs and NS CV-MSCs groups rather than the VCAM-1$^-$CV-MSCs and PBS groups (Fig. 3b–i). H & E staining revealed that the new outgrowth contained erythrocytes and the smooth muscle layer (Fig. 3b ii, iii). Moreover, vessel densities in the VCAM-1$^+$CV-MSCs and NS CV-MSCs groups were significantly higher than in the VCAM-1$^-$CV-MSCs and PBS groups (10.66 ± 0.67 and 11.84 ± 1.23 per mm^2 vs. 0.36 ± 0.24 and 0.27 ± 0.19 per mm^2, n = 3, p <0.0001; Fig. 3c). However, the vessel density in the VCAM-1$^+$CV-MSCs and NS CV-MSCs groups was similar (p >0.05). Besides that, a larger vessel lumen was observed in the VCAM-1$^+$CV-MSCs group rather than in the NS CV-MSCs group, which could be related to a higher VCAM-1$^+$CV-MSC proportion in the transplanted cells. Moreover, immunostaining of vWF and α-SMA revealed that the fresh blood vessels contained endothelial cells (labeled with anti-vWF antibodies) and smooth muscle cells (labeled with anti-α-SMA antibodies; Fig. 3d), indicating that the vessel structures were intact and mature.

VCAM-1$^+$CV-MSC^{CM} effectively promoted endothelial cell proliferation and migration

To explore the paracrine activities of VCAM-1$^+$CV-MSCs and VCAM-1$^-$CV-MSCs, we collected their CMs and performed endothelial cell proliferation and scratch wound healing assay. Our data revealed that compared with the VCAM-1$^-$CV-MSC^{CM}, VCAM-1$^+$CV-MSC^{CM} significantly promoted endothelial cell proliferation during 48 hours (n = 3, p <0.01), with the most significant point at 24 hours (n = 3, p <0.001). But this pro-proliferative effect was not significant after 72 hours (n = 3, p >0.05; Fig. 4a). The reason for this might be the exhaustion of angiogenic cytokines. In addition, scratch assay that mimicked the wound healing process in vitro was used to evaluate the pro-migratory effects. After incubation for 18 hours, we surprisingly found that endothelial cells cultured in VCAM-1$^+$CV-MSC^{CM} reached confluence again. Representative photographs were taken under × 40 magnification and the percentage of area repopulation was calculated by Image J software (NIH, USA) (Fig. 4b). VCAM-1$^+$CV-MSC^{CM} significantly increased the cleared area recovery compared with VCAM-1$^-$CV-MSC^{CM} (80.58 ± 6.88 vs. 56.36 ± 4.23, n = 3, p <0.01; Fig. 4c), indicating that VCAM-1$^+$CV-MSC^{CM} was richer in pro-migratory cytokines than VCAM-1$^-$CV-MSC^{CM}. To figure out the paracrine mechanism of VCAM-1$^+$CV-MSCs, we performed VEGF and

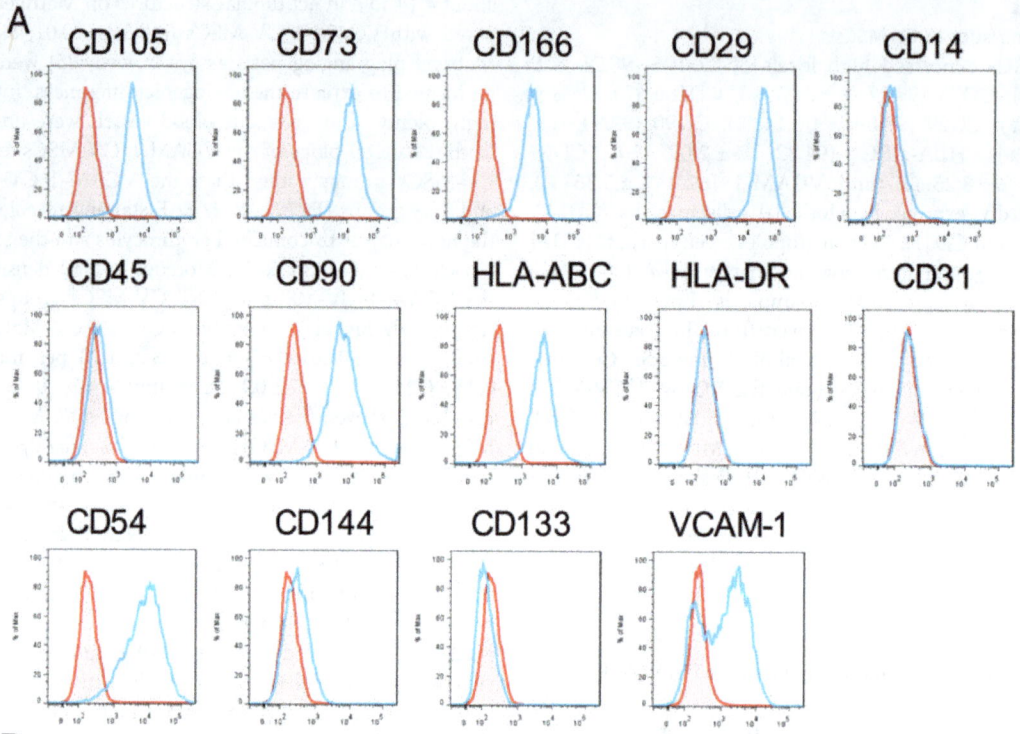

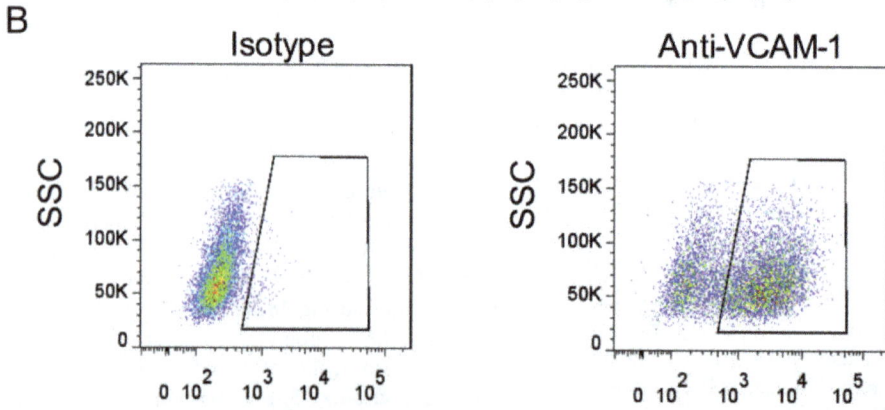

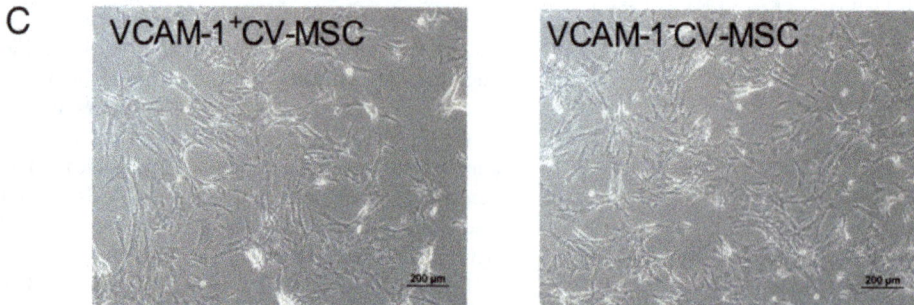

Fig. 1 (See legend on next page.)

(See figure on previous page.)
Fig. 1 Phenotype of CV-MSCs and flow cell sorting. **a** Surface markers of CV-MSCs were evaluated by FACS analysis. CV-MSCs positively expressed CD105, CD73, CD166, CD29, CD90, HLA-ABC, CD54, and VCAM-1, and hardly expressed CD14, CD45, CD31, CD144, CD133 and HLA-DR. A representative sample is shown. **b** VCAM-1+CV-MSCs and VACM-1−CV-MSCs were separated by the BD Aria III cell sorting system. **c** Morphology of VCAM-1+CV-MSCs and VCAM-1−CV-MSCs (scale bar = 200 μm). *CV* chorionic villi, *MSC* mesenchymal stem cell, *SSC* side scatter, *VCAM-1* vascular cell adhesion molecule 1

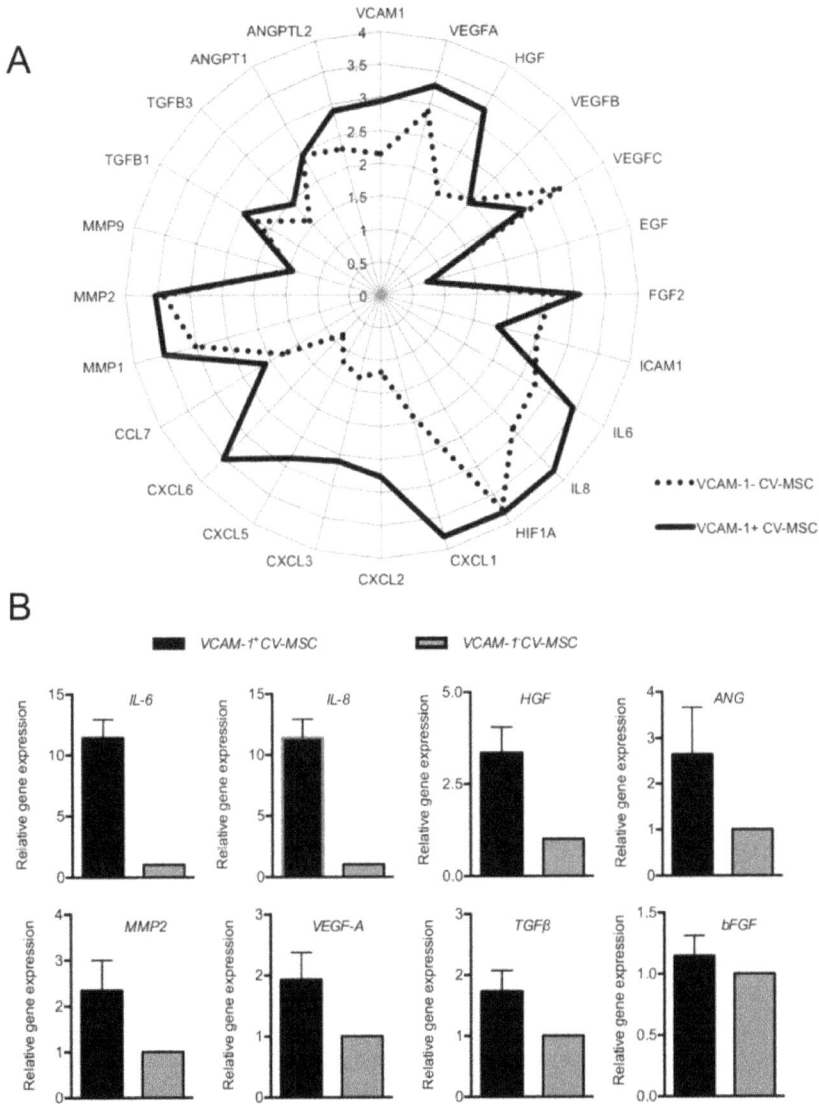

Fig. 2 Angiogenic genes were upregulated in VCAM-1+CV-MSCs. **a** Gene expression profile of VCAM-1+CV-MSCs and VCAM-1−CV-MSCs determined using Affymetrix oligoarray, with the angiogenic genes valued and expressed in $\log_{10}$. **b** Several raised angiogenic genes in VCAM-1+CV-MSCs were confirmed by real-time PCR, including IL-6, IL-8 [23], HGF, ANG, MMP2, VEGF-A, TGFβ, and bFGF (*n* = 3–5). *ANG* angiogenin, *ANGPT2* angiopoietin-2, *ANGPTL2* angiopoietin-like protein 2, *BFGF* basic fibroblast growth factor, *CCL* Chemokine (C-C motif) ligand, *CV* chorionic villi, *CXCL* chemokine (C-X-C motif) ligand, *EGF* epidermal growth factor, *HGF* hepatocyte growth factor, *HIF* hypoxia-induced factor, *IL* interleukin, *MMP* matrix metalloproteinase, *MSC* mesenchymal stem cell, *TGF* transforming growth factor, *VCAM-1* vascular cell adhesion molecule 1, *VEGF* vascular endothelial cell growth factor

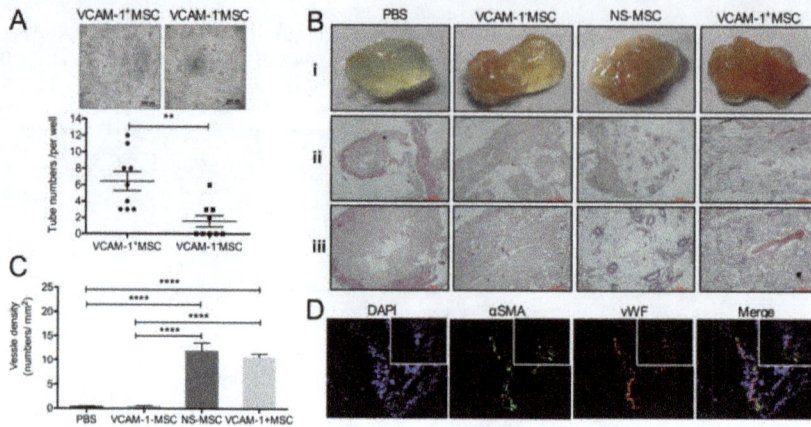

Fig. 3 VCAM-1+CV-MSCs revealed vasculoangiogenic potential by angiogenesis analysis with Matrigel in vitro and vivo. **a** VCAM-1+CV-MSCs spontaneously formed much more intact tube-structures on Matrigel than VCAM-1−CV-MSCs (n = 3, ** p <0.01), indicating that VCAM-1+CV-MSCs possessed vasculogenic potential. Representative images are shown (scale bar = 500 μm). Each sample was performed in triplicate. **b** Macroscopic and microscopic view of Matrigel plugs. The Matrigel plug was harvested 21 days later; macroscopic vessels were seen in the Matrigel plug in the VCAM-1+CV-MSCs and NS CV-MSCs groups **i**. H & E staining was performed to reveal the vessel density in Matrigel plug (scale bar: **ii** = 500 μm; **iii** = 200 μm). **c** Vessel densities in the VCAM-1+CV-MSCs and NS CV-MSCs groups were much greater than in the PBS and VCAM-1−CV-MSCs groups (n = 3, **** p <0.0001). **d** New outgrowth in Matrigel plug was immunostained with vWF and α-SMA antibodies to indicate the endothelial cells and smooth muscle cells, respectively. Photographs were taken under × 20 (bottom) and × 60 (upper) magnifications. α-SMA alpha smooth muscle actin, CV chorionic villi, DAPI 4',6-diamidino-2-phenylindole, MSC mesenchymal stem cell, NS nonseparated, PBS phosphate-buffered saline, VCAM-1 vascular cell adhesion molecule 1, vWF von Willebrand factor

sVCAM-1 ELISAs. Results showed that the VEGF concentration in VCAM-1+CV-MSC^{CM} was 200 pg/ml, 3.6-fold higher than VCAM-1−CV-MSC^{CM} (n = 4, p <0.0001; Fig. 4d), while the sVCAM-1 concentration was <20 pg/ml (Additional file 1: Figure. S1). The fact that VEGF can induce endothelial cell proliferation and migration [7] may partially explain the pro-proliferative and pro-migratory differences between VCAM-1+CV-MSC^{CM} and VCAM-1−CV-MSC^{CM} on endothelial cells.

Cytokine antibody array revealed the angiogenic secretome of VCAM-1+CV-MSCs

To systematically study the secretome of VCAM-1+CV-MSCs and VCAM-1−CV-MSCs, we performed the human cytokine antibody array (AAH-CYT-G1000). Our data revealed that: the differential cytokines from the two donors were similar between VCAM-1+CV-MSCs and VCAM-1−CV-MSCs (Fig. 5a); the VCAM-1+CV-MSC secretome contained a significantly higher level of angiogenic cytokines (Fig. 5b, Additional file 1: Table S4), including IL-1β (6.57-fold), hematopoietic colony-stimulating factor 2 (CSF2)/granulocyte–macrophage colony-stimulating factor (GM-CSF, 5.73-fold), CSF3/granulocyte colony-stimulating factor (G-CSF; 2.03-fold), IL-8 (1.70-fold), CXCL1 (Growth regulated oncogene-α (GRO-α), 1.48-fold), osteoprotegerin (OPG, 1.46-fold), urokinase-type plasminogen activator receptor (μPAR, 1.69-fold), IL-1α (1.6-fold), angiopoietin-2 (ANGPT2, 1.35-fold), HGF (1.33-fold), ANG (1.33-fold), monocyte

chemotactic protein-3 (MCP-3/CCL-7, 1.31-fold), and cutaneous T-cell attracting chemokine (CTACK/CCL27, 1.30-fold), some of those differential cytokines consistent with our gene profile results; and the secretion of RANTES (0.68-fold) and TARC (0.74-fold) was lower in VCAM-1+CV-MSCs than in VCAM-1−CV-MSCs (Fig. 5b). Because of a signal value less than 200, VEGF was not the principal cytokine secreted by CV-MSCs in normal conditions.

VCAM-1+CV-MSCs exerted therapeutic efficacy on hind limb ischemia

To investigate the therapeutic neovascularization of VCAM-1+CV-MSCs, we constructed a vascular ischemia animal model and intramuscularly injected 10^6 VCAM-1+/−CV-MSCs into the ischemic limbs within 6 hours post surgery. PBS served as a negative control. To estimate the therapeutic effect, we classified mice into three outcomes: limb salvage, foot necrosis, and limb loss. Different percentage distributions of outcomes among three groups were calculated and Fisher's exact test (Freeman–Halton) was used to analyze this result (p = 0.10, n = 11; Fig. 6a). From the data, mice in the PBS group suffered the maximal amputation rate (54.5 %) and foot necrosis rate (27 %). The amputation rate in the VCAM-1−CV-MSCs group was much higher than in the VCAM-1+CV-MSCs group (36.4 % vs. 9 %), while the foot necrosis rate in both of them was 18.2 %. Semiquantitative scores of ischemia damage and

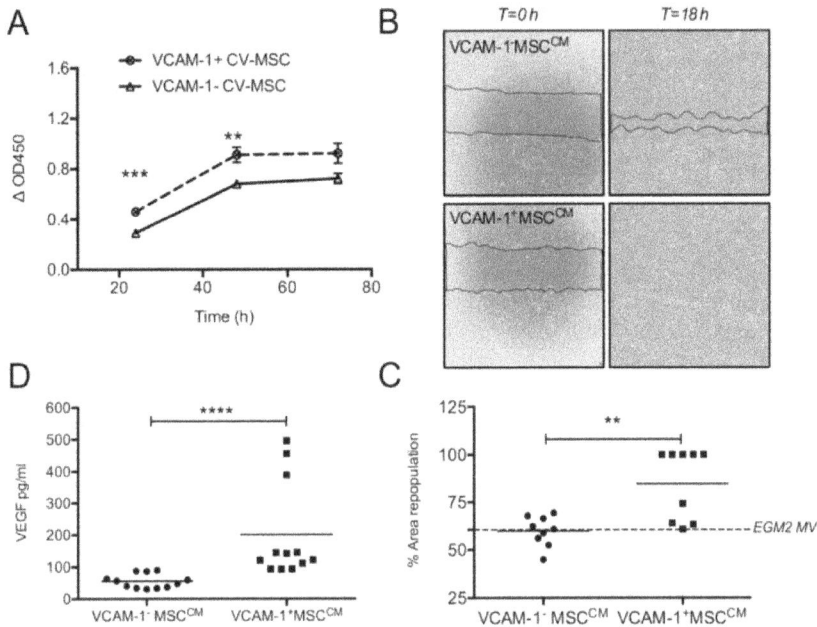

Fig. 4 VCAM-1$^+$CV-MSC^{CM} exerted angiogenic paracrine effects on endothelial cells. **a** Endothelial cell proliferation assay was used to study the pro-proliferative activity of VCAM-1$^+$-MSCs and VCAM-1$^-$CV-MSCs. By comparison with VCAM-1$^-$CV-MSC^{CM}, VCAM-1$^+$CV-MSC^{CM} significantly promoted endothelial cell proliferation during 48 hours, but this effect was not significant at 72 hours ($n = 3$, **p <0.01, ***p <0.001). Each sample was done in quadruplicate. **b** Wound healing assay was performed to study the pro-migratory effect of VCAM-1$^{+/-}$CV-MSCs. Representative photographs were shown at 0 and 18 hours under × 40 magnification. **c** Result of area repopulation (%) indicating that VCAM-1$^+$CV-MSC^{CM} efficiently accelerated the endothelial cell wound healing process compared with VCAM-1$^-$CV-MSC^{CM} ($n = 3$, **p <0.01). Each sample was performed in triplicate. **d** VEGF concentration in VCAM-1$^+$CV-MSC^{CM} and VCAM-1$^-$CV-MSC^{CM} measured by ELISA ($n = 4$, ****p <0.0001). Each sample was tested in triplicate. *CM* conditioned medium, *CV* chorionic villi, *MSC* mesenchymal stem cell, *VCAM-1* vascular cell adhesion molecule 1, *VEGF* vascular endothelial cell growth factor

ambulatory impairment were used to assess ischemic states and physiological function of ischemic limbs. Results indicated that VCAM-1$^+$CV-MSCs significantly alleviated the ischemia damage and ambulatory impairment (0.77 ± 0.37 and 0.59 ± 0.24), much better than the PBS group (2.77 ± 0.52 and 1.82 ± 0.33, $n = 11$, p <0.05), while VCAM-1$^-$CV-MSCs showed a slight improvement compared with PBS treatment (1.86 ± 0.57 and 1.18 ± 0.36, $n = 11$, p >0.05; Fig. 6b, c).

In addition, blood perfusion detected by the PeriCam PSI System was utilized to evaluate ischemia restoration on days 0, 7 and 20 post surgery (Fig. 6d). The blood perfusion ratio in ischemic and healthy limbs was calculated and compared by ANOVA Bonferroni's multiple test. Data showed a significant blood flow increased on day 7 and day 20 post surgery in the VCAM-1$^+$CV-MSCs group (0.79 ± 0.07 and 0.92 ± 0.07), much higher than the VCAM-1$^-$CV-MSCs group (0.48 ± 0.08 and 0.65 ± 0.08, $n = 11$, *p <0.05) and the PBS group (0.50 ± 0.07 and 0.57 ± 0.07, $n = 11$, *p <0.05, **p <0.01; Fig. 6e). By comparison, VCAM-1$^-$CV-MSCs did not show similar therapeutic effects as VCAM-1$^+$CV-MSCs (p >0.05; Fig. 6e). To study collateral vessel

development at the ischemic site, mice underwent angiography and in vivo images on day 20 post surgery. Representative photographs are shown in Fig. 6f. The angiography score that described vessel density was used to analyze the neovascularization. Results demonstrated that VCAM-1$^+$CV-MSCs significantly augmented the generation of collateral vessels at the ischemia site ($n = 3$–5, *p <0.05; Fig. 6g), whose angiography score was 1.52-fold and 1.28-fold higher than the PBS and VCAM-1$^-$CV-MSCs groups, respectively. H & E staining further confirmed that the vessel density in the VCAM-1$^+$CV-MSCs group was 6.3-fold and 2.17-fold more than the PBS and VCAM-1$^-$CV-MSCs groups ($n = 7$, *p <0.05, ***p <0.001, ****p <0.0001; Fig. 6h, i).

Discussion

Previous studies have reported that MSCs displayed remarkable therapeutic properties on vascular ischemic diseases such as myocardial infarction, stroke, and perivascular ischemic diseases [12]. However, the mechanisms of therapeutic angiogenesis induced by MSCs have not yet been well defined. Several investigators have proposed that paracrine factors secreted from MSCs, including a core of

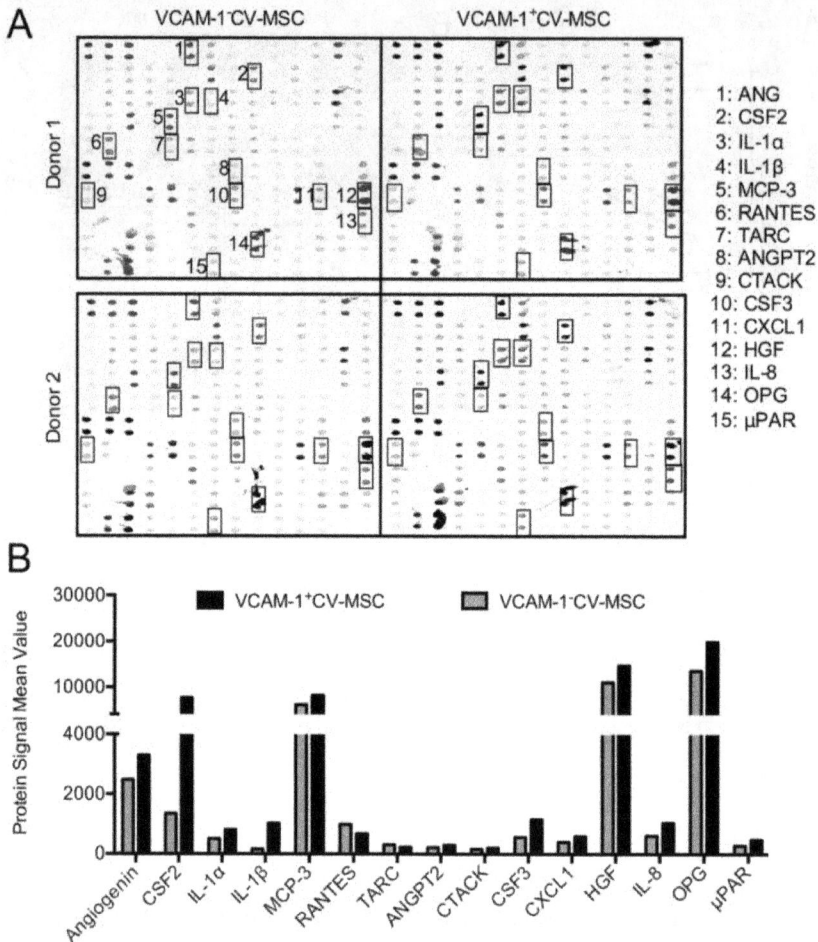

Fig. 5 Human cytokine antibody array displayed the angiogenic secretome of VCAM-1⁺CV-MSCs. **a** Expression of 120 cytokines in SN of VCAM-1⁺CV-MSCs and VCAM-1⁻CV-MSCs was determined by human cytokine antibody array (AAH-CYT-G1000). VCAM-1⁺CV-MSC and VCAM-1⁻CV-MSC SN derived from two healthy donors were used. Each sample was performed in duplicate. The differential angiogenic cytokines between VCAM-1⁺CV-MSCs and VCAM-1⁻CV-MSCs were similar in two healthy donors. **b** Cytokine signal >200 was analyzed, and ratio of cytokine signal in VCAM-1⁺CV-MSCs to VCAM-1⁻CV-MSCs was calculated. This was statistically significant if the cytokine signal ratio was >1.3 or <0.75. Data revealed that VCAM-1⁺CV-MSCs secreted more abundant angiogenic cytokines than VCAM-1⁻CV-MSCs, including HGF, IL-8, ANG, ANGPT2, μPAR, CXCL1, IL-1β, IL-1α, CSF2, CSF3, MCP-3, CTACK, and OPG. *CV* chorionic villi, *MSC* mesenchymal stem cell, *VCAM-1* vascular cell adhesion molecule 1. See Abbreviations for cytokine definitions

angiogenic cytokines (i.e., VEGF, HGF, IL-8, TGFβ), exosomes [26], and microvesicles [27], might be the major contributors [28]. Gnecchi et al. [29] reported that injection with the CM of Akt-modified MSCs abundant with VEGF, bFGF, HGF, and TB4 significantly improved cardiac performance after induced myocardial infarction. Recent studies using cell labeling [30] and single cell technology [31] also supported the major status of paracrine action in MSC-mediated angiogenesis.

In this study, we have firstly demonstrated that the VCAM-1⁺CV-MSC subpopulation displayed a potent angiogenic property and exerted enhanced therapeutic efficacy on regeneration after ischemia in comparison with the VCAM-1⁻CV-MSC subpopulation.

We then wanted to know why VCAM-1⁺CV-MSCs possessed superior pro-angiogenic activities than VCAM-1⁻CV-MSCs. We were interested to note a superior angiogenic secretome from VCAM-1⁺CV-MSCs, including HGF, IL-8, ANG, ANGPT2, CXCL1/GRO-α, μPAR, IL-1β, IL-1α, CSF2/GM-CSF, CSF3/G-CSF, MCP-3, CTACK/CCL27, and OPG. Previous studies have shown that HGF potently stimulated endothelial cell motility and growth [32]. IL-8 promoted angiogenesis via directly enhancing endothelial cell proliferation, survival, and MMP production [33]. ANG potently induced new blood vessel formation [34]. ANGPT-2 potentiated the effects of other angiogenic cytokines in vivo and initiated neovascularization [35]. CXCL1 enhanced microvascular endothelial cell

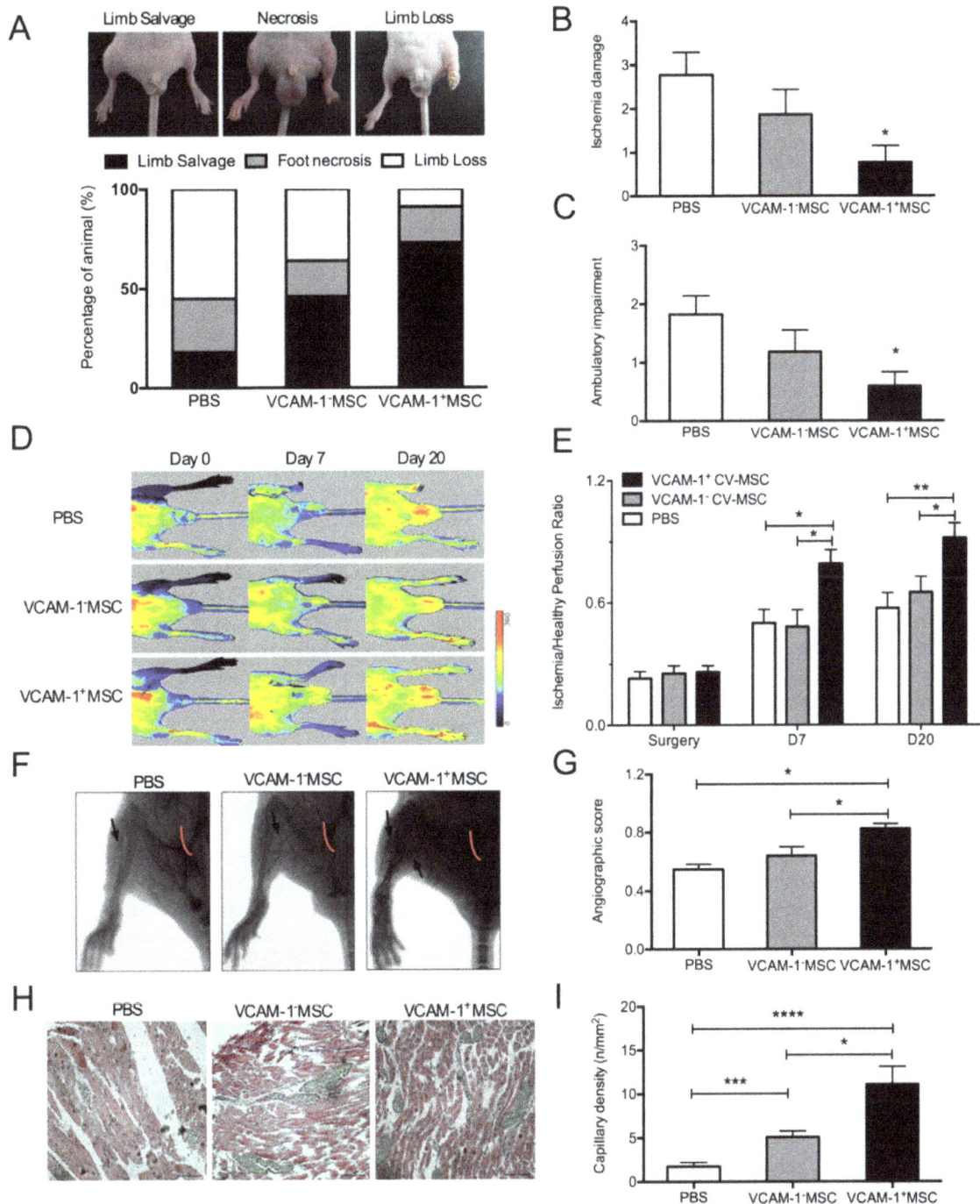

Fig. 6 (See legend on next page.)

(See figure on previous page.)
Fig. 6 Transplantation of VCAM-1[+]CV-MSCs significantly enhanced the blood perfusion and the generation of collateral vessels in the ischemic sites. **a** VCAM-1[+/−]CV-MSCs or PBS were injected into the ischemic site of nude mice. Percentage distributions of limb salvage, foot necrosis, and limb loss in the three groups are shown and analyzed by the Fisher's exact test ($n = 11$, $p = 0.102$). Ischemia damage and physiological function of ischemic limbs were semiquantified by ischemia scores **b** and ambulatory impairment scores **c** ($n = 11$, *$p < 0.05$). **d** Blood perfusion in ischemic/healthy limb was detected by the Kodak In-Vivo FX Pro Imaging System on days (D) 0, 7, and 20. Different colors indicate different blood perfusion. Blood flow increased from *dark blue* to *red*. **e** Blood perfusion ratio in ischemic to healthy limbs was used to quantitatively analyze the blood flow restoration in ischemic limbs ($n = 11$, *$p < 0.05$, **$p < 0.01$). **f** Angiography was performed to assess the collateral vessel generation in the ischemic site. *Red curves* indicated the site of the femoral arteries incision; *black arrows* showed the collateral vessels in the ischemic hind limb. **g** Angiography score indicated that VCAM-1[+]CV-MSCs were superior to VCAM-1[−]CV-MSCs in augmenting collateral vessels ($n = 3–5$, *$p < 0.05$). **h** H & E staining was performed to study the vessel density in ischemia limbs. Pictures showed that the blood vessels were full of barium sulfate (*silver*, scale bar = 100 μm). **i** Vessel density in VCAM-1[+]CV-MSC or VCAM-1[−]CV-MSCs group was significantly greater than the PBS group ($n = 7$, ***$p < 0.001$, ****$p < 0.0001$). Furthermore, VCAM-1[+]CV-MSC transplantation apparently promoted the vessel generation compared with the VCAM-1[−]CV-MSCs group (**$p < 0.05$). *CV* chorionic villi, *MSC* mesenchymal stem cell, *PBS* phosphate-buffered saline, *VCAM-1* vascular cell adhesion molecule 1 (Color figure online)

migration and tube formation [36]. μPAR induced endothelial cell invasion and proliferation in the initial period of angiogenesis [37]. IL-1β [38], IL-1α [39], GM-CSF [40], and G-CSF [41] were reported to initiate angiogenesis by stimulating VEGF production or activating the angiogenesis-related pathway. MCP-3 stimulated the migration of circulating angiogenic cells and angiogenesis partially via the chemokine (C-X-C motif) receptor 1 (CCR1) [42]. CTACK/CCL27 was reported to accumulate the CD34[+] bone marrow cells (expressing CCR10) to participate in skin wound healing and repair [43]. OPG was a positive regulator of microvessel formation in vivo and could activate endothelial colony-forming cells [44]. In addition, we have performed the endothelial cell differentiation assay in vitro and have not found significant differences between VCAM-1[+]CV-MSCs and VCAM-1[−]CV-MSCs (seen by immunostaining of vWF) under a confocal microscope (Additional file 1: Figure. S2). Based on these studies, we believed that paracrine action rather than differentiation was the principal mechanism of the therapeutic angiogenesis induced by MSCs. Besides, the superior angiogenic effect of VCAM-1[+]CV-MSCs could be a result of a synergic effect of multiple angiogenic factors secreted by cells.

To date, the identification of MSC still relies on the minimal criteria specified in 2006 (plastic adhesion, expressing a set of membrane antigens and tridifferentiation capacities) [45]. Besides these properties, the trait of MSCs varies among different origins and individuals; that is, the paracrine actions [46], and immunomodulatory [23] and hematopoietic support capacities [47]. In addition, MSCs isolated from the same tissue also comprised a heterogeneous population. A variety of markers (i.e., Stro-1, SSEA-4, CD271, CD146) have hence been adopted to investigate the potential of particular MSC subpopulations [14]. Psaltis et al. [48] reported that stro-1[+] bone marrow-derived MSCs possessed unique cardiovascular paracrine activities. Interestingly, Gronthos et al. [49] employed VCAM-1 as a coexpressed maker to enrich stro-1[+] MSCs. Our data agree with Psaltis et al.'s study, which verified the consistent angiogenic

potentials of VCAM-1[+] MSCs. Most recently, Wang et al. [50] reported that MSCs pretreated with IL-1β and tumor necrosis factor alpha could enhance the therapeutic efficacy on cardiovascular ischemia via upregulating VCAM-1 expression. Consistently, our study demonstrated the presence of a natural VCAM-1[+] MSC subpopulation in vivo in placenta CV that exerted excellent paracrine action. Additionally, it has been shown that placenta CV and bone marrow abundant with capillaries contained many more VCAM-1[+] MSCs (68 % and 13 %) than adipose tissue and umbilical cord (0.24 % and 4 %) [23], suggesting that VCAM-1[+] MSCs might play important roles in the physiological vasculogenesis and angiogenesis.

Conclusion

Our comparative studies at multiple levels on the angiogenic properties of VCAM-1[+]CV-MSCs and VCAM-1[−]CV-MSCs showed that VCAM-1 could be used as a surface marker to select a MSC subpopulation with superior pro-angiogenic activity. Moreover, the exciting therapeutic efficacy of VCAM-1[+]CV-MSCs on ischemic nude mice not only provided a novel strategy for cell-based therapy of ischemic diseases, but also a hint for banking appropriate MSCs for clinical usage.

Additional file

Additional file 1: is Table S1 presenting primers for real-time reverse transcription PCR, Table S2 presenting the description of the human cytokine antibody array (AAH-CYT-G1000), Table S3 presenting the phenotype of three donor-derived CV-MSCs, Table S4 presenting differential angiogenesis cytokines of VCAM-1[−]CV-MSCs and VCAM-1[+]CV-MSCs, Fig. S1 showing sVCAM-1 concentration in 48-hour CM of VCAM-1[+]CV-MSCs and VCAM-1[−]CV-MSCs measured by ELISA, and Fig. S2 showing the endothelial-like cells derived from VCAM-1[+]CV-MSCs and VCAM-1[−]CV-MSCs harvested after in vitro endothelial induction and immunostained by anti-vWF antibodies to evaluate their endothelial differentiation capacities. (PDF 524 kb)

Abbreviations

α-SMA: Alpha-smooth muscle actin; ANG: Angiogenin; ANGPT2: Angiopoietin-2; ANGPTL2: Angiopoietin-like protein 2; ANOVA: Analysis of variance; bFGF: Basic fibroblast growth factor; CCL: Chemokine (C-C motif) ligand; CCR: Chemokine (C-C motif) receptor; CLI: Critical limb ischemia; CM: Conditioned medium;

CSF: Colony-stimulating factor; CTACK: Cutaneous T-cell attracting chemokine; CV: Chorionic villi; CXCL: Chemokine (C-X-C motif) ligand; EGF: Epidermal growth factor; ELISA: Enzyme-linked immunosorbent assay; FACS: Fluorescence-activated cell sorting; FBS: Fetal bovine serum; FITC: Fluorescein isothiocyanate; G-CSF: Granulocyte-colony stimulating factor; GM-CSF: Granulocyte–macrophage colony-stimulating factor; GRO-α: Growth regulated oncogene-α; H & E: Hematoxylin and eosin; HGF: Hepatocyte growth factor; HIF: Hypoxia-induced factor; HUVEC: Human umbilical vein endothelial cell; ICAM-1: Intercellular cell adhesion molecule-1; IL: Interleukin; MCP: Monocyte chemotactic protein; MMP: Matrix metalloproteinase; MSC: Mesenchymal stem cell; NS: Nonseparated; OPG: Osteoprotegerin; PAD: Peripheral arterial disease; PBS: Phosphate-buffered saline; PE: Phycoerythrin; RANTES: Regulated on activation, normal T cell expressed and secreted; SN: Supernatants; sVCAM-1: Soluble vascular cell adhesion molecule 1; TARC: Thymus and activation regulated chemokine; TGFβ: Transforming growth factor beta; μPAR: Urokinase-type plasminogen activator receptor; VCAM-1: Vascular cell adhesion molecule 1; VEGF: Vascular endothelial cell growth factor; vWF: Von Willebrand factor.

Competing interests
The authors declare that they have no competing interests.

Authors' contributions
WD and Z-CH carried out the conception, designed the experiment, and drafted the manuscript. WD, XL, YC, and SY isolated CV-MSCs, and performed FACS analysis and cell sorting. Z-BH and ZY performed the Affymetrix oligoarrays. WD and ZY carried out the real-time PCR. WD and SY performed the Matrigel angiogenesis assay in vitro and in vivo. WD performed proliferation and the scratch wound healing assay. Z-BH performed the cytokine array and analyzed the data. WD, XL, YC, TM, BS, JC, LL, JL, FC, SL, and JT built the vascular ischemia model, performed cell transplantation, evaluated the ischemia scores, detected the blood perfusion, carried out the angiography, and performed the histological staining. Z-CH, Z-BH, FM, ZL, and XF participated in data analysis and manuscript writing. All authors revised their corresponding content and approved the final manuscript.

Acknowledgements
This study was supported by The National Basic Research Program of China (2011CB964802), The National Science and Technology Support Program (2013BAI01B09), and The Natural Science Foundation of China (81330015 and 31470951).

Author details
[1]The State Key Laboratory of Experimental Hematology, Institute of Hematology and Hospital of Blood Diseases, Chinese Academy of Medical Sciences & Peking Union Medical College, No.288, Nanjing Road, Heping District, Tianjin 300020, China. [2]Beijing Institute of Health and Stem Cells, No.1 Kangding Road, BDA, Beijing 100176, China. [3]National Engineering Research Center of Cell Products, No.80, Fourth Avenue, TEDA, Tianjin 300457, China.

References
1. Ouriel K. Peripheral arterial disease. Lancet. 2001;358(9289):1257–64. doi:10.1016/S0140-6736(01)06351-6.
2. Kinnaird T, Stabile E, Burnett MS, Shou M, Lee CW, Barr S, et al. Local delivery of marrow-derived stromal cells augments collateral perfusion through paracrine mechanisms. Circulation. 2004;109(12):1543–9. doi:10.1161/01.CIR.0000124062.31102.57.
3. Xu Y, Meng H, Li C, Hao M, Wang Y, Yu Z, et al. Umbilical cord-derived mesenchymal stem cells isolated by a novel explantation technique can differentiate into functional endothelial cells and promote revascularization. Stem Cells Dev. 2010;19(10):1511–22. doi:10.1089/scd.2009.0321.
4. Kim SW, Zhang HZ, Kim CE, An HS, Kim JM, Kim MH. Amniotic mesenchymal stem cells have robust angiogenic properties and are effective in treating hindlimb ischaemia. Cardiovasc Res. 2012;93(3):525–34. doi:10.1093/cvr/cvr328.
5. Hongyan Tao ZH, Zhong Chao Han, and Zongjin Li. Proangiogenic features of mesenchymal stem cells and their therapeutic applications. Stem Cells Int. 2016;2016:1314709. doi:10.1155/2016/1314709.
6. Wu Y, Chen L, Scott PG, Tredget EE. Mesenchymal stem cells enhance wound healing through differentiation and angiogenesis. Stem Cells. 2007; 25(10):2648–59. doi:10.1634/stemcells.2007-0226.
7. Shizukuda Y, Tang S, Yokota R, Ware JA. Vascular endothelial growth factor-induced endothelial cell migration and proliferation depend on a nitric oxide-mediated decrease in protein kinase Cdelta activity. Circ Res. 1999;85(3):247–56.
8. Ding S, Merkulova-Rainon T, Han ZC, Tobelem G. HGF receptor up-regulation contributes to the angiogenic phenotype of human endothelial cells and promotes angiogenesis in vitro. Blood. 2003;101(12):4816–22. doi:10.1182/blood-2002-06-1731.
9. Li A, Dubey S, Varney ML, Dave BJ, Singh RK. IL-8 directly enhanced endothelial cell survival, proliferation, and matrix metalloproteinases production and regulated angiogenesis. J Immunol. 2003;170(6):3369–76.
10. Mahmoud M, Upton PD, Arthur HM. Angiogenesis regulation by TGFbeta signalling: clues from an inherited vascular disease. Biochem Soc Trans. 2011;39(6):1659–66. doi:10.1042/BST20110664.
11. Rundhaug JE. Matrix metalloproteinases and angiogenesis. J Cell Mol Med. 2005;9(2):267–85.
12. Salem HK, Thiemermann C. Mesenchymal stromal cells: current understanding and clinical status. Stem Cells. 2010;28(3):585–96. doi:10.1002/stem.269.
13. Tolar J, Le Blanc K, Keating A, Blazar BR. Concise review: hitting the right spot with mesenchymal stromal cells. Stem Cells. 2010;28(8):1446–55. doi:10.1002/stem.459.
14. Lv FJ, Tuan RS, Cheung KM, Leung VY. Concise review: the surface markers and identity of human mesenchymal stem cells. Stem Cells. 2014;32(6): 1408–19. doi:10.1002/stem.1681.
15. Rice GE, Bevilacqua MP. An inducible endothelial cell surface glycoprotein mediates melanoma adhesion. Science. 1989;246(4935):1303–6.
16. Castrechini NM, Murthi P, Gude NM, Erwich JJ, Gronthos S, Zannettino A, et al. Mesenchymal stem cells in human placental chorionic villi reside in a vascular Niche. Placenta. 2010;31(3):203–12. doi:10.1016/j.placenta.2009.12.006.
17. Kwee L, Baldwin HS, Shen HM, Stewart CL, Buck C, Buck CA, et al. Defective development of the embryonic and extraembryonic circulatory systems in vascular cell adhesion molecule (VCAM-1) deficient mice. Development. 1995;121(2):489–503.
18. Koch AE, Halloran MM, Haskell CJ, Shah MR, Polverini PJ. Angiogenesis mediated by soluble forms of E-selectin and vascular cell adhesion molecule-1. Nature. 1995;376(6540):517–9. doi:10.1038/376517a0.
19. Vanderslice P, Munsch CL, Rachal E, Erichsen D, Sughrue KM, Truong AN, et al. Angiogenesis induced by tumor necrosis factor-agr; is mediated by alpha4 integrins. Angiogenesis. 1998;2(3):265–75.
20. Ding YB, Chen GY, Xia JG, Zang XW, Yang HY, Yang L. Association of VCAM-1 overexpression with oncogenesis, tumor angiogenesis and metastasis of gastric carcinoma. World J Gastroenterol. 2003;9(7):1409–14.
21. Byrne GJ, Ghellal A, Iddon J, Blann AD, Venizelos V, Kumar S, et al. Serum soluble vascular cell adhesion molecule-1: role as a surrogate marker of angiogenesis. J Natl Cancer Inst. 2000;92(16):1329–36.
22. Hemmerlein B, Scherbening J, Kugler A, Radzun HJ. Expression of VCAM-1, ICAM-1, E- and P-selectin and tumour-associated macrophages in renal cell carcinoma. Histopathology. 2000;37(1):78–83.
23. Yang ZX, Han ZB, Ji YR, Wang YW, Liang L, Chi Y, et al. CD106 identifies a subpopulation of mesenchymal stem cells with unique immunomodulatory properties. PLoS One. 2013;8(3):e59354. doi:10.1371/journal.pone.0059354.
24. Zhou B, Bi YY, Han ZB, Ren H, Fang ZH, Yu XF, et al. G-CSF-mobilized peripheral blood mononuclear cells from diabetic patients augment neovascularization in ischemic limbs but with impaired capability. J Thromb Haemost. 2006;4(5):993–1002. doi:10.1111/j.1538-7836.2006.01906.x.
25. Malinda KM. In vivo matrigel migration and angiogenesis assay. Methods Mol Biol. 2009;467:287–94. doi:10.1007/978-1-59745-241-0_17.
26. Bian S, Zhang L, Duan L, Wang X, Min Y, Yu H. Extracellular vesicles derived from human bone marrow mesenchymal stem cells promote angiogenesis in a rat myocardial infarction model. J Mol Med (Berl). 2014;92(4):387–97. doi:10.1007/s00109-013-1110-5.
27. Zhang HC, Liu XB, Huang S, Bi XY, Wang HX, Xie LX, et al. Microvesicles derived from human umbilical cord mesenchymal stem cells stimulated by hypoxia promote angiogenesis both in vitro and in vivo. Stem Cells Dev. 2012;21(18):3289–97. doi:10.1089/scd.2012.0095.
28. Liang X, Ding Y, Zhang Y, Tse HF, Lian Q. Paracrine mechanisms of mesenchymal stem cell-based therapy: current status and perspectives. Cell Transplant. 2014;23(9):1045–59. doi:10.3727/096368913X667709.

29. Gnecchi M, He H, Noiseux N, Liang OD, Zhang L, Morello F, et al. Evidence supporting paracrine hypothesis for Akt-modified mesenchymal stem cell-mediated cardiac protection and functional improvement. FASEB J. 2006;20(6):661–9. doi:10.1096/fj.05-5211com.

30. Suga H, Glotzbach JP, Sorkin M, Longaker MT, Gurtner GC. Paracrine mechanism of angiogenesis in adipose-derived stem cell transplantation. Ann Plast Surg. 2014;72(2):234–41. doi:10.1097/SAP. 0b013e318264fd6a.

31. Yao Y, Huang J, Geng Y, Qian H, Wang F, Liu X, et al. Paracrine action of mesenchymal stem cells revealed by single cell gene profiling in infarcted murine hearts. PLoS One. 2015;10(6):e0129164. doi:10.1371/journal.pone.0129164.

32. Bussolino F, Di Renzo MF, Ziche M, Bocchietto E, Olivero M, Naldini L, et al. Hepatocyte growth factor is a potent angiogenic factor which stimulates endothelial cell motility and growth. J Cell Biol. 1992;119(3):629–41.

33. Li A, Varney ML, Valasek J, Godfrey M, Dave BJ, Singh RK. Autocrine role of interleukin-8 in induction of endothelial cell proliferation, survival, migration and MMP-2 production and angiogenesis. Angiogenesis. 2005;8(1):63–71. doi:10.1007/s10456-005-5208-4.

34. Wiedlocha A. Following angiogenin during angiogenesis: a journey from the cell surface to the nucleolus. Arch Immunol Ther Exp (Warsz). 1999;47(5):299–305.

35. Asahara T, Chen D, Takahashi T, Fujikawa K, Kearney M, Magner M, et al. Tie2 receptor ligands, angiopoietin-1 and angiopoietin-2, modulate VEGF-induced postnatal neovascularization. Circ Res. 1998;83(3):233–40.

36. Wang D, Wang H, Brown J, Daikoku T, Ning W, Shi Q, et al. CXCL1 induced by prostaglandin E2 promotes angiogenesis in colorectal cancer. J Exp Med. 2006;203(4):941–51. doi:10.1084/jem.20052124.

37. Del Rosso M. uPAR in angiogenesis regulation. Blood. 2011;117(15):3941–3. doi:10.1182/blood-2011-02-337733.

38. Sola-Villa D, Camacho M, Sola R, Soler M, Diaz JM, Vila L. IL-1beta induces VEGF, independently of PGE2 induction, mainly through the PI3-K/mTOR pathway in renal mesangial cells. Kidney Int. 2006;70(11):1935–41. doi:10.1038/sj.ki.5001948.

39. Salven P, Hattori K, Heissig B, Rafii S. Interleukin-1alpha promotes angiogenesis in vivo via VEGFR-2 pathway by inducing inflammatory cell VEGF synthesis and secretion. FASEB J. 2002;16(11):1471–3. doi:10.1096/fj.02-0134fje.

40. Zhao J, Chen L, Shu B, Tang J, Zhang L, Xie J, et al. Granulocyte/macrophage colony-stimulating factor influences angiogenesis by regulating the coordinated expression of VEGF and the Ang/Tie system. PLoS One. 2014;9(3):e92691. doi:10.1371/journal.pone.0092691.

41. Shojaei F, Wu X, Qu X, Kowanetz M, Yu L, Tan M, et al. G-CSF-initiated myeloid cell mobilization and angiogenesis mediate tumor refractoriness to anti-VEGF therapy in mouse models. Proc Natl Acad Sci U S A. 2009;106(16): 6742–7. doi:10.1073/pnas.0902280106.

42. Bousquenaud M, Schwartz C, Leonard F, Rolland-Turner M, Wagner D, Devaux Y. Monocyte chemotactic protein 3 is a homing factor for circulating angiogenic cells. Cardiovasc Res. 2012;94(3):519–25. doi:10.1093/cvr/cvs140.

43. Inokuma D, Abe R, Fujita Y, Sasaki M, Shibaki A, Nakamura H, et al. CTACK/CCL27 accelerates skin regeneration via accumulation of bone marrow-derived keratinocytes. Stem Cells. 2006;24(12):2810–6. doi:10.1634/stemcells.2006-0264.

44. Benslimane-Ahmim Z, Heymann D, Dizier B, Lokajczyk A, Brion R, Laurendeau I, et al. Osteoprotegerin, a new actor in vasculogenesis, stimulates endothelial colony-forming cells properties. J Thromb Haemost. 2011;9(4):834–43. doi:10.1111/j.1538-7836.2011.04207.x.

45. Dominici M, Le Blanc K, Mueller I, Slaper-Cortenbach I, Marini F, Krause D, et al. Minimal criteria for defining multipotent mesenchymal stromal cells. The International Society for Cellular Therapy position statement. Cytotherapy. 2006;8(4):315–7. doi:10.1080/14653240600855905.

46. Hsiao ST, Asgari A, Lokmic Z, Sinclair R, Dusting GJ, Lim SY, et al. Comparative analysis of paracrine factor expression in human adult mesenchymal stem cells derived from bone marrow, adipose, and dermal tissue. Stem Cells Dev. 2012;21(12):2189–203. doi:10.1089/scd.2011.0674.

47. Peltzer J, Montespan F, Thepenier C, Boutin L, Uzan G, Rouas-Freiss N, et al. Heterogeneous functions of perinatal mesenchymal stromal cells require a preselection before their banking for clinical use. Stem Cells Dev. 2015;24(3): 329–44. doi:10.1089/scd.2014.0327.

48. Psaltis PJ, Paton S, See F, Arthur A, Martin S, Itescu S, et al. Enrichment for STRO-1 expression enhances the cardiovascular paracrine activity of human bone marrow-derived mesenchymal cell populations. J Cell Physiol. 2010; 223(2):530–40. doi:10.1002/jcp.22081.

49. Gronthos S, Zannettino AC, Hay SJ, Shi S, Graves SE, Kortesidis A, et al. Molecular and cellular characterisation of highly purified stromal stem cells derived from human bone marrow. J Cell Sci. 2003;116(Pt 9):1827–35.

50. Wang CM, Guo Z, Xie YJ, Hao YY, Sun JM, Gu J, et al. Co-treating mesenchymal stem cells with IL1beta and TNF-alpha increases VCAM-1 expression and improves post-ischemic myocardial function. Mol Med Rep. 2014;10(2):792–8. doi:10.3892/mmr.2014.2236.

15

The hepatocyte growth factor-expressing character is required for mesenchymal stem cells to protect the lung injured by lipopolysaccharide in vivo

Shuling Hu, Jinze Li, Xiuping Xu, Airan Liu, Hongli He, Jingyuan Xu, Qihong Chen, Songqiao Liu, Ling Liu, Haibo Qiu and Yi Yang[*]

Abstract

Background: Acute respiratory distress syndrome (ARDS) is a life-threatening condition in critically ill patients. Recently, we have found that mesenchymal stem cells (MSC) improved the permeability of human lung microvascular endothelial cells by secreting hepatocyte growth factor (HGF) in vitro. However, the properties and functions of MSC may change under complex circumstances in vivo. Here, we sought to determine the role of the HGF-expressing character of MSC in the therapeutic effects of MSC on ARDS in vivo.

Methods: MSC with HGF gene knockdown (MSC-ShHGF) were constructed using lentiviral transduction. The HGF mRNA and protein levels in MSC-ShHGF were detected using quantitative real-time polymerase chain reaction and Western blotting analysis, respectively. HGF levels in the MSC culture medium were measured by enzyme-linked immunosorbent assay (ELISA). Rats with ARDS induced by lipopolysaccharide received MSC infusion via the tail vein. After 1, 6, and 24 h, rats were sacrificed. MSC retention in the lung was assessed by immunohistochemical assay. The lung wet weight to body weight ratio (LWW/BW) and Evans blue dye extravasation were obtained to reflect lung permeability. The VE-cadherin was detected with inmmunofluorescence, and the lung endothelial cell apoptosis was assessed by TUNEL assay. The severity of lung injury was evaluated using histopathology. The cytokines and HGF levels in the lung were measured by ELISA.

Results: MSC-ShHGF with markedly lower HGF expression were successfully constructed. Treatment with MSC or MSC carrying green fluorescent protein (MSC-GFP) maintained HGF expression at relatively high levels in the lung at 24 h. MSC or MSC-GFP decreased the LWW/BW and the Evans Blue Dye extravasation, protected adherens junction VE-cadherin, and reduced the lung endothelial cell apoptosis. Furthermore, MSC or MSC-GFP reduced the inflammation and alleviated lung injury based on histopathology. However, HGF gene knockdown significantly decreased the HGF levels without any changes in the MSC retention in the lung, and diminished the protective effects of MSC on the injured lung, indicating the therapeutic effects of MSC on ARDS were partly associated with the HGF-expressing character of MSC.

Conclusions: MSC restores lung permeability and lung injury in part by maintaining HGF levels in the lung and the HGF-expressing character is required for MSC to protect the injured lung.

Keywords: Acute lung injury, Mesenchymal stem cell, Hepatic growth factor, Lung vascular permeability

* Correspondence: yiyiyang2004@163.com
Department of Critical Care Medicine, Zhongda Hospital, Southeast
University School of Medicine, No.87 Dingjiaqiao Road, Nanjing 210009,
Jiansu, P.R. China

Background

Acute lung injury (ALI) and its severe form, acute respiratory distress syndrome (ARDS), are life-threatening conditions in critically ill patients [1]. Although substantial progress has been made in the treatment of ALI/ARDS, especially in the development of lung-protective ventilation and a fluid-conservation strategy, mortality remains high [2, 3]. Novel therapies for ALI/ARDS are therefore needed.

ALI/ARDS are characterized by diffuse injury to the lung endothelial and epithelial cells, which leads to an increase in alveolar-capillary permeability and alveolar pulmonary edema [4]. The lung endothelium is the first barrier to prevent cells and proteins in the blood from infiltrating into the interstitial space and the alveoli of the lung. Therefore, it is essential to restore injured endothelial cells for the treatment of ALI/ARDS.

Bone marrow-derived mesenchymal stem cells (MSC) are one of several types of stem cells that may have therapeutic effects on damaged organs. MSC are capable of differentiating into different types of cells, secreting paracrine soluble factors, releasing microvesicles (MV), homing to injury sites, and so forth. Many studies show that MSC are beneficial in ALI/ARDS [5–7]. They can restore lung protein permeability, reduce inflammation, and improve survival in lipopolysaccharide- (LPS) or *Escherichia coli*-induced lung injury in mice. However, the underlying mechanism by which MSC restores lung protein permeability in vivo has not yet been fully clarified.

Recently, the strong paracrine property of MSC has been considered to be the principal mechanism for maintaining function in damaged organs [8, 9]. The pulmotrophic factor, hepatocyte growth factor (HGF), which plays an important role in protecting vascular permeability, is a major paracrine soluble factor of MSC. Many studies have detected HGF protein in the cell culture medium of MSC [10, 11]. Furthermore, HGF expression increased when MSC were under hypoxic conditions or stimulated with LPS [10]. In our previous in vitro study [12], we found that, by secreting HGF, MSC improved protein permeability across the human lung microvascular endothelial cell (HMVEC) monolayer and repaired intercellular junctions, including the tight junctions and the adherens junctions of lung endothelial cells. However, in in-vivo conditions, the circumstances for MSC are different due to the unchecked inflammation, severe hypoxemia, and the increased pulmonary alveolar and capillary membrane leakage of ARDS. These all may affect the properties and functions of MSC. Hence, we sought to determine whether the HGF-expressing character of MSC would also be critical for MSC to maintain lung permeability and protect the lung from injury in vivo.

Methods

Ethics statement

Male wild-type Sprague-Dawley (SD) rats (Laboratory Animal Center, Shanghai, China) were maintained under specific pathogen-free conditions. The Animal Care and Use Committee of Southeast University approved all experiments involving the use of animals.

Cell culture

Rat MSC, isolated from the bone marrow of SD rats, and the MSC culture medium were purchased from Cyagen Bioscience, Inc. (Guangzhou, China). 293 T cells were supplied by Cell Bank of Chinese Academy of Sciences. The supplier identified MSC according to cell surface phenotypes and multipotency. Fluorescence-activated cell sorting (FACS) analysis characterized surface phenotypes using the following markers: CD90+, CD44+, CD29+, CD34–, CD45–, and CD11b/c–, and the capacities to differentiate into the adipogenic, osteogenic and chondrogenic lineages were determined by staining with oil red-O, alizarin red or alcian blue, respectively, after culturing in adipogenic, osteogenic or chondrogenic differentiation media (Cyagen Bioscience, Inc., Guangzhou, China) for 2–3 weeks (See Additional file 1).

MSC were cultured in the MSC culture medium made from SD rat MSC basal medium containing 10 % SD rat MSC-qualified fetal bovine serum, 1 % penicillin-streptomycin and 1 % glutamine. The culture medium was changed every 2–3 days and the cells were split when they achieved 90 % confluency. The MSC used for in-vitro studies had been passaged 10 times after lentiviral transduction. MSC with the total passage number <9 were used in the in-vivo experiments. 293 T cells were cultured in Dulbecco's modified Eagle's medium (DMEM) containing 10 % fetal bovine serum (Wisent, Inc., St-Bruno, Quebec, Canada), 1 % L-glutamine, and 1 % penicillin-Streptomycin, and incubated at 37 °C in a humidified atmosphere of 5 % CO_2. Additional materials were provided in Additional file 2.

Lentiviral vector-mediated HGF gene knockdown in MSC

MSC with passage number <6 were used for this experiment. Briefly, the rat HGF knockdown constructs expressing short-hairpin RNA targeting endogenous HGF (ShRNA HGF) were encoded into a lentivirus-based ShRNA vector pGLV3/H1/GFP + Puro (LV3) driven by the H1 promoter containing green fluorescent protein (GFP) and puromycin. Target sequences were designed and selected with software Designer 3.0 provided by GenePharma. Additionally, LV3 containing nonspecific ShRNA (LV3-GFP) was used as a negative control. The recombinant vectors were integrated and replicated in *E. coli* Top10 (GenePharma, Shanghai, China).

The recombinant plasmid DNAs were extracted from *E. coli* Top10 and purified using the Plasmid Preparation Kit (GenePharma, Shanghai, China). The purity of the DNA was assessed with a spectrophotometer (Tecan, Switzerland). A260/A280 nm absorbance ratios of 1.8–2.2 suggested a pure DNA sample. Theses plasmids were then separately co-transfected with three packaging plasmids (pGag/Pol, pRev, pVSV-G) into 293 T cells using RNAi-mate (Genepharma, Shanghai, China) according to the manufacturer's instruction. The lentiviral particles were collected and stored at −80 °C for future use. Titer was obtained by GFP expression assay [13].

MSC were seeded and cultured in six-well plates for 24 h. The lentiviral vectors (carrying LV3-GFP or LV3-GFP ShRNA HGF) were then added to the wells at a multiplicity of infection (MOI) value of 100:1 and cultured with MSC for 24 h. After 24 h, the culture medium was changed, and puromycin was added at the minimal lethal concentration (1.5 µg/ml) for transfected MSC. The puromycin-resistant cells were then collected.

RNA isolation and quantitative real-time polymerase chain reaction (qRT-PCR)

MSCs treated with LV3-GFP (MSC-GFP) or LV3-GFP-ShRNA HGF (MSC-ShHGF) were collected, respectively. Total RNA was isolated from MSC, MSC-GFP or MSC-ShHGF using TRIzol reagent (Takara Bio, Inc., Kyoto, Japan) according to the manufacturer's protocol. The quality of the RNA was assessed with a spectrophotometer (Tecan, Switzerland). 260/280 nm absorbance ratios of 1.8–2.2 suggested a pure RNA sample. The RT-PCR primers for rat glyceraldehyde-3-phosphate dehydrogenase (GAPDH) and rat HGF (Table 1) were provided by GenePharma (Shanghai, China). RT-PCR assays were performed following the One-Step RT-PCR protocol described by Funglyn Biotech Inc. (Shanghai, China).

Western blotting analysis

MSC, MSC-GFP, and MSC-ShHGF were collected after transduction with lentiviral vector. Total cellular protein from either MSC, MSC-GFP, or MSC-ShHGF was extracted and separated using SDS-PAGE gels (10 %), as previously described [14]. Protein was then incubated with primary antibodies to HGF (1:600 dilution; Santa

Cruz Biotechnology, Inc., Santa Cruz, CA, USA) or β-actin (1:10,000 dilution; Abcam Ltd., Cambridge, UK). The blots were washed three times and then incubated with goat anti-rabbit IgG conjugated with horseradish peroxidase (HRP; Zhongshan Golden Bridge Biotechnology Co., Ltd, China). Immunoreactive complexes were visualized using chemiluminescence reagents (Thermo Scientific).

Evaluation of HGF levels by ELISA

MSC, MSC-GFP, and MSC-ShHGF were seeded in a 12-well plate at a density of 1×10^5 cells per well. After 12 h the culture medium was changed, and MSC were cultured in an incubator at 37 °C, 5 % CO_2 for 24 h. The culture medium was then collected and HGF protein levels in the culture medium were quantified using an enzyme-linked immunosorbent assay (ELISA) kit (ExCellBio, Shanghai, China) according to the manufacturer's instructions.

LPS-induced ALI in rats

To induce ALI, 6- to 8-week-old wild-type SD rats received an intratracheal instillation of LPS (2 mg/kg, *E. coli* 0111:B4; Sigma-Aldrich, St. Louis, MO, USA) dissolved in 100 µl phosphate-buffered saline (PBS; Wisent, Inc., St-Bruno, Quebec, Canada) as described previously [15]. PBS, MSC, MSC-GFP, or MSC-ShHGF (5×10^6 cells resuspended in 100 µl PBS) were injected into the tail vein 5 h after LPS challenge. Rats without LPS challenge were injected with PBS as a control. Rats were sacrificed at 1, 6 and 24 h after MSC injection, and the lung lobes were collected for further analysis.

Measurement of lung edema

Lung wet weight to body weight (LWW/BW) ratios, which reflected the severity of lung vascular permeability and lung edema, were obtained from the control, ALI, MSC, MSC-GFP, and MSC-ShHGF group.

Evans blue dye leakage

For the control, ALI, MSC-GFP, and MSC-ShHGF group (n = 6 per group, see Additional file 3 for sample size calculation), Evans blue dye (20 mg/kg in 1 ml saline; Sigma-Aldrich, St. Louis, MO, USA) was injected into the tail vein of the rats. After 30 min, the right ventricle of the heart was perfused with 100 ml heparinized saline

Table 1 The primer sequence of genes

Gene	Primer	Primer sequence	PCR amplified products (bp)
GAPDH	Forward primer	5' > GTGCTGAGTATGTCGTGGAGTCT < 3'	104
	Reverse primer	5' > GGAAGGGGCGGAGATGA < 3'	
HGF	Forward primer	5' > GCACCTCCTCCTGCTTCC < 3'	288
	Reverse primer	5' > CCAAACCCTTTTTTCACTCCA < 3'	

bp base pair, *GAPDH* glyceraldehyde-3-phosphate dehydrogenase, *HGF* hepatocyte growth factor, *PCR* polymerase chain reaction

to clean up the dye remaining in the lung vascular system. When all the dye had been cleared from the lung vascular system, the whole lung was collected. Lung tissue (100 mg) from the right lobe was then incubated in formamide (Sigma-Aldrich, St. Louis, MO, USA) for 24 h at 60 °C, and the concentration of Evans blue dye was measured using a spectrophotometer at 630 nm (Tecan, Switzerland).

Immunohistochemical staining

Immunohistochemical analysis was performed to determined expression of GFP in the lung. Sections (5 μm thick) were cut from paraffin-embedded tissues. After being routinely deparaffinized and dehydrated, sections were subjected to antigen retrieval by microwave treatment in boiling 0.01 M citrate buffer (pH 6.0) for 20 min. Then the sections were blocked in 1 % bovine serum albumin (BSA) for 1 h at 37 °C and incubated with anti-GFP primary antibody (Abcam Ltd., Cambridge, UK) overnight at 4 °C, followed by HRP conjugated goat anti-rabbit IgG (Santa Cruz Biotechnology, Inc., Santa Cruz, CA, USA). After washing with 0.1 M PBS, the sections were reacted with a staining solution containing 0.03 % 3,3′-diaminobenzidine tetrahydrochloride (Nanjing Lufei Biotechnology Co., Ltd., Nanjing, China) for 5–10 min at room temperature. Mayer's hematoxylin was used for counterstaining. The sections were examined with an Olympus IX71 microscope (Olympus Co., Tokyo, Japan). Cells with GFP were counted by a pathologist based on five randomly selected high-power fields (×400).

Immunofluorescence

To measure co-localization of VE-cadherin in the lung, rat lungs were collected gently and then fixed in 4 % paraformaldehyde at 4 °C. Twenty four hours later, lung tissue was frozen in optimal cutting temperature medium (OCT; Sakura Finetek USA, Inc., Torrance, CA, USA) and cut into 5-μm-thick sections. Then the sections were stained with anti-VE-cadherin antibody (Santa Cruz Biotechnology, Inc., Santa Cruz, CA, USA), followed by Fluorescence (FITC)-AffiniPure Donkey Anti-Rabbit IgG (H + L; Jackson ImmunoResearch Inc., PA, USA), and mounted with 4,6-diamidino-2-phenylindole (DAPI; Sigma-Aldrich). Fluorescence was monitored with an Olympus IX71 microscope (Olympus Co., Tokyo, Japan). The mean fluorescence intensity of VE-cadherin of five randomly chosen high-power fields per lung lobe section per rat was assessed and calculated.

TUNEL assay

To assess the apoptosis of endothelial cells in the lung, the right lobe was collected and fixed in 4 % paraformaldehyde at 4 °C for 24 hours, and the lung sections were stained with a TUNEL Apoptosis Assay kit (Nanjing Lufei

Biotechnology Co., Ltd., Nanjing, China). The sections were observed with an Olympus IX71 microscope (Olympus Co., Tokyo, Japan). Endothelial cells were quantified by a pathologist based on five randomly selected high-power fields (×400). The number of apoptotic lung endothelial cells and total lung endothelial cells were recorded, and the apoptosis index of lung endothelial cells, which was defined as the number of apoptotic lung endothelial cells divided by the total number of lung endothelial cells, was used to assess the severity of lung endothelial cell apoptosis.

Lung histopathology

To examine the severity of lung injury, the right lobes were collected at 1, 6 and 24 h after MSC infusion and then fixed in 4 % paraformaldehyde. After fixation, the lungs were embedded in paraffin and cut into 5-μm sections. The lung sections were stained using a hematoxylin and eosin staining kit purchased from the Beyotime Institute of Biotechnology. The severity of lung injury was quantified and assessed using a total lung injury score as we previously described [15].

HGF expression in the lung

To detect the HGF levels in the lung, the left lung was collected and stored at −80 °C. The lung tissue was homogenized thoroughly in PBS solution and then centrifuged at 3000 rpm for 20 min at 4 °C. The levels of HGF in the supernatant were then measured using an ELISA kit (ExCellBio, Shanghai, China), according to the manufacturer's instructions.

Measurement of cytokines in the lung

The left lobe was snap-frozen and later processed for lung homogenization. Then the lung homogenate was centrifuged at 3000 rpm for 20 min at 4 °C and the supernatant was collected. The concentrations of interleukin (IL)-1β and IL-10 in the supernatant were evaluated using ELISA kits (NeoBiosicence, Shenzhen, China), according to the manufacturer's instructions.

Statistical analysis

Data from in vitro experiments are shown as the mean ± standard deviation. Data from in vivo experiments are shown as individual point + median. Statistical analyses were performed using SPSS 16.0. For comparisons between multiple groups, one-way analysis of variance (ANOVA) followed by Bonferroni's post-hoc test was used. p values <0.05 were considered to be statistically significant.

Results
The efficiency of lentiviral vector-mediated HGF gene knockdown
HGF mRNA levels were detected by qRT-PCR. The result showed that HGF mRNA expression was significantly lower in the MSC-ShHGF group than in the MSC ($p < 0.05$) and MSC-GFP group ($p < 0.05$). However, there was no significant difference between the MSC and MSC-GFP group ($p > 0.05$) (Fig. 1a) in the HGF mRNA expression. The result from Western blotting analysis (Fig. 1c and d) showed that the HGF protein expression in the cytoplasm was decreased in the MSC-ShHGF group compared with the MSC ($p < 0.05$) and MSC-GFP group ($p < 0.05$). The HGF protein levels in the cell culture medium of the MSC-ShHGF group were also significantly lower those that in the MSC ($p < 0.05$) and MSC-GFP groups ($p < 0.05$) (Fig. 1b). However, there was no significant difference in HGF protein levels between the MSC and MSC-GFP group in either the cytoplasm or the cell culture medium. Taken together, these results suggested that lentiviral-mediated HGF gene knockdown was efficient.

MSC retention in the lung at 24 h
To measure the MSC retention in the lung tissue after cell infusion, immunohistochemical assays were performed for the GFP carried by MSC. As Fig. 2 shows, no significant differences were found in the retention of MSC with GFP in the lung tissue between the MSC-GFP and MSC-ShHGF groups ($p > 0.05$), demonstrating that HGF gene knockdown in MSC had no significant impact on MSC homing to the lung in respond to lung injury.

HGF levels in the lung after MSC treatment
To examine the effect of MSC infusion on HGF levels in ALI rats, we measured the HGF concentration in the lung. As Fig. 3 shows, HGF protein levels in the lung increased at 1 h after LPS challenge and then gradually decreased at 6, 24 and 48 h ($p < 0.05$). MSC treatment significantly increased HGF levels at 6, 24 and 48 h ($p < 0.05$). When the HGF gene was knocked down in MSC, the HGF level in the lung decreased significantly compared with the MSC or MSC-GFP group at 24 and 48 h ($p < 0.05$). Since no differences were found in the retention of MSC with GFP in the lung tissue between the MSC-GFP and MSC-ShHGF groups, these results suggest that MSC regulated HGF level in the injured lung in part through secreting HGF. Additional results were provided in Additional file 2.

The effect of MSC on pulmonary vascular permeability
In this study, LWW/BW and Evans blue dye extravasation were used to evaluate the role of the HGF-expressing

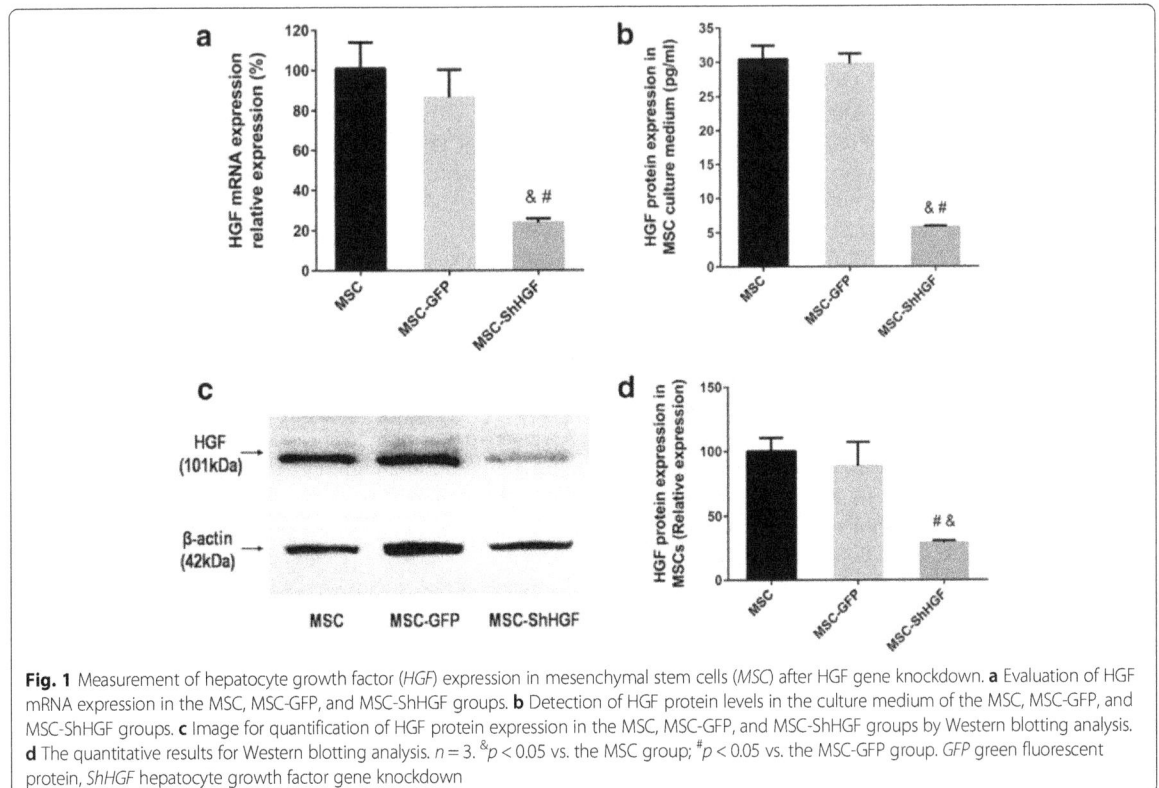

Fig. 1 Measurement of hepatocyte growth factor (HGF) expression in mesenchymal stem cells (MSC) after HGF gene knockdown. a Evaluation of HGF mRNA expression in the MSC, MSC-GFP, and MSC-ShHGF groups. b Detection of HGF protein levels in the culture medium of the MSC, MSC-GFP, and MSC-ShHGF groups. c Image for quantification of HGF protein expression in the MSC, MSC-GFP, and MSC-ShHGF groups by Western blotting analysis. d The quantitative results for Western blotting analysis. $n = 3$. $^\&p < 0.05$ vs. the MSC group; $^\#p < 0.05$ vs. the MSC-GFP group. GFP green fluorescent protein, ShHGF hepatocyte growth factor gene knockdown

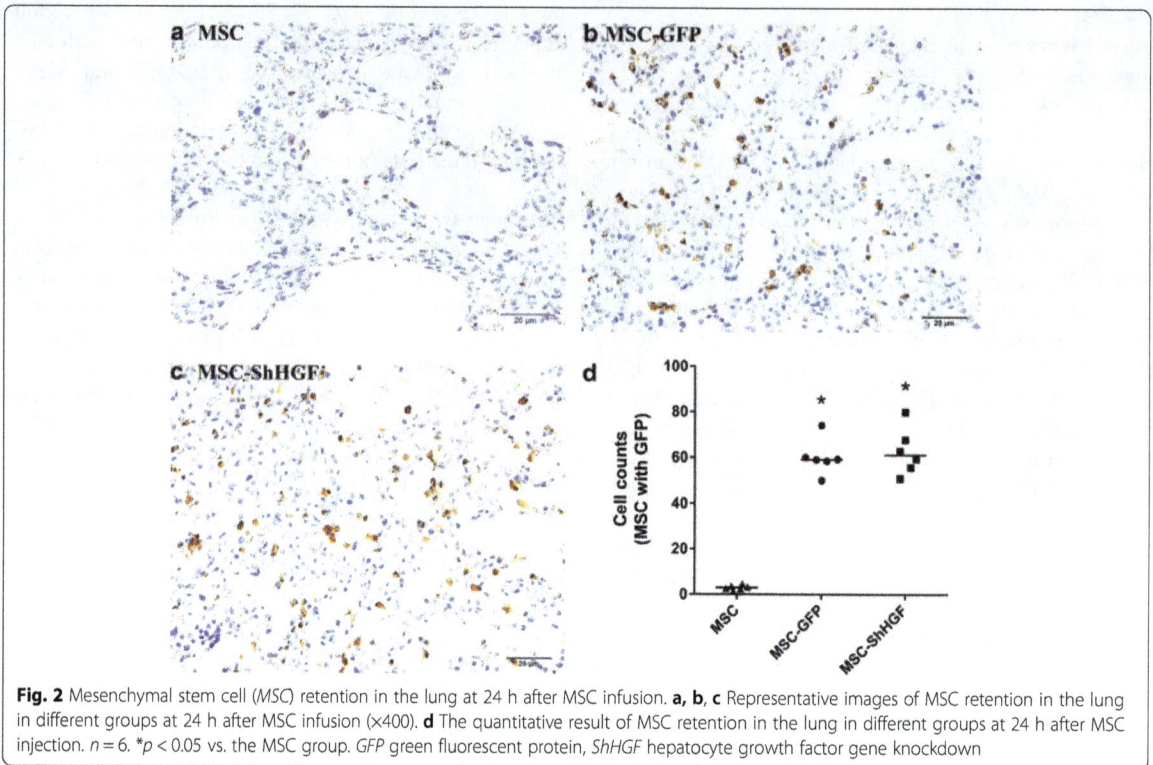

Fig. 2 Mesenchymal stem cell (*MSC*) retention in the lung at 24 h after MSC infusion. **a, b, c** Representative images of MSC retention in the lung in different groups at 24 h after MSC infusion (×400). **d** The quantitative result of MSC retention in the lung in different groups at 24 h after MSC injection. $n = 6$. *$p < 0.05$ vs. the MSC group. *GFP* green fluorescent protein, *ShHGF* hepatocyte growth factor gene knockdown

character of MSC in the therapeutic effect of MSC on lung vascular permeability. The results (Fig. 4a and b) showed that LPS stimulation increased LWW/BW markedly at 1, 6 and 24 h ($p < 0.05$). After treatment with MSC, MSC-GFP, or MSC-ShHGF, the LWW/BW was decreased significantly at 24 h in the ALI rats ($p < 0.05$). More interestingly, the LWW/BW in rats treated with MSC with HGF gene knockdown was significantly higher than that in rats treated with MSC or MSC-GFP ($p < 0.05$). Furthermore, as Fig. 4c and d show, LPS stimulation also increased Evans blue dye extravasation from the lung vascular to the lung interstitial space and lung alveoli at 1, 6 and 24 h ($p < 0.05$). With MSC, MSC-GFP, or MSC-ShHGF treatment, the Evans blue dye extravasation decreased markedly at 6 and 24 h ($p < 0.05$) but, in the MSC-ShHGF group, the Evans blue dye extravasation was significantly higher compared with the MSC or MSC-GFP group ($p < 0.05$). These results indicate that MSC improved vascular permeability in lung tissue injured by LPS, and that the HGF-expressing character of MSC was required for MSC to restore this lung vascular permeability.

VE-cadherin expression in the lung following MSC injection

To investigate the effect of MSC with HGF gene knockdown on intercellular junctions in the lungs of rats with ALI, we observed the expression of the adherens junction protein VE-cadherin in the rat lung by

immunofluorescence at 24 h. As Fig. 5 shows, the VE-cadherin expression in lung tissue was decreased markedly after intratracheal instillation of LPS at 24 h ($p < 0.05$). After treatment with MSC or MSC with HGF gene knockdown, the VE-cadherin expression was increased in the lung compared with the ALI group ($p < 0.05$). However, the increase in VE-cadherin expression in the MSC-ShHGF group was markedly lower than that in the MSC group ($p < 0.05$). These results suggest that MSC had a protective effect on the adherens junction of cells in the lung, and that this protective effect required the HGF expressed by MSC.

Assessment of endothelial cell apoptosis in the injured lung after MSC treatment by TUNEL assay

To evaluate the effect of the HGF-expressing character of MSC on endothelial apoptosis in the ALI lung, we carried out a TUNEL assay 24 h after MSC treatment. As Fig. 6 shows, the apoptosis index of the lung endothelium was increased after LPS challenge, whereas it decreased to a large degree after MSC treatment. However, in MSC with gene knockdown, the apoptosis index of the lung endothelium was significantly higher than that in MSC or the MSC-GFP group. These results indicate that MSC treatment was able to protect lung endothelial cells from apoptosis in the lung of rats with ALI, and that this

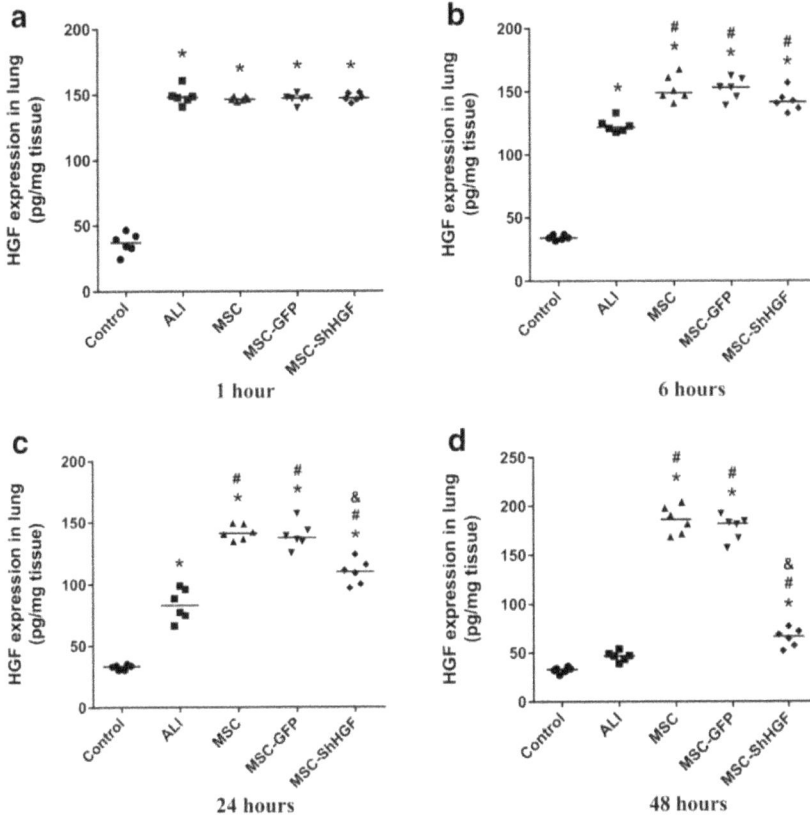

Fig. 3 Hepatocyte growth factor (*HGF*) expression in injured lung after mesenchymal stem cell (*MSC*), MSC-GFP, or MSC-ShHGF treatment. HGF protein levels in the lung tissue of different groups at **a** 1, **b** 6, **c** 24, and **d** 48 h after MSC delivery were measured by ELISA. $n = 6$. $*p < 0.05$ vs. the control group; $^{#}p < 0.05$ vs. the ALI group; $^{&}p < 0.05$ vs. the MSC group. *GFP* green fluorescent protein, *ShHGF* hepatocyte growth factor gene knockdown

protective effect was related to the ability of MSC to release the HGF protein.

Histological evaluation of the therapeutic potential of MSC in ALI rats

As Fig. 7 shows, histopathology indicated extensive inflammatory infiltrates, marked inter-alveolar septal thickening, and diffuse interstitial and alveolar edema at 1, 6 and 24 h in the ALI group. The lung injury score was 8.23 at 1 h, 10.6 at 6 h and 15.15 at 24 h in the ALI group. The administration of MSC, MSC-GFP, and MSC-ShHGF significantly attenuated lung injury at 24 h ($p < 0.05$). However, the lung injury score in the MSC-ShHGF group (9.53) was significantly higher than that in the MSC (5.23) and MSC-GFP group (4.98) ($p < 0.05$). There was no significant difference in lung injury score between the MSC and MSC-GFP group. These results suggest that MSC had a

protective effect on lung injury, and that this effect was diminished when the HGF gene was knocked down in MSC.

The effect of MSC on cytokine levels in the injured lung after LPS-induced ALI in rats

In this study, we also investigated the effect of MSC-GFP and MSC with HGF gene knockdown on IL-10 and IL-1β production in the lung. As Fig. 8a shows, the IL-1β level in the lung was markedly increased after LPS challenge at 1, 6 and 24 h ($p < 0.05$). After treatment with MSC, MSC-GFP, and MSC-ShHGF, IL-1β levels in the lung were significantly decreased at 6 and 24 h ($p < 0.05$). There was no significant difference in the lung IL-1β levels between the MSC-ShHGF, MSC, and MSC-GFP group at 6 h. However, there was a significant difference between the MSC-ShHGF and the MSC or MSC-GFP group at 24 h ($p < 0.05$).

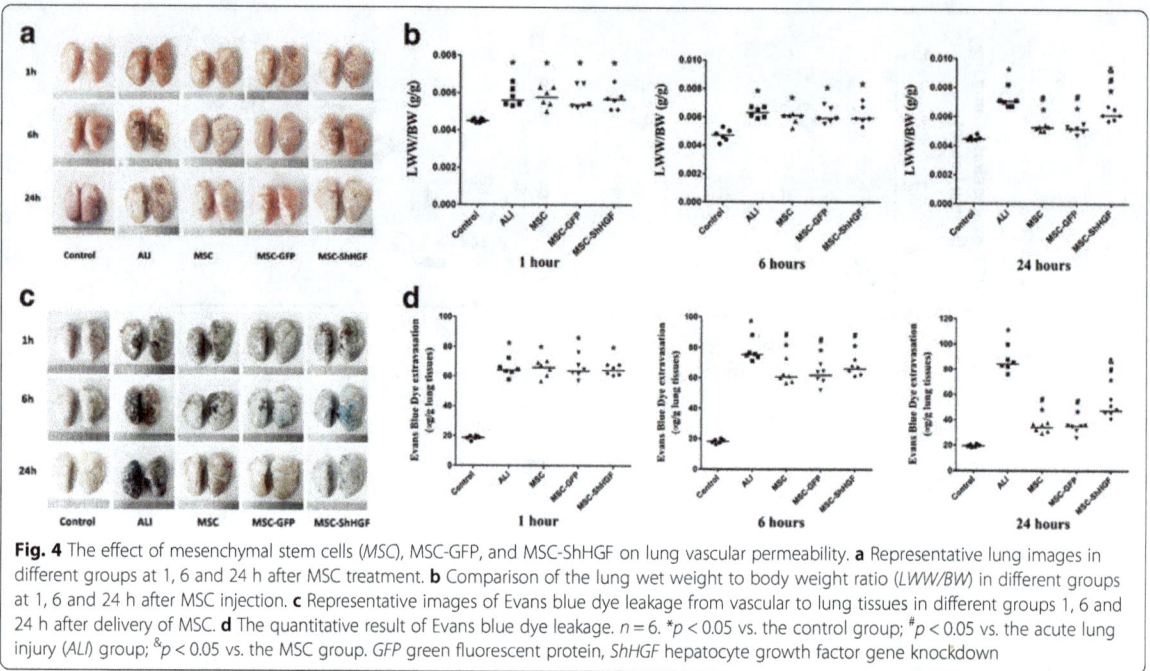

Fig. 4 The effect of mesenchymal stem cells (*MSC*), MSC-GFP, and MSC-ShHGF on lung vascular permeability. **a** Representative lung images in different groups at 1, 6 and 24 h after MSC treatment. **b** Comparison of the lung wet weight to body weight ratio (*LWW/BW*) in different groups at 1, 6 and 24 h after MSC injection. **c** Representative images of Evans blue dye leakage from vascular to lung tissues in different groups 1, 6 and 24 h after delivery of MSC. **d** The quantitative result of Evans blue dye leakage. $n = 6$. *$p < 0.05$ vs. the control group; #$p < 0.05$ vs. the acute lung injury (*ALI*) group; &$p < 0.05$ vs. the MSC group. *GFP* green fluorescent protein, *ShHGF* hepatocyte growth factor gene knockdown

The IL-10 level in the lung was significantly decreased after LPS stimulation at 1, 6 and 24 h ($p < 0.05$) (Fig. 8b). However, it was markedly increased after MSC, MSC-GFP, or MSC-ShHGF treatment at 6 and 24 h. In addition, compared with the MSC or MSC-GFP group, the IL-10 level in the lung were significantly lower in the MSC-ShHGF group. These results indicate that MSC reduced LPS-induced inflammation in the lung, and that the HGF-expressing characteristic of MSC played an important part in this effect.

Discussions

Our previous in vitro studies showed that, by secreting the protective soluble factor HGF, MSC are able to restore the permeability of the HMVEC monolayer following LPS-induced injury [12]. By producing HGF, MSC may maintain the integrity of the injured endothelial monolayer by restoring endothelial intercellular junctions, decreasing caveolin-1 protein expression, and inducing proliferation in endothelial cells. Here, we knocked down HGF in MSC to examine the role of the HGF-expressing character of MSC in the therapeutic effect of MSC on lung permeability and lung injury in vivo.

Firstly, we successfully constructed a stable and long-term MSC cell line with low HGF expression by employing a lentiviral vector-mediated HGF gene knockdown technique. Our data showed that MSC-ShHGF at the tenth passage had notably low HGF mRNA and protein expression. Moreover, HGF levels in the cell culture medium of MSC-ShHGF also markedly decreased. The

successful construction of MSC-ShHGF facilitates our further investigations into the role of the HGF-expressing character of MSC in the protective effect of MSC on lung endothelial permeability and lung injury in vivo.

According to our data from the in vivo study, the HGF level in the lung was markedly elevated after LPS challenge. However, it began to decrease at 6 h and was further reduced to approximately half by 24 h. After MSC treatment, the HGF level increased significantly. Interestingly, when the HGF gene was knocked down in MSC, the HGF protein levels in the injured lung were significantly decreased at 24 h but were still higher than those in the ALI group. However, there was no significant difference in MSC retention in the injured lung between the MSC-GFP and MSC-ShHGF groups as shown by the immunohistochemical assay. Taken together, these results suggested that the HGF-expressing character of MSC was required for MSC to increase the HGF levels in the lung, which then exerted a protective effect on the injured lung. There are studies showing that the intravenous or the intra-organ administration of HGF reduced organ injury [12, 16–18]. Recently, studies [19–21] have also demonstrated that MSC overexpressing HGF elevated the HGF level in injured organs and delivered a stronger therapeutic effect compared with normal MSC. Here, our results are consistent with these previous findings.

In this study, data from the LWW/BW and Evans blue dye extravasation assay showed that the HGF-expressing character of MSC played a modest role in the protective effect of MSC on lung endothelial permeability. HGF-

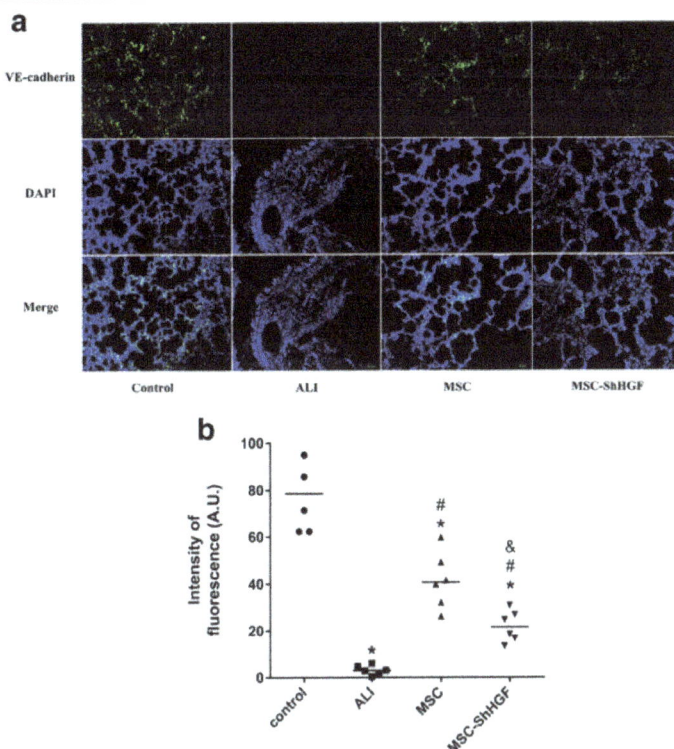

Fig. 5 Detection of the changes of VE-cadherin in the lung tissue after mesenchymal stem cell (*MSC*) injection by immunofluorescence. **a** Intercellular junctions were evaluated by detection of the adherens junction protein VE-cadherin using fluorescence microscopy 24 h after MSC-ShHGF treatment (×200; *blue*, DAPI; *green*, VE-cadherin). **b** The quantitative result of VE-cadherin expression in the lung in different groups at 24 h after MSC injection. $n = 6$. *$p < 0.05$ vs. the control group; #$p < 0.05$ vs. the ALI group; &$p < 0.05$ vs. the MSC group. *ALI* acute lung injury, *AU* arbitrary units, *DAPI* 4,6-diamidino-2-phenylindole, *ShHGF* hepatocyte growth factor gene knockdown

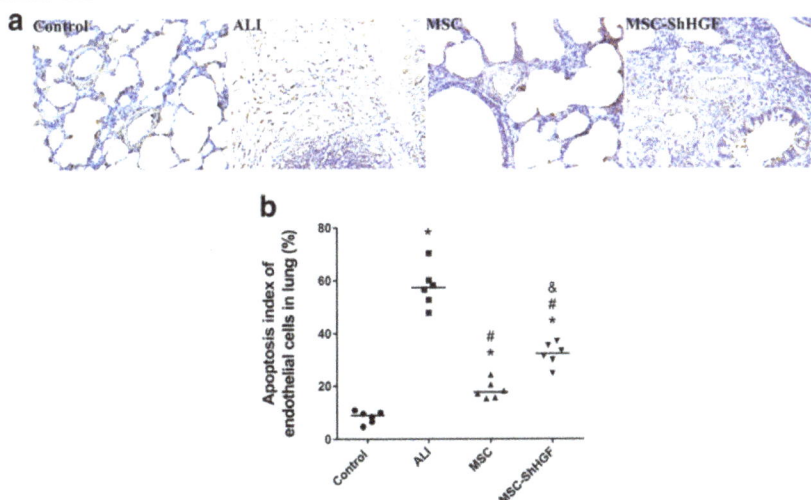

Fig. 6 The evaluation of vascular endothelial cell apoptosis in the lung. **a** Representative images of lung endothelial cell apoptosis in different groups 24 h after mesenchymal stem cell (*MSC*) treatment (×400). **b** The quantitative result of the apoptosis index of lung endothelial cells in different groups at 24 h after MSC injection. $n = 6$. *$p < 0.05$ vs. the control group; #$p < 0.05$ vs. the ALI group; &$p < 0.05$ vs. the MSC group. *ALI* acute lung injury, *ShHGF* hepatocyte growth factor gene knockdown

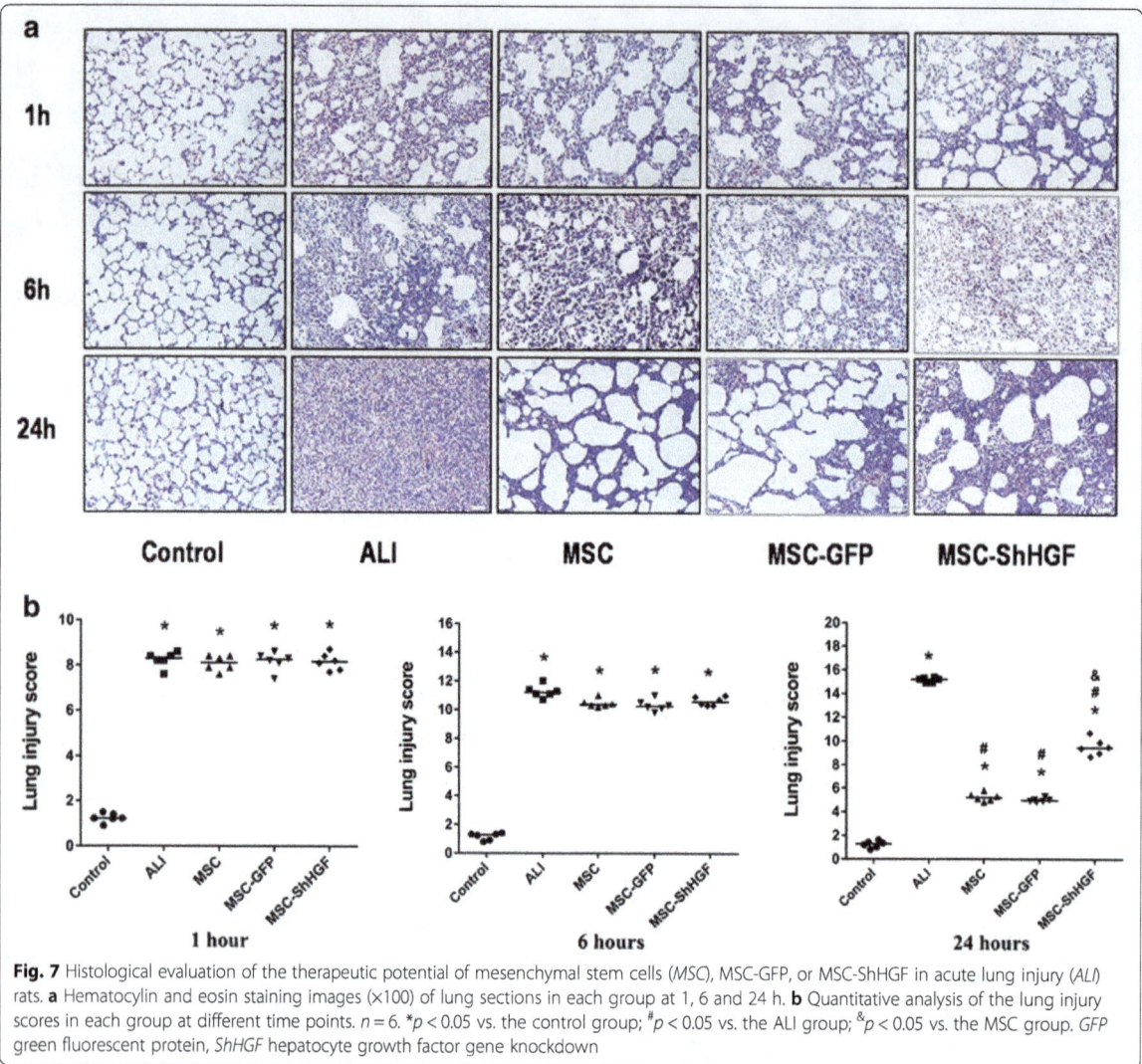

Fig. 7 Histological evaluation of the therapeutic potential of mesenchymal stem cells (*MSC*), MSC-GFP, or MSC-ShHGF in acute lung injury (*ALI*) rats. **a** Hematocylin and eosin staining images (×100) of lung sections in each group at 1, 6 and 24 h. **b** Quantitative analysis of the lung injury scores in each group at different time points. $n = 6$. *$p < 0.05$ vs. the control group; #$p < 0.05$ vs. the ALI group; &$p < 0.05$ vs. the MSC group. *GFP* green fluorescent protein, *ShHGF* hepatocyte growth factor gene knockdown

expressing MSC have a beneficial effect on maintaining the integrity of the lung endothelium. This may be related to the ability of HGF itself to protect intercellular junctions, as our data demonstrated, preventing endothelial cell apoptosis. Previous studies have demonstrated that HGF had a strong angiogenic effect [22–24]. This could facilitate endothelial cell proliferation [25], prevent apoptosis [26, 27], and restore intercellular junctions and cytoskeleton structures [28, 29] in vitro. HGF administration could also improve the prognosis of lung injury caused by inflammation, oxidative stress, radiation, fibrosis, and so forth [17, 30, 31], which may be partly related to the effect of HGF on repairing the lung endothelium.

In addition to restoring lung endothelial permeability, the HGF-expressing character also plays a positive role in the beneficial effect of MSC on lung inflammation

induced by LPS. As shown in Fig. 8, IL-1β levels were significantly higher and IL-10 levels were lower in the MSC-ShHGF group compared with the MSC or MSC-GFP group. This finding suggested that the ability of MSC to control lung inflammation was diminished after the HGF gene in MSC was knocked down. Possible explanations for this result include: 1) HGF has a direct anti-inflammatory effect as demonstrated by previous studies [32, 33]; or 2) the decreased infiltration of immunocytes from the blood to the interstitial space and alveoli of the lung due to the improvement of endothelial permeability related to HGF.

Finally, in the current study, we also assessed lung injury using histopathology and lung injury score (Fig. 7). The improvements in histopathology and lung injury score typically occurred 24 h after MSC treatment.

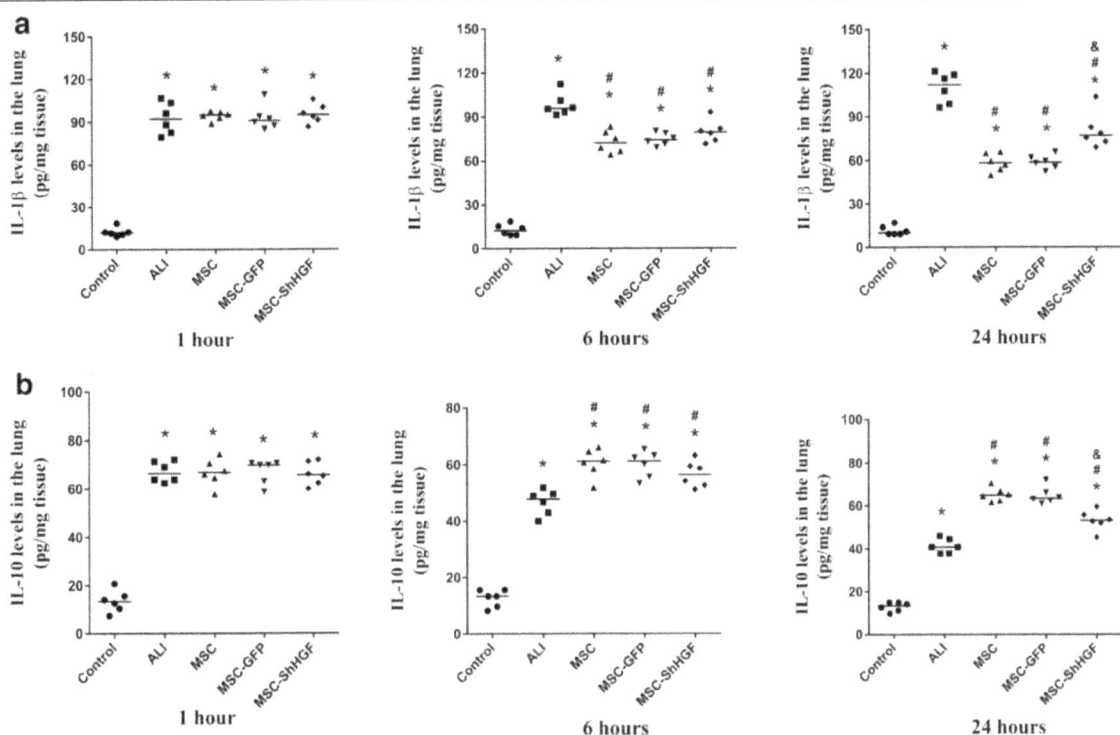

Fig. 8 The effect of mesenchymal stem cell (*MSC*), MSC-GFP, and MSC-ShHGF on cytokine levels in the lung. **a** IL-1β levels in lung tissues in different groups at 1, 6 and 24 h after MSC treatment measured by ELISA. **b** IL-10 levels in lung tissues in different groups at 1, 6 and 24 h after MSC treatment measured by ELISA. $n = 6$. *$p < 0.05$ vs. the control group; #$p < 0.05$ vs. the ALI group; &$p < 0.05$ vs. the MSC group. *ALI* acute lung injury, *GFP* green fluorescent protein, *IL* interleukin, *ShHGF* hepatocyte growth factor gene knockdown

However, in the MSC-ShHGF group this improvement was relatively small compared with the MSC or MSC-GFP group. These results suggested that HGF secretion was required for MSC to exert a better therapeutic effect in ALI. Recently, several studies demonstrated that HGF gene modification (overexpression) in MSC could enhance the therapeutic effect of MSC on injured organs such as the heart [34], liver [35], intestine [33], and lung [30]. Here, we inhibited HGF expression in MSC with a method using lentiviral vector and found out that the therapeutic effects of MSC on lung injury were diminished to some degree. This result is consistent with the previous studies and indicates a modest role of the HGF-expressing character of MSC in the therapeutic effects of MSC on ALI. In summary, our results showed that the HGF-expressing property of MSC was required for MSC-based therapies in lung injury.

There are some limitations to this study. Firstly, adherens junctions were the only type of intercellular junction we examined in this study. Therefore, further studies are needed to clarify the effect of HGF protein production in MSC on the endothelial tight junctions

and cytoskeleton in rats with ALI. Secondly, we did not investigate the cellular origin of HGF in the lung, which may help us to understand the underlying mechanism of the beneficial effect of the HGF-expressing character of MSC in ALI more clearly. Moreover, we did not explore the effect of MSC with HGF overexpression on ALI. This would also provide some additional evidence for this work. Finally, we only focused on the short-term effect (24 h) of MSC-ShHGF on ALI. The long-term effects remain unknown and merit further investigation.

Conclusion

In conclusion, the results presented here suggest that MSC restores lung vascular permeability, which might be associated with the protective effects of MSC on the adherens junction protein VE-cadherin and lung vascular endothelial cell apoptosis, reducing inflammation, and attenuating lung injury in LPS-induced ALI in rats in part by maintaining the HGF level in the injured lung. Moreover, the HGF-expressing character is required for MSC to protect the injured lung.

Abbreviations

ALI: acute lung injury; ARDS: acute respiratory distress syndrome; ELISA: enzyme-linked immunosorbent assay; GFP: green fluorescent protein; HGF: hepatocyte growth factor; HMVEC: human lung microvascular endothelial cell; HRP: horseradish peroxidase; IL: interleukin; LPS: lipopolysaccharide; LV3-GFP: lentivirus carrying GFP; LV3-ShRNA HGF: lentivirus carrying green fluorescent protein and hepatocyte growth factor gene knockdown; LWW/BW: lung wet weight to body weight ratio; MSC: mesenchymal stem cells; MSC-GFP: mesenchymal stem cells carrying green fluorescent protein; MSC-ShHGF: mesenchymal stem cells with hepatocyte growth factor gene knockdown; PBS: phosphate-buffered saline; qRT-PCR: quantitative real-time polymerase chain reaction; SD: Sprague-Dawley; Sh: short-hairpin.

Competing interests

The authors declare that they have no competing interests.

Authors' contributions

SH participated in the study design, performed laboratory work and statistical analysis, prepared the drafts of the manuscript and revised it according to advice from the other authors. JL, XX and HH participated in the laboratory work, performed statistical analysis and drafted the manuscript. AL, JX, QC, SL and LL participated in the study design, and assisted in statistical analysis and drafting the manuscript. HQ and YY were responsible for the study design and revised the manuscript for important intellectual content. All the authors have given final approval of the version to be published and agree to be accountable for all aspects of this work.

Acknowledgements

We all thank Yanli An for assistance with the experimental techniques. We are also thankful to Pablo Gabatto and Fengyun Xu. The authors are grateful for the financial support of the National Natural Science Foundation of China (Grant No. 81170057).

References

1. Rubenfeld GD, Caldwell E, Peabody E, Weaver J, Martin DP, Neff M, et al. Incidence and outcomes of acute lung injury. N Engl J Med. 2005;353(16):1685–93. doi:10.1056/NEJMoa050333.

2. The Acute Respiratory Distress Syndrome Network. Ventilation with lower tidal volumes as compared with traditional tidal volumes for acute lung injury and the acute respiratory distress syndrome. N Engl J Med. 2000; 342(18):1301–8. doi:10.1056/NEJM200005043421801.

3. Buregeya E, Fowler RA, Talmor DS, Twagirumugabe T, Kiviri W, Riviello ED. Acute respiratory distress syndrome in the global context. Glob Heart. 2014; 9(3):289–95. doi:10.1016/j.gheart.2014.08.003.

4. Matthay MA, Ware LB, Zimmerman GA. The acute respiratory distress syndrome. J Clin Invest. 2012;122(8):2731–40. doi:10.1172/JCI60331.

5. Lee JW, Fang X, Gupta N, Serikov V, Matthay MA. Allogeneic human mesenchymal stem cells for treatment of E. coli endotoxin-induced acute lung injury in the ex vivo perfused human lung. Proc Natl Acad Sci U S A. 2009;106(38):16357–62. doi:10.1073/pnas.0907996106.

6. Lee JW, Krasnodembskaya A, McKenna DH, Song Y, Abbott J, Matthay MA. Therapeutic effects of human mesenchymal stem cells in ex vivo human lungs injured with live bacteria. Am J Respir Crit Care Med. 2013;187(7):751–60. doi:10.1164/rccm.201206-0990OC.

7. Matthay MA, Goolaerts A, Howard JP, Lee JW. Mesenchymal stem cells for acute lung injury: preclinical evidence. Crit Care Med. 2010;38(10 Suppl): S569–73. doi:10.1097/CCM.0b013e3181f1ff1d.

8. Liang X, Ding Y, Zhang Y, Tse HF, Lian Q. Paracrine mechanisms of mesenchymal stem cell-based therapy: current status and perspectives. Cell Transplant. 2014;23(9):1045–59. doi:10.3727/096368913X667709.

9. Matthay MA. Therapeutic potential of mesenchymal stromal cells for acute respiratory distress syndrome. Ann Am Thorac Soc. 2015;12 Suppl 1:S54–7. doi:10.1513/AnnalsATS.201406-254MG.

10. Crisostomo PR, Wang Y, Markel TA, Wang M, Lahm T, Meldrum DR. Human mesenchymal stem cells stimulated by TNF-alpha, LPS, or hypoxia produce growth factors by an NF kappa B- but not JNK-dependent mechanism. Am J Physiol Cell Physiol. 2008;294(3):C675–82. doi:10.1152/ajpcell.00437.2007.

11. Gnecchi M, He H, Liang OD, Melo LG, Morello F, Mu H, et al. Paracrine action accounts for marked protection of ischemic heart by Akt-modified mesenchymal stem cells. Nat Med. 2005;11(4):367–8. doi:10.1038/nm0405-367.

12. Chen QH, Liu AR, Qiu HB, Yang Y. Interaction between mesenchymal stem cells and endothelial cells restores endothelial permeability via paracrine hepatocyte growth factor in vitro. Stem Cell Res Ther. 2015;6(1):44. doi:10. 1186/s13287-015-0025-1.

13. Sastry L, Johnson T, Hobson MJ, Smucker B, Cornetta K. Titering lentiviral vectors: comparison of DNA, RNA and marker expression methods. Gene Ther. 2002;9(17):1155–62. doi:10.1038/sj.gt.3301731.

14. Liu AR, Liu L, Chen S, Yang Y, Zhao HJ, Liu L, et al. Activation of canonical wnt pathway promotes differentiation of mouse bone marrow-derived MSCs into type II alveolar epithelial cells, confers resistance to oxidative stress, and promotes their migration to injured lung tissue in vitro. J Cell Physiol. 2013;228(6):1270–83. doi:10.1002/jcp.24282.

15. He H, Liu L, Chen Q, Liu A, Cai S, Yang Y, et al. Mesenchymal stem cells overexpressing angiotensin-converting enzyme 2 rescue lipopolysaccharide-induced lung injury. Cell Transplant. 2015;24(9):1699-1715. doi:10.3727/096368914X685087.

16. Adachi E, Hirose-Sugiura T, Kato Y, Ikebuchi F, Yamashita A, Abe T, et al. Pharmacokinetics and pharmacodynamics following intravenous administration of recombinant human hepatocyte growth factor in rats with renal injury. Pharmacology. 2014;94(3–4):190–7. doi:10.1159/000363412.

17. Meng F, Meliton A, Moldobaeva N, Mutlu G, Kawasaki Y, Akiyama T, et al. Asef mediates HGF protective effects against LPS-induced lung injury and endothelial barrier dysfunction. Am J Physiol Lung Cell Mol Physiol. 2015; 308(5):L452–63. doi:10.1152/ajplung.00170.2014.

18. Wang X, Li Q, Hu Q, Suntharalingam P, From AH, Zhang J. Intra-myocardial injection of both growth factors and heart derived Sca-1+/CD31- cells attenuates post-MI LV remodeling more than does cell transplantation alone: neither intervention enhances functionally significant cardiomyocyte regeneration. Plos One. 2014;9(2):e95247. doi:10.1371/journal.pone.0095247.

19. Li J, Zheng CQ, Li Y, Yang C, Lin H, Duan HG. Hepatocyte growth factor gene-modified mesenchymal stem cells augment sinonasal wound healing. Stem Cells Dev. 2015;24(15):1817-1830. doi:10.1089/scd.2014.0521.

20. Seo KW, Sohn SY, Bhang DH, Nam MJ, Lee HW, Youn HY. Therapeutic effects of hepatocyte growth factor-overexpressing human umbilical cord blood-derived mesenchymal stem cells on liver fibrosis in rats. Cell Biol Int. 2014;38(1):106–16. doi:10.1002/cbin.10186.

21. Kim MD, Kim SS, Cha HY, Jang SH, Chang DY, Kim W, et al. Therapeutic effect of hepatocyte growth factor-secreting mesenchymal stem cells in a rat model of liver fibrosis. Exp Mol Med. 2014;46:e110. doi:10.1038/emm.2014.49.

22. Awada HK, Johnson NR, Wang Y. Dual delivery of vascular endothelial growth factor and hepatocyte growth factor coacervate displays strong angiogenic effects. Macromol Biosci. 2014;14(5):679–86. doi:10.1002/mabi.201300486.

23. Yang X, Zhang XF, Lu X, Jia HL, Liang L, Dong QZ, et al. MicroRNA-26a suppresses angiogenesis in human hepatocellular carcinoma by targeting hepatocyte growth factor-cMet pathway. Hepatology. 2014;59(5):1874–85. doi:10.1002/hep.26941.

24. Ding S, Merkulova-Rainon T, Han ZC, Tobelem G. HGF receptor up-regulation contributes to the angiogenic phenotype of human endothelial cells and promotes angiogenesis in vitro. Blood. 2003;101(12):4816–22. doi:10.1182/blood-2002-06-1731.

25. Yu F, Lin Y, Zhan T, Chen L, Guo S. HGF expression induced by HIF-1alpha promote the proliferation and tube formation of endothelial progenitor cells. Cell Biol Int. 2015;39(3):310–7. doi:10.1002/cbin.10397.

26. Panganiban RA, Day RM. Hepatocyte growth factor in lung repair and pulmonary fibrosis. Acta Pharmacol Sin. 2011;32(1):12–20. doi:10.1038/aps.2010.90.

27. Gazdhar A, Fachinger P, van Leer C, Pierog J, Gugger M, Friis R, et al. Gene transfer of hepatocyte growth factor by electroporation reduces bleomycin-

induced lung fibrosis. Am J Physiol Lung Cell Mol Physiol. 2007;292(2):L529–36. doi:10.1152/ajplung.00082.2006.

28. Higginbotham K, Tian Y, Gawlak G, Moldobaeva N, Shah A, Birukova AA. Hepatocyte growth factor triggers distinct mechanisms of Asef and Tiam1 activation to induce endothelial barrier enhancement. Cell Signal. 2014; 26(11):2306–16. doi:10.1016/j.cellsig.2014.07.032.

29. Martin TA, Mansel RE, Jiang WG. Antagonistic effect of NK4 on HGF/SF induced changes in the transendothelial resistance (TER) and paracellular permeability of human vascular endothelial cells. J Cell Physiol. 2002;192(3): 268–75. doi:10.1002/jcp.10133.

30. Gazdhar A, Grad I, Tamo L, Gugger M, Feki A, Geiser T. The secretome of induced pluripotent stem cells reduces lung fibrosis in part by hepatocyte growth factor. Stem Cell Res Ther. 2014;5(6):123. doi:10.1186/scrt513.

31. Chakraborty S, Chopra P, Hak A, Dastidar SG, Ray A. Hepatocyte growth factor is an attractive target for the treatment of pulmonary fibrosis. Expert Opin Investig Drugs. 2013;22(4):499–515. doi:10.1517/13543784.2013.778972.

32. Molnarfi N, Benkhoucha M, Funakoshi H, Nakamura T, Lalive PH. Hepatocyte growth factor: a regulator of inflammation and autoimmunity. Autoimmun Rev. 2015;14(4):293–303. doi:10.1016/j.autrev.2014.11.013.

33. Wang H, Sun RT, Li Y, Yang YF, Xiao FJ, Zhang YK, et al. HGF gene modification in mesenchymal stem cells reduces radiation-induced intestinal injury by modulating immunity. Plos One. 2015;10(5):e124420. doi: 10.1371/journal.pone.0124420.

34. Loghmanpour NA, Druzdzel MJ, Antaki JF. Cardiac Health Risk Stratification System (CHRiSS): a Bayesian-based decision support system for left ventricular assist device (LVAD) therapy. Plos One. 2014;9(11): e111264. doi:10.1371/journal.pone.0111264.

35. Zhang J, Zhou S, Zhou Y, Feng F, Wang Q, Zhu X, et al. Hepatocyte growth factor gene-modified adipose-derived mesenchymal stem cells ameliorate radiation induced liver damage in a rat model. Plos One. 2014;9(12): e114670. doi:10.1371/journal.pone.0114670.

Permissions

The contributors of this book come from diverse backgrounds, making this book a truly international effort. This book will bring forth new frontiers with its revolutionizing research information and detailed analysis of the nascent developments around the world.

We would like to thank all the contributing authors for lending their expertise to make the book truly unique. They have played a crucial role in the development of this book. Without their invaluable contributions this book wouldn't have been possible. They have made vital efforts to compile up to date information on the varied aspects of this subject to make this book a valuable addition to the collection of many professionals and students.

This book was conceptualized with the vision of imparting up-to-date information and advanced data in this field. To ensure the same, a matchless editorial board was set up. Every individual on the board went through rigorous rounds of assessment to prove their worth. After which they invested a large part of their time researching and compiling the most relevant data for our readers.

The editorial board has been involved in producing this book since its inception. They have spent rigorous hours researching and exploring the diverse topics which have resulted in the successful publishing of this book. They have passed on their knowledge of decades through this book. To expedite this challenging task, the publisher supported the team at every step. A small team of assistant editors was also appointed to further simplify the editing procedure and attain best results for the readers.

Apart from the editorial board, the designing team has also invested a significant amount of their time in understanding the subject and creating the most relevant covers. They scrutinized every image to scout for the most suitable representation of the subject and create an appropriate cover for the book.

The publishing team has been an ardent support to the editorial, designing and production team. Their endless efforts to recruit the best for this project, has resulted in the accomplishment of this book. They are a veteran in the field of academics and their pool of knowledge is as vast as their experience in printing. Their expertise and guidance has proved useful at every step. Their uncompromising quality standards have made this book an exceptional effort. Their encouragement from time to time has been an inspiration for everyone.

The publisher and the editorial board hope that this book will prove to be a valuable piece of knowledge for researchers, students, practitioners and scholars across the globe.

List of Contributors

Fernanda Gubert, Ana B. Decotelli, Igor Bonacossa-Pereira, Fernanda R. Figueiredo, Luísa Hoffmann, Turan P. Urmenyi, Marcelo F. Santiago and Rosalia Mendez-Otero1
Instituto de Biofísica Carlos Chagas Filho, Centro de Ciências da Saúde, Sala G2-028, Universidade Federal do Rio de Janeiro, Cidade Universitária, RJ 21941-902 Rio de Janeiro, Brazil.

Camila Zaverucha-do-Valle
Evandro Chagas National Institute of Infectious Diseases (INI), Oswaldo Cruz Foundation, Avenida Brasil 4365Maguinhos RJ 21040-900 Rio de Janeiro, Brazil.

Fernanda Tovar-Moll
Institute of Biomedical Sciences and National Center of Structural Biology and Bioimaging, CENABIO, Centro de Ciências da Saúde, Universidade Federal do Rio de Janeiro, Cidade Universitária, RJ 21941-902 Rio de Janeiro, Brazil
Instituto D'Or de Pesquisa e Educação (IDOR), Rua Diniz Cordeiro 30Botafogo RJ 22281-100 Rio de Janeiro, Brazil

Ieva Bruzauskaite, Edvardas Bagdonas, Jaroslav Denkovskij and Eiva Bernotiene
Department of Regenerative Medicine, State Research Institute Centre for Innovative Medicine, Vilnius, Lithuania
Department of Pathology, Forensic Medicine and Pharmacology, Vilnius University, Faculty of Medicine, Vilnius, Lithuania

Daiva Bironaite
Department of Regenerative Medicine, State Research Institute Centre for Innovative Medicine, Vilnius, Lithuania
Department of Pathology, Forensic Medicine and Pharmacology, Vilnius University, Faculty of Medicine, Vilnius, Lithuania

Vytenis Arvydas Skeberdis
Institute of Cardiology, Lithuanian University of Health Sciences, Kaunas, Lithuania

Tomas Tamulevicius
Institute of Materials Science, Kaunas University of Technology, Kaunas, Lithuania

Valentinas Uvarovas
Clinic of Rheumatology, Orthopedic and Traumatology and Reconstructive Surgery, Faculty of Medicine, Vilnius University, Vilnius, Lithuania

Marcelo Ezquer, Scarleth Montecino, Karla Leal, Paulette Conget and Fernando E
Centro de Medicina Regenerativa, Facultad de Medicina Clínica Alemana-Universidad del Desarrollo, Av. Las Condes 12438, Lo Barnechea, Santiago 7710162, Chile

Cristhian A. Urzua
Departamento de Oftalmología, Facultad de Medicina, Universidad de Chile, Av. Independencia 1027, Santiago, Chile

Simone Riis, Jeppe Emmersen, Cristian Pablo Pennisi, Vladimir Zachar and Trine Fink
Department of Health Science and Technology, Laboratory for Stem Cell Research, Aalborg University, Fredrik Bajers Vej 3B, Aalborg 9220, Denmark

Allan Stensballe and Svend Birkelund
Department of Health Science and Technology, Laboratory for Medical Mass Spectrometry, Aalborg University, Aalborg, Denmark

Ming Liu, Dunqiang Ren, Baodan Yu, Lixia Zheng and Jun Xu
1State Key Laboratory of Respiratory Diseases, Guangzhou Institute of Respiratory Diseases, The First Affiliated Hospital of Guangzhou Medical University, Guangzhou Medical University, Guangzhou, P. R. China

Xiansheng Zeng
State Key Laboratory of Respiratory Diseases, Guangzhou Institute of Respiratory Diseases, The First Affiliated Hospital of Guangzhou Medical University, Guangzhou Medical University, Guangzhou, P. R. China

Junli Wang and Zhiping Fu
State Key Laboratory of Respiratory Diseases, Guangzhou Institute of Respiratory Diseases, The First Affiliated Hospital of Guangzhou Medical University, Guangzhou Medical University, Guangzhou, P. R. China

Jinsong Wang
State Key Laboratory of Respiratory Diseases, Guangzhou Institute of Respiratory Diseases, The First Affiliated Hospital of Guangzhou Medical University, Guangzhou Medical University, Guangzhou, P. R. China
Shenzhen Beike Cell Engineering Research Institute, Shenzhen, P. R. China

Muyun Liu and Xiang Hu
Shenzhen Beike Cell Engineering Research Institute, Shenzhen, P. R. China

Wei Shi
Developmental Biology and Regenerative Medicine Program, Department of Surgery, The Saban Research Institute of Children's Hospital Los Angeles, University of Southern California Keck School of Medicine, Los Angeles, CA, USA

Mahmoud Yousefifard and Mansoor Keshavarz
Electrophysiology Research Center, Tehran University of Medical Sciences, Tehran, Iran
Department of Physiology, School of Medicine, Tehran University of Medical Sciences, Tehran, Iran

Farinaz Nasirinezhad
Physiology Research Center, Iran University of Medical Sciences, Tehran, Iran
Department of Physiology, Iran University of Medical Sciences, Tehran, Iran

Homa Shardi Manaheji
Department of Physiology, Shahid Beheshti University of Medical Sciences, Tehran, Iran
Neuroscience Research Center, Shahid Beheshti University of Medical Sciences, Tehran, Iran

Atousa Janzadeh
Physiology Research Center, Iran University of Medical Sciences, Tehran, Iran

Mostafa Hosseini
Department of Epidemiology and Biostatistics, School of Public Health, Tehran University of Medical Sciences, Tehran, Iran
Pediatric Chronic Kidney Disease Research Center, Childrens Hospital Medical Center, Tehran
University of Medical Sciences, Tehran, Iran

Jeremy John Mathan, Salim Ismail, Jennifer Jane McGhee, Charles Ninian John McGhee and Trevor Sherwin
Department of Ophthalmology, New Zealand National Eye Centre, Faculty of Medical and Health Sciences, The University of Auckland, Private Bag 92019, Auckland 1010, New Zealand

Ambreen Shaikh, Deepa Bhartiya, Sona Kapoor and Harshada Nimkar
Stem Cell Biology Department, National Institute for Research in Reproductive Health (ICMR), Jehangir Merwanji Street, Parel, Mumbai 400 012, India

Jifeng Liu
Laboratory of Molecular Developmental Biology, Shanghai Jiao Tong University School of Medicine, Shanghai 200025, China
Key Laboratory of Stem Cell Biology, Institute of Health Sciences, Shanghai Institutes for Biological Sciences, Chinese Academy of Sciences, Shanghai Jiao Tong University School of Medicine, New Life Science Building A, Room 1328, 320 Yue Yang Road, Shanghai 200032, China

Xinlong Luo
Laboratory of Molecular Developmental Biology, Shanghai Jiao Tong University School of Medicine, Shanghai 200025, China KU Leuven Department of Development and Regeneration, Stem Cell Institute Leuven, Herestraat 49, 3000, Leuven, Belgium

Yanli Xu
Laboratory of Molecular Developmental Biology, Shanghai Jiao Tong University School of Medicine, Shanghai 200025, China

Junjie Gu, Fan Tang, Ying Jin and Hui Li
Laboratory of Molecular Developmental Biology, Shanghai Jiao Tong University School of Medicine, Shanghai 200025, China Key Laboratory of Stem Cell Biology, Institute of Health Sciences, Shanghai Institutes for Biological Sciences, Chinese Academy of Sciences, Shanghai Jiao Tong University School of Medicine, New Life Science Building A, Room 1328, 320 Yue Yang Road, Shanghai 200032, China

Beatriz Bravo, Alba Puente-Bedia, Javier Hernández, Alicia Ballester and Sara Ballester1
Instituto de Salud Carlos III, Unidad Funcional de Investigación en Enfermedades Crónicas, Laboratory of Gene Regulation, Carretera de Majadahonda-Pozuelo Km 2, 28220 Madrid, Spain

Marta I. Gallego, Elena García-Zaragoza
Instituto de Salud Carlos III, Unidad Funcional de Investigación en Enfermedades Crónicas, Laboratory
of Mammary Gland Pathology, Carretera de Majadahonda-Pozuelo Km 2, 28220 Madrid, Spain

Ana I. Flores and Paz de la Torre
Grupo de Medicina Regenerativa, Instituto de Investigación Hospital 12 de Octubre, Avda. Córdoba s/n, 28041 Madrid, Spain

Rafael Bornstein
Hospital Central de Cruz Roja, Servicio de Hematología y Hemoterapia, Avenida de Reina Victoria 24, 28003 Madrid, Spain

Raquel Perez-Tavarez
Instituto de Salud Carlos III, Unidad Funcional de Investigación en Enfermedades Crónicas, Histology Core Unit, Carretera de Majadahonda-Pozuelo Km 2, 28220 Madrid, Spain

Jesús Grande
Grupo de Medicina Regenerativa, Instituto de Investigación Hospital 12 de Octubre, Avda. Córdoba s/n, 28041 Madrid, Spain

Joice Fülber Fernanda Agreste and Raquel Y. Arantes Baccarin
Department of Internal Medicine, School of Veterinary Medicine and Animal Science, University of São Paulo (USP), Avenida Prof. Orlando Marques de Paiva, 87, 05508-270 São Paulo, SP, Brazil.

Durvanei A. Maria
Laboratory of Biochemistry and Biophysics, Butantan Institute, Avenida Vital Brasil 1500, São Paulo
05503-900SP, Brazil

Luis Cláudio Lopes Correia da Silva
Department of Surgery, School of Veterinary Medicine and Animal Science, University of São Paulo (USP), Avenida Prof. Orlando Marques de Paiva, 87, SP 05508-270SP, Brazil

Cristina O. Massoco
Department of Pathology, School of Veterinary Medicine and Animal Science, University of São Paulo
(USP), Avenida Prof. Orlando Marques de Paiva, 87, São Paulo 05508-270SP, Brazil

Aastha Singh and Abhishek Singh
School of Bio Sciences and Technology, VIT University, Vellore, India
Cellular and Molecular Therapeutics Laboratory, Centre for Biomaterials, Cellular and Molecular Theranostics (CBCMT), VIT University, Vellore 632014, Tamil Nadu, India

Dwaipayan Sen
School of Bio Sciences and Technology, VIT University, Vellore, India
Cellular and Molecular Therapeutics Laboratory, Centre for Biomaterials, Cellular and Molecular Theranostics (CBCMT), VIT University, Vellore 632014, Tamil Nadu, India

Nadeeka Bandara
School of Biomedical Sciences, Charles Sturt University, Wagga Wagga, NSW 2650, Australia
O'Brien Institute Department, St. Vincent's Institute of Medical Research, Fitzroy, VIC 3065, Australia

Saliya Gurusinghe and Shiang Y. Lim
O'Brien Institute Department, St. Vincent's Institute of Medical Research, Fitzroy, VIC 3065, Australia
Department of Surgery, St. Vincent's Hospital, University of Melbourne, Melbourne, VIC 3002, Australia

Padraig Strappe
School of Biomedical Sciences, Charles Sturt University, Wagga Wagga, NSW 2650, Australia

Haiying Chen and Shuangfeng Chen
Central laboratory and key Laboratory of Oral and Maxillofacial-Head and Neck Medical Biology, Liaocheng People's Hospital, Liaocheng 252000, PR China.

Le-xin Wang
School of Biomedical Sciences, Charles Sturt University, Wagga Wagga, NSW 2650, Australia

Wenjing Du, Xue Li, Ying Chi, Fengxia Ma, Shaoguang Yang, Baoquan Song, Junjie Cui, Juanjuan Li, Jianjian Tian, Zhouxin Yang, Xiaoming Feng Fang Chen, Zhi-Bo Han and Shihong Lu
The State Key Laboratory of Experimental Hematology, Institute of Hematology and Hospital of Blood Diseases, Chinese Academy of Medical Sciences & Peking Union Medical College, No.288, Nanjing Road, Heping District, Tianjin 300020, China

Zongjin Li and Lu Liang
Institute of Health and Stem Cells, No.1 Kangding Road, BDA, Beijing 100176, China

Tao Ma
National Engineering Research Center of Cell Products, No.80, Fourth Avenue, TEDA, Tianjin
300457, China

Zhong-Chao Han
The State Key Laboratory of Experimental Hematology, Institute of Hematology and Hospital of Blood Diseases, Chinese Academy of Medical Sciences & Peking Union Medical College, No.288, Nanjing Road, Heping District, Tianjin 300020, China
Beijing Institute of Health and Stem Cells, No.1 Kangding Road, BDA, Beijing 100176, China
National Engineering Research Center of Cell Products, No.80, Fourth Avenue, TEDA, Tianjin
300457, China

Shuling Hu, Jinze Li, Xiuping Xu, Airan Liu, Hongli He, Jingyuan Xu, Qihong Chen, Songqiao Liu, Ling Liu, Haibo Qiu and Yi Yang
Department of Critical Care Medicine, Zhongda Hospital, Southeast University School of Medicine, No.87 Dingjiaqiao Road, Nanjing 210009, Jiansu, P.R. China

Index

www.ingramcontent.com/pod-product-compliance
Lightning Source LLC
Chambersburg PA
CBHW061939190326
41458CB00009B/2782